# TECHNICAL Writing

## A READER-CENTERED APPROACH

**PAUL V. ANDERSON**
*Miami University, Ohio*

**HARCOURT BRACE JOVANOVICH, PUBLISHERS**
San Diego   New York   Chicago   Austin   Washington, D.C.
London   Sydney   Tokyo   Toronto

*For my family—*
*Margie, Christopher, Rachel*

# Preface

I wrote *Technical Writing: A Reader-Centered Approach* in the belief that, as teachers, we all desire to prepare our students to become confident, flexible, and resourceful writers. At work they will encounter such a wide variety of writing situations that no single formula or short series of recipes that we might teach will serve them well. Instead, we must equip them with strategies, together with the skills, knowledge, and attitudes needed to put those strategies to good use.

## MAJOR FEATURES OF THIS BOOK

I have designed this book to teach students to think for themselves: to size up a situation, create a writing plan, and then carry out that plan with skill and flair. These are the book's major features:

- *Reader-Centered Approach.* Fundamental to this book is a strategy that students can use in the myriad writing situations they will encounter at work. That strategy is to create lively, individual portraits of their readers. Using these portraits at every step of the writing process, students can accurately predict their readers' thoughts, feelings, needs, and desires— and write accordingly. Every chapter increases the students' ability to create and use these portraits.

- *Organization around Guidelines.* To focus attention on the *principles* of effective technical writing, I have distilled the advice in most chapters into short lists of easily understood, easily remembered guidelines. At the same time, students learn that the guidelines are not inflexible rules: to write effectively, writers must follow the guidelines creatively and thoughtfully.

- *Focus on Real-World Writing.* Nothing is tougher for students than figuring out how the writing principles they learn in the classroom apply on the job. Consequently, throughout the book students learn what readers expect, desire, and need in various typical on-the-job situations. In addition, students find numerous examples and anecdotes, and much advice about the practical problems of writing at work.

## COMPREHENSIVE COVERAGE OF KEY TOPICS

This book covers both the process and the products of writing on the job. The first 25 chapters are devoted to the writing *process*. Besides presenting a thorough and fresh treatment of such basic activities as defining purpose, drafting prose, and creating visual aids, they provide detailed advice about the following important topics:

- *Finding Out What's Expected.* Chapter 6 tells students how to verify their assumptions about what their employers and readers require and expect concerning length, format, style, and content.
- *Generating Ideas about What To Say.* Chapter 8 explains ten techniques for rhetorical invention that are especially useful in on-the-job writing.
- *Writing Beginnings and Endings.* Two chapters (13 and 14) discuss strategies students can use for writing these crucial elements.
- *Designing Page Layouts.* Chapter 21 provides detailed advice about creating multicolumn page designs, which are particularly useful in instructions, reference manuals, and technical brochures.
- *Evaluating Drafts.* Three chapters (22–24) tell students how to check drafts over themselves, derive the full value from reviews by other people, and conduct formal tests. The chapter on reviewing helps students participate effectively in peer review sessions in the classroom as well as in the workplace.

In addition, the book devotes eight chapters to the *products* of technical writing. Chapter 26 describes formats for memos, letters, and such booklike communications as formal reports and instruction manuals. Chapter 27 treats reference lists, footnotes, and bibliographies. Chapters 28–33 teach students how to prepare six frequently written types of communication: instructions, proposals, and four types of reports. These chapters on types of communication are carefully coordinated with the chapters on the writing process so that students learn how to adapt conventional formats to particular writing situations. Students thereby avoid the mistake of treating the conventional forms as models to be imitated blindly.

Finally, Chapters 34 and 35 present a "listener-centered" approach to oral presentations that builds on the "reader-centered" lessons in the rest of the text.

## LEARNING AIDS

Several features of the book will help students learn quickly and thoroughly. These include chapter previews and summaries, two overview chapters (2 and 3) that introduce the basic strategies and principles that are explained in more detail later in the book, worksheets that students can use when planning and reviewing their communications in the classroom or on the job, and marginal annotations that point out the key features of sample passages and communications.

To further aid learning, I've tried to use a supportive tone in which I speak directly to students and their concerns, attempting always to be helpful, respectful, and encouraging.

## TEACHER SUPPORT

To provide you with ideas about how to use this book in your classes, I have prepared an *Instructor's Manual*. The manual covers such topics as course design, classroom activities, and grading criteria. Also, Appendix I of this book includes seven writing assignments designed to support and reinforce the principles taught here.

## CONCLUSION

With the help of many other instructors, I have tested this book in the classroom over the several years of its development. This classroom use and the advice of my colleagues have strengthened every aspect of the book. I am confident that you will find it to be very teachable and that your students will agree that it is informative, realistic, and helpful.

Paul V. Anderson
Oxford, Ohio

# Acknowledgments

I take special pleasure in this opportunity to thank the many people who have helped me with this book. First, I would like to express my gratitude to my own teachers, James W. Souther and Myron White, of the University of Washington. Many of the basic insights that underlie this book come from their classes and from Jim's book (coauthored by Mike in the second edition): *Technical Report Writing* (New York: Wiley-Interscience, 1977). As a teacher and writer, I would feel that I have achieved an extraordinary success if I could manage to make as much difference to just one student as they have made not only to me but also to numerous others.

Three other technical writing textbooks have also taught me a great deal about what and how to teach; people familiar with these books will see their influence in the pages that follow. The books are Kenneth Houp and Thomas E. Pearsall's *Reporting Technical Information* (New York: Macmillan, 1984), J. C. Mathes and Dwight W. Stevenson's *Designing Technical Reports* (Indianapolis: Bobbs-Merrill, 1978), and Gordon Mills and John Walter's *Technical Writing* (New York: Holt Rinehart and Winston, 1986).

Other publications that have significantly influenced the substance or presentational strategies of this book are David L. Carson's "Audience in Technical Writing: The Need for Greater Realism in Identifying the Fictive Reader," in *Teaching Technical Writing: Audience Analysis and Adaptation,* ed. Paul V. Anderson (Morehead, Kentucky: Association of Teachers of Technical Writing, 1980), 24–31; Theodore Clevenger's *Audience Analysis* (Indianapolis: Bobbs-Merrill, 1966); Donald A. Daiker, Andrew Kerek and Max Morenberg's *The Writer's Options: Combining to Composing* (New York: Harper & Row, 1986); Linda Flower's *Problem-Solving Strategies for Writing* (New York: Harcourt Brace Jovanovich, 1985); E. D. Hirsh, Jr.'s *Philosophy of Composition* (Chicago: University of Chicago Press, 1977); William Strunk, Jr., and E. B. White's *The Elements of Style* (New York: Macmillan, 1979); Stephen Toulmin, Richard Rieke, and Allan Janik's *An Introduction to Reasoning* (New York: Macmillan, 1984); Joseph M. Williams' *Style: Ten Lessons in Clarity & Grace* (Oakland, New Jersey: Scott Foresman, 1984); and Richard E. Young, Alton L. Becker and Kenneth L. Pike's *Rhetoric: Discovery and Change* (New York: Harcourt Brace Jovanovich, 1970). (In this paragraph and the preceding one, I cite the current editions of books, but for those that have had more than one edition my debt originates earlier.)

While writing this book, one of my major goals has been to devise a framework that will help us as teachers to integrate our instruction in the standard forms of technical communication with the rhetorical and process-oriented approach that many of us take. Five works that helped me in that effort are Lloyd F. Bitzer's "The Rhetorical Situation," *Philosophy and Rhetoric* 1(1968):1–14; Teun A. van Dijk's *Macrostructures* (Hillsdale, New Jersey: Erlbaum, 1980); Carolyn R. Miller's "Genre as Social Action," *Quarterly Journal of Speech* 70(1984):151–67; Vickie M. Winkler's "The Role of Models in Technical and Scientific Writing," In *New Essays in Technical and Scientific Communication: Research, Theory, Practice,* ed. Paul V. Anderson, R. John Brockmann, and Carolyn R. Miller (Farmingdale, New York: Baywood, 1983), 111–22; and James P. Zappen's "A Rhetoric for Research in Sciences and Technologies," in *New Essays in Technical and Scientific Communication: Research, Theory, Practice,* ed. Paul V. Anderson, R. John Brockmann, and Carolyn R. Miller (Farmingdale, New York: Baywood, 1983), 123–38.

The following people helped me immensely by reading through a nearly final draft of my manuscript and then offering detailed—and excellent—advice about it: Barbara Couture (Wayne State University), John DiGaetani (Hofstra University), William H. Hardesty (Miami University, Ohio), and Carolyn R. Miller (North Carolina State University). I have also benefited from the helpful comments and suggestions offered by Mary C. Cosgrove (University of Cincinnati), Ruth Falor (Ohio State University), AEleen Frisch (University of Pittsburgh), Patricia Harkin (University of Akron), Joseph C. Mancuso (North Texas State University), and Gerald J. Schiffhorst (University of Central Florida), who read and commented on earlier drafts.

Several members of Miami University's English department have helped me develop and refine the book by using drafts of it in their classes. The reactions they reported and the suggestions that they and their students made have been most helpful to me. These colleagues are John H. Crow, Jim Clark, David N. Dobrin (now with Lexicom, Cambridge, Massachusetts), Jean Alexander Coakley, William H. Hardesty, Jim Flavin (now at Shawnee State University, Ohio), John Heyda, Thomas Kent (now at Iowa State University), Becky Lukens, Jean A. Lutz, Jack Selzer (now at Pennsylvania State University), and C. Gilbert Storms. I also want to thank the many graduate students who have used drafts in classes they have taught.

For introducing me to the approach to designing pages that I describe in Chapter 21, I am grateful to Professor Joseph M. Cox III here at Miami University (Ohio). I also want to express my gratitude to Professor Thomas Effler for his suggestions concerning Chapter 21. Other Miami faculty whom I would like to thank for their assistance are Tony Esposito, T. William Houk, and Alan Mills. I also extend my thanks to Miami's president, Paul G. Pearson, whose memo about computer policies inspired one of the model communications presented in this book.

I am grateful, too, for the aid provided by several people in industry: Norma Allen (General Electric), Roy E. and Gladys M. Anderson (Mobile Satellite Corporation), C. Mike Bickford (Monarch Marking Systems, Incorporated), David Bradford (IBM), Kent Bradshaw (Armco Steel), Thomas Collins (Oxford Associates, Oxford, Ohio), Joseph R. Davis (WCI Machine Tools &

Systems), David Feitlowitz-Anderson (Medical Economics, Inc.), Neil Herbkersman (Sinclair Community College Office of Grants Development), Tom Linxweiler (Monsanto), Thomas R. Milligan (O'Neil & Associates), and Terri Parker (Committed to Results, Cincinnati).

Many present and former students have contributed to this book by doing a variety of things, such as providing samples and ideas, working in the library, gathering writing samples from industry, and copyediting. These students include Ann Blakeslee, Mary Brady, Kelly Blose, James P. Conway, Jr., Kirk J. Donnan, Thomas E. Esposito, James (Jay) Farlow, Cecilia Franz, Cindy Griffin, Allen J. Hines, Sheryl Hofmann, Scott Houck, Carol Millard Kosarko, Karen Levine, Kelly McBride, Mary Mason, William C. Morton, Caroline S. Neal, Stephen Oberjohn, Sally C. Rustin, Cindy Silletto, Phyllis Smallwood, Heather Strasburg, Stephanie Swartz, Peter Ronald Toivonen, Curt Walor, Greg Wickliff, Carla Wojnaroski, and Karen Kinsey Zink.

Trudi Nixon did an immense amount of work preparing the final manuscript, and Jackie Kearns, Tammy Forester, and Betty Marak helped greatly also.

I want to give special thanks to the people at Harcourt Brace Jovanovich who worked with me. Bill McLane has been particularly helpful and encouraging. Others who contributed greatly include Janice Anderson, Bruce Daniels, Lesley Lenox, Becky Lytle, Diane Pella, and Karl Yambert.

For various forms of help and encouragement while I was working on this book, I am deeply indebted to Dick and Sandy Huggins.

My gratitude to my family is of another sort than that which I have expressed above—and very deeply felt.

# Contents

Preface     iii

Acknowledgments     vii

PART I     INTRODUCTION     1

Chapter 1     **Writing, Your Career, and This Book**     2

Sharing Ideas Is a Major Activity at Work     3
Your Two Roles at Work     4
Writing is a Form of Action     6
Writing is a Form of *Social* Action     6
The Main Advice of This Book: Write to People     7
Knowledge and Common Sense about People at Work     8
Guidelines for Writing     8
The Organization of This Book     9
Improving Your Writing by Improving Your Writing Process     9
Summary 10     Exercises 11

Chapter 2     **A Reader-Centered Approach to Writing**     12

*General Guideline 1*     When Writing, Talk *with* Your Readers     13
*General Guideline 2*     Write in a Respectful Voice     19
*General Guideline 3*     Write in a Professional Manner     21
*General Guideline 4*     Help Your Readers Focus on the Key Information     26
*General Guideline 5*     Write Sentences Your Readers Can Understand Easily     31
*General Guideline 6*     Help Your Readers Follow the Flow of Thought from Sentence to Sentence     33
*General Guideline 7*     Help Your Readers Build Hierarchies of Meaning     37
*General Guideline 8*     Tell Your Readers the Significance *to Them* of Your Message     41
*General Guideline 9*     Use Visual Aids     42
Summary 43     Exercises 44

**Chapter**

**3**

**Examples of Reader-Centered Writing: Resumes and Letters of Application**    48

The Persuasive Aim of This Chapter    49
Writing Your Resume    50
Writing Your Letter of Application    66
Summary 73    Exercises 74


**PART II    DEFINING YOUR OBJECTIVES**    75


**Chapter**

**4**

**Defining Your Purpose**    76

Two Kinds of Purpose    77
*Guideline 1*    Describe the Final Result You Want Your Communication To Bring About    77
*Guideline 2*    Describe the Tasks Your Readers Will Try to Perform while Reading    78
*Guideline 3*    Tell How You Want Your Communication To Alter Your Readers' Attitudes    84
Relationship between Guidelines 2 and 3    88
When To Use Your Definition of Your Communication's Purpose    88
Worksheet for Defining Purpose    89
Summary 89    Exercises 93


**Chapter**

**5**

**Understanding Your Readers**    94

Making Predictions about Readers    95
Readers at Work Compared with Readers at School    95
*Guideline 1*    Create Imaginary Portraits of Your Readers in the Act of Reading    96
*Guideline 2*    Portray Your Readers as People Who Ask Questions while They Read    97
*Guideline 3*    Learn Who Your Readers Will Be    98
*Guideline 4*    Find Out Whether Your Readers Will Be Decision-Makers, Advisers, or Implementers    103
*Guideline 5*    Find Out How Familiar Your Readers Are with Your Specialty    108
*Guideline 6*    Find Out Your Readers' Reasons for Holding Their Present Attitudes    108
*Guideline 7*    Find Out If Your Readers Have Any "Special" Characteristics    109
*Guideline 8*    Find Out about the Setting in Which Your Readers Will Read Your Communication    110

*Guideline 9*    Aggressively Seek Information about Your Readers    110
Worksheet for Defining Objectives    113
Summary 113    Exercises 114

Chapter

**6**

**Finding Out What's Expected**    115

The Main Advice of This Chapter    116

*Guideline 1*    Ask Your Boss, Readers, and Other Knowledgeable
People What's Expected    118

*Guideline 2*    Obtain and Study Any Written Regulations
Concerning Your Writing    118

*Guideline 3*    Look at Similar Communications Written by Other
People    119

*Guideline 4*    Find Out Your Deadline and Make a Schedule For
Meeting It    119

*Guideline 5*    Find Out What Length Is Expected    120

*Guideline 6*    Find Out What Your Communication Should Look
Like    120

*Guideline 7*    Find Out What Style You Should Use    120

*Guideline 8*    Find Out What You are Expected To Say—Or Not
Say    121

*Guideline 9*    Find Out What Policies Apply to the
Communication You Are Preparing    121

*Guideline 10*    Find Out Whether and When You Will Need to
Have Your Drafts Reviewed    122

A Final Note 122    Summary 122

**PART III    PLANNING**    125

Chapter

**7**

**Deciding What To Say**    130

Your Two Activities When Deciding What To Say    131

*Guideline 1*    List the Questions Your Readers Will Ask, and
Answer Them    132

*Guideline 2*    List the Additional Things Your Readers Will Need
To Know    137

*Guideline 3*    List the Persuasive Arguments You Will Have To
Make    138

*Guideline 4*    Generate Ideas Freely before You Evaluate Them    140

*Guideline 5*    Review and Revise the Contents of Your
Communication throughout Your Work on It    140

Worksheet for Deciding What To Say    141

Summary 141    Exercises 142

Chapter
8
**Ten Techniques for Generating Ideas about What To Say**     144

Structured versus Unstructured Lists     145
When To Use These Techniques     145
Ask Your Readers     146
Talk with Someone Else     146
Look at Successful Communications Used in Similar Situations     147
Look at Conventional Superstructures     148
Brainstorm     148
Write a Throw-Away Draft     149
Make an Idea Tree     154
Make an Outline     155
Draw a Flow Chart     158
Make a Matrix     158
Exercises 160

Chapter
9
**Organizing**     163

Various Approaches to Organizing     164
Importance of Thinking about Your Purpose and Readers     164
*Guideline 1*   Arrange Your Material into Groups Meaningful to Your Readers     167
*Guideline 2*   Build a Hierarchy by Combining Your Groups into Clusters Meaningful to Your Readers     170
*Guideline 3*   Arrange the Parts of Your Hierarchy into a Reader-Centered Sequence     173
*Guideline 4*   Plan Your Visual Aids     173
*Guideline 5*   Organize All the Parts of Your Communication, Small as Well as Large     174
*Guideline 6*   Adapt Organizational Patterns Used Successfully by Other Writers     175
Summary 176     Exercises 177

Chapter
10
**Using Outlines**     180

Outlining versus Organizing     181
An Outline Is a Sketch     182
The Value of Outlining     182
*Guideline 1*   Outline To Find the Best Content and Organization for Your Communication     183
*Guideline 2*   Outline To Check the Content and Organization of Your Draft     184
*Guideline 3*   Outline To Obtain Approval of Your Writing Plan     185
*Guideline 4*   Outline When You Are on a Writing Team     185

*Guideline 5*    Use a Sentence Outline To Sharpen the Focus of
                Your Writing                                        186
*Guideline 6*    Change Your Outline When Circumstances Warrant      190
*Guideline 7*    Don't Outline When Outlining Won't Help You         190
Summary 191    Exercises 192


PART IV    **DRAFTING PROSE ELEMENTS**                               195

Chapter    **Writing Segments**                                     198

**11**     Recognizing Segments                                     199
           *Guideline 1*    At the Beginning of Each Segment, Tell What It Is
                           About                                     202
           *Guideline 2*    At The Beginning of Each Segment, Tell How It Is
                           Organized                                 207
           *Guideline 3*    At the Beginning of Each Segment, Generalize about
                           Its Contents in a Reader-Centered Way     208
           Relationship among Guidelines 1 through 3                 211
           *Guideline 4*    Put the Most Important Information First  212
           *Guideline 5*    Answer Your Readers' Questions Where They Arise  213
           *Guideline 6*    Provide Transitional Statements between Segments  216
           Summary 217    Exercises 218


Chapter    **Seven Types of Segments**                              222

**12**     Benefits of Learning and Using Conventional Patterns     223
           Four Important Points before You Read Further             224
           The Organization of This Chapter                          225
           Classification                                            225
           Partition: Description of an Object                       231
           Segmentation: Description of a Process                    235
           Comparison                                                242
           Persuasion                                                245
           Cause and Effect                                          251
           Problem and Solution                                      255
           Exercises 258


Chapter    **Writing the Beginning of a Communication**            263

**13**     The Two Functions of a Beginning                         264
           *Guideline 1*    Imagine Your Readers' Thoughts, Feelings, and
                           Expectations as They First Pick Up Your
                           Communication                             264

*Guideline 2*    Tell Your Readers How Your Communication Is
Relevant to Them                                                267
*Guideline 3*    Tell How Your Communication Is Organized           272
*Guideline 4*    Tell the Scope of Your Communication               273
*Guideline 5*    Tell Your Main Points                              274
*Guideline 6*    Provide the Background Information Your Readers
Will Need To Understand Your Communication     275
Applying These Guidelines                                          275
Summary 279    Exercises 280

**Chapter**

**14**

## Writing the Ending of a Communication      282

The Three Aims of an Ending                                        283
About the Guidelines                                               284
*Guideline 1*    After You've Made Your Last Point, Stop           284
*Guideline 2*    Repeat Your Main Point                            285
*Guideline 3*    Summarize Your Key Points                         285
*Guideline 4*    Focus on a Key Feeling                            285
*Guideline 5*    Tell Your Readers How To Get Assistance or More
Information                                       287
*Guideline 6*    Tell Your Readers What To Do Next                 287
*Guideline 7*    Identify Any Further Study That Is Needed         288
*Guideline 8*    Refer to a Goal Stated Earlier in Your
Communication                                     288
*Guideline 9*    Follow the Social Conventions That Apply to Your
Situation                                         289
Summary 289    Exercises 291

**Chapter**

**15**

## Mapping Your Communication      293

Your View versus Your Readers' View                                294
The Focus of This Chapter                                          296
*Guideline 1*    Use Headings Often                                298
*Guideline 2*    Phrase Each Heading To Tell Specifically What
Follows                                           305
*Guideline 3*    Write Each Heading Independently, Using Parallel
Structure among Headings Where It Is Helpful      305
*Guideline 4*    Adjust the Size and Location of Your Headings To
Indicate Organizational Hierarchy                 307
*Guideline 5*    Adjust the Location of Your Prose To Indicate
Organizational Hierarchy                          311
*Guideline 6*    Provide a Table of Contents If It Is Useful       318
*Guideline 7*    Provide an Index If It Is Useful                  318
Summary 319    Exercises 321

Chapter **16** **Writing Sentences** 325

*Guideline 1* Eliminate Unnecessary Words 326
*Guideline 2* Keep Related Words Together 328
*Guideline 3* Use the Active Voice Unless There Is Good Reason To Use the Passive Voice 330
*Guideline 4* Put Main Actions in Verbs 332
Summary 334    Exercises 334

Chapter **17** **Guiding Your Readers** 338

Relating New Information to Old 339
Your Readers' View versus Your View 339
*Guideline 1* Avoid Needless Shifts in Topic from One Sentence to the Next 340
*Guideline 2* Use Transitional Words and Phrases at the Beginning of Your Sentences 343
*Guideline 3* Use Echo Words at the Beginning of Your Sentences 344
*Guideline 4* Use Paragraph Breaks To Signal Major Shifts in Topic 347
Summary 350    Exercises 350

Chapter **18** **Choosing Words** 354

The Main Advice of This Chapter 355
*Guideline 1* Use Concrete, Specific Words 356
*Guideline 2* Use Words Accurately 359
*Guideline 3* Choose Plain Words over Fancy Ones 361
*Guideline 4* Avoid Unnecessary Variation of Terms 363
*Guideline 5* Use Technical Terms when Your Readers Will Understand and Expect Them—but Not Otherwise 365
*Guideline 6* Explain Technical Terms That Are Not Familiar To Your Readers 366
Summary 368    Exercises 368

**PART V    DRAFTING VISUAL ELEMENTS** 371

Chapter **19** **Using Visual Aids** 372

More than Just Aids 373
*Guideline 1* While Planning Your Communication, Look for Places To Use Visual Aids 376
*Guideline 2* Choose Visual Aids Appropriate To Your Purpose 379

*Guideline  3*    Choose Visual Aids Your Readers Know How
To Read      384
*Guideline  4*    Make Your Visual Aids Easy To Use      388
*Guideline  5*    Make Your Visual Aids Simple      388
*Guideline  6*    Label the Important Content Clearly      389
*Guideline  7*    Provide Informative Titles      392
*Guideline  8*    Introduce Your Visual Aids in Your Prose      394
*Guideline  9*    State the Conclusions You Want Your Readers To
Draw from Your Visual Aids      396
*Guideline 10*    Put Your Visual Aids Where Your Readers Can Find
Them Easily      396
Summary 397

**Chapter 20 Twelve Types of Visual Aids**      399

The Organization of This Chapter      400
A Note about Computers and Visual Aids      400
A Note about Titles      401
Tables      401
Bar Graphs      410
Pictographs      414
Line Graphs      417
Pie Charts      421
Choosing the Best Visual Display for Numerical Data      422
Photographs      423
Drawings      429
Choosing between Photographs and Drawings      433
Diagrams      433
Flow Charts      436
Organizational Charts      438
Schedule Charts      439
Budget Statements      440
Exercises 442

**Chapter 21 Designing Pages**      447

The Importance of Good Page Design      448
Single-Column and Multicolumn Designs      449
*Guideline 1*    Create a Grid To Serve as the Visual Framework for
Your Pages      454
*Guideline 2*    Use the Grid to Coordinate Related Visual Elements      456
*Guideline 3*    Use the Same Design for All Pages That Contain the
Same Types of Information      462
Practical Procedures for Designing Pages      466
Summary 469    Exercises 469

## PART VI  EVALUATING 481

### Chapter 22  Checking 485

Guideline 1  Play the Role of Your Readers 486
Guideline 2  Consider the Consequences of Your Communication
for Your Employer 486
Guideline 3  Let Time Pass before You Check 488
Guideline 4  Read Your Draft More Than Once, Concentrating on
Different Aspects of Your Writing Each Time 488
Guideline 5  Use Checklists 489
Guideline 6  Read Your Draft Aloud 490
Guideline 7  When Checking for Spelling, Focus on Individual
Words 490
Guideline 8  Double-Check Everything You Are Unsure About 492
Guideline 9  Use Computer Aids To Find, but Not To Cure,
Possible Problems 493
Summary 494    Exercises 496

### Chapter 23  Reviewing 499

Reviewing in the Workplace 500
Reviewing in the Classroom 501
Importance of Good Relations between Writers and Reviewers 504
Introduction to Guidelines for Having Your Writing Reviewed 504
Guideline 1  Think of Your Reviewers as Your Partners 505
Guideline 2  Tell Your Reviewers about the Purpose and Readers
of Your Communication 505
Guideline 3  Tell Your Reviewers What You Want Them To Do 505
Guideline 4  Stifle Your Tendency To Be Defensive 506
Guideline 5  Ask Your Reviewers To Explain the Reasons for Their
Suggestions 506
Guideline 6  Take Notes on Your Reviewers' Suggestions 507
Summary of Guidelines for Having Your Writing Reviewed 507
Introduction to Guidelines for Reviewing the Writing of Other
People 508
Guideline 1  Think of the Writers as People You Are Helping, Not
People You Are Judging 508
Guideline 2  Ask the Writers about the Purpose and Readers of
Their Communication 510
Guideline 3  Ask the Writers What They Want You To Do 510
Guideline 4  Review from the Point of View of the Readers 510
Guideline 5  Begin Your Suggestions by Praising Strong Points in
the Communication 511

*Guideline 6*   Distinguish Matters of Substance from Matters of
Personal Taste ........................................................................ 511
*Guideline 7*   Explain Your Suggestions Fully ........................................ 512
*Guideline 8*   Determine Which Revisions Will Most Improve the
Communication ...................................................................... 512
Summary of Guidelines for Reviewing the Writing of Other People ... 513
Exercises 514


Chapter
**24**

## Testing ........................................................................................... 516

Basic Strategy of Testing ............................................................... 517
Two Major Uses of Test Results .................................................... 518
When Should You Test a Communication? ...................................... 519
Three Elements of a Test .............................................................. 520
*Guideline 1*   Ask Your Test Readers To Use Your Draft in the
Same Ways Your Target Readers Will ............................ 520
*Guideline 2*   Learn How Your Draft Affects Your Test Readers'
Attitudes ............................................................................... 526
*Guideline 3*   Learn How Your Test Readers Interact with Your
Draft while Reading ............................................................. 528
*Guideline 4*   Choose Test Readers From Your Target Audience ...... 531
*Guideline 5*   Use Enough Test Readers To Determine Typical
Responses ............................................................................. 531
*Guideline 6*   Use Drafts That Are as Close as Possible to Final
Drafts .................................................................................... 532
Summary 532    Exercises 533


**PART VII    REVISING** ........................................................................ 535


Chapter
**25**

## Revising ......................................................................................... 536

When Revising, You Are an Investor ............................................. 537
To Revise Well, Follow the Guidelines for Writing Well ............... 537
*Guideline 1*   Determine How Good Your Communication Needs
To Be ..................................................................................... 538
*Guideline 2*   Determine How Good Your Communication Is Now,
Before You Revise It ........................................................... 541
*Guideline 3*   Determine Which Revisions Will Improve Your
Communication Most ........................................................... 541
*Guideline 4*   Plan To Make The Most Important Revisions First ..... 543
*Guideline 5*   Plan When To Stop Revising ............................................ 545
*Guideline 6*   Revise To Become a Better Writer ................................. 545
A Caution 546    Summary 546

**PART VIII    USING CONVENTIONAL FORMATS**                       549

Chapter    **Formats for Letters, Memos, and Books**            550
**26**
           Importance of Using Conventional Formats             551
           Formats and Employers' Style Guides                  552
           Choosing the Best Format for Your Communication      552
           Prepare Neat, Legible Communications                 553
           Organization of the Rest of This Chapter             554
           Using the Letter Format                              554
           Using the Memo Format                                564
           Using the Book Format                                568

Chapter    **Formats for References, Footnotes,**
**27**     **and Bibliographies**                               594
           Purposes of Documentation                            595
           Deciding What To Acknowledge                         595
           Choosing a Format for Documentation                  596
           Using Author–Year Citations Combined with a Reference List    597
           Using Numbered Citations Combined with a Reference List       605
           Using Footnotes                                      605
           Using Bibliographies                                 613
           Exercises 619

**PART IX    USING CONVENTIONAL SUPERSTRUCTURES**                621

Chapter    **Reports: General Superstructure**                  623
**28**
           How To Use the Advice in Chapters 28 through 31      624
           Varieties of Report-Writing Situations               624
           Your Readers Want to Use the Information You Provide  625
           The Questions That Readers Ask Most Often            626
           General Superstructure for Reports                   626
           A Note about Summaries                               631
           Sample Reports 632    Writing Assignments 632

Chapter    **Empirical Research Reports**                       640
**29**
           Typical Writing Situations                           641
           The Questions Readers Ask Most Often                 642
           Superstructure for Empirical Research Reports        643
           An Important Note about Headings                     651
           Sample Research Report 652    Writing Assignment 652

Chapter **Feasibility Reports**                                          671

**30**
Typical Writing Situation                                               672
The Questions Readers Ask Most Often                                    672
Superstructure for Feasibility Reports                                  673
Sample Feasibility Report 683    Writing Assignment 683


Chapter **Progress Reports**                                            690

**31**
Typical Writing Situations                                              691
The Readers' Concern with the Future                                    692
The Questions Readers Ask Most Often                                    692
Superstructure for Progress Reports                                     693
Tone in Progress Reports                                                697
Sample Progress Report 698    Exercise 698
   Writing Assignment 699


Chapter **Proposals**                                                   703

**32**
Variety of Proposal-Writing Situations                                  704
Proposal Readers Are Investors                                          706
The Questions Readers Ask Most Often                                    706
Strategy of the Conventional Superstructure for Proposals               707
Conventional Superstructure for Proposals                               710
Sample Proposal 720    Writing Assignment 720


Chapter **Instructions**                                                726

**33**
The Variety of Instructions                                             727
Three Important Points to Remember                                      727
Conventional Superstructure for Instructions                            730
Conclusion                                                              744
Exercises 747


**PART X    MAKING ORAL PRESENTATIONS**                                 761


Chapter **Preparing Your Talk**                                         764

**34**
Organization of This Chapter                                            765
*Guideline 1*    Define Your Objectives                                 765
*Guideline 2*    Select the Form of Oral Delivery
                 Best Suited to Your Purpose and Audience               768
*Guideline 3*    Integrate Visual Aids into Your Presentation           770
*Guideline 4*    Select the Visual Medium Best Suited to Your
                 Purpose, Audience, and Situation                       772

*Guideline 5*   In Your Presentation, Talk *with* Your Listeners   776
*Guideline 6*   Strongly Emphasize Your Main Points   778
*Guideline 7*   Make the Structure of Your Talk Evident   779
*Guideline 8*   Prepare for Interruptions and Questions   780
*Guideline 9*   Rehearse   780
Summary 781   Exercises 782

Chapter
## 35 Presenting Your Talk   783

Concentrate on Communicating with Each Individual in
  Your Audience   784
*Guideline 1*   Arrange Your Stage so You and Your Visual Aids Are
    the Focus of Attention   785
*Guideline 2*   Look at Your Audience   786
*Guideline 3*   Speak in a Natural Manner, Using Your Voice To
    Clarify Your Message   787
*Guideline 4*   Exhibit Enthusiasm and Interest in Your Subject   788
*Guideline 5*   Display Each of Your Visual Aids Only when It
    Supports Your Words   788
*Guideline 6*   Make Purposeful Movements, but Otherwise
    Stand Still   788
*Guideline 7*   Respond Courteously to Interruptions, Questions,
    and Comments   789
*Guideline 8*   Learn To Accept and Work with Your Nervousness   789
Summary 790   Exercises 791

## PART XI APPENDIXES   793

### Appendix I Writing Assignments   794

### Appendix II Sample Unsolicited Recommendation   802

Index   809

# Introduction

The three chapters in Part 1 introduce you to the aims and approach of this book. You will learn about the kind of writing you will do on the job and its importance to you. You will learn nine general guidelines for writing, and you will learn how to use your understanding of your reader to guide you through the various steps of writing at work.

## CONTENTS OF PART I

**Chapter 1**
**Writing, Your Career, and This Book**

**Chapter 2**
**A Reader-Centered Approach to Writing**

**Chapter 3**
**Examples of Reader-Centered Writing: Resumes and Letters of Application**

# Writing, Your Career, And This Book

## PREVIEW OF CHAPTER

**Sharing ideas is a major activity at work**

**Your two roles at work**

**Writing is a form of action**

**Writing is a form of *social* action**

**The main advice of this book: write to people**

**Knowledge and common sense about people at work**

**Guidelines for writing**

**The organization of this book**

**Improving your writing by improving your writing process**

While writing this book, I have pictured you in my mind. I imagine that you are a capable, creative person who often has valuable insights and good ideas. You take your studies seriously, especially in your major department. And you look forward to graduation, to the challenges and pleasures of full-time work. Perhaps you are working now, so that you already have the opportunity to use your education on the job.

I have also imagined what your days at work will be like. You will spend much of your time using the special knowledge and skills you learned in college. You will search out answers to questions your co-workers have asked, develop recommendations your boss has requested, and solve problems your customers are facing. Furthermore, you will generate many good ideas on your own, without being asked to do so by anyone else. Looking around, you will notice ways to make things work better and find ways to do them less expensively. You will discover ways to overcome difficulties that have stumped others.

Finally, I have imagined your excitement as you tell your fellow employees and customers about your work. You will be eager to help them by explaining what you know, found out, or created. You will look forward to receiving their recognition for your accomplishments. To enjoy those pleasures, however, you must *share* your thoughts and knowledge by talking or writing.

This book is about that sharing. It offers many suggestions that will help you communicate effectively on the job—so your ideas are understood, your recommendations are accepted, your information is welcomed and used, and your readers are pleased.

## SHARING IDEAS IS A MAJOR ACTIVITY AT WORK

Sharing ideas will be one of your major activities at work. Numerous surveys indicate that if you are at all like the typical college graduate, you will spend a large part of your time at work writing. For example, 245 people listed in *Engineers of Distinction* reported to a researcher that on the average they devote 24% of their work time to writing.[1] That's more than one day out of every five-day work week.

In another survey, 200 people selected to represent a cross-section of college-educated people in the United States indicated that on the average they spend over 20% of their time at work writing.[2] Similar experiences were reported by 836 graduates of seven departments that send students to technical writing courses—departments ranging from systems analysis and chemistry to home economics and office administration.[3] In fact, as Figure 1-1 shows, 15% of the graduates reported spending more than 40% of their time at work writing.

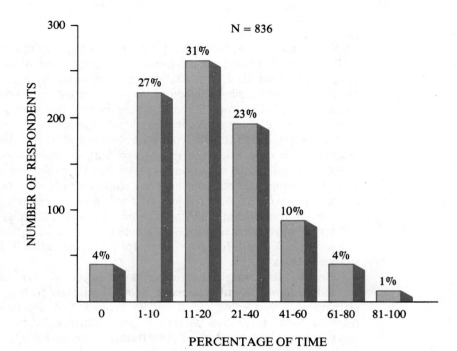

**Figure 1-1.**   Percentage of Time Spent Writing by Graduates of Seven Departments That Send Students to Technical Writing Courses.

From Paul V. Anderson, "What Survey Research Tells Us about Writing at Work," in *Writing in Nonacademic Settings,* ed. Lee Odell and Dixie Goswami (New York: Guilford Press, 1985), 18.

The college graduates responding to these and other surveys also indicate that if you are typical, you will find that writing is important to your success on the job. For instance, 94% of the graduates from the seven departments that send students to technical writing courses reported that the ability to "write well" (not just to write, but to *write well*) is of at least "some" importance to them; 58% said that it is of "great" or "critical" importance. (See Figure 1-2.) In the survey of people listed in *Engineers of Distinction,* 89% said that writing ability is considered when a person is being evaluated for advancement, and 96% said that the ability to communicate on paper has helped their own advancement.

The overwhelming conclusion of these and many other surveys is that writing will almost certainly be one of your major activities at work.

## YOUR TWO ROLES AT WORK

Why will writing be so important to you on the job? Two major reasons have already been mentioned. First, you will write to win recognition for your hard, creative work. Beyond that, your employer will *expect* you to communicate extensively and effectively. How come? Because all the facts, all the theories, all the skills that you have at your command will be useless to your employer unless you share them with someone else.

Consider, for example, the situation of Larry, a recent graduate in metallurgy. He has just analyzed a group of pistons that broke when used in an experimental automobile engine. No matter how skillfully he conducted this analysis, Larry's work can help his employer only if Larry communicates the results to someone else, such as the engineer who must redesign the pistons.

Sarah, a newly hired dietitian, must also communicate to make her work valuable to her employer, a large hospital. She has devised a way to reorganize the hospital's kitchen that will save money and provide better service to the patients. However, her insight will benefit her employer only if she

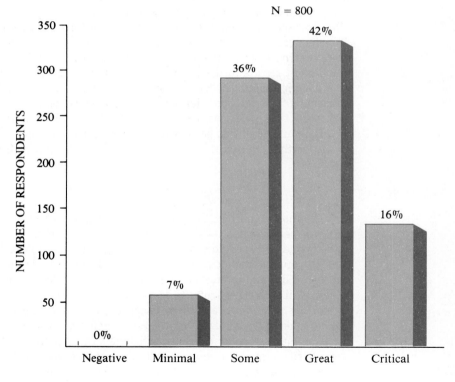

**Figure 1-2.** Importance of Writing to Graduates of Seven Departments That Send Students to Technical Writing Courses. Percentages do not add to exactly 100 because of rounding.

From Paul V. Anderson, "What Survey Research Tells Us about Writing at Work," in *Writing in Nonacademic Settings*, ed. Lee Odell and Dixie Goswami (New York: Guilford Press, 1985), 19.

communicates her recommendations to someone else, such as the kitchen director, who has the power to implement them.

Like Larry and Sarah, you will be able to make your work valuable to your employer only if you communicate effectively. Consequently, you will perform two separate roles on the job. First, you will be a *specialist*. Using your knowledge and skills in your specialty, you will generate information and ideas that will be potentially useful. Second, you will be a *communicator*. In this role, you will share the results of your specialized activities with co-workers, customers, and other people who rely on you.

## WRITING IS A FORM OF ACTION

Your two roles—specialist and communicator—have something important in common. In both you perform an *action*. You exert your powers to bring about a specific result. For example, consider Larry's actions. As a metallurgical specialist, his action is to test the pistons to learn why they failed. As a metallurgical communicator, his action is to construct tables of data, write sentences, and perform a broad range of other writing activities to present his test results to the engineer who will redesign the pistons. Similarly, as a dietetics specialist, Sarah's action is to devise a plan for improving the efficiency of the kitchen. Her action as a dietetics communicator is to write a proposal that persuades the kitchen director to adopt her plan.

The purpose of this book is to help you learn to act effectively when you write—so that your writing brings about the results you desire.

## WRITING IS A FORM OF *SOCIAL* ACTION

The most important thing for you to remember about the "writing acts" you will perform at work is that they will be *social* actions. They will be interchanges between you (a person) and your readers (other individual people). In your interactions, you and your readers will be influenced by all the considerations and interests that affect any human interchange. You and your readers will have social roles to play—you will be the supervisor telling your co-workers what you want them to do, the adviser trying to persuade your boss to make a certain decision, the expert helping a worker operate a certain piece of equipment. Your reader will be an experienced employee who has some hesitancies about your request, a supervisor who has been educated to ask certain questions when making decisions, a machine operator who has a particular sense of personal dignity and a specific amount of knowledge of the equipment to be operated.

In all these relationships, many social conventions will apply, various personal interests will be at stake, and a wide array of personal feelings will be touched. Although it's true that your interactions with your readers will be defined to a great extent by the particular kinds of social relationships found in the workplace, at bottom there always will be those individuals: you and your readers.

## THE MAIN ADVICE OF THIS BOOK: WRITE TO PEOPLE

From the observation that writing is a social action follows the main advice of this book: *when writing, think constantly about your readers.* Think about what they want from you—and why. Think about the ways you want to affect them and the ways they will react to what you have to say. Think about them in all the ways you would if they were standing right there in front of you while you talked together.

You may be surprised that this book emphasizes the personal dimension of writing more than such important characteristics of writing as clarity and correctness. Although clarity and correctness are important, they cannot—by themselves—ensure that something you write at work will succeed.

To succeed, the things you write at work must affect in a certain, specific way the individual people you are writing to. You can see why if you think about Sarah's proposal for reorganizing the hospital's kitchen. If her proposal explains the problems created by the present organization in a way that her *reader* finds compelling, if it addresses the kinds of objections her *reader* will have to her recommendations, if it reduces her *reader's* sense of threat at having a new employee suggest improvements to a system that the reader set up, then it may succeed.

On the other hand, if Sarah's proposal fails to persuade her reader that the present organization of the kitchen creates problems that are significant from his point of view, if it leaves some key objections unanswered, if it makes her seem like a pushy person who has overstepped her appropriate role, it will not succeed. It won't matter how "correct" the writing is, or how "clear." A communication can be perfectly clear, perfectly correct, and yet be utterly unpersuasive, utterly ineffective.

Similarly, consider Larry's report on the faulty pistons. If he writes it so the engineer who must redesign the pistons can find the information she needs quickly and in a readily usable form, his report will achieve its purpose. It will do what Larry created it to do. However, if the engineer has to hunt through the report for the needed information and must convert it into another form before she can use it, then Larry's report will be ineffective—no matter how clearly and correctly it is written.

Of course, you will often write communications that will be read by many people, not just one individual. Even Sarah's proposal and Larry's report are likely to be read by many of their co-workers. How do you write to your "individual reader" when your communication will have *many* readers? In later chapters you will find much advice about how to write effectively in those situations. For now, the important point for you to remember is that you should take a "reader-centered" approach to writing. As you write, think constantly about your readers.

## KNOWLEDGE AND COMMON SENSE ABOUT PEOPLE AT WORK

If you have not had any work experience, you may be wondering how you can succeed in using a textbook that seems to place so much emphasis upon knowing what people do at work—and why and how they do it.

First, this book provides you with the basic information you will need. For example, Chapter 5 describes the basic roles that readers play at work, and Chapter 4 describes the basic reasons that people will have for reading what you have written. Further, all the advice in this book takes into account the way things are done on the job.

You can gain additional information about people in the working world from your teacher and classmates who have work experience. Also, you can talk with instructors in your major department and people outside of college, such as your parents and friends who have graduated.

However, even without using any of those resources you already know the most important things about the people you will be writing to. Although people's attitudes and interests are shaped somewhat by their jobs, they remain people. Hence, you already know such crucial things as what makes them feel good, what offends them, what they are likely to do when they are frustrated, and how they are likely to search for answers to the questions they have. If you use that knowledge when you write, you will have taken the single most important step towards writing effectively at work.

This book provides many detailed suggestions about ways you can use your extensive knowledge of people as you write.

## GUIDELINES FOR WRITING

So that these suggestions will be as easy as possible for you to remember and use, they are presented as *guidelines*. Each chapter offers between three and ten. You may want to skim through them all now.

As you read the guidelines—whether now or later—you should keep two things in mind.

First, the guidelines look deceptively simple. Indeed, you could probably memorize them all in an hour or so. What's difficult about learning to write effectively isn't remembering the guidelines. It's learning how to apply them in particular situations. That takes concentrated thought and much practice. To aid you, the book explains the reasons for each guideline and presents examples of their use. In addition, you will gain much practice in using them while working on the exercises and projects your instructor assigns.

Second, as you read the guidelines keep in mind that they are merely that: guidelines. They are not rules. Although you can usually succeed by following them, every one has exceptions.

For example, one guideline reads, "Put the most important information first." That's good advice, for instance, when you are responding to your boss' request for a recommendation. The most important information is going to be the recommendation itself, and your boss will want to read it right away.

However, when you are writing a recommendation to someone who *hasn't* asked for your advice, you might be more successful if you withhold your recommendation until you have explained why your reader should agree with you that some kind of change is desirable.

Thus, you will need to use good judgment to apply the guidelines successfully. You will find plenty of help in this book. It discusses all the guidelines in a way that will assist you in developing and refining your ability to use—and ignore—them wisely.

## THE ORGANIZATION OF THIS BOOK

As you look through this book, you will notice that most of the chapters focus on one or another of the tasks you perform when you write: deciding what to say, organizing your material, phrasing your sentences, and so on. Accordingly, as you use this book in your writing class, you will systematically study each of these crucial activities. In addition, if you find that you need help with any of these tasks, you can turn directly to the chapter that provides the advice you need.

You will notice also that the chapters are gathered into parts that suggest a general process for you to follow when you write. The following paragraphs describe the five steps in the process.

- *Define Your Objectives (Part II).* First, you decide exactly what you want to accomplish, and determine what limitations there are on the way you can pursue that goal.
- *Plan (Part III).* You create a strategy for accomplishing your objectives. When doing that, you decide such things as what material you will include, how you will organize it, and what visual aids you will use.
- *Draft (Parts IV and V).* You put your plan into action by drafting your prose and visual aids.
- *Evaluate (Part VI).* You look carefully at your draft, preferably with help from other people, to see if it is likely to achieve the objectives you defined in the first step of the writing process.
- *Revise (Part VII).* You improve your draft by making the changes suggested by your evaluation. When you have finished these improvements, you are ready to deliver your communication to your readers.

## IMPROVING YOUR WRITING BY IMPROVING YOUR WRITING PROCESS

If you don't already follow this process, you can probably improve your writing a great deal just by organizing your work in the way it suggests. For example, you will generally write most effectively if you define your objectives and plan your strategy for achieving those objectives *before* you begin to draft your prose and visual aids. Likewise, you can usually improve a communication by carefully evaluating and revising your draft before you deliver it to your readers.

At the same time, you should avoid following the five steps in the process in rigid order. It's natural for you to skip around among the steps. When you are defining the objectives for a communication, you may think of some good ideas about how to write it. Hang onto the ideas so you can use them when you are planning and drafting. When you are drafting, you may discover a way to improve the outline you originally planned. Rather than stick with the old, less effective outline, change it in light of your discovery. While you are evaluating or revising, you may learn something important about your readers. Act on that insight rather than send a communication that is aimed slightly off the target.

Is there any particular pattern you should follow when skipping among the steps? Not really. Approaches to writing vary widely, even among good writers. Find one that works for you. Just be sure to follow the general *spirit* of the process. Create communications in which your plans and drafts reflect a careful definition of your objectives. Use those same objectives to evaluate and revise your work before giving it to your readers.

## SUMMARY

At work, writing will consume a substantial portion of your time, and it will be important to your success in your career. It will provide you with a way to show off your good work, and it will be a major way you will make your work valuable to other people.

When writing on the job, you are performing an *action*. That is, you exert your powers to bring about a specific result. The most important feature of this action is that it is a *social* action, involving you and each of your readers as individuals. The most important piece of advice this book has to offer you, the fundamental recommendation that underlies all its other suggestions, is this: as you write, think constantly about your readers as individual people.

This book presents its specific suggestions about writing in the form of guidelines. However, you cannot become a successful writer merely by memorizing these guidelines. You must also learn how to apply them effectively. That requires practice, practice that is accompanied by knowledgeable and sensitive reactions to your efforts.

Such reactions, unfortunately, are something a textbook cannot give. To learn whether or not your way of applying the guidelines is working, you must turn to other people. In fact, the value to you of this book will depend to a large extent upon your openness to insights and advice provided by your instructor—and by your classmates, if your instructor asks you and your classmates to review one another's work.

In your efforts to prepare yourself for the writing you will do at work, I wish you good luck. And I wish you the sense of satisfaction and self-confidence that comes from knowing that you can write well, that you can communicate your good ideas and your special knowledge in ways that will make good things happen for you and your readers.

## NOTES

1. Richard M. Davis, "How Important is Technical Writing?—A Survey of Opinions of Successful Engineers," *The Technical Writing Teacher* 4(1977):83–8.
2. Lester Faigley and Thomas P. Miller, "What We Learn from Writing on the Job," *College English* 44(1982):557–69.
3. Paul V. Anderson, "What Survey Research Tells Us about Writing at Work," in *Writing in Nonacademic Settings,* ed. Lee Odell and Dixie Goswami (New York: Guilford Press, 1985), 3–85.

## EXERCISES

1. Imagine two situations in which you will write on the job. For each, explain what purposes you will have for writing and what purposes your readers will have for reading.
2. Find a communication that has been written by someone who has the kind of job you want. Explain its purposes from the points of view of both the writer and the readers. Describe some of the writing strategies the writer has used to achieve those purposes.
3. Imagine that another student in your major has asked you the following questions about writing on the job:
   - Will writing be an important part of *my* job?
   - In what ways will my writing affect my success?
   - How much will I write?
   - Why will I write?
   - What kinds of communications will I write?
   - Who will my readers be?

   Help this student by calling or visiting one or two people who have the kind of job that you want to hold. Then write your friend a memo in which you report your results.

   Here are some alternative ways to answer those questions, if your instructor approves:
   - Ask an instructor in your major department. If possible, find an instructor who has worked off-campus in the kind of job you want.
   - If you have already worked in the kind of position you want as an intern, co-op student, or regular employee, base your answers on what you observed.

# A Reader-Centered Approach to Writing

## PREVIEW OF GUIDELINES

1. When writing, talk *with* your readers.
2. Write in a respectful voice.
3. Write in a professional manner.
4. Help your readers focus on the key information.
5. Write sentences your readers can understand easily.
6. Help your readers follow the flow of thought from sentence to sentence.
7. Help your readers build hierarchies of meaning.
8. Tell your readers the significance *to them* of your message.
9. Use visual aids.

Imagine that you are sitting at your desk, preparing to write something. Maybe it's a report for your employer or an assignment for class. You know the result you want your writing to bring about: to persuade someone to accept your recommendation, to enable someone to perform a new procedure, to help someone learn the answer to a question. Furthermore, you remember reading in the first chapter of this book that you should take a reader-centered approach to writing, one in which you think constantly about your readers.

But you feel a little frustrated. You ask, "How am I supposed to *use* that advice? What, exactly, am I supposed to think about my readers, and how will my thoughts about my readers help me put words down on paper?"

The rest of this book answers those questions. It presents many guidelines for thinking about your readers and for using the knowledge you gain about them to help you write effectively. The guidelines provide detailed advice about all of the various activities you perform when writing, from defining your objectives to organizing your message and choosing your words.

Before you study the detailed advice contained in these guidelines, however, you may find it helpful to consider nine *general* guidelines that summarize the reader-centered approach to writing. This chapter describes those general guidelines.

## INTRODUCTION TO GENERAL GUIDELINES 1 THROUGH 3

The first three general guidelines focus on the relationship you establish in your communication with your readers. One involves a way you can keep that relationship vivid in your mind as you write. The others concern the way your writing should "sound" so that your readers will feel receptive to you and (consequently) to your message.

General
Guideline

**1**

### When Writing, Talk *with* Your Readers

As you read in the first chapter, writing is a social activity in which you, a particular person, communicate to other individuals, your readers. Appropriately, then, the first general guideline is this: when writing, talk *with* those other people.

The key word in this guideline is *with*. Don't talk *to* your readers. Talk *with* them. When you talk *to* other people, you are like an actor reciting a speech: you stick to your script without regard to how your audience is reacting to your words.

In contrast, when you talk *with* others, you adjust your statements to fit their reactions. Does someone squeeze his brows in puzzlement? You explain the point more fully. Does someone twist her hands impatiently? You abbreviate your message. Are your listeners unpersuaded by one argument? Then you abandon it and try another. Do they ask to hear more? You provide additional information.

When you write, you should respond to your readers' reactions in much the same way—paying attention to their moment-by-moment reactions to your communication and shaping your message accordingly. How do you do that when you are writing rather than speaking face-to-face? You use your imagination. You imagine your readers in the act of reading the words you are writing and then write accordingly. Shortly you'll read some examples that show how to do that.

## A Reader's Moment-by-Moment Reactions

Why should you think about your readers' reactions during *each moment* they are reading your communications? The following demonstration will show you.

Imagine that you are Dwight Gardner, manager of the Personnel Department in a factory. A few days ago you met with Donna Pryzblo, manager of the Data Processing Department, to discuss a problem: recently, the company's computer has been issuing some payroll checks for the wrong amount.

The two of you discussed this problem because your department and Ms. Pryzblo's department work together to create each week's payroll. First, your clerks collect a timesheet for each employee. Then they transfer the information on those sheets to time-tickets, which they forward to the clerks in Ms. Pryzblo's department. Her clerks enter the information into a computer, which calculates each employee's pay and prints the payroll checks. The whole procedure is summarized in the following diagram:

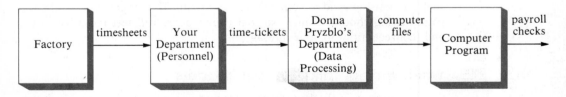

When meeting with Ms. Przyblo, you (Dwight Gardner) proposed a solution that she did not like. Because you and Ms. Pryzblo are both at the same level in your company's organization (you report to the same boss), neither one of you has the power to impose your suggestions on the other. In today's mail, you find a memo from her.

Your task in this demonstration will be to read that memo, but not at your normal reading speed. You should read the memo *very slowly*—so slowly that you can focus your attention on something that normally happens too rapidly for you to notice, namely the way you react, moment-by-moment, to each statement you read.

Here is the procedure you should use to slow down your reading. First, cover the memo entirely with a sheet of paper. Then slide the paper down the page, stopping as you expose each new sentence. Once you have read that sentence—and before going on to the next one—write down your thoughts and feelings (in your role as Dwight Gardner). In this demonstration, there are no "right" or "wrong" answers. You are not being tested.

Ready? Then take out a sheet of paper, slip it quickly over Figure 2-1 and begin to slide it down Ms. Pryzblo's memo, reading and then recording your reaction to each sentence. When you are finished, turn back here and read the next section.

## Discussion of the Demonstration

Finished? As you have probably guessed, Ms. Pryzblo's memo did not work. That is, Ms. Pryzblo wanted Mr. Gardner to accept the recommendation she makes in the final sentence. He did not do so. Why? We cannot know for sure. You can make some reasonable guesses, however, by looking back at your notes on your own reactions to the memo as you read it in your role as Dwight Gardner.

If you reacted in one of the ways that most people do, your reactions will illustrate two points about how people read that are basic to the advice in this book.

1. *As readers, we interact with the things we read on a moment-by-moment basis.* When we are reading a humorous novel, we all chuckle as we read the sentence that tells the punch line. We don't wait to laugh until we have finished the book. Likewise, on the job, people react to each part of a memo or other business communication as they come to that part.

   For example, most people who participate in the Dwight Gardner demonstration react to the quotation marks that surround the word *errors* in the first sentence. And they react immediately; they don't wait until they have finished reading the entire memo. The word *insinuated* in the second sentence likewise draws an immediate reaction from most readers. (They laugh if they forget to play the role of Dwight Gardner; most cringe if they remember to play the role.)

2. *Our reaction during one moment of reading shapes our reactions in subsequent reading moments.* That is, our reaction to one sentence affects our reaction to the sentences we read afterwards. Consider, for example, what happens as most readers read through Donna Pryzblo's memo, playing their role as Dwight Gardner. They start to become defensive the moment they see the quotation marks around the word *errors,* or at least by the time they come to the word *insinuated.* After they have read Ms. Pryzblo's third paragraph, their defensiveness has hardened into a grim determination to resist any suggestion that Ms. Pryzblo might make.

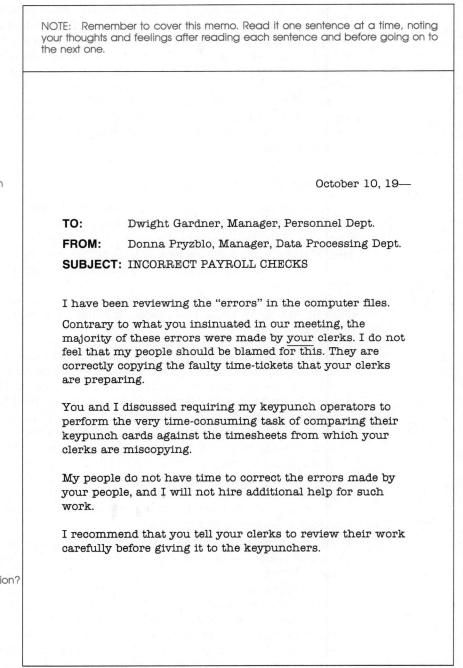

Make notes on your thoughts and feelings here

October 10, 19—

**TO:**      Dwight Gardner, Manager, Personnel Dept.

**FROM:**    Donna Pryzblo, Manager, Data Processing Dept.

**SUBJECT:** INCORRECT PAYROLL CHECKS

I have been reviewing the "errors" in the computer files.

Contrary to what you insinuated in our meeting, the majority of these errors were made by your clerks. I do not feel that my people should be blamed for this. They are correctly copying the faulty time-tickets that your clerks are preparing.

You and I discussed requiring my keypunch operators to perform the very time-consuming task of comparing their keypunch cards against the timesheets from which your clerks are miscopying.

My people do not have time to correct the errors made by your people, and I will not hire additional help for such work.

I recommend that you tell your clerks to review their work carefully before giving it to the keypunchers.

Would you (as Dwight Gardner) accept this recommendation? Why? Or why not?

**Figure 2-1.** Memo for Demonstration.

Thus, they have little interest in accepting the recommendation she makes in her last sentence.

A smaller group of readers is more even tempered. Instead of becoming defensive, they become skeptical. Their skepticism is triggered by the first two sentences, where they interpret the quotation marks around the word *errors* and the use of the word *insinuated* as signs that Donna Pryzblo is behaving emotionally. Consequently, they decide to evaluate her statements very carefully. When they read the accusation that their clerks (that is, Dwight Gardner's) are "miscopying," they immediately want to know what evidence Ms. Pryzblo has to support that claim. When they read the next sentence, which fails to provide that evidence, they are disappointed and also disinclined to go along with any suggestions she might make.

Thus, even though different readers react to the early sentences of this memo in different ways, their early reactions shape their reactions to later sentences, including the recommendation. No reader I've met feels like following this recommendation, at least until Ms. Pryzblo provides more information than she does here.

As you read Ms. Pryzblo's memo, you may have wondered about her purpose in writing. What if her purpose was not to persuade Mr. Gardner to accept her recommendation but rather to inflame him into making a rash and foolish response that would get him in trouble with their mutual boss? Or, what if she had some entirely different purpose? Then her memo might have worked very well. Even in those cases, however, the basic points of this demonstration would remain unchanged: readers react to the things they read on a moment-by-moment basis and their reactions in one moment shape their responses to what they read next.

## How to Talk with Your Readers

How can you make practical use of this knowledge about how people read? You can imagine that your communications are conversations in which you make a statement and your readers respond. Then, with the image of your readers' reactions foremost in your mind, write each of your sentences, each of your paragraphs, each of your chapters, to create the interaction—the conversation—between you and your readers that you think will bring about the final result you desire.

Consider how much Ms. Pryzblo could have benefited from "talking" with Mr. Gardner as she drafted her memo. After writing the first sentence, she would have seen him wrinkle his forehead as he saw the quotation marks around *errors*. After writing the second sentence, she would have seen his jaw tighten and heard him exclaim, "What proof do you have that my clerks are making the mistakes?" Seeing these things, she would have known that she could persuade Mr. Gardner to accept her recommendation only if she changed her draft. In the same way, you can benefit from imagining your readers' reactions as you draft your communications.

In fact, you can also benefit from thinking about your talk with your readers when you are still planning your communication. You are probably already experienced at planning important conversations in advance, for instance, when you want to talk with your instructor about a grade or when you want to ask a large favor of a friend.

If Ms. Pryzblo had planned in this way, she probably would have written even her first draft much differently. As she planned, she might have said, "I want to begin this memo in a way that will make Mr. Gardner feel open-minded about my recommendation." Then she could have chosen among the various ways of beginning her memo that would promote open-mindedness, such as beginning with a friendly statement (not accusations) and then saying that she wants to work with Mr. Gardner to solve the problem of the payroll checks in a way that will best serve the interests of their mutual employer.

Finally, you can benefit from this technique of talking with your readers when you have completed a draft. As *you* read through the draft, you can determine the likelihood that your entire communication, taken together, will affect your readers in the way you desire as *they* read through it.

By imagining that you are talking with your readers, you can help yourself write any kind of communication, not just ones (like Ms. Pryzblo's) that are intended to be persuasive. Here, for instance, is an account of how a young engineer, Marty, improved one of his instructions for calibrating an instrument that tests steel girders. Marty's original instruction read like this:

15. Check the reading on Gauge E.

Marty then imagined how a typical member of the audience for his set of instructions would react after reading that instruction. The reader looked up at Marty and asked, "What should the reading be?" So Marty told the reader to look for the correct reading in the Table of Values. He then imagined his reader asking, "Where is that table?" So he added the location of the table. Then Marty imagined that when the reader looked at the Table of Values, the reader discovered that the value on Gauge E was incorrect. The reader asked, "What do I do now?" So, Marty revised again. In the end, his instruction read as follows:

15. Check the reading on Gauge E to see if it corresponds with the appropriate value listed in the Table of Values (see Appendix IV, page 38).

   • If the value is *incorrect,* follow the procedure for correcting imbalances (page 27).
   • If the value is *correct,* proceed to step 16.

## When Your Communication Will Be Read by Many Readers

At work, you will often write things that will be read by many people. How can you "talk" with your readers then? The answer depends upon your situation. If all of your readers will be using your communication for the same purpose, and if all bring to it essentially the same knowledge of your subject matter, then you can do what Marty did: imagine a *typical* reader.

Many times, however, you will write communications that are read by many people who have different purposes for reading. For example, if you write a proposal to a customer on behalf of your employer, various people in the customer's organization may read it for different reasons. A technical specialist might read it to see if your proposal makes sense technically. A management specialist might read it to see if your plan for managing the project is sound. And an accountant might read it to see if your budget is reasonable. In these situations, you need to imagine several readers, thinking about how each of them will read your communication. When you write, you will have to devise ways to present your message so that all are satisfied. Later chapters will give you numerous ideas for doing that.

**General Guideline 2**

## Write in a Respectful Voice

As you were reading about the first general guideline ("When writing, talk with your readers"), you may have been wondering what you should *sound* like when you "talk" with your readers. The question is an important one. As you read in the first chapter, writing is a social interaction between individuals. Like all people, your readers will like to be spoken to in certain ways, but not in others. Further, your readers will make judgments about you based upon the voice you use when you write. These judgments will affect their reactions to what you say.

You can demonstrate to yourself the importance of voice by thinking once again about Donna Pryzblo's memo. In it, she speaks in an angry, accusing voice. Typically, readers react much more strongly to the voice Ms. Pryzblo uses than to the information or the recommendation she presents. If she had chosen to speak in a different voice—for example, in the voice of a reasonable problem-solver or a helpful co-worker—then her memo might have been much more effective.

### How To Choose Your Voice

Given that the voice Donna Pryzblo uses is usually ineffective on the job, what voice will work? There is no pat answer to that question. The answer depends on many things, including the following:

- Your professional relationship with your readers—for example, whether you are a customer, supervisor, or subordinate.

- Your purpose—for example, whether you are asking for something, apologizing for an oversight, providing advice, ordering your readers to do something, or granting a request.
- Your subject—for example, whether you are writing about a routine matter or an urgent problem.
- Your personality.
- Your readers' personalities.

### Are You Being Insincere if You Choose Your Voice?

Perhaps you are hesitating right now because you are thinking that you might be acting in an insincere or artificial way if you were to take so many things into account when you write on the job. After all, you will be the same person every time you speak. Shouldn't you write in the same voice each time?

You can answer that question by imagining that you are walking across campus. On the way, you stop to talk with three people: a close friend, the instructor in a course that you have just begun, and an interviewer from a company you would like to work for. The words you choose, the subjects you discuss, even the way you stand are likely to be different in each of these three conversations. In fact, each of these individuals would think it strange if you didn't adjust your conversation to him or her. What would the interviewer think if you spoke to her in the same casual way you speak to your close friend? And what would your friend think if you were to address him in the way you would speak to your instructor?

These adjustments in the way you present yourself are all perfectly natural. You have different relationships with each person, and you interact with each of them accordingly. On the job, you will make the same kinds of adjustments. You will probably talk one way with the person who works next to you and another way with your boss's boss, one way with someone in the department down the hall and another way with a potential customer.

On the other hand, no matter how much you adjust your voice to suit the situation, you will write most successfully if the voice you use is *respectful*. In addition to the ethical arguments for acting respectfully toward others, there is also a practical one: your readers will be more likely to cooperate with you if you write respectfully than if you don't. They will probably become resentful and uncooperative if they feel you are not respectful.

To let your readers know that you respect them, you can use your common sense, show that you understand your readers' point of view, and acknowledge your readers' abilities and areas of authority.

### Use Your Common Sense

The first and most important strategy for writing in a respectful voice is to use your common sense about how to show your regard for other people.

Imagining that you are talking face-to-face with your readers will help you call your instincts into play.

### Show that You Understand Your Readers' Point of View

You can show your understanding in many ways. In communications where you are attempting to persuade, you can couch your arguments in terms of goals that you know your readers have. You can state and carefully discuss the points you know your readers would raise, even when you disagree with those points. When rejecting some request or proposal, you can mention the reasonableness of the request or the good points of the proposal and *explain* your reasons for rejecting it.

### Acknowledge Your Readers' Abilities and Areas of Authority

For example, when writing instructions, you should avoid telling your readers things they surely know—as if they were not competent to do their jobs. When writing recommendations, you should avoid implying that your readers *must* agree with you—as if they had no room for making a decision.

General
Guideline
3

## Write in a Professional Manner

Besides showing your respect for your readers, you should also strive to win your readers' respect for you. To a large extent, you can win their respect by the way you perform the work you do *before* you write: by using your specialized knowledge to produce important information, solid ideas, and sound advice. You must keep in mind, however, that your readers' opinion of you will be shaped not only by the quality of your information, ideas, and advice, but also by the manner in which you present those things when you write.

There are many strategies you can use to show that you are writing about your subject in a professional manner. Four important ones are discussed in the following paragraphs.

### Discuss Your Subject from the Point of View of Your Professional Role

You will be hired to perform a specific professional role. By playing that role you make your special contribution to the overall cooperative effort that brings you and your readers together in the first place. Maybe you will be a computer analyst working with your fellow employees to make your employer's company run more efficiently. Or maybe you will be a research biologist working with fellow scientists around the world to increase the understanding of human genetics. In any event, you and your readers will be pursuing common goals. You should write from the point of view of your professional role and those common goals. If you abandon your role and discuss your subject from the point of view of your personal feelings or private advantage, your readers will believe that you are writing unprofessionally.

To see the difference between an unprofessional and a professional treatment of the same subject, compare the two drafts of a memo written by Terry Prindle. Mr. Prindle is a marketing specialist who wishes to persuade his boss, Frieda Van Husen, to give him an office of his own.

In the first draft of his memo (Figure 2-2), Mr. Prindle seems to be concerned mainly with explaining how a new office would help him. Though the decision to change the office arrangements is Ms. Van Husen's, this draft of the memo fails to provide her with any reasons why she, acting in her professional role, should want to alter them.

In the revised memo (Figure 2-3), Mr. Prindle treats his subject in a more objective and professional manner by explaining how a new office for him would help the organization achieve its goals.

Because Mr. Prindle clearly will benefit from a new office, you may be wondering whether or not his revised memo is manipulative. As a general rule, things written at work are not manipulative if they argue from the point of view of the shared professional goals and values of the reader and writer. Mr. Prindle is expected to present his case in those terms, and Ms. Van Husen is expected to evaluate it in those same terms. Mr. Prindle's memo would be manipulative only if he suggested that his reader use some other standard for evaluation, such as her personal interests. For instance, he would be manipulative if he appealed to Ms. Van Husen's vanity by suggesting that a person as important as she is should not have her receptionist's office cluttered up with his things. As he has written it, Mr. Prindle's revised memo is more persuasive than his original draft, but it is not manipulative.

## Choose Words Commonly Used in Your Professional Setting

The second action you can take to write in a professional manner is to use the vocabulary of your professional setting—the specialized terms your readers are used to using. By doing so, you show that you are a member of that professional community, that you are a person who shares your readers' words.

At the same time, you must avoid using specialized terms when you are writing to people who won't understand them. Readers often feel resentful when they face a communication whose vocabulary they can't understand.

## Write Grammatically Correct Sentences and Spell Words Correctly

At work, readers expect correct grammar and spelling. They are much harsher in their judgment of slips in these areas than are most college

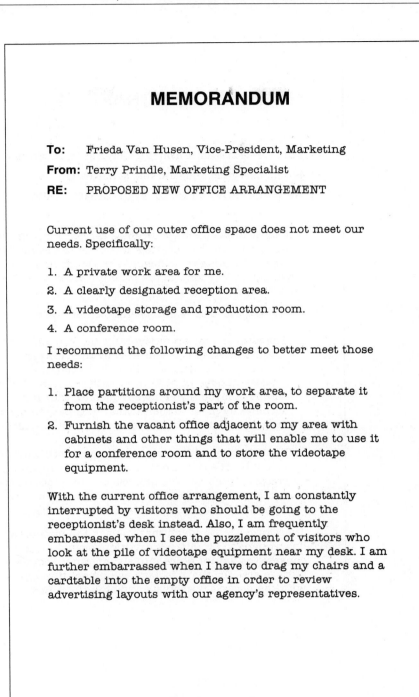

# MEMORANDUM

**To:**   Frieda Van Husen, Vice-President, Marketing

**From:** Terry Prindle, Marketing Specialist

**RE:**   PROPOSED NEW OFFICE ARRANGEMENT

Current use of our outer office space does not meet our needs. Specifically:

1. A private work area for me.
2. A clearly designated reception area.
3. A videotape storage and production room.
4. A conference room.

I recommend the following changes to better meet those needs:

1. Place partitions around my work area, to separate it from the receptionist's part of the room.
2. Furnish the vacant office adjacent to my area with cabinets and other things that will enable me to use it for a conference room and to store the videotape equipment.

With the current office arrangement, I am constantly interrupted by visitors who should be going to the receptionist's desk instead. Also, I am frequently embarrassed when I see the puzzlement of visitors who look at the pile of videotape equipment near my desk. I am further embarrassed when I have to drag my chairs and a cardtable into the empty office in order to review advertising layouts with our agency's representatives.

**Figure 2-2.**   A First Draft Written in an Unprofessional Manner.

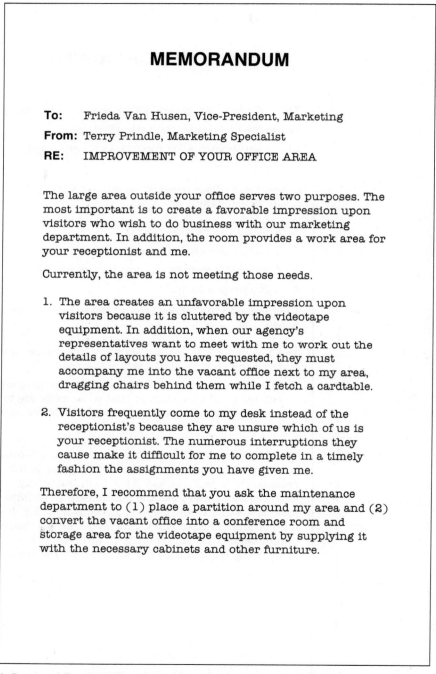

# MEMORANDUM

**To:** Frieda Van Husen, Vice-President, Marketing

**From:** Terry Prindle, Marketing Specialist

**RE:** IMPROVEMENT OF YOUR OFFICE AREA

The large area outside your office serves two purposes. The most important is to create a favorable impression upon visitors who wish to do business with our marketing department. In addition, the room provides a work area for your receptionist and me.

Currently, the area is not meeting those needs.

1. The area creates an unfavorable impression upon visitors because it is cluttered by the videotape equipment. In addition, when our agency's representatives want to meet with me to work out the details of layouts you have requested, they must accompany me into the vacant office next to my area, dragging chairs behind them while I fetch a cardtable.

2. Visitors frequently come to my desk instead of the receptionist's because they are unsure which of us is your receptionist. The numerous interruptions they cause make it difficult for me to complete in a timely fashion the assignments you have given me.

Therefore, I recommend that you ask the maintenance department to (1) place a partition around my area and (2) convert the vacant office into a conference room and storage area for the videotape equipment by supplying it with the necessary cabinets and other furniture.

**Figure 2-3.** A Revised Draft Written in a More Professional Manner.

instructors—even English instructors. Errors in spelling and grammar will cause your readers to believe that you are not performing up to professional standards.

## Prepare Neat Communications

When you first begin work, you may be struck by how neatly people prepare written communications. Neatness is expected. Sloppiness and carelessness are considered unprofessional.

In sum, to write in a professional manner, you must discuss your subject from the point of view of your professional role, use words that are common in your professional setting, write grammatically and spell correctly, and prepare neat communications.

# INTRODUCTION TO GENERAL GUIDELINES 4 THROUGH 9

Whereas the first three general guidelines for a reader-centered approach to writing focus on your relationship with your readers, the next six concern the mental processes your readers perform when reading.

Fundamental to these guidelines is considerable research into the way we all derive meaning from what we read. This research has been conducted by cognitive psychologists (who study the way people think), social psychologists, linguists, reading theorists, and specialists in artificial intelligence (who study ways to make machines "think"—not just to send information from one machine to another).

## Readers Create Meaning

These researchers agree that instead of *receiving* meaning when we read, we interact with the text to *create* meaning. As one researcher says, reading depends at least as much on what we bring to the page as it does on what is printed on the page.[1]

Consider, for instance, the amount of knowledge that we must possess and apply to interpret even so simple a sentence as, "It's a dog."

To begin with, remember that the words themselves *(it's, a, dog)*—those black marks against the white background of the page—have no meaning at all to most children under five years of age.

Remember, too, that the word *dog* has several meanings, so that we cannot know the meaning of the sentence without knowing its context. Is it the answer to the question, "What kind of animal is Jim's new pet?" or to the question, "What did you think of the new movie?"

## Other Examples of Readers' Creativity

When reading, we create meaning not only from individual words and sentences but also from whole communications and the major parts of them. Here's a quick way to demonstrate that to yourself. Imagine that you have been asked by someone who has not read this book to explain the meaning of the first general guideline ("When writing, talk with your readers"). First, write down a sentence that provides that explanation. Then, try to find a sentence in this book that matches yours exactly. Most likely, you will not be able to find one. The sentence you wrote is not one that you memorized; it expresses a meaning you *created* through your interaction with the text.

## Readers' Purposes and Feelings

When we read, we usually seek to do much more than merely determine the meaning of a communication. Especially at work, we read with a *purpose*. We want to use the information we gain from the communication to do something—to make some decision, perform some action, or complete some other activity. To do that, we must not only understand a communication, but also relate its meaning to other things we know. To make the decision about whether to go to a movie, for example, we not only must understand what our friend means by the words she uses to evaluate the movie, but we must also interpret her evaluation in light of our attitudes toward her taste in films.

Furthermore, when we interact with something we read, we not only gather information from it but also *experience feelings*. The demonstration involving Donna Pryzblo's memo shows how important those feelings can be in shaping readers' responses to something they read.

## Significance to You of This Discussion

Why is all this important to you as a writer? By learning how your readers create meaning when they read, you can also learn a great deal about how to write to them effectively. The next six general guidelines provide advice based upon the study of how people read. In the discussion of each you will be introduced to some of the more specific guidelines that are explained at greater length in later chapters.

General
Guideline

**4**

## Help Your Readers Focus on the Key Information

The first step in reading is to focus our eyes on the page. We can't read what we aren't looking at. On the job, people focus their eyes purposefully. They look for information that is useful to them. One of your most important tasks as a writer is to help them find it.

To do that you must have thought enough about your readers to know *what* they will want from the communications you write. You must have thought enough about them to know also *how* they will look for it, so you can assist them in their search.

Readers search for the information they want in two ways. Sometimes they read sequentially from the front to the back of a communication. But they read sequentially only when they think that's the most efficient way to find what they want. And they will continue sequential reading only if they find key information very quickly.

At other times, readers dip into a communication, hoping to find the information that's important and relevant to them without becoming bogged down in unimportant information. They do that when reading such things as company policy books and computer reference manuals. But they also do it when reading most other types of communication—including memos, letters, reports, and proposals. Often they feel they don't have time to read the entire communication sequentially. Often they want to review some specific part of a communication they first read sequentially.

Six ways you can help your readers find the key information in the things you write are to include everything that's needed, but nothing more; put the most important information first; use topic sentences; use headings; use lists; and provide indexes and tables of contents. These strategies are discussed below and are illustrated in Figure 2-4.

## Say Everything Needed, but Nothing More

To help your readers find the information they seek, you must first be sure that you include it. Then you must be sure that it isn't buried in a lot of information that is irrelevant to your readers.

You can determine what's relevant and what isn't by thinking carefully about your purpose and readers. What will your readers want and what else do you need to tell them in order to achieve your objectives?

Consider, for instance, the memo shown in Figure 2-4. In it, the writer, Frank Thurmond, provides information about some tests he ran to see what would happen if a new plastic were used to make bottles for one of his employer's products, a fluid that cleans ovens. Frank has carefully selected the particular results that will be relevant to his reader, a person in the marketing department who would like to use the plastic for packaging the oven cleaner. Frank omits any discussion of many other results. If he were writing to someone else—say, another researcher who wants to understand the chemical interaction between the cleaner and the plastic—he would have included a different set of test results.

In the same way, you should tailor the contents of your communications to your readers and purpose, including everything that's needed but nothing more.

**Summary of Key Points** →

# PAIGETT HOUSEHOLD PRODUCTS

**Intracompany Correspondence**

| | | | |
|---|---|---|---|
| TO: | Herman Wyatt | DATE: | October 12, 19-- |
| FROM: | Frank Thurmond | SUBJECT: | Tests of Salett 321 Bottles for StripIt |

We have completed the tests you requested to find out whether we can package StripIt Oven Cleaner in bottles made of the new plastic, Salett 321. We conclude that we cannot, chiefly because StripIt attacks and begins to destroy the plastic at 100°F and 125°F. We also found other significant problems.

**Heading** →

Test Methods

**Topic Sentence** →

To test Salett 321, we used two procedures that are standard in the container industry. First, we tested the storage performance of filled bottles by placing 24 of them in a chamber for 28 days at 73°F. We stored other sets of 24 bottles at 100°F and 125°F for the same period. Second, we tested the response of filled bottles to environmental stress by exposing 24 of them for 7 days to varying humidities and varying temperatures up to 140°F.

We simultaneously subjected glass bottles filled with StripIt to each of these test conditions.

**Heading** →

Results and Discussion

**Topic Sentences** →

In the 28-day storage tests, we found three major problems:

**List (begins with the most important item)** →

- The StripIt attacked the bottles at 100°F and 125°F. At 125°F, this attack was particularly serious, causing localized but very severe deformation in several bottles. Most likely, the attack was by the ketone solvents present in StripIt. The deformed bottles leaned, which could cause them to fall off displays in retail stores.

**Figure 2-4.** Strategies for Helping Readers Focus on Key Information.

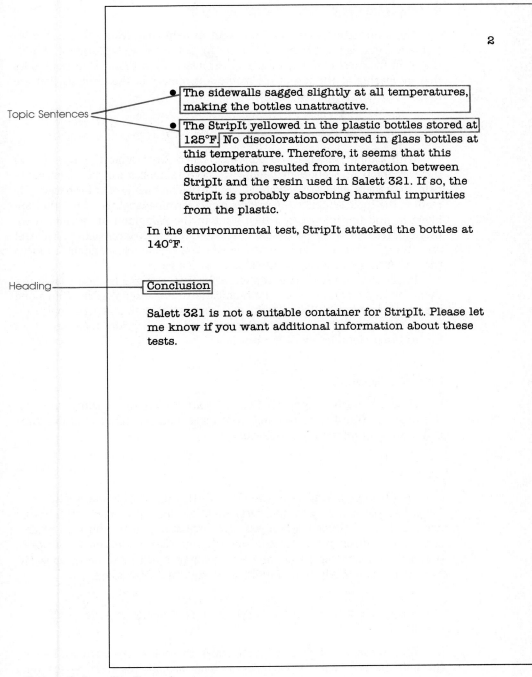

Figure 2-4. *Continued*

## Put the Most Important Information First

The information that you most want to help your reader find is the information that is most important. You can help your reader locate it by putting it in the most obvious and prominent place: first. Frank Thurmond does that by placing at the top of his memo a summary of the findings that are most important to his reader.

## Begin Your Sections and Paragraphs with Topic Sentences

With a topic sentence you tell your readers what subject you discuss in the paragraph or section that follows. These sentences can be of immense help to readers who are searching for particular pieces of information. By reading the topic sentence the readers can determine whether or not they are likely to find the information they want in the paragraph or section. If so, they can read all of the paragraph or section. If not, they can skip to the next one. Topic sentences also help people who are reading sequentially to focus their attention on the key points in each of the parts.

In addition to using topic sentences for your paragraphs, you should use them for larger parts of your communications. For example, Mr. Thurmond begins his section entitled "Results and Discussion" with a general topic sentence ("We found three major problems") and then provides topic sentences for his discussion of each of the problems.

## Use Headings

Headings are like signposts that tell your readers where things are. Notice how Mr. Thurmond's headings help readers quickly find any of the three major kinds of information he includes.

## Use Lists

Lists help your readers in two ways. First, they provide information in a form that makes it easy to find. Lists stand out from the other material in a communication. Second, when you place material in lists, you help readers scan quickly through the items in the list. Your lists might consist of words, phrases, complete sentences, or even paragraphs—like the paragraphs Mr. Thurmond uses in his section entitled "Results and Discussion."

## In Longer Communications, Provide Indexes and Tables of Contents

Your readers will find it much easier to spot a key term by looking through the few pages of an index or table of contents than by scanning through many pages of text.

You will find additional advice about how to help your readers focus on the key information in Chapters 7, 11, and 13.

## Write Sentences Your Readers Can Understand Easily

Once your readers focus their eyes and attention on some part of your communication, they must extract the meaning from the sentences they find there. You can help your readers quickly grasp the meaning of your sentences by keeping related words together, eliminating unnecessary words, using the active voice, and using words familiar to your readers.

### Keep Related Words Together

As you may know, our eyes can focus on only a very small area at a time. That's because the fovea (the region of sharpest focus on the retina) is large enough to see only about two dozen letters and spaces—not any more. Furthermore, as our eyes move from one small patch of words to the next, we cannot see. We can see only when our eyes stop to focus.

To understand a sentence, we must grasp the information provided in small groups of words and then piece that information together into the meaning of the sentence. This work is done by the short-term memory, which has one characteristic that is absolutely crucial to you as a writer: it can hold only a few bits of information at a time. Most psychologists agree that the number of items that can be stored in the short-term memory is between five and nine.

To see the significance of this fact, consider the following sentence.

> Carlson's report, because it showed that the gear teeth were damaged by continuous stress (not by sudden, jolting forces), demonstrated the value of our new quality control procedures.

If you are like most people, you had difficulty understanding that sentence the first time you read it. Your difficulty arose from the limitations of short-term memory. By the time you got to the verb *(demonstrated)* your short-term memory had forgotten the subject *(Carlson's report)*. If the sentence were rewritten in the following way, you could readily grasp its meaning because the subject and verb are close enough together to meet the needs of short-term memory.

> Carlson's report demonstrated the value of our new quality control procedures because it showed that the gear teeth were damaged by continuous stress (not by sudden, jolting forces).

To keep the related parts of your sentences together, you must keep your subject close to your verb and your modifiers close to the words or phrases they modify. By doing so you will avoid situations in which your readers' short-term memories have forgotten one of the related parts by the time they read the other.

## Eliminate Unnecessary Words

Every word requires work by the short-term memory. By eliminating words not needed to convey your meaning, you will reduce the burden your writing imposes on your readers. The following example shows how you can cut away unnecessary words, making your sentences lean and easy to understand.

> **Before:** Because *of the fact that* altocumulus clouds usually appear when high pressure areas are approaching, they generally *may be taken as a sign that* good weather is coming. (28 words)

> **After:** Because altocumulus clouds usually appear when high pressure areas are approaching, they generally mean good weather is coming. (18 words)

## Use the Active Voice

The verb is the heart of a sentence. Readers understand sentences written in the active voice more rapidly than they understand sentences written in the passive voice.

To write in the active voice, place the *actor*—the person or thing performing the action—in the *subject* position. Your verb will then tell the *action* the actor performed. In the active voice, the subject acts:

$$\begin{array}{cc} \text{Subject} & \text{Verb} \\ \downarrow & \downarrow \end{array}$$

The *technician mixed* the fluids.

$$\begin{array}{cc} \uparrow & \uparrow \\ \text{Actor} & \text{Action} \end{array}$$

In the passive voice, the subject is acted upon:

$$\begin{array}{cc} \text{Subject} & \text{Verb} \\ \downarrow & \downarrow \end{array}$$

The *fluids were mixed* by the *technician*.

$$\begin{array}{cc} \uparrow & \uparrow \\ \text{Action} & \text{Actor} \end{array}$$

As you can see, active verbs help readers see immediately what action is described and who performed it. Frank Thurmond's memo (Figure 2-4) provides a good example of the use of the active voice. Look for instance, at these verbs: *have completed, conclude, used, exposed,* and *subjected.*

## Use Words Familiar to Your Readers

If you use an unfamiliar word, your readers will have to devote the limited capacity of their short-term memories to figuring out what the word means. When that happens, the readers' short-term memories will forget the other words in the sentence and the readers will have to reread material to pick up the thread of thought again.

This does not necessarily mean that you should avoid using technical terms. When you are writing to a person familiar with such terms, the technical terms are probably the best way to convey your meaning. But when you are writing to an audience unfamiliar with those terms, you should look for more ordinary words.

In Chapters 16 and 18, you will find additional advice about writing lean, active, easily understandable sentences.

General
Guidelines

6

## Help Your Readers Follow the Flow of Thought from Sentence to Sentence

To read, we must not only put the parts of a sentence together to grasp its meaning, but we must relate the meaning of one sentence to the meaning of another. To help your readers follow the flow of thought from one sentence to another, you can use transitional words and phrases, and you can use echo words. These devices are discussed below, and their use is illustrated in Figure 2-5.

## Use Transitional Words and Phrases

One way to help your readers see how your sentences fit together is to tell them explicitly. In that way you eliminate the need for the readers to guess, and you eliminate the chance that your readers will guess incorrectly.

There are transitional words and phrases for many different kinds of relationships, such as the following:

| | |
|---|---|
| time | after, before, simultaneously, then |
| space | above, below, inside |
| cause and effect | because, consequently |
| similarity | just as, like, similarly |
| contrast | despite, however, nevertheless |

**PAIGETT HOUSEHOLD PRODUCTS**

**Intracompany Correspondence**

TO:    Herman Wyatt        DATE:      October 12, 19--

FROM:    Frank Thurmond    SUBJECT:    Tests of Salett 321
Bottles for StripIt

We have completed the tests you requested to find out
whether we can package StripIt Oven Cleaner in bottles
made of the new plastic, Salett 321. We conclude that we
cannot, chiefly because StripIt attacks and begins to destroy
the plastic at 100°F and 125°F. We also found other
significant problems.

Test Methods

To test Salett 321, we used two procedures that are
standard in the container industry. First, we tested the
storage performance of filled bottles by placing 24 of them
in a chamber for 28 days at 73°F. We stored other sets of 24
bottles at 100°F and 125°F for the same period. Second, we
tested the response of filled bottles to environmental stress
by exposing 24 of them for 7 days to varying humidities
and varying temperatures up to 140°F.

We simultaneously subjected glass bottles filled with StripIt
to each of these test conditions.

Results and Discussion

In the 28-day storage tests, we found three major problems:

- The StripIt attacked the bottles at 100°F and 125°F.
  At 125°F, this attack was particularly serious,
  causing localized but very severe deformation in
  several bottles. Most likely, the attack was by the
  ketone solvents present in StripIt. The deformed
  bottles leaned, which could cause them to fall off
  displays in retail stores.

**Figure 2-5.**   Strategies for Providing Transitions.

*(Margin labels:)* Echo Word (associated word); Transitional Word; Echo Words (associated words); Transitional Word; Echo Words (repeated words); Transitional Word

2

Echo Word
(synonym)

Transitional
Phrase

- The sidewalls sagged slightly at all temperatures, making the bottles unattractive.

- The StripIt yellowed in the plastic bottles stored at 125°F. No discoloration occurred in glass bottles at this temperature. Therefore, it seems that this discoloration resulted from interaction between StripIt and the resin used in Salett 321. If so, the StripIt is probably absorbing harmful impurities from the plastic.

In the environmental test, StripIt attacked the bottles at 140°F.

Conclusion

Salett 321 is not a suitable container for StripIt. Please let me know if you want additional information about these tests.

**Figure 2-5.**    *Continued*

## Use Echo Words

Echo words help readers see how the sentence they are reading now relates to the preceding sentence by recalling to mind some topic mentioned in the preceding sentence. Consider the following example.

> The mechanic placed a small warning device on each brake. The *device* "chirps" when the brake pad is worn dangerously low.

The repetition of the word *device* in the second sentence tells readers that the second sentence is about the device mentioned in the first one. (The writer could also have used the pronoun *it.*)

There are many kinds of echo words, including the following:

- *Repeated Words*. For example, the word *device* in the preceding example.
- *Pronouns*. For example, *he, she, it, this, that.*
- *Associated Words*. Words closely associated with each other, including synonyms. In the following example, the first three words of the second sentence are echo words:

> The industrial design department is developing a totally new style for ovens, refrigerators, and dishwashers. *Large kitchen appliances* have looked basically the same for several decades now.

## Place Transitions and Echo Words Near the Beginning of Sentences

Transitions and echo words can help your reader most if you place them early in your sentences, so that your readers come to them before their short-term memories forget what the preceding sentence was about. Compare the following pair of sentences.

> **Bad Example:** During the past six months, almost all of our agricultural products have sold well. Workers at the the Trumbley plant seem more willing to negotiate their contract demands before the strike deadline next month because of the additional overtime pay resulting from this increased production.

> **Revised:** During the past six months, almost all of our agricultural products sold well. *Because of the additional overtime pay resulting from this increased production,* workers at the Trumbley plant seem more willing to negotiate their contract demands before the strike deadline next month.

In Chapter 17 you will find detailed advice about helping readers follow the flow of thought from one sentence to the next.

General
Guideline

7

## Help Your Readers Build Hierarchies of Meaning

To understand the larger meanings of the things we read, we must fit the bits of information we derive from individual sentences into larger frameworks. We do this by building hierarchies; that is, we group small pieces of related information together, and then we gather those groups into still larger ones. Each of these groups is associated with some generalization that characterizes its contents or expresses its meaning. By building hierarchies, we strive to provide a meaningful place in our understanding for every piece of information in the communication. Figure 2-6 shows the hierarchy of information in Frank Thurmond's memo.

You can help your readers by telling them in advance about the hierarchy you have constructed for your communication. You can do that by using topic sentences, using forecasting statements, and stating your generalizations explicitly *before* providing your detailed information. Figure 2-7 shows how Frank Thurmond uses these strategies in his memo.

### Use Topic Sentences

When you begin a paragraph with a topic sentence, you tell your readers something about the way the contents of the paragraph fit together: they all

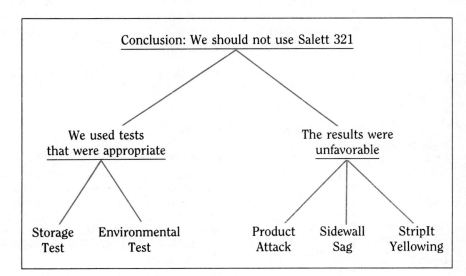

**Figure 2-6.**    A Hierarchy of Information from the Memo shown in Figures 2-4 and 2-5.

## PAIGETT HOUSEHOLD PRODUCTS

### Intracompany Correspondence

TO:        Herman Wyatt        DATE:        October 12, 19--

FROM:     Frank Thurmond     SUBJECT:     Tests of Salett 321
                                          Bottles for StripIt

Beginning explains relevance to reader

Summary focuses on significance to reader

We have completed the tests you requested to find out whether we can package StripIt Oven Cleaner in bottles made of the new plastic, Salett 321. We conclude that we cannot, chiefly because StripIt attacks and begins to destroy the plastic at 100°F and 125°F. We also found other significant problems.

Test Methods

Topic sentence includes forecasting

To test Salett 321, we used two procedures that are standard in the container industry. First, we tested the storage performance of filled bottles by placing 24 of them in a chamber for 28 days at 73°F. We stored other sets of 24 bottles at 100°F and 125°F for the same period. Second, we tested the response of filled bottles to environmental stress by exposing 24 of them for 7 days to varying humidities and varying temperatures up to 140°F.

We simultaneously subjected glass bottles filled with StripIt to each of these test conditions.

Results and Discussion

Topic Sentence includes forecasting

In the 28-day storage tests, we found three major problems:

- The StripIt attacked the bottles at 100°F and 125°F. At 125°F, this attack was particularly serious, causing localized but very severe deformation in several bottles. Most likely, the attack was by the ketone solvents present in StripIt. The deformed

Significance explained explicitly

bottles leaned, which could cause them to fall off displays in retail stores.

**Figure 2-7.**   Strategies for Telling Readers the Significance of Information.

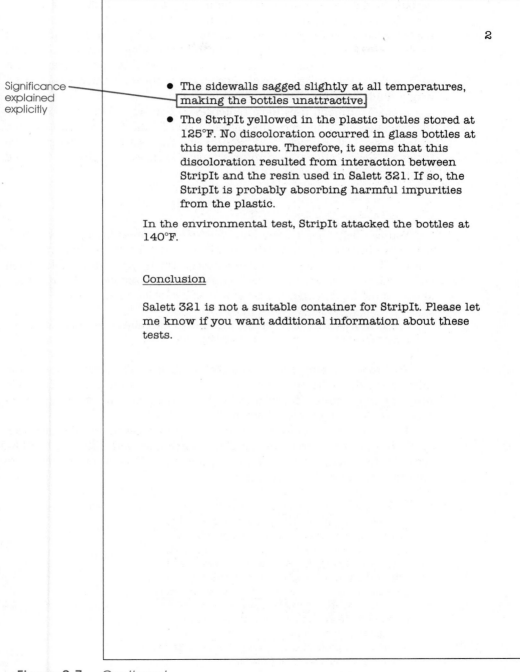

Significance explained explicitly

2

- The sidewalls sagged slightly at all temperatures, making the bottles unattractive.
- The StripIt yellowed in the plastic bottles stored at 125°F. No discoloration occurred in glass bottles at this temperature. Therefore, it seems that this discoloration resulted from interaction between StripIt and the resin used in Salett 321. If so, the StripIt is probably absorbing harmful impurities from the plastic.

In the environmental test, StripIt attacked the bottles at 140°F.

Conclusion

Salett 321 is not a suitable container for StripIt. Please let me know if you want additional information about these tests.

**Figure 2-7.**   *Continued*

pertain to the topic you mention. Similarly, when you begin a section or other larger part of your communication with a topic sentence, you tell your readers what its major parts have in common. By using topic sentences at all levels of your communication, you build an interlocking system of statements that describe the overall structure of your message. As your readers begin each of the parts—large or small—they learn about the framework for organizing the information they are going to read.

Frank Thurmond uses topic sentences in that way in his memo. In the first sentence, he announces his overall topic. In the first sentences of the sections on "Test Methods" and "Results and Discussion" he announces the topics of those sections. He also uses topic sentences to begin the first and third items in his list of results.

## Use Forecasting Statements

You can help your readers build hierarchies of meaning by telling them not only what subjects they are about to read about but also how you have organized your discussion of those subjects. Often, you will be able to tell your readers about your organization in your topic sentences. Here, for example, is the topic sentence for one part of a booklet on allergies:

> People with allergies often wonder what causes their allergic reactions and why those reactions are more severe at certain times.

That topic sentence enables readers to anticipate the organization of the information that follows: first the booklet will discuss the causes of allergic reactions and then it will explain why those causes produce more severe reactions at some times than at others.

However, you can't always use your topic sentences to provide your readers with a preview of your organization. Sometimes you will need to add a separate statement, called a forecasting statement. Here is an example from an article addressed to veterinarians.

Today, veterinarians need a practical guide to chemotherapy. —— Topic Sentence

If you have not yet been asked to provide chemotherapy to a pet, you probably will be asked soon. The general public is becoming aware that chemotherapy is being used to improve and extend the lives of people with inoperable cancer. As a result, many animal owners are asking for similar treatment for their pets. This article describes the drugs that are available for chemotherapy, their mechanisms of action, and their dosages. —— Forecasting Statement

It also discusses the costs of chemotherapy, its undesirable side effects, and suggested protocols.

## State Your General Points Explicitly *before* You Provide Details

Some topic sentences are more specific than others. Often, they merely indicate the subject you are about to discuss—saying, in effect, this is a paragraph about optical gyroscopes, capacitor leakage, or the mechanics of roof design. However, when you are beginning a section in which you are trying to support an important generalization, you should state the generalization explicitly in your topic sentence. Consider, for instance, the following paragraph.

> Since he joined our company, we have noticed several things about Michael's abilities. As he showed in his work last year on the Weingarten project, he can create novel and ingenious engineering designs. He also understands materials very well. At the same time, his work on the Weingarten project showed that he has trouble communicating instructions and motivating other people to do their best work.

Readers of that paragraph are left asking, "So what?" To figure out the point of the paragraph, they must guess at the general point the writer wants to make. Here is a revised version of that same paragraph in which the writer makes that general point explicitly in the topic sentence.

> Michael is a talented and knowledgeable engineer, but he does not have the qualities needed by a project manager. As he showed in his work last year on the Weingarten project, he can create novel and ingenious engineering designs. He also understands materials very well. At the same time, his work on the Weingarten project showed that he has trouble communicating instructions and motivating other people to do their best work.

In Chapters 9, 11, 15, and 20, you will find more advice about helping your readers build hierarchies of meaning.

General Guideline

8

## Tell Your Readers the Significance *to Them* of Your Message

At work, people read purposefully. They won't read things that don't seem useful to them. Accordingly, one of your tasks as a writer is to help your readers understand the significance of the material you are presenting. Here are some actions you can take to help your readers understand the significance of the things you are saying.

### Begin by Telling Your Readers Why They Should Read What You Have Written

An excellent way to explain the significance of a communication is to tell your readers explicitly at the outset why your communication is important to them. Your explanation of its relevance might be as brief as the phrase, "As you asked." For example, Frank Thurmond opens his memo by saying, "We have completed the tests *you requested.*" In the rest of that sentence he reminds the reader of the reason why he asked for those tests: "to find out whether we can package StripIt Oven Cleaner in bottles made of the new plastic, Salett 321."

In long communications, writers sometimes devote as much as an entire chapter to explaining the relevance to their readers of the topic they are addressing.

### Explicitly State the Significance of the Information You Present

When writing, you should do more than simply list facts. For example, if you are writing a report or proposal, tell your readers the implications and consequences of the facts and findings you discuss. If you are writing instructions that have warnings, tell what might happen if the warnings are ignored. In every communication you write, be sure that your readers know the significance to them of what you are saying.

In his memo on the plastic bottles, Frank Thurmond states the significance of the test results in several places. For instance, he not only reports that StripIt attacks plastic bottles, but he also explains why the attack is a problem: it deforms the bottles so that they might fall off the shelves in retail stores.

### Begin with Summaries that Present the Most Significant Information Succinctly

By beginning with a summary of your most important information, you can help your readers understand the general significance of your material before reading any of your specific details. In a relatively short communication, like Frank Thurmond's memo, you can do that with a sentence or two in the opening paragraph. Frank Thurmond does that in his second sentence, where he states the overall conclusion and the main reason for it.

In longer communications, you can write a separate section called a "summary" or "abstract" at the beginning of the entire communication. In it, you would emphasize the points most relevant to your readers.

In Chapters 11 and 13, you will find additional advice about ways to help your readers see the significance to them of your communications.

General
Guideline

9

## Use Visual Aids

The first eight general guidelines for reader-centered writing focus on prose. At work, however, you will often find that you can communicate some of your material much more effectively in visual aids, such as tables, graphs, or drawings.

The first step in using visual aids is to look for places where they can help you achieve your communication objectives. These places include the following:

- *Where you describe something physical.* A drawing or photograph can be much more effective than prose in helping your readers "see" the object.
- *Where you discuss the organization of something that has several parts.* For instance, a flow chart can provide readers with a quick understanding of a process, and an organizational chart can provide a ready understanding of a company's structure.
- *Where you want your readers to be able to locate specific pieces of data among many pieces.* In such situations, tables can be very helpful.
- *Where you focus on the relationship among separate pieces of data.* For example, you may have sales data for each of the past five years. Using a line graph, you can show how sales have changed from year to year. Or you may have data from two or more sources, such as people with and without college degrees. Using a bar graph, you can help your readers readily see how the data from one source compare with those from the other source.

Chapters 19 and 20 provide detailed advice about ways you can use visual aids in your communications—whether one-page memos or one-hundred-page reports.

## SUMMARY

This chapter presented nine general guidelines for a reader-centered approach to writing on the job. The first three concern the personal relationship you establish with your readers.

1. *When writing, talk with your readers.* Imagine that your communications are conversations in which you make a statement and your readers respond. In this way, you can anticipate your readers' reactions and adjust your statements to fit those reactions.
2. *Write in a respectful voice.* Three ways to show respect for your readers are to use your common sense about how to show respect, show that you understand your readers' point of view, and acknowledge your readers' abilities and areas of authority.
3. *Write in a professional manner.* Discuss your subject from the point of view of your professional role, not your personal feelings. Choose words that are common in your professional setting. Write grammatically correct sentences that are free from misspellings and typographical errors.

The other six general guidelines concern ways you can help your readers read efficiently.

4. *Help your readers focus on the key information.* You can help your readers focus on essential information by eliminating irrelevant material and by putting the most important information first. You can also use topic sentences, headings, lists, indexes, and tables of contents.

5. *Write sentences your readers can understand easily.* Keep related words together, eliminate unnecessary words, use the active voice, and use words familiar to your readers.

6. *Help your readers follow the flow of thought from sentence to sentence.* Use transitional words and phrases, as well as echo words. These devices are especially helpful when you place them at the beginnings of your sentences.

7. *Help your readers build hierarchies of meaning.* In a hierarchy, the various details in a communication are related to one another through an interrelated series of groups and subgroups. You can help your readers see these relationships by using topic sentences and forecasting statements, as well as by stating your generalizations explicitly before providing details.

8. *Tell your readers the significance to them of your message.* You can increase the effectiveness of your communications by explicitly telling your readers how your information is useful to them.

9. *Use visual aids.* You can often communicate your information much more effectively through visual aids than through prose.

## NOTE

1. Frank Smith, *Understanding Reading: A Psycholinguistic Analysis of Reading and Learning to Read,* 3rd ed. (New York: Holt, Rinehart and Winston, 1982).

## EXERCISES

1. A. Find a piece of writing that you believe to be ineffective. You might look for a poor set of instructions or an unpersuasive advertisement for some business or technical product. Then write a brief analysis of three or four "reading moments" in which you interact with the text in a way that inhibits the communication from achieving the result the author desires.

   B. Do the same for an effective piece of writing. This time, write about three or four "reading moments" in which you interact with the text in a way that helps to bring about the desired result.

2. A. Using your common sense about how people react to one another, rewrite the memo by Donna Pryzblo so that it is more likely than hers to persuade Dwight Gardner to follow her recommendation. You may assume

that she knows that Mr. Gardner's clerks are miscopying because she examined the timesheets, time-tickets, and computer files associated with 37 incorrect payroll checks; in 35 cases the errors were made by Mr. Gardner's clerks. Remember to take into account the initial reaction you expect Mr. Gardner to have at finding a memo from Ms. Pryzblo in his mail; make sure that the first sentence of your memo addresses a person in that frame of mind, and that your sentences lead effectively from there. Leave the last sentence unchanged.

B. In the margin of your memo or on a separate sheet of paper, tell what you expect Dwight Gardner's reaction to be after reading each of the individual sentences.

3. Playing the role of Dwight Gardner, write two replies to Donna Pryzblo. In the first, write in a respectful, professional manner. In the other, do not.

4. A. The purpose of this exercise is to help you appreciate the great extent to which readers create the meaning of the things they read. Scan this book and find five words that would mean something different than they do here if they were in some other sentence. In each case, write down the sentence from this book and another in which the word's meaning is different.

B. In the first part of this book find one paragraph whose main point is explicitly stated in it and one whose main point is plain to you even though it isn't stated explicitly. How, exactly, do you think you derived the meaning of the latter paragraph? Write out your answer.

5. The purpose of this exercise is to give you an opportunity to apply several of the general guidelines presented in this chapter. The exercise has two parts, both of which use the notes on page 46.

A. Imagine that you are the assistant to the Production Engineer of a large manufacturing plant. The plant is having excessive problems with its forklift truck, which it purchased ten years ago from a firm that went defunct shortly thereafter. Consequently, your plant is unable to obtain replacement parts for the forklift. Now, the forklift's hydraulic system must be rebuilt. The Plant Manager has therefore decided to buy a new forklift.

To be sure that he spends the company's money wisely, the Plant Manager has asked your boss to recommend a forklift for the company. Your boss, in turn, has asked you to gather information about the two forklifts he believes are the best candidates.

You write up the information you have gathered in the form of notes on the present situation, the electric forklift, and the gasoline forklift. The notes are given below. You now sit down to write your report to your boss from your notes.

- List the specific questions you believe your boss will want your report to answer. Note: your boss does *not* want you to include a recommendation in your report.
- Underline the facts in your notes (given below) that you would include in your report, and put an asterisk by those you would emphasize.
- Decide how you would organize the report.
- Tell how you would apply Guidelines 3, 4, and 8 when writing this report.

B. For the second part of this exercise, you are promoted to the position of Production Engineer. You must now design the report that you will send to the Plant Manager. It must contain your recommendation. Tell what your report will look like.

### Notes for Exercise 5

**Present Situation.** The present forklift, which is red, moves raw material from the loading dock to the beginning of the production line, and it takes finished products from the Packaging Department back to the loading dock. When it moves raw materials, the forklift transports pallets weighing six-hundred pounds, which it must hoist to a platform eight feet high, so that the raw materials can be emptied into a hopper. When transporting finished products, the forklift moves pallets weighing two hundred pounds, which it picks up and delivers at ground level. For both jobs, the forklift moves between stations at 10 miles per hour, although some improvements in the production line will increase that rate to 15 mph in the next two months. The present forklift is easy to operate, so that no injuries and very little damage have been associated with its use.

**Electric Forklift.** The electric forklift carries loads of up to 1000 pounds at speeds up to 30 mph. Although the electric forklift can hoist materials only six feet, a two-foot ramp could be built beneath the hopper platform in three days (perhaps over a long weekend, when the plant would be closed). During construction of the ramp, production would have to stop. The ramp itself would cost $600. The electric forklift costs $17,250, and a special battery charger costs $1500 more. The lift would use about $2000 worth of electricity each year. Preventive maintenance costs would be about $300 per year, and repair costs would be about $800 per year. While operating, the electric forklift emits no harmful fumes whatever. Parts are available from a warehouse 500 miles away. They are ordered by phone and delivered the next day. Records indicate that the electric forklift is operated with very little damage to goods and with no injuries at all. It comes in blue and red.

**Gasoline Forklift.** The gasoline forklift, green in color, carries loads of up to one ton as rapidly as 40 mph. It can hoist materials twelve feet high. Because this forklift is larger than the one presently being used, the company would have to widen a doorway in the cement wall separating the Packaging Department and the loading dock. This alteration would stop production for two days and would cost $800. The gasoline forklift costs $19,000, but needs no auxiliary equipment. However, under regulations established by the Occupational Safety and Health Administration, the company would have to install a ventilation fan to carry the exhaust fumes away from the hopper area. The fan costs $780. The gasoline forklift would use about $1800 of fuel per year. Preventive maintenance would cost an additional $400 per year, and repair costs would be about $600 per year. Repair parts are available from the factory, which is seventeen miles from our plant. Other owners of the forklift have incurred no damage or injuries during the operation of this gasoline forklift.

# Examples of Reader-Centered Writing: Resumes and Letters of Application

## PREVIEW OF CHAPTER

The persuasive aim of this chapter

Writing your resume

    Defining your objectives

    Planning

    Drafting

    Evaluating

    Revising

Writing your letter of application

    Defining your objectives

    Planning

    Drafting

    Evaluating

    Revising

Writing is like building a home. You start with nothing except a need or desire: you want to house yourself or to communicate with someone. Then you carefully assess your aims and materials, make your plans, and construct a structure that serves your purposes.

In the previous chapter, you learned some general guidelines for "constructing" a successful communication at work: for example, "Talk with your readers," "Write in a professional manner," and "Help your readers focus on the key information." These guidelines are like general principles of home building: "Consider the lifestyle of the people who will live in the home," "Group the activity areas away from the sleeping areas," and so on.

Such guidelines are very helpful—even essential—to your success. Yet, you need something more than guidelines and principles to build either a house or a communication. You need a procedure. This chapter discusses the overall procedure for creating reader-centered communications on the job.

You already know the most important feature of that procedure: as you perform each step, think about your readers. Further, in the first chapter you learned what the basic steps are: defining your objectives, planning, drafting, evaluating, and revising.

In this chapter, you will learn how these steps fit together and how to put your thoughts about your readers to good, *practical* use as you perform each of them.

In addition, as this chapter guides you through the writing process, it will *demonstrate* each step by showing how you can perform it while creating actual communications. The two communications chosen for this demonstration are your resume and letter of application for employment.

## THE PERSUASIVE AIM OF THIS CHAPTER

Why were these particular communications chosen for the demonstration? Partly to be helpful to you. They are likely to play an important role in your career. You may even need them soon to look for a summer job, a part-time job, or the full-time position you want to have waiting for you after you graduate.

Beyond that, the letter and resume were chosen because with them you should be able to see quite readily the value of thinking constantly about your readers. With resumes and letters of application it is especially evident that your success depends entirely upon the impression they make upon your readers.

So, when you finish studying this chapter you should know two things: how to write a resume and employment letter and how to follow a reader-centered writing process. In addition, you should see why it is worthwhile for you to use a reader-centered writing process, one in which you keep your readers—not yourself, not your subject matter—primarily in mind at every step.

# WRITING YOUR RESUME

This chapter guides you through the writing process *twice*. The first time, it explains and illustrates each step of your work on your resume. The second time, it leads you through the process much more rapidly, focusing on your letter of application.

## Defining Your Objectives

When writing your resume, as when writing any communication related to your job, the first—and essential—step is to define your objectives. Your objectives form the basis for all later steps in the writing process. They are the foundation you build on, the goal you strive for, the target you aim at.

To define your objectives, you must do three things:

- Understand your readers.
- Define your communication's purpose.
- Find out about any expectations or limitations (such as page limits) that may restrict your choices about the way you will write.

### Understanding Your Readers

Before you can understand your readers, you must determine who they will be. Typically, employers recruit new employees through a process that has two stages, each with a different group of readers.

In the first stage, employers try to attract applications from as many qualified people as possible. They do this by maintaining a personnel office that accepts and screens applications; by interviewing students on college campuses; and by placing advertisements in job catalogs (like the *College Placement Annual*), professional journals, and newspapers. At this stage of recruiting, resumes are usually read by people who work in the personnel office. Their main objective is to screen out applicants who are obviously unqualified.

As suggested in the preceding chapter, when you write you may find it very helpful to draw an imaginary portrait of a typical reader, so that you can think of that person's reactions to your communication and write accordingly. For this first stage of recruiting, you might imagine your reader to be a man who has just sat down at his desk to read the stack of twenty-five to fifty new applications that arrived in today's mail. With jacket off and sleeves rolled up, he begins looking for the two or three persons who might be worth additional consideration. Only occasionally does he read a full resume, because he quickly finds reasons for disqualifying most applicants. Throughout your work on your resume, you should keep in mind that it must quickly attract and hold this man's attention.

In the second stage of recruiting, employers look very carefully at the qualifications of the most promising applicants. Often, this involves visits by the candidates to the employer's site. The readers of your resume during this second stage of recruiting include the various people you would meet on such a visit, including the manager of the department you would work for, other people in that and other departments, and maybe even some of the employer's clients.

To represent your readers in this stage, you might imagine the head of the department you wish to work for. This person is shorthanded, and hopes very much to fill one or more openings. She has been waiting for a week for the personnel department to forward resumes from the most promising applicants. When they finally arrive, she clears a space on her crowded desk and hopes that no one will knock on the door or call on the phone until she has had a chance to read them. Her needs are very specific, and she knows just the kind of qualifications she seeks. Throughout your work on your resume, you should keep in mind that it must persuade this reader that you have the specific abilities, education, and other experience that she is looking for.

## Defining Your Communication's Purpose

While you are seeking to understand the readers of your resume, you should also be thinking about the resume's purpose. More precisely, you should define the way you want your resume to affect your readers *while they are reading it*. To determine how you want your resume—or any other communication—to affect your readers, you can think about two things:

- The way you want it to alter your readers' attitudes.
- The task you want to help your readers perform while they read.

### Altering Your Readers' Attitudes

As the first step in defining the way you want a communication to alter your readers' attitudes, you should determine how your readers feel now, *before* reading what you are writing, and how you want them to feel *after* reading it. When you write a resume, you might say that your readers' attitude toward you now, before reading your resume, is neutral because the readers don't know anything about you. You want those people as they read to come to think that you are highly qualified for the job you are seeking. Once you have described your readers' present attitudes and the attitudes you hope they will hold after reading your resume, you should learn what characteristics your readers value, what qualities they look for in the people they hire. It's by persuading your readers that you have these qualities that your resume will alter their attitude about you.

As common sense will tell you, employers want to hire people who can be described with the following adjectives:

- *Capable.* Through their previous training and through their ability to learn additional material, the applicants must be qualified to perform the specific tasks the employer will assign them.
- *Responsible.* The applicants must be trustworthy enough to use their training and abilities in ways that will benefit the organization.
- *Pleasant.* The applicants must be personally compatible with the other employees with whom they will work on the job.

Of course, these qualities are stated only generally. The readers of your resume will ask them in much more specific terms. Instead of asking, "Is this applicant capable," they will ask, "Can this applicant perform differential equations?" "Program computers in COBOL?" "Manage a clothing store?" and so on.

What can you do to learn exactly what qualities your readers will be looking for? You might ask professors in your major department, or you might visit people who work in or manage the kind of job you desire.

Information you obtain about the qualities your readers want in new employees is information you can *use*. It tells you what things you can say about yourself that will persuade your reader that *you* are sufficiently capable, responsible, and pleasant for the specific job you want.

### Helping Your Readers Perform Their Tasks

While part of your resume's purpose is to alter your readers' attitudes, another part is to enable your readers to perform the *work* of reading. Different kinds of communications involve different tasks. When you know what tasks your particular readers will perform, you can write in a way that will help them read efficiently.

When reading your resume, your readers' primary task will be to seek the answers to the questions they ask when determining whether or not to hire someone.

The readers' first question will be:

- What, exactly, do you want to do?

Then, to determine whether you are capable and responsible enough for the particular job you want, your reader will ask such questions as:

- Do you have the required knowledge?

- Do you have experience in this or a similar job?
- What responsibilities have you been given or have you undertaken on your own initiative?

To determine whether you are pleasant, your readers may ask:

- Do your interests show that you would be a pleasant and compatible employee?

Finally, if your readers are persuaded that you are capable, responsible, and pleasant, your readers will ask:

- How can I get more information about your qualifications?

Knowing that your readers will be looking for the answers to these questions tells you a great deal about what to include in your resume. Knowing that your readers will devote only a brief time to reading your resume initially lets you know that you should help your readers find that information quickly.

## Finding Out What's Expected

At work, your readers or other people will often have expectations that influence or dictate the way you will write. Sometimes these expectations are expressed as requirements—such as deadlines, page limits, and regulations about what you have to include and how your communication should look. An important step in writing is to learn about any expectations your readers will have.

The readers of your resume will have several expectations, mostly shaped by the many conventions about resumes: resumes should be typed; they should have the applicant's name, address, and telephone number at the top; they should be on white, buff, or light gray paper; and so on.

Some of these conventions make good sense. They serve your needs and those of your readers. For instance, by including your name, address, and phone number, you ensure that your readers can get in touch with you. Because this information is always placed at the top, readers always know where to look for it.

Other customs make no particular sense. For instance, there seems to be no special reason why resumes should be printed on white, buff, or light gray paper, but the convention is so strong that you risk seeming very odd if you ignore it.

Because some of the conventions about resumes can help you plan and draft effectively, they are discussed later in this chapter. In fact, that's the purpose of defining your objectives—to learn things that can assist you in later steps of writing.

## Planning

When you *plan,* you use your understanding of your purpose, your readers, and your readers' expectations to decide what to say and how to organize your material.

### Deciding What To Say

As you think about what you want to say in your resume or any other communication, you can begin in any of three places.

- You can list all the things you have to say on your subject. When planning your resume, for instance, you can list all the things that you think show your qualifications for the job you seek.
- You can begin with the questions you know your readers have about your subject. With your resume, those questions would be the ones listed above, in the section on "Helping Your Readers Perform Their Tasks."
- You can begin by finding out what other people have said in similar situations. With your resume, that would mean looking at resumes written by other people to see what their writers have said.

There is no need for you to use one strategy alone. You will often benefit from using all three. With your resume, a reasonable way to begin your list of topics is to see what other people have said in theirs. That's because some fairly standard patterns have developed for writing resumes, patterns that help both readers and writers in job-hunting situations. The topics conventionally covered in resumes closely match the questions employers usually ask when evaluating job applications, as the following list indicates. The readers' questions are those listed earlier in this chapter:

| Conventional Topics | Usual Questions by Readers |
|---|---|
| Professional Objective | What, exactly, do you want to do? |
| Education<br>Honors | Do you have the required knowledge? |
| Work Experience<br>Activities | Do you have experience in this or a similar job? What responsibilities have you been given or have you undertaken on your own initiative? |
| Interests | Do your interests show that you would be a pleasant and compatible employee? |
| References | How can I get more information about your qualifications? |

Later in this chapter, you will find ideas about things you might say about each topic.

## Organizing Your Material

Most writers organize their resumes around the topics just listed. That organization helps readers find the answers to their questions, and it enables writers to present their information persuasively because it makes evident the relationship of their qualifications to their readers' concerns. In Figures 3-1 and 3-2 you will find two resumes organized in this way.

However, some individuals choose to organize a substantial part of their material around their accomplishments and abilities. Such a resume is sometimes called a *functional* resume because it emphasizes the functions the applicant can perform. Generally, this kind of resume is used by people who have extensive work experience so that they have on-the-job accomplishments to talk about. Usually college students who do not have extensive work records use the more traditional arrangement of their material.

Whichever organizational pattern you use, you still must decide the order in which you will present your material. By convention, your name should go at the top. Then you should state your professional objective. By knowing what job you want, your readers can tell how to judge your qualifications as they read the rest of your resume. After your professional objective, you can follow one of the basic strategies for writing at work: put the most important information *first*. In a resume, the most important information is that which describes your most impressive qualification. If you are like most college students, that will be your education. However, if you have had substantial work experience, that experience may eclipse your education and so should go first. Of course, if you are writing a functional resume, you are trying to emphasize your accomplishments and skills, so they should lead off.

## Drafting

When you draft, you transform your plans—your notes, outlines, ideas—into a communication. With your resume, as with the communications you will write at work, you must draft not only the prose but also the visual arrangement or appearance of your message.

## Drafting the Prose

While drafting your resume, call upon your imaginary portraits of your readers. As you write, keep those people in mind. Remember that your purpose is to enable them to find the answers to their questions about you (listed above) and to provide those answers in a way that will persuade the readers that you are capable, responsible, and pleasant. The following paragraphs suggest some ways you can achieve that purpose.

PATRICIA NORMAN
Box 88 Wells Hall
University of Washington
Seattle, Washington 98195
(206) 529-5097

<u>PROFESSIONAL OBJECTIVE</u>

To join an executive training program leading to a management position in merchandising with an innovative and growing retailer.

<u>EDUCATION</u>

University of Washington, Seattle, Washington, B.S. in Home Economics Retailing, May 19—. 3.6 accumulative grade-point average.

Earned 23 credit hours in marketing management, obtaining a working knowledge of the factors motivating today's consumer. Also learned how a product is marketed and distributed to the consumer. Took eight credit hours of study focused specifically on principles and problems of retail management.

<u>HONORS</u>

Dean's Merit List seven semesters. Mortar Board, senior women's national honorary. Held position of Editor, responsible for a summer newsletter and all publicity on campus. Phi Upsilon Omicron, home economics national honorary. Held position of Treasurer, responsible for receipt and disbursement of all money in the chapter.

<u>WORK EXPERIENCE</u>

Danzig's Department Store, Dallas, Texas, Summers 19— through 19—. Responsibilities included selling to customers, creating displays, and managing inventory within the men's department.

**Figure 3-1.** Resume.

Patricia Norman                                                    2

## ACTIVITIES

Commander (President) for Angel Flight, campus volunteer service organization. Responsibilities included overseeing all flight activities by delegating responsibility to appropriate people, authorizing legislation, presiding at all regular and special meetings, and seeing that all ends were tied together. Also held positions as Membership-Selection Chairman and as representative to regional and national conclaves.

## PERSONAL INTERESTS

Art history, sewing, water skiing, and reading.

## REFERENCES

Derek Yoder, Store Manager
Danzig's Department Store
11134 Longhorn Drive
Dallas, Texas 75220
(214) 394-4875

Dr. Lydia Zelasko, Chair
Home Economics Department
Putman Hall
University of Washington
Seattle, Washington 98195
(206) 579-3589

Professor C. Gregory Yule
Marketing Department
Pinehurst Hall
University of Washington
Seattle, Washington, 98195
(206) 579-9481

**Figure 3-1.**    *Continued.*

ROBERT C. MACDONALD

<u>Address Until December 15</u>
37 W. Collinsworth, Apt. 3
Oxford, Ohio 45056
(513) 529-1538

<u>Permanent Address</u>
7932 Houghton Drive
Toledo, Ohio 43606
(419) 474-4708

PROFESSIONAL
OBJECTIVE

To design production processes for a
heavy equipment manufacturer.

EDUCATION

Miami University, Oxford, Ohio. B.S. in
Manufacturing Engineering, with a concentration in computerized manufacturing operations. December 19—.

SPECIALIZED
COURSES

Important engineering coursework includes (credit hours shown):

| | |
|---|---|
| Basic and Advanced Electronics | 12 |
| Materials and Manufacturing | 20 |
| Instrumentation and Controls | 12 |
| Electro-Mechanical Systems | 8 |

Important computer science courses include:

| | |
|---|---|
| Assembler Language Programming | 3 |
| Computer Assisted Manufacturing | 8 |
| Analysis of Manufacturing Systems | 4 |

WORK
EXPERIENCE

<u>Wallace King Amusement Park, Toledo, Ohio</u>
Worked as maintenance man. Responsibilities included keeping shop clean and
safe, and repairing and servicing eight
vehicles.

<u>Ballentine's Texaco, Toledo, Ohio</u>
Worked as serviceman. Responsibilities
included waiting on customers, closing
station at night, cleaning premises, and
repairing engines.

**Figure 3-2.**   Resume.

2

ACTIVITIES   Member of the Society of Manufacturing Engineers.

INTERESTS   Water skiing, boating, rock climbing.

REFERENCES   Professor Milton Pultinas, Manufacturing Engineering Department Miami University, Oxford, Ohio 45056. (513) 529-2650

Professor Janet Casalano, Computer Science Department, Miami University, Oxford, Ohio 45056. (513) 529-5928

Wallace King, Jr., President, Wallace King Amusement Park, 118 Adams Avenue, Toledo, Ohio 43694. (419) 462-5312.

**Figure 3-2.**   *Continued*

### Professional Objective

When you state your professional objective, you answer your readers' question, "What, exactly, do you want to do?" Your answer to this question can be extremely important to the success of your resume. In response to a survey, personnel officers of the 500 largest corporations in the United States reported that the most serious problem they find with resumes and letters of application is the applicant's failure to indicate career and specific job objectives.[1]

That feeling by personnel officers may surprise you. "After all," you may reason, "once I present my qualifications, shouldn't an employer be able to match me to an appropriate opening?" The answer to that question lies in your imaginary portrait of your reader—the harried man trying to work his way rapidly through a large number of applications. That person does not have time to act as your personal employment agency, trying to match you with a job. For every opening that an organization has, it will receive more than enough resumes from applicants who want that particular job—applicants who do not have to be asked, "Do you think you might like this job?" Furthermore, when compared to those applicants, people who do not describe their employment objectives may seem aimless—pleasant enough, talented enough, but not motivated enough to compete successfully with applicants who have defined their goals more precisely.

What should your professional objective look like? By convention, such statements are one or two sentences long. Also by convention, the statements are usually general enough that the writer can send them, without alteration, to many prospective employers. If you follow this convention, you would not say (for example), "I want to work in the process control department of the Fort Lauderdale plant of Channing Electronics." Instead, you would make a more general statement, such as "I want to work in process control for a mid-sized electronics company located in the southeastern United States."

So, the challenge you face when writing your professional objective is this: to avoid being too general while not being too specific. To do that, you must describe your objective so that if you sent your resume to several organizations, each would feel that you want to work in:

- their particular kind of organization (a manufacturing plant, a pharmaceutical company), and
- some particular department (quality control, new product development).

To see how two students described their professional objectives, look at the sample resumes written by Patricia Norman (Figure 3-1) and Robert MacDonald (Figure 3-2). Both students achieved the desired level of specificity.

What should you do if you are genuinely interested and qualified for two different types of jobs—such as engineering design and engineering sales? Write two resumes, one for each type.

## Education

When describing your education, you have a chance to provide evidence that you are capable of performing the job you have applied for. The basic evidence is your college degree, so you should tell where you are going to college, what degree you will earn, and when that degree will be conferred. But there are also several other persuasive things you may be able to say.

- If your grades are good, mention them.
- If you earned any academic honors, tell what they are.
- If you had any special educational experience, such as a co-op assignment or internship, describe it.

Also, keep in mind that the title of your degree does not necessarily tell the readers very much about your specific qualifications for the particular job you want. For that reason, you may want to describe some of the courses or kinds of courses you have taken that relate directly to that job.

By looking at the resumes of Patricia Norman (Figure 3-1) and Robert MacDonald (Figure 3-2), you can see how two very different students elaborated upon the way their educations qualified them for the jobs they wanted. Interested in a management position in merchandising, Patricia used this section to state that she had taken considerable course work in two crucial areas: marketing management and retail management. Also, because her grades were high, she mentioned her accumulative grade-point average under the heading "Education." By devoting a separate section to her honors, she made them more prominent than they would have been if she had merely included them with the other material under the heading "Education."

In contrast to Patricia, Robert was on academic probation almost every term. Naturally, he did not mention his accumulative grade-point average, but he still managed to give a favorable impression of his capabilities by providing an extensive list of courses he has taken that relate to his employment objective of working in the production engineering department of a heavy-equipment manufacturer. To emphasize that list, he provided it with a separate heading.

## Work Experience

By describing your work experience, you can present further evidence about your capabilities, and you can also show that you are responsible. However, you will miss the opportunity to fully display your qualifications if you merely name your former employers and give your job titles.

You can show your capabilities by mentioning any duties you performed that resemble duties you will have on the kind of job you are currently seeking. Notice, for instance, how Patricia Norman, who is seeking a job as a manager in merchandising, describes her duties during her summer work in a department store:

Responsibilities included selling to customers, creating displays, and managing inventory within the men's department.

However, even if your previous work has been unrelated to your career objectives, you can still describe it in a way that will show that you are so reliable that others have entrusted you with significant responsibility. You can do that by asking yourself, "What did I do in that job that showed that others relied upon my sense of responsibility?" By asking himself that question, Robert MacDonald found that even his duties as a maintenance man could be turned to advantage, if described properly:

Worked as maintenance man. Responsibilities included keeping shop clean and safe, and repairing and servicing eight vehicles.

By the way, if you paid a large portion of your college expenses with your earnings, be sure to mention that fact.

Usually, applicants list their jobs in reverse chronological order, putting the most recent one first. This arrangement will work very well for you if your most recent job was also the one that best displays your qualifications for the job you now want. However, if it doesn't, you may want to devise some other way of presenting your jobs that gives the most important one the proper emphasis. One way to do that is to separate that job out and give it a special heading, such as "Related Job Experience"—related, that is, to the job you are seeking. Then you might describe the less impressive jobs under the heading "Other Job Experience."

### Activities

At the very least, your participation in group activities will indicate that you are a pleasant person who gets along with others. Beyond that, your activities may also show that you developed some of the specific abilities that are important in the job you want. Notice, for instance, how Patricia Norman describes one of her activities in a way that shows her ability to handle a management position:

Commander (President) for Angel Flight, campus volunteer service organization. Responsibilities included overseeing all flight activities by delegating responsibility to appropriate people, authorizing legislation, presiding over all regular and special meetings, and seeing that all ends were tied together.

This statement suggests that Patricia has been entrusted with considerable responsibility by people who know her well.

In considering material you might include in your resume, be sure not to overlook ways in which you have demonstrated unusual abilities of any sort, such as being selected for a high school choir that toured Europe or playing varsity sports at your college.

### Interests

Although it is not necessary for you to tell about your interests in a resume, by doing so you can provide your reader with evidence that you are a pleasant person—one who can talk about something other than studies and jobs. Of course, if you are engaged in extracurricular activities, your description of them under the "Activities" heading may provide plenty of information about your interests, making a separate section on the matter unnecessary.

### Personal Data

Federal law prohibits employers from discriminating upon the basis of sex, religion, race, age, or national origin. It also makes it illegal for employers to ask you if you are married or plan to be married. Many students welcome these restrictions because the restrictions prohibit employers from asking about matters the students consider to be personal or irrelevant to their qualifications.

On the other hand, federal law does not prohibit you from giving employers information of this sort, if you wish. And many students do.

Guidelines for effective writing cannot provide you with much help in deciding whether or not you should include personal information in a resume. Your understanding of your readers can help you predict whether or not any of this information might help you obtain the job you want, but it cannot determine your personal values for you.

### References

Your references are people who agree to tell employers about your qualifications. Some students decide not to include the names of their references on their resumes. However, there are two good reasons why you should.

First, by listing your references you make it easier for employers to contact them. Otherwise, employers will have to write to you and then await your reply before making contact.

Second, the mere appearance of your references' names can strengthen your resume if you make your references seem as important as possible. After all, the more important the people who will speak on your behalf, the more important you will seem in the eyes of your readers.

Because you want your readers to see what important individuals are serving as your references, you must take great care in the way you give information about them. Too many resume writers write references like these:

William Houk
Culler Hall
Miami University
Oxford, Ohio
(513) 529-5265

Walter Williamson
6050 Busch Boulevard
Columbus, Ohio 43220
(614) 376-2554

For all an employer will know, Culler Hall might be a dormitory and Mr. Houk might be the applicant's roommate. Likewise, a reader might think that Mr. Williamson is the applicant's neighbor on Busch Boulevard. To prevent such misunderstandings, you should give the title and full business address of your references:

Dr. William Houk, Chairman
Department of Physics
Culler Hall
Miami University
Oxford, Ohio 45056
(513) 529-5625

Mr. Walter Williamson, Director
Corporate Services Division
Orion Oil Company
6050 Busch Boulevard
Columbus, Ohio 43220
(614) 376-2554

Before you include anyone as a reference in your resume, be sure to check with that person. Asking can benefit you in two ways. First, you can assure yourself that the person feels that he or she knows you well enough to give a good, supportive account of you to employers. Second, it helps the people prepare for calls from employers. If you are naming a teacher who had you in class a year ago or an employer who hired you two summers ago, those people may have momentary trouble recalling who you are if they receive an unexpected phone call about you. Third, when you speak to the persons you want to serve as your references, you can provide them with a copy of your resume, so that they will have at hand some specific information about you if they receive any inquiries.

## Designing the Appearance

One of the most important things to keep in mind when you are designing the appearance of your resume is that you want to help your very busy readers find *quickly* the specific information they want.

You can do that by following one of the conventional layouts shown in Figures 3-1 and 3-2. Notice that both make extensive use of headings to help readers locate the specific kinds of information that interest them. These resumes also have ample amounts of white space to help the reader see the divisions between blocks of information.

Because resumes are usually sent to more than one company, they may be printed, either by copy machine or offset press. In most organizations, recruiters expect resumes to be prepared on white, 8½ × 11-inch paper; students who want theirs to be a little different sometimes use a buff or light gray. A more startling color is likely to look not simply different but jarringly odd to most readers. To readers in some professions, however, oddity can be effective; for example, recruiters for marketing firms are used to getting resumes printed in all kinds of unique ways—as the outsides of boxes, as messages on balloons, in three-dimensional models. Naturally, your resume should be neat and carefully proofread.

## Evaluating

Throughout your work at drafting, you concentrate on developing writing strategies that you think will affect your readers in the way you desire. But will your strategies really work? An important step in writing involves stepping back from your role as a writer and attempting to see your communication as your readers will see it so you can use the insights you gain to strengthen your message before you send it.

Unfortunately, as writers we all have difficulty seeing our own work in just the way our readers will see it. When we look at our pages, we often see what we intended—which is not always the same as what we accomplished. To compensate for that difficulty, you can show a draft of your communication to other people, asking for their reactions and advice. When you are writing a resume, various people can help you. If your instructor asks you and your classmates to share your drafts with one another, you will get (and give) helpful advice. Likewise, both your writing instructor and instructors in your major department can provide valuable assistance.

## Revising

When you revise, you improve your communication in light of the advice and insights you gain from your evaluation. To do that you will probably have to consider not only matters of right and wrong but also matters requiring your sensitivity, insight, and good judgment.

For example, the people you ask for help in evaluating drafts of your resume may identify problems but not be able to tell you how to overcome them. In addition, some people may give you advice that contradicts the advice you get from others. This chapter for instance, suggests that you include your references in your resume. From other people, you may hear the more traditional advice to leave them out. When you are writing a resume as an assignment in your writing course, good sense dictates that you follow the advice given by your instructor. But outside the course, you are going to be on your own.

How can you choose then? You need to think back to your objectives: to your descriptions of your readers and purpose. Then, using your own good sense and creativity, you are going to have to revise in the way that you think most likely to match your communication to its readers and purpose.

In the preceding discussion of how to write your resume, the most important thing for you to notice is the great amount of help you get from taking a reader-centered approach—from thinking constantly about your readers. You define your purpose in terms of your readers' tasks and attitudes so that you will know how you want your resume to affect your readers as they read it. You create a mental portrait of your readers so that you can imagine what they will want from your resume and how they are likely to react while reading it. Then, you use your understanding of your reader and your reader-centered definition of purpose to guide you as you draft, evaluate, and revise your resume.

The rest of this chapter shows how you can use this same approach when writing your letter of application.

## WRITING YOUR LETTER OF APPLICATION

Your letter of application is a companion to your resume. It goes to the same readers, and it has the same *general* purpose: to get you the job you want. But the differences between the specific purposes of resumes and letters of application, though apparently slight, are great enough to make the two quite distinct from one another. Consequently, besides providing you with practical advice about writing letters of application, the following paragraphs show why you should think very carefully about your purpose and readers whenever you write at work: even small differences in purpose or audience can have a large effect on the way you should write.

### Defining Your Objectives

How does the purpose of your letter of application differ from the purpose of your resume? To begin with, your resume is a versatile document. You give it to employers whether you first contact them in person or through the mails. Furthermore, the people involved with hiring you will use it throughout the various steps of the hiring process—from their initial assessment of your qualifications through your visit to their facilities.

In contrast, you send a letter *only* when you first contact employers through the mail. Unlike your resume, your letter becomes a much less important document to your readers after they have met you in an interview. That's because your letter of application is, in essence, a substitute for a personal meeting. Therefore, it must serve, by means of the written word, several of the purposes of an initial interview.

Accordingly, one way to understand what readers look for in letters of application is to think about what they look for in interviews. The most important of these things, beyond doubt, is some sense of your personality. Your readers want to know what you are like as an individual, what you would be like as an employee, a co-worker. That is information you cannot readily provide in a resume, which is more like a table of data than an expression of your personal characteristics.

In addition, from your letter employers will want to learn how you feel your education and other experiences relate to the *particular* job you have applied for. In fact, in the survey (mentioned above) of personnel directors of the 500 largest corporations in the United States, 88% either "agreed" or "strongly agreed" that a letter should show how the applicant's education and experience fit the requirements of the particular job he or she has applied for.

Employers can't get that information from your resume because the convention of the resume is to present your qualifications in only a general way so that you can send it to several employers. From your letter, however, readers will want to learn how your specific qualifications match the requirements of the particular job you are applying for.

Also, employers will want your letter to tell them why you chose *their* organizations to apply to. In that same survey, 81% of the personnel directors either "agreed" or "strongly agreed" that a letter should tell why the applicant applied for *this* particular job.

Thus, you might say that the one objective of your letter is to answer your readers' questions about your personality, your interest in them, and your perception of the relationship between your qualifications and the particular job they offer. Of course, your letter also aims to answer all those questions in a way that persuades your readers that you are a top candidate for their job.

## Planning

With such an understanding of the purpose of your letter, and keeping in mind the facts about your readers that were discussed earlier in this chapter, you are ready to think about the kinds of material you should include in your letter.

The best way to begin is to tell your readers why you are writing. First, you can explain why you are writing by telling them what you like about their organization and what job you desire. You thus answer your readers' questions about what you want and why you are applying to *their* organization rather than to someone else's.

Furthermore, through your explanation of your reason for writing, you can show yourself to be likable and pleasant. One thing that makes us like someone else is the sense that the other person knows and likes us. Consequently, one of the most persuasive messages you can include in a letter of application is, "I know about you and like you." In their first paragraphs, both Patricia Norman and Robert MacDonald show that they have a direct, personal knowledge of the organizations they are writing to. And they explain why they like the organizations. (See Figures 3-3 and 3-4.)

Box 88
Wells Hall
University of Washington
Seattle, Washington 98195
February 12, 19—

Kevin Mathews, Director
Corporate Recruiting
A.L. Lambert Department Stores, Inc.
Fifth and Noble Streets
San Diego, California 92103

Dear Mr. Mathews:

Writer demonstrates knowledge of company.

At about the time that I saw A.L. Lambert's advertisement in the *College Placement Annual*, I read the article about your company in the July issue of *Retail Management*. I was very impressed by your success at opening free-standing specialty shops within the stores of your chain. Since I would very much like to work for a company where I could help develop creative marketing strategies, I hope that you will consider me for an opening in your Executive Development Program.

Writer tells what job she wants.

In June, I will graduate from the University of Washington's retailing program, where I have focused my study on marketing management. I have learned a great deal about consumer behavior, advertising, and innovative sales techniques. Furthermore, I have gained a thorough overview of the retailing industry, and I have studied successful and unsuccessful retailing campaigns through the case-study method.

Writer highlights special features of her education.

In addition to my educational qualifications, I have experience in both retail sales and in managing volunteer organizations. While working in a Dallas department store for the past four summers, I had many opportunities to apply the knowledge and skills that I have learned in college. Likewise, in my extracurricular activities, I have gained experience working and communicating with people.

Writer explains importance of her summer jobs and college activities.

**Figure 3-3.**   Letter of Application.

2

For instance, I have been the Commander of Angel Flight, a volunteer service organization at the University of Washington. Like a manager, I supervised many of the organization's activities. Similarly, while holding offices in two honorary societies, I have developed my senses of organization and responsibility.

Writer explains importance of her summer jobs and college activities.

I would appreciate an opportunity to talk with you in person about my qualifications. Is there any way that can be arranged?

Sincerely yours,

Patricia Norman

**Figure 3-3.**    *Continued*

37 W. Collins
Oxford, Ohio 4505
October 12, 19—


Daniel J. Grice, Manager
Search and Placement
Stark Equipment Company
17 Springwood Road
Buchanan, Michigan 49109

Dear Mr. Grice:

There may be no better way to learn about a company than to work closely with its products. Although I have long known your company by name, I really got to know about the Stark Equipment Company last summer. At that time, I worked with a Stark forklift truck, one of eight vehicles I was responsible for while employed as a maintenance man by a large amusement park.

As a result of seeing the craftsmanship and engineering skill that went into that forklift, I decided that Stark is exactly the kind of firm for which I would like to work. Therefore, I hope that you will be interested in interviewing me for a position in your Production Engineering Department on November 10, when I will be in the Buchanan area.

If it turns out that you have an opening for me, I will be able to begin work in December, after I complete my B.S. in Manufacturing Engineering at Miami University. My field of concentration is computerized manufacturing operations, so that I have supplemented my work in a broad variety of engineering subjects with considerable study in computer science courses that have direct application in the manufacture of heavy equipment. In the enclosed resume, I provide additional information about my education, as well as the other experiences through which I have prepared myself for a career in industry.

I am eagerly waiting to learn whether you will be able to talk with me on November 10.

Sincerely,


Robert C. MacDonald

*Annotations (left margin):*

Writer demonstrates knowledge of company.

Writer tells why he is applying to this company.

Writer tells what job he wants.

Writer highlights his special qualifications for the job.

**Figure 3-4.** Letter of Application.

You can't tell why you are applying to a particular organization unless you first *know* something about it. If you don't have the kind of prior knowledge of the company that Robert MacDonald had, you might ask the staff in your campus placement center or your college librarians to help you locate information about it. The work involved will be well rewarded.

You should also take the trouble to learn the name of the person you are writing to. Some companies publish recruiting information about themselves without giving the name of any of their individual recruiters, so you may need to call the company itself (that is, its switchboard) to learn the name of some particular person to whom you can address your letter.

After opening your letter in a way that explains your interest in your readers' organization and tells what job you want, you can discuss your qualifications. When doing that, you should do more than merely repeat the highlights of your resume. You should link your education and experiences together in a way that explains how they are relevant to the job you want—just as you would do in an interview. By doing that, you answer one of the major questions, mentioned above, that your readers will want you to answer in your letter.

How do you answer your readers' questions about your personality? One way is through the kinds of things you choose to say about your qualifications and reasons for applying. When you tell what appeals to you about the organization, you reveal some of your personality. You do the same through the qualifications you choose to emphasize and the way you explain their relevance to the employer's position.

Notice, for instance, how the first paragraph of Patricia Norman's letter shows her to be an observant, curious person who is able to appreciate and warmly praise good work done by others. Notice, also, how her third paragraph shows her to be serious and well-organized—aspects of her personality that she thought would appeal to her readers.

## Drafting

Even when you have made excellent and thorough plans for a letter of application, you will probably find it a challenge to write. The most difficult thing for most college students to achieve in such a letter is an appropriate tone. Yet, in your letter nothing is more important than tone. In the survey mentioned above, 90% of the personnel directors either "agreed" or "strongly agreed" that the tone of the letter of application is important.

Why is tone so important? As much as anything else in your letter or resume, it can reveal your personality. Or, it can *hide* your personality. If you try to control the tone of your letter by using standard phrases that you have seen in other resumes, your readers will know it. Rather than being impressed, however, they will be uninterested.

Many people have difficulty achieving a tone in their letters of application that is appropriate to both the occasion and their personalities. This difficulty is often especially great for people who haven't written such letters before. Here are some things that are particularly troublesome to college students.

## Achieving a Self-Confident Tone

Some students have difficulty indicating an appropriate level of self-confidence. In your letter, you want to seem self-assured. However, you do not want to appear brash or overconfident. Thus, you want to avoid statements like the following:

> I am sure that my excellent education qualifies me for a position in your Ball Bearing Department.

There are two problems with that sentence. The first is the offensive "I am sure," and the second is that the sentence pushes the readers out of consideration by asserting that the writer has already performed for them their job of evaluating the writer's qualifications. To improve the sample sentence, you might write:

> I hope that you will find that my education qualifies me for a position in your Ball Bearing Department.

## Achieving a Good Tone in the Conclusion

In addition, some students also have problems achieving the proper tone in the conclusion of their letters. You should avoid conclusions like this:

> I would like to meet with you at your earliest convenience. Please let me know when this is possible.

or:

> I am available for interviews in the Columbus area on April 15. I will be expecting to hear from you soon.

To sound less demanding, you might revise the second sentence of the first sample to say, "Please let me know whether this is possible." Likewise, you might revise the second sentence of the second sample to read, "I look forward to hearing from you."

## Format

Finally, when drafting your letter, you should use one of the several standard formats for business letters. You are not likely to impress your readers favorably by using any other. Standard formats are described in Chapter 26, and used in the sample letters shown in this chapter.

### Evaluating

You will undoubtedly find that writing your letter of application is a challenging undertaking. The chief reason is that no model can tell you how to do it. The personality you need to reveal is your own. If you merely imitate one of the sample letters out of this or any other book, you will be revealing someone else's personality. If you are like most college students, writing letters of application is also difficult because you have had very little experience in writing about yourself in a business (as distinct from a personal) situation.

Because of the difficulty and the novelty of writing these letters, it will probably take you many drafts to arrive at an effective draft. Furthermore, it would be wise for you to ask other people to read your draft over. Ask them to play the role of the recruiter or the department head you hope to impress. Good choices for these role-playing readers would be the instructor of your writing course and instructors in your major department. Both can tell you whether the letter you have written is likely to affect your readers in the way you intend.

### Revising

When revising your letter of application, you are likely to encounter some of the same difficulties you encounter when revising your resume. The people who help you evaluate your letter may identify problems but not tell you how to solve them. They may give you contradictory advice. To revise well, you have to use your own creativity and good judgment. And you need to think constantly about the way your readers will react, moment by moment, to what you have written.

## SUMMARY

This chapter has had three major purposes:

1. To show you how a reader-centered approach can help you make some of the most important decisions involved with writing a resume and a letter of application.

2. To provide you with an overview of the writing process so that you can see how its various stages are related to one another. In particular, this chapter has emphasized the importance of developing a thorough understanding of your readers and a thoughtful sense of your purpose. Those things will provide you with sound guidance throughout the later stages of the writing process.

3. To show how completely the design of a communication depends upon your particular purpose and readers, by showing how the differences between two similar communications—the resume and the letter of application—are attributable to small differences in their purposes.

Along the way, this chapter has also given you practical advice about how to prepare an effective resume and letter of application. Keep in mind that the reader-centered strategy described here for writing those two documents is fundamentally the same as the strategy you should use when writing any other kind of communication related to your job. That strategy is explained in more detail in the rest of this book.

## NOTE

1. Baron Wells, Nelda Spinks, and Janice Hargrave, "A Survey of the Chief Personnel Officers in the 500 Largest Corporations in the United States to Determine Their Preferences in Job Application Letters and Personal Resumes," *ABCA Bulletin* 14(June 1981):3–7.

## EXERCISES

1. Find a sample resume by looking in handouts from your college's job placement center or in books about resume writing that are available in your library. Evaluate that sample from the point of view of its intended readers.

2. An assignment that involves writing a resume and letter of application is included in Appendix I, "Writing Assignments."

# Defining Your Objectives

Todd recently graduated with a degree in chemical engineering. For three months he has been working in the research division of a paint company. Todd's first assignment has been to investigate ways of making car paint that will chip less easily than conventional paint. This morning, Todd's boss said that some upper-level managers want Todd to send them a report on his progress.

## A WRITER'S QUESTIONS

"How should I write the report?" Todd asks himself. "What should I tell the managers?"

To be able to answer those questions—and all others he will face as he writes—Todd needs to figure out what he is trying to accomplish in his report. He needs to define his objectives.

If Todd states his objectives clearly and precisely, his statement can guide him throughout his work at writing. It can help him decide what to say and how to say it. It can help him decide what to present in prose and what in tables and illustrations, how to design his pages and how to structure his paragraphs and sentences—how, in fact, to handle every aspect of his report.

Like Todd, you will benefit in many ways from carefully defining your objectives before you begin to write.

## THREE ACTIVITIES OF DEFINING OBJECTIVES

In order to construct a truly helpful definition of your objectives, you need to do three things:

- *Define Your Purpose* by deciding how you want your communication to affect your readers.
- *Understand Your Readers* so you can determine which writing strategies will cause them to respond in the ways you desire.
- *Find Out What's Expected* about such things as your deadline and the length, style, appearance, and other features of the communication you are about to write.

The next three chapters present guidelines that will help you perform each of these activities.

## CONTENTS OF PART II

**Chapter 4**
**Defining Your Purpose**
**Chapter 5**
**Understanding Your Readers**
**Chapter 6**
**Finding Out What's Expected**

# Defining Your Purpose

## PREVIEW OF GUIDELINES

1. Describe the final result you want your communication to bring about.
2. Describe the tasks your readers will try to perform while reading.
3. Tell how you want your communication to alter your readers' attitudes.

On page 75, you read about Todd, who has been investigating ways to make car paint that will chip less easily than conventional paints. This morning, he was asked to write a report on his progress for upper-level managers.

Todd has wisely decided to begin his work on this report by defining its purpose. By taking the time to figure out exactly what he wants to accomplish through his report, he can provide himself with a sound basis for deciding how to write it.

Consider, for example, how Todd's definition of purpose can help him decide what to include in his report. Different purposes suggest different contents. Suppose Todd determines that the managers want to understand the technical information about paint chipping that he has gathered so far. Then he should write a long report explaining what he has read and what he has learned in the laboratory. On the other hand, suppose Todd determines that the managers are trying to decide whether or not to assign him a helper because they are worried that he cannot complete his research by their deadline. Then he should write a very brief report telling how much he has accomplished, how much he has left to do, and how much time it will take him to do it.

Todd's definition of purpose will also help him with all the other aspects of his report: its organization, page design, tone of voice, and so on. Similarly, by defining the purposes of the communications you write at work, you can provide yourself with a sound, helpful guide for writing.

This chapter presents three guidelines for defining purpose.

## TWO KINDS OF PURPOSE

While reading about the guidelines, keep in mind that the statements of purpose described here are for *your* use when writing communications. They are not the same as the statements of purpose you would include in a communication to help your readers read and use it effectively. You will find advice about ways of telling your readers the purpose of a communication in Chapter 13, "Writing the Beginning of a Communication."

As you read the next few pages, you will see the reason for distinguishing between the purpose you define for yourself and the purpose you explain to your readers.

### Describe the Final Result You Want Your Communication To Bring About

Guideline

1

As explained in Chapter 1, writing is action. When you write, you exert your powers to bring about a specific result. The easiest way to begin defining your purpose is to tell what you want that result to be. For example:

- You want your boss to accept your proposal for beginning a new project.
- You want the twenty-seven people in the department you manage to adhere to a new policy you have established.
- You want the employees in three departments to be able to run a computer program that you wrote.

In these three examples, it's obvious what result you desire. In some situations, however, you may need to make a special effort to find out what final result your communication should bring about. Sometimes managers ask new employees to write things (like Todd's progress report) without telling them why. If that happens to you, you could guess at what result is wanted. But guessing involves a great risk: you might guess incorrectly. If you write a communication that will achieve one final result excellently, but your boss has another result in mind, you are not likely to succeed with that communication. When you don't know what final result is wanted, you should ask.

To develop a truly helpful definition of purpose, however, you need to know more than the final result your communication should bring about. Why? Because the final result is something that happens *after* your readers have finished reading. If you are to succeed in bringing about that result, you must focus your attention on what you want to happen *while* your readers are reading. The next two guidelines will help you do that.

<table>
<tr><td>Guideline</td><td rowspan="2">

### Describe the Tasks Your Readers Will Try To Perform while Reading.

</td></tr>
<tr><td>

**2**

</td></tr>
</table>

Whenever you write at work, you are trying to alter the world, in at least some small way. You attempt to change things from the way they are now to the way you want them to be.

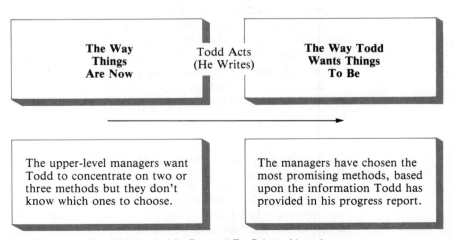

Figure 4-1.   The Change Todd Wants His Report To Bring About.

Imagine, for instance, that Todd's purpose is to help the upper-level managers decide which methods for producing chip-resistant paint they want Todd to concentrate his research upon in the coming weeks. Then, Todd's aim would be to bring about the change shown in Figure 4-1.

Figure 4-2 shows some other types of changes that you might want to bring about through the writing you do on the job.

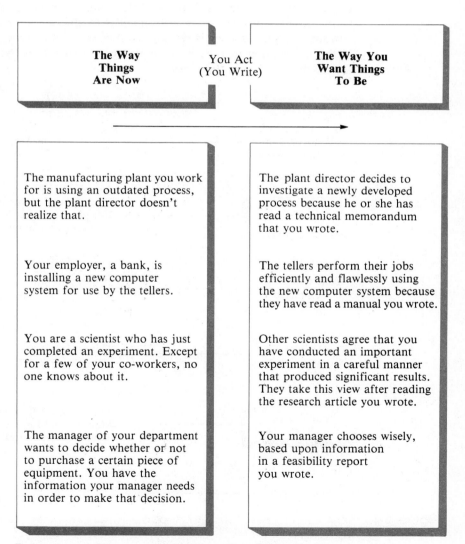

| The Way Things Are Now | You Act (You Write) | The Way You Want Things To Be |
|---|---|---|
| The manufacturing plant you work for is using an outdated process, but the plant director doesn't realize that. | | The plant director decides to investigate a newly developed process because he or she has read a technical memorandum that you wrote. |
| Your employer, a bank, is installing a new computer system for use by the tellers. | | The tellers perform their jobs efficiently and flawlessly using the new computer system because they have read a manual you wrote. |
| You are a scientist who has just completed an experiment. Except for a few of your co-workers, no one knows about it. | | Other scientists agree that you have conducted an important experiment in a careful manner that produced significant results. They take this view after reading the research article you wrote. |
| The manager of your department wants to decide whether or not to purchase a certain piece of equipment. You have the information your manager needs in order to make that decision. | | Your manager chooses wisely, based upon information in a feasibility report you wrote. |

**Figure 4-2.**   Examples of Changes that Writers Can Bring About.

## Focus on Your Readers in the Act of Reading

Of course, your writing can bring about the final result you desire only if it affects your readers in a certain way while they are reading it. It certainly can't bring about any result before they read it, and any effects it creates indirectly afterwards will depend upon what happened *while* your readers were reading (Figure 4-3.) That's why you should define a communication's purpose not only by the final result you want the communication to achieve, but also by the way you want it to affect your readers while they read it.

To focus your attention on the way you want a communication to affect your readers while they read, think of the communication as a tool, an instrument. One purpose of this instrument is to enable your readers to perform some specific task while reading. To succeed, it must provide them with (1) the information they need (2) in a manner that makes the information easy for them to use in their task.

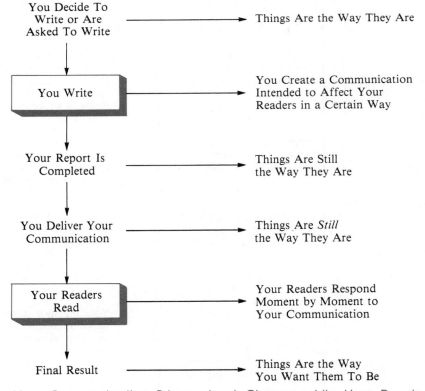

You Decide To
Write or Are               → Things Are the Way They Are
Asked To Write

You Write                  → You Create a Communication
                             Intended to Affect Your
                             Readers in a Certain Way

Your Report Is             → Things Are Still
Completed                    the Way They Are

You Deliver Your           → Things Are *Still*
Communication                the Way They Are

Your Readers               → Your Readers Respond
Read                         Moment by Moment to
                             Your Communication

Final Result               → Things Are the Way
                             You Want Them To Be

**Figure 4-3.**   Your Communication Brings about Change while Your Reader is Reading it.

## Typical Reader Tasks

How can you identify the specific tasks that your readers will try to perform while reading? You can do that by following the first of the general guidelines given in Chapter 2: when you write, talk with your readers. In practice, that means that you must imagine your readers in the act of reading, moment by moment, the communication you are writing. What questions are they asking as they read that paragraph? What are they trying to do with the information you provide?

As you try to imagine your readers in the act of reading, you may find it helpful to keep in mind Figure 4-4, which lists five reading tasks people often perform at work. The figure includes sample strategies you can use to help your readers perform each of the tasks.

| Readers' Tasks | Sample Writing Strategies |
|---|---|
| *Compare various pieces of information,* as when making a decision or conducting an evaluation. | Determine the specific points upon which your readers will make the comparison. Organize the relevant facts around those points. |
| *Locate specific facts,* as when using reference books (such as computer documentation) or when reviewing particular parts of a report. | Provide a route to specific facts by means of an index and table of contents. Use prominent headings. |
| *Understand,* as when reading scientific reports and other conceptual material. | Move from the familiar to the unfamiliar. Move from the general to the particular. |
| *Follow step-by-step instructions* by reading a step, doing a step, and so on. | Number the steps and present them in lists to help readers find their place again after performing each one. |
| *Apply information,* as when reading policies and regulations or communications presenting a general procedure. | Focus on actions that readers might take. Give examples of people doing specific things in particular situations. |

**Figure 4-4.** Five Tasks Readers Often Perform and Writing Strategies Suited to Them.

## Overlap of Tasks

Of course, these tasks overlap. For example, to perform any of the other tasks, a reader must *understand* what you have written.

In addition, many communications are divided into sections, each devoted to helping readers perform a different task. For example, many manuals for computer programs begin with a general introduction that helps readers *understand* the uses and general strategy of the program. They move to a tutorial section, which helps readers *follow the step-by-step instructions* that tell how to use the program, and conclude with a reference section, which helps readers *locate specific pieces of information* about the program and its operation. For such communications, you need to think about the tasks your readers will be trying to perform in each part.

## Example

To see how thinking about your readers' tasks can help you, consider Lorraine's situation. Lorraine works for a steel mill that has decided to build a new blast furnace. She has been asked to prepare a report on the two types of furnace that the mill is thinking of constructing. The ultimate objective of her communication is to help the president of the mill decide which furnace to buy.

Having gathered all the relevant information about these furnaces, she finds herself unable to make one of the basic decisions about her report: how to organize the 150 pages of material she has amassed. She realizes she could use either of the organizational patterns shown in Figure 4-5.

In the divided pattern, Lorraine would devote the first half of her report to one blast furnace and the second half to the other. In the alternating pattern, she would discuss the two furnaces side by side in terms of one factor, then another.

How can she decide which pattern is best? *Logic* won't help. From the point of view of logic, the two patterns are equally correct. Consideration of *length* won't help. Even though the outline for the alternating pattern is longer than the outline for the divided pattern, the patterns would yield reports of the same length. Her *statement of the final result* she hopes to achieve won't help either. Both patterns will provide the president of the mill with the information he needs to make his decision. From the point of view of *ease of writing,* the divided pattern is probably the easiest. When conducting her investigation, Lorraine organized her notes into two big piles, one about blast furnace A and the other about blast furnace B. But then ease of writing is a writer-centered consideration, not a reader-centered one.

What is a reader-centered consideration? It's one that involves thinking not about the work the writer will do, but about the work *the reader* will do *while reading.* By thinking about her reader's work, Lorraine will have a sound basis for choosing between these two organizational patterns.

| **Divided Pattern** | **Alternating Pattern** |
|---|---|
| Furnace A<br>  Cost<br>  Efficiency<br>  Construction Time<br>  Air Pollution<br>  Etc. | Cost<br>  Furnace A<br>  Furnace B |
| | Efficiency<br>  Furnace A<br>  Furnace B |
| Furnace B<br>  Cost<br>  Efficiency<br>  Construction Time<br>  Air Pollution<br>  Etc. | Construction Time<br>  Furnace A<br>  Furnace B |
| | Air Pollution<br>  Furnace A<br>  Furnace B |
| | Etc.<br>  Furnace A<br>  Furnace B |

**Figure 4-5.**   Two Organizational Patterns Lorraine Can Use in Her Report.

Suppose, for the sake of this example, that the mill president intends to compare the two furnaces in detail in terms of specific criteria: cost, efficiency, construction time, and so on. To make such a point-by-point comparison using a report organized according to the divided pattern, he would have to search the first part of the report for information about cost (for example) of the first furnace and then search the second part of the report for information on the cost of the second furnace. Then he would have to flip back and forth between the two locations in order to compare the two costs. Furthermore, he would have to continue this tedious and time-consuming process for *every* one of the criteria he had in mind.

In contrast, the mill president could make these comparisons much more easily if Lorraine used the alternating pattern. That pattern puts in one place the information about the cost of both furnaces, the efficiency of both furnaces, and so on. For other purposes, the divided pattern—or some other pattern—may be a better way to organize a communication. But in this situation, the alternating pattern will make the reader's work easier.

Of course, there are other ways Lorraine could help the mill president perform this task of comparing the two blast furnaces point by point. For instance, she could open her report with a summary that makes a point-by-point comparison. Such strategies might work very well. What's important for you to remember from this example, however, is *not* that the alternating order is a good way of helping readers make point-by-point comparisons. The key point is this: you can gain important insights about how to write if you think about the tasks your readers will try to perform while reading. These insights are valuable because your communication must enable your readers to perform those tasks if it is to bring about the final result you desire.

## What Else You Need To Know

As you think about your readers in the act of reading, however, you should think not only about the tasks they are trying to perform, but also about the attitudes they are developing. The next guideline explains how to do that.

Guideline

3

## Tell How You Want Your Communication To Alter Your Readers' Attitudes

When you listen to someone talk, you respond in two ways. Imagine that you are listening to your instructor give an assignment. First, you respond by performing the mental tasks required to understand and remember what he or she is saying. Second, you respond emotionally. Perhaps you feel eager because the assignment sounds interesting, or perhaps you feel worried because you already have several projects to complete for other classes.

The same things happen when you read. You think about the writer's message, performing the mental tasks necessary to understand and use it. And you develop and change your attitudes.

Because people respond to the things they read with feelings as well as thought, you should always plan the way you want your communications to affect your readers' attitudes. You can do that by describing to yourself two things: the way your readers feel *before* reading your communication and the way you want them to feel *after* reading it. An important part of your communication's purpose is to bring about that change *while* your readers are reading.

As an example, consider again the situation of Todd, the new college graduate who must write a progress report for some upper-level managers about his investigations into ways of making chip-resistant paint. Especially because this is the first time that these managers will read anything of his, Todd is very concerned about the way it will make them feel toward him and his abilities. Thus, he might say that he wants his communication to alter his readers' attitudes as shown in Figure 4-6.

To succeed in altering his readers' attitudes in this way, Todd must write a report that leads the readers to the desired attitudes—through the moment-by-moment responses it elicits from them.

## What Readers' Attitudes Are About

In the example given in the preceding paragraph, Todd is primarily concerned about the way that his progress report will affect his readers' attitudes

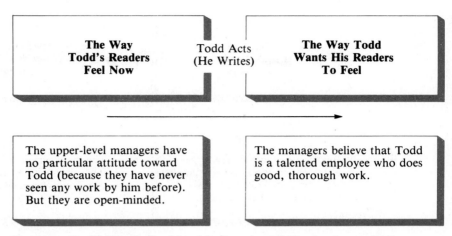

| | | |
|---|---|---|
| **The Way Todd's Readers Feel Now** | Todd Acts (He Writes) | **The Way Todd Wants His Readers To Feel** |
| The upper-level managers have no particular attitude toward Todd (because they have never seen any work by him before). But they are open-minded. | | The managers believe that Todd is a talented employee who does good, thorough work. |

**Figure 4-6.**   The Way Todd Wants To Alter His Readers' Attitudes toward Him.

toward him. Often, writers at work are also concerned about their readers' attitudes toward other things as well.

For instance, they are often concerned about their readers' attitudes toward their subject matter. That happens, for instance, when they are writing recommendation reports and proposals. After all, the major aim of these communications is to persuade the readers to take a certain attitude toward the action the writer is suggesting.

Consider Todd's situation, for example. Imagine that, based upon his research, he believes that one of the methods he has examined is much more likely than the others to produce highly chip-resistant paint. Further, imagine that he is confident that his readers would welcome a recommendation from him as they decide which method they will ask him to concentrate upon in his future work. Then, Todd might say that a major part of his purpose is to alter his readers' attitudes as shown in Figure 4-7.

Sometimes, writers are concerned with altering their readers' attitudes toward the readers themselves. Such a concern often arises, for example, when the writers are preparing instruction manuals. For instance, many people who have not used computers before fear that they will embarrass themselves when they first try. They may even fear that they are not capable of learning to use a computer at all. Accordingly, people who write instructions for these people might say that they want to alter their readers' attitudes as shown in Figure 4-8.

In sum, on the job you may be concerned with your readers' attitudes toward any of the following things: you, your subject matter, or themselves.

Figure 4-9 shows some other examples of ways you might want to alter your readers' attitudes when you write on the job.

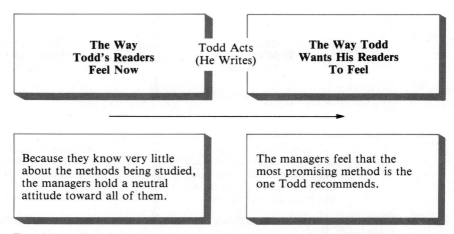

| The Way Todd's Readers Feel Now | Todd Acts (He Writes) | The Way Todd Wants His Readers To Feel |

Because they know very little about the methods being studied, the managers hold a neutral attitude toward all of them.

The managers feel that the most promising method is the one Todd recommends.

**Figure 4-7.** The Way Todd Wants To Alter His Readers' Attitudes toward His Subject.

## How Attitudes Can Change

As you think about the ways you want to change your readers' attitudes, keep in mind that you can change them in more than one way. For example, you might want your readers to reverse an attitude they presently hold. That is, you might want to persuade them to like something they now dislike. Or

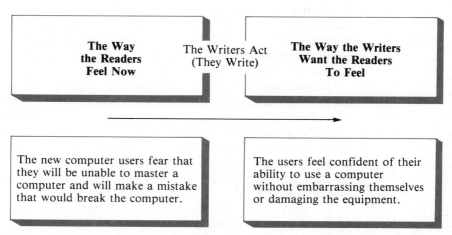

| The Way the Readers Feel Now | The Writers Act (They Write) | The Way the Writers Want the Readers To Feel |

The new computer users fear that they will be unable to master a computer and will make a mistake that would break the computer.

The users feel confident of their ability to use a computer without embarrassing themselves or damaging the equipment.

**Figure 4-8.** A Way that Writers Sometimes Want Their Readers To Alter Their Attitudes about Themselves.

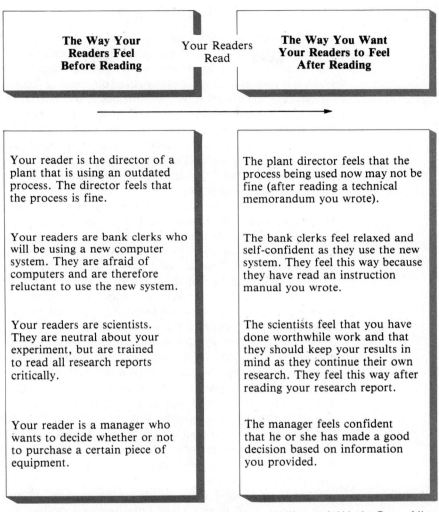

| The Way Your Readers Feel Before Reading | Your Readers Read | The Way You Want Your Readers to Feel After Reading |
|---|---|---|
| Your reader is the director of a plant that is using an outdated process. The director feels that the process is fine. | | The plant director feels that the process being used now may not be fine (after reading a technical memorandum you wrote). |
| Your readers are bank clerks who will be using a new computer system. They are afraid of computers and are therefore reluctant to use the new system. | | The bank clerks feel relaxed and self-confident as they use the new system. They feel this way because they have read an instruction manual you wrote. |
| Your readers are scientists. They are neutral about your experiment, but are trained to read all research reports critically. | | The scientists feel that you have done worthwhile work and that they should keep your results in mind as they continue their own research. They feel this way after reading your research report. |
| Your reader is a manager who wants to decide whether or not to purchase a certain piece of equipment. | | The manager feels confident that he or she has made a good decision based on information you provided. |

**Figure 4-9.** Examples of Ways that Communications Written at Work Can Alter Attitudes.

you might want to persuade them to feel that there is a problem where they presently feel everything is fine.

On the other hand, you might want your readers to hold one of their present attitudes even more firmly than before. For example, your readers might already think that you are a good employee who shows a lot of promise for a management position. You might aim to write your communication so that it persuades your readers to feel even more favorably impressed by your qualifications.

## RELATIONSHIP BETWEEN GUIDELINES 2 AND 3

Whenever they read, readers both think and feel. Accordingly, every time you write, you should be sure to follow both Guideline 2 ("Describe the tasks your readers will try to perform while reading") *and* Guideline 3 ("Tell how you want your communication to alter your readers' attitudes").

Writers sometimes forget to think about one or the other of these guidelines. For instance, when they are writing a recommendation or a proposal, they may concentrate so much on their readers' attitudes that they forget about the mental tasks the readers will try to perform while reading. Similarly, when writing instructions or a procedure manual, they may concentrate so much on the tasks that the readers will perform that they forget to think about the readers' attitudes

When you are writing, your readers' tasks and attitudes are *both* important. When you write a recommendation report, of course you want to focus on persuading your readers to feel that your suggestion is a good one. Nevertheless, you must keep in mind that as your readers read, they will be performing specific mental tasks, such as comparing various pieces of information and trying to find the answers to certain questions. If you write a report that frustrates your readers as they try to perform these tasks, you will reduce your chances of persuading them to adopt the attitude you desire, namely that your suggestion is worth taking.

Similarly, when writing an instruction manual, you must work hard at enabling your readers to perform the various steps of the procedure you are describing. But you must also concern yourself with your readers' attitudes. As our everyday experience tells us, people are not necessarily eager to read instructions. Many would just as soon toss them aside. If your readers do that, they may devise alternative ways of performing the task at hand—ways that for one reason or another you may not want them to employ. Perhaps the alternatives are less efficient than your procedure, or more dangerous. Usually, part of your purpose when preparing instructions will be to persuade your readers to use them.

For these reasons, when you define the purpose of your communications, keep in mind both your readers' tasks *and* your readers' attitudes.

## WHEN TO USE YOUR DEFINITION OF YOUR COMMUNICATION'S PURPOSE

The guidelines provided in this chapter will help you define your purpose in a way that will help you make many decisions as you write. As pointed out at the beginning of this chapter, you usually cannot include this definition of purpose in the communication you are writing. It's true, of course, that you can help your readers read efficiently if you begin by telling them what your communication is intended to do. In most cases, however, you will want to explain your purpose to your readers in a somewhat different way than you explain it to yourself.

Imagine, for example, how your readers would respond if you wrote, "One part of my purpose in this report on my current project is to persuade you that I have the qualifications needed to become a manager." Such a *statement* would be entirely inappropriate, although such a *purpose* would not be. Usually, the statement of purpose you address to your readers is different—often very different—from the statement you create to help yourself write well.

You will find advice concerning ways to tell your readers about the purpose of a communication in Chapter 4.

## WORKSHEET FOR DEFINING PURPOSE

Figure 4-10 shows a worksheet that will help you follow the three guidelines for defining purpose. Figures 4-11 and 4-12 show how two writers—a student and an employee—filled out the worksheet.

## SUMMARY

To make good decisions about how to write something at work, you need to know exactly what you want your communication to accomplish. In part, that means that you must identify the final result you want the communication to bring about: approval of the plan you proposed, success in teaching your reader how to perform some procedure, adherence to the policy you have established.

You should keep in mind, however, that your communications work toward that final result only while your readers are reading them. Even if the readers remember and react to your communication later, the memories themselves will depend upon the readers' responses while reading. Therefore, when defining your communication's purpose, you should also tell how you want your readers to respond while engaged in the act of reading.

The three guidelines presented in this chapter will assist you in defining your purpose in a useful way.

1. *Describe the final result you want your communication to bring about.* If you write on your own, *you* will decide what this result is. Usually, if you write at someone else's request, the person who gives you the assignment has already determined what he or she wants this final result to be.

2. *Describe the tasks your readers will try to perform while reading.* At work, five common tasks are:

   ● Comparing facts (as when making a decision).

## Worksheet
## DEFINING PURPOSE

**PROJECT:** _____ **DUE DATE:** _____

**READERS:**

**FINAL RESULT:** What final result do you want to achieve?

**READERS' TASK:** What will your readers try to do while reading?

_____ **Compare Facts.** What criteria will your readers use?

_____ **Locate Specific Facts.** What will these facts be?

_____ **Understand.** What are the key points for readers to understand?

_____ **Follow Step-by-Step Instructions.**

_____ **Apply Information.** In what situations will your readers apply it?

_____ **Other.** What are the important parts of this task?

**READERS' ATTITUDES.** How do you want to change your readers' attitudes? (Tell what the readers' attitudes are now and what you want them to be.)
**About you:**

**About your subject:**

**About themselves:**

**Figure 4-10.**    Worksheet for Stating Purpose. (An expanded version of this worksheet is presented in Figure 5-4.)

## Worksheet
## DEFINING PURPOSE

**PROJECT:** <u>Instruction Manual for Using a Graphics Printer</u>   **DUE DATE:** <u>11/15</u>

**READERS:** Students using computer lab

**FINAL RESULT:** What final result do you want to achieve? I want students to be able to use the graphics printer to draw graphs for their courses without needing assistance from the lab supervisors.

**READERS' TASK:** What will your readers try to do while reading?

_____ **Compare Facts.** What criteria will your readers use?

\_\_x\_\_ **Locate Specific Facts.** What will these facts be? Students may want to locate the specific instructions for using any of the printer's functions.

\_\_x\_\_ **Understand.** What are the key points for readers to understand? Students will want to understand the capabilities of the printer and how they can use it.

\_\_x\_\_ **Follow Step-by-Step Instructions.** This will be the students' main task.

\_\_x\_\_ **Apply Information.** In what situations will your readers apply it? Students will want to apply the instructions to the particular projects they have.

_____ **Other.** What are the important parts of this task?

**READERS' ATTITUDES.** How do you want to change your readers' attitudes? (Tell what the readers' attitudes are now and what you want them to be.)
**About you:** They have no attitude toward me now. While they are reading my instructions, I want them to feel that I am a reliable source of information.

**About your subject:** They may not know about graphics printers now. I want them to be curious and a little excited.

**About themselves:** They feel confident using computers, but may be unsure about their ability to work a graphics printer. I want them to feel confident in themselves.

**Figure 4-11.**   Worksheet for Stating Purpose that Was Filled Out by a Student.

## Worksheet
## DEFINING PURPOSE

**PROJECT:** Feasibility Study of a New Way of Assigning
Braille Translations _____    **DUE DATE:** April 14 _____

**READERS:** Mrs. Land, Head of Braille, Division of Kansas City Society for the Blind and
two of her assistants.

**FINAL RESULT:** What final result do you want to achieve? I want Mrs. Land to adopt a
new method of assigning work to volunteers who translate for the blind. The most
urgent requests would go to the most reliable volunteers.

**READERS' TASK:** What will your readers try to do while reading?

__X__ **Compare Facts.** What criteria will your readers use? Mrs. L. will compare proposed system with present one in terms of cost, speed of producing translations,
amount of work required of her, effect on volunteer morale.

_____ **Locate Specific Facts.** What will these facts be?


__X__ **Understand.** What are the key points for readers to understand? Will want to
understand, in detail, how the proposed system would work.


__X__ **Follow Step-by-Step Instructions.** Will want specific instructions about how to
implement the proposed plan if it is adopted.


_____ **Apply Information.** In what situations will your readers apply it?


_____ **Other.** What are the important parts of this task?


**READERS' ATTITUDES.** How do you want to change your readers' attitudes? (Tell what the
readers' attitudes are now and what you want them to be.)

**About you:** Mrs. L. thinks I don't see the various issues involved with administering
this work. I want her to feel confident that I do. Her assistants are more open to new
ideas; I want them to like this one.

**About your subject:** Mrs. L. worries that a systematic method would increase com-
petitiveness and decrease morale among the volunteers. I want her to feel the office
can work more efficiently without losing volunteers.

**About themselves:** She gives the impression of being very sure of herself, but feels
threatened by suggestions for change. She's worked here for 20 years.

**Figure 4-12.**    Worksheet for Stating Purpose that Was Filled Out by an Employee.

- Locating facts (as when using a reference document).
- Understanding (as when reading a scientific article).
- Following step-by-step instructions (as when reading a step, doing a step, reading a step, and so on).
- Applying information (as when trying to follow general regulations).

3. *Tell how you want your communication to alter your readers' attitudes.* Tell what your readers' attitudes are now and what you want them to be by the time they have finished reading. These attitudes might be about you or your subject matter, or they might be the readers' attitudes toward themselves.

## EXERCISES

1. Study the completed worksheet for defining objectives shown in Figure 4-11 or 4-12. Describe some of the writing strategies that the author could use to achieve the purpose set out there. Try to think of at least one strategy to match each of the answers that the writer provides.

2. Find an example of a communication that you might write in your career. After studying that communication, fill out the worksheet shown in Figure 4-10 in a way that describes its purpose.

3. Think of a project you might like to do in your writing class. Fill out the worksheet for defining objectives (Figure 4-10) for it.

# Understanding Your Readers

## PREVIEW OF GUIDELINES

1. Create imaginary portraits of your readers in the act of reading.
2. Portray your readers as people who ask questions while they read.
3. Learn who your readers will be.
4. Find out whether your readers will be decision-makers, advisers, or implementers.
5. Find out how familiar your readers are with your specialty.
6. Describe your readers' reasons for holding their present attitudes.
7. Find out if your readers have any "special" characteristics.
8. Find out about the setting in which your readers will read your communication.
9. Aggressively seek information about your readers.

In the preceding chapter, you learned how to develop reader-centered statements of the purpose for communications you write at work. Such statements tell not only what you want to happen *after* your readers have finished reading but also what you want to happen *while* your readers are reading: what tasks you want to enable your readers to perform while reading and how you desire to shape their attitudes. To be able to prepare a communication that will achieve those objectives, you must have some very specific information about the particular individuals you are writing to.

Why? Because reading is a highly individual, highly personal act. We are all used to the idea that if any two of us read a particular novel, essay, or story, we will each react to it in a slightly different way. The same is true of communications written at work. A research report will create one effect upon one reader and another upon someone else. So will an instruction manual or proposal for a new project. In fact, a communication that creates one effect upon one reader now might very well create another effect upon the same reader if it were instead read next year.

## MAKING PREDICTIONS ABOUT READERS

Nevertheless, it is possible for you to make very reasonable predictions about the ways that your readers will react to the things you write at work. That's partly because people's reactions are shaped by their professional responsibilities. Research scientists, production engineers, and middle managers are expected to look for certain kinds of information—and to react in certain ways to what they find. People's responsibilities are also shaped by their education and their professional backgrounds. Furthermore, you possess a commonsense understanding of other people that can help you predict the ways that your readers will respond to the things you write.

This chapter explains nine guidelines for developing the kind of understanding of your readers that you will need if you are to write effectively on the job.

## READERS AT WORK COMPARED WITH READERS AT SCHOOL

Before you read these guidelines, you may find it helpful to think about the differences between the people you write to at school and the people you will write to on the job.

At school, you probably write primarily for your instructors. They ask you to write because they believe you can learn about a subject by writing a report or essay on it. And they ask you to write so that they can see what you know about a certain subject—so they can evaluate your performance in their courses.

In contrast, at work you will write mostly for people who want to *use* the information you provide. Your information will be important to them. They will need it to do their jobs. Your report will enable them to make a crucial decision. Your instructions will enable them to perform a necessary procedure.

This difference in the reasons that your college instructors and your readers on the job have for reading is very important to you as a writer. It means that the kind of writing that you have done in many of your courses probably won't succeed on the job. At work, instead of telling everything you know, you should usually tell only what your readers want to learn. Instead of organizing according to a *logical* plan, you should organize according to a *useful* one.

The following guidelines for understanding readers will help you see more precisely what factors shape the way people read and respond on the job.

Guideline

1

## Create Imaginary Portraits of Your Readers in the Act of Reading

To be able to predict the ways that the particular readers of each communication you write will read and respond, you need to know a great deal about your readers. This guideline suggests a strategy that will help you search for that information about your readers and organize it in a way you will find useful as you write. The strategy is to create imaginary portraits of your readers reading the things you are about to write. Using these portraits, you will be able to anticipate your readers' reactions and write accordingly. You will be able to "talk" with your readers in the way that is described in Chapter 2.

A variation of this strategy of creating imaginary portraits of your readers was used in an interesting way by the writer of an instruction manual that tells owners of very small businesses how to use a computer program designed to help them with their bookkeeping. The writer placed over his desk a photograph of an elderly couple standing behind the counter of the neighborhood grocery that they owned. "Whenever I sat down to write," this writer said, "I looked at that photograph so that I would always start out by thinking of the people who would be using my manual."

Sometimes you will find it easy to create helpful mental portraits of your readers, for instance when you are writing to people you work with constantly. In other situations, as when writing to a customer you have never met, you will have to rely heavily upon your creativity and common sense.

And sometimes, you will have to create more than one portrait for a single communication. Many of the things you write at work will be read by more than one person—perhaps dozens, perhaps even thousands. If these people all have essentially the same purpose for reading and if they all bring essentially the same knowledge of your subject matter, then you can imagine a single typical reader, as was done by the writer of the computer manual. Often, however, you will write communications that will be read by groups of people who have various reasons for reading and who possess various levels of knowledge of your subject—a point that will be discussed more fully later in this chapter. When that happens, you will have to create portraits not just of one reader but of several.

Whether you need to portray one reader or many, your mental pictures of your readers can help you immensely throughout all your work at writing a communication.

Guideline

2

## Portray Your Readers as People Who Ask Questions while They Read

As you create your portraits, keep in mind that you want to imagine your readers *in action,* not just in still lifes. You want to be able to imagine their reactions to your message so that you can anticipate those reactions and write accordingly.

One good way to put your imaginary portraits into action is to think of your readers as people who ask questions while they read. The questions they ask will indicate what information they need and want as they perform their reading tasks. Their questions will also indicate what their attitudes are and what you must do to persuade them to change those attitudes in the way you desire.

By thinking of your readers as people who are looking for answers to questions, you can gain insight into their state of mind at any point in their reading of your communication. For example, if you ask yourself what questions your readers will have as they start reading, you'll have a clue as to their state of mind at that point and, consequently, a clue about the best way to begin your communication. If you suspect, for instance, that your readers will come to your communication asking urgently for some particular piece of information, you will know that you would probably be wise to put that information at the beginning. After all, if your readers cannot find the answers to their questions in the first few sentences, they may begin skimming; if they don't soon find the desired information by skimming, they may quit reading altogether. In either case, many of your words will never have been read at all, never have had a chance to affect your readers in any way.

As this example shows, your readers' state of mind begins to change as soon as they start to read. Each sentence, each paragraph, each section affects the readers' thoughts and feelings. By determining what questions your readers will probably ask as they reach any particular point in your communication, you can learn things that will help you shape your communication to affect the readers in the way you desire. You've identified a question at the end of the seventh paragraph? Then you should consider answering it in the eighth. Or, if you would rather that the readers didn't ask that question, at least not at that place, you should rewrite paragraph seven so that it does not arouse the unwanted question.

In sum, by thinking about your readers as people who ask questions, you can predict not only the general nature of your readers' reading activity but also how your readers will be reading any particular part of your communication, from the very first sentence to the very last.

Guideline

3

## Learn Who Your Readers Will Be

Before you can create imaginary portraits of your readers, you must learn who they will be. That can be a challenging activity, particularly when you are new on the job.

Perhaps the most important thing for you to remember as you set out to identify your readers is that many of the things you write at work will be read by more than one person. One researcher asked 122 managers and professional people in a large organization to supply him with a typical report they had written and to tell him how many people had received the report.[1] He found that the reports were received by an average of 28.6 readers: 10 readers above the writer's level in the organization, 11 at the same level as the writer, and 7.6 below the writer's level.

Why are there so many readers for the things people write at work? In most organizations, the working relationships among individuals and groups is necessarily so close that almost any decision made by one individual or group will affect the activities of several others.

Imagine, for instance, that you have developed an idea about how to improve one of your company's products. If your idea is approved, designers may have to redesign the product, the production department may have to alter its production process, the purchasing department may need to obtain different supplies, the marketing department may have to revise its sales materials and strategies, and the service department may have to learn new maintenance and repair procedures.

Even when only a few people are affected by a decision, the decision is often made by many individuals. Many employers expect widespread consultation and advice on decisions.

### Complex Audiences

One of the chief characteristics of audiences for reports, proposals, and many other communications that are read by more than one person is that each reader will read from a different point of view. Each has his or her own professional role and areas of expertise, and each will play that role and apply that special knowledge when reading. When you write to such a group of people, you are addressing a *complex* audience.

Writing to a complex audience is much more challenging than writing to a single reader or to an audience composed of many people who will all read in essentially the same way (as might happen when you write a set of instructions, for instance). Instead of writing a communication that will affect one reader (or one typical reader) in some particular way, you must write a communication that will affect a variety of readers—each in a different but particular way. To do that, you must begin by identifying each of the readers or types of readers your communication will have.

## Phantom Readers

Your efforts to find out who your readers are may sometimes be complicated by the presence of "phantom readers." Phantom readers are the real users of communications addressed to someone else. You will most often encounter phantom readers when you are writing things that require some sort of decision.

One clue to their presence is that the person you are addressing is not high enough in the organizational hierarchy to make the kind of decisions that your communication requires. Perhaps the decision will directly affect more parts of an organization than are managed by the person addressed, or perhaps it involves more money than the person addressed is likely to control. In such situations, there are probably phantom readers at higher levels who will make the decision.

Many of the things you write to your own boss may actually be used by phantom readers. For instance, your boss may be one of those managers who accomplish the work they have been assigned by asking their assistants to perform the necessary task or parts of it. Thus, when you turn in your report, your boss may first check it over and then pass it along to his or her superiors.

After people have worked at a job for a while, they usually learn which communications will be passed up the organizational hierarchy, but many a new employee has been chagrined to discover that his or her hastily written memo has been read by people very high in the organizational structure. To avoid these kinds of embarrassments, you need to be able to identify phantom readers. Also, you need to identify phantom readers so that you can be sure that you meet their needs, not just the needs of the less influential person you are addressing. Thus, both you and your real readers benefit if you are able to recognize phantom readers.

## Procedures for Identifying Readers

How can you identify your readers, phantom and otherwise? When you are writing at someone else's request, you can ask that person. Sometimes, though, that won't be possible or the person won't be sure. Then you need some other strategy.

### Brainstorming

Employees who have worked for an organization long enough to know thoroughly its ways of operating can simply brainstorm. They can ask themselves, "Who in my own department will read this communication?" "Who in other departments that work closely with mine?" and, "Who in more remote parts of the organization?"

However, taking such an approach involves the risk of missing someone, particularly when used by someone fairly new on the job. For that reason it is often best to use a more systematic procedure, one that minimizes the chances of overlooking an important reader.

### Using Organizational Charts

One such procedure is to use your employer's organizational chart. Such a chart, like the one shown in Figure 5-1, names all the units of activity within an organization. Thus every individual (every potential reader) in the organization is represented in it. To find the readers for your communications, go from block to block within the chart asking yourself whether anyone in the group represented by each block would have any reason to read what you have written.

As you look through an organizational chart, however, you must remember that you are not really trying to identify groups that might read what you are going to write, but to identify your individual readers, so that you can learn the characteristics of each. Therefore, once you have identified a block that is likely to have a potential reader (or readers), you should try to find out who that person is.

Of course, there is an added complication when you communicate to complex audiences outside your own organization. You may not have organizational charts for those other organizations. In such situations you may nevertheless be able to discover your probable readers by asking yourself who would read the communication if it came to your employer's organization. In this way, you can determine what types of people in the other organization will probably read your communication.

### Looking for Indirect Readers

For some communications, there is an additional element of complexity. These communications may have readers who are not directly involved in the present decision at all. The readers may be people who are working on a problem similar to the one addressed in the present communication. They will hope to gain some ideas about solving their own problem by seeing how someone else solved a similar one. Furthermore, these readers may be people who won't even begin working on their problem until sometime in the future. In fact, it is possible that the authors of a communication may become its readers at some later time—as the chemists in an industrial research center found out.

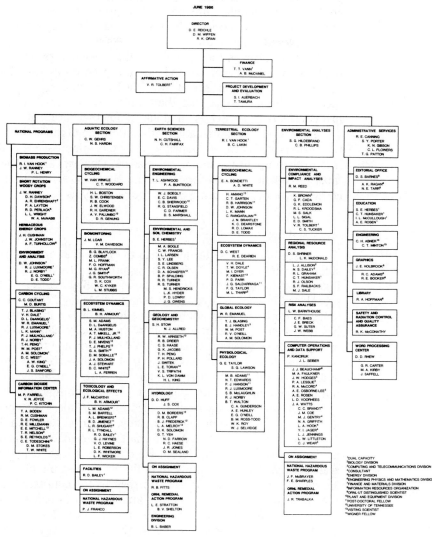

**Figure 5-1.** Typical Organizational Chart.

Martin Marietta Energy Systems, Inc., Oak Ridge National Laboratory.

These chemists worked in a department that analyzed plants from around the world to see whether the plants contained any compounds known to have medicinal value. Such analyses can be extremely complex. For instance, the chemists once searched through some plant samples for a certain compound whose presence was hidden by certain others. The chemists finally succeeded in finding the compound by using an ingenious procedure. They dutifully placed a record of the results in the company's files. Three years later, they encountered a similar problem with a plant obtained from another part of the world. Remembering their earlier adventure, they went to the files to see what they had done in the past.

To their dismay, they discovered that they had recorded only the results of their ingenious analysis and not the procedures they had used. Consequently, they had to spend two weeks re-creating—by a hit and miss method—the procedure that they could have imitated in a day if only they had considered themselves an audience for their original communication.

Thus, even the systematic method of using organizational charts to identify readers needs to be supplemented by common sense and imagination if it is to turn up all the important readers. One way or another, however, you must learn who your readers are if you are to have any real chance of writing communications that bring about the results you desire.

## Example: Identifying Readers

To see how one writer identified the readers of one communication, consider the efforts of Mr. McKay as he prepared to write the letter shown in Figure 5-2 (page 104). In the letter, Mr. McKay addresses Mr. Fulton, vice-president for sales at a company that sold McKay's firm an air sampler that failed to work properly. Although the letter makes it appear that Mr. McKay has only a single reader, Mr. Fulton, he actually had a wide phantom readership in three different organizations: the company that employs Mr. McKay (Midlands Research), the company that supplied the air sampler (Aerotest), and the federal agency that paid for the work that Midlands Research could not complete on time (Environmental Protection Agency).

When Mr. McKay set out to identify his readers, he could obtain an organizational chart for his own company, Midlands Research. Figure 5-3 (page 106) shows how he marked that chart to designate the people within Midlands who would read his letter. He circled the legal department because he realized that Aerotest's failure to fulfill his request might lead to legal action. He circled the field teams because they had encountered the problem with the air sampler and would want to know what was being done about it.

To identify the likely readers within Aerotest, Mr. McKay used the same chart, thinking that every reader in Midlands Research would probably have a counterpart in Aerotest: lawyer for lawyer, vice-president for vice-president. But Mr. McKay thought that in Aerotest there would be some other readers as well; to identify them he had to use his common sense. He thought that the departments that engineered and then manufactured the faulty equipment would read the letter, since their work was being criticized. And he figured that some of Aerotest's sales force would read it also, especially the sales engineer who sold the product to Midland.

Mr. McKay also planned to send the Environmental Protection Agency a copy of his letter, as a way of explaining the delays in Midlands' work. He guessed that at the agency the letter would not travel beyond the contracting officer and those most directly affected by the delay in obtaining the test results.

After he identified the many readers of his letter, Mr. McKay asked himself what questions each would want answered. He found that his readers would be playing a wide variety of roles and that, accordingly, they would have a wide variety of questions. For instance, the lawyer within Mr. McKay's own company would read the letter (before it was sent) to answer this question, "Has McKay said anything that will abrogate Midlands' right to sue Aerotest?" In contrast, the lawyer for Aerotest would ask the same question, but hope to find a different answer. Likewise, the people in the Aerotest engineering division would ask, "What evidence does McKay have that the problems encountered by Midlands were caused by poor work on our part?" To answer that question, Mr. McKay drew up the two detailed accounts of those problems. Similarly, the Aerotest repair shop would ask, "Did the repairs made by Midlands really cost $1500?" Anticipating that question, Mr. McKay prepared the detailed statement of repair expenses.

As you can see, the readership of a communication can be quite large and diverse. Yet all of these readers are important people with important questions the writer must answer. Thus, whether you use a systematic or an unsystematic method, you must be careful to find out exactly who will read what you write.

# INTRODUCTION TO GUIDELINES 4 THROUGH 7

Once you know who your readers will be, you are ready to gather the information about them that you will need to have so that you can create accurate imaginary portraits of them. The next four guidelines tell you what kinds of information about your readers will help most.

## Find Out Whether Your Readers Will Be Decision-Makers, Advisers, or Implementers

Guideline

4

When a person is hired by a business, industry, or government organization, that person is expected to perform a certain role, such as the role of assistant process engineer, manager of the credit department, or auditor for electronic data processing operations. When the person is promoted, retires, or leaves that position for any reason, the role itself is likely to remain, to be filled by another person.

It's important for you to know your readers' roles because the kinds of questions people ask when reading are determined largely by their roles. For instance, when reading a report on the toxic emissions from a factory, a chemical engineer working for the factory might ask, "How can these

# MIDLANDS RESEARCH INCORPORATED

Mr. Robert Fulton
Vice-President for Sales
Aerotest Corporation
485 Connie Avenue
Sea View, California 94024

Dear Mr. Fulton:

In August, Midlands Research Incorporated purchased a Model Bass 0070 sampling system from Aerotest. Our Environmental Monitoring Group has been using—or trying to use—that sampler to fulfill the conditions of a contract that MRI has with the Environmental Protection Agency to test for toxic substances in the effluent gases from thirteen industrial smokestacks in the Cincinnati area. However, the manager of the Environmental Monitoring Group reports that his employees have had considerable trouble with the sampler while trying to use it on the first two smokestacks.

These difficulties have prevented MRI from fulfilling some of its contract obligations on time. Thus, besides frustrating our Environmental Monitoring Group, particularly the field technicians, these problems have also troubled Mr. Bernard Gordon, who is our EPA contracting officer, and the EPA enforcement officials who have been awaiting data from us.

I am enclosing two detailed accounts of the problems we have had with the sampler. As you can see, these problems arise from serious design and construction flaws in the sampler itself. Because of the strict schedule contained in our contract with EPA, we have not had time to return our sampler to you for repair. Therefore, we have had to correct the flaws ourselves, using our Equipment Support Shop, at a cost of approximately $1500. Since we are incurring that additional expense only because of the poor engineering and construction of your sampler, we

**Figure 5-2.**   Letter to Complex Audience. (Enclosures not shown.)

2

hope that you will be willing to reimburse us, at least in part, by supplying without charge the replacement parts listed on the enclosed page. We will be able to use those parts in future work.

Thank you for your consideration in this matter.

Sincerely yours,

*Thomas McKay*

Thomas McKay
Vice-President
Environmental Division

Enclosures: 2 Accounts of Problems
1 Statement of Repair Expenses
1 List of Replacement Parts

**Figure 5-2.** *Continued*

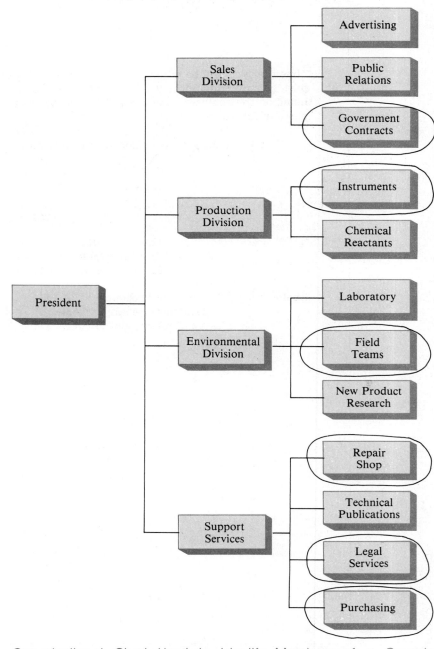

**Figure 5-3.** Organizational Chart Used to Identify Members of a Complex Audience.

emissions be reduced?" The corporate attorney might ask, "Will we be fined by the Environmental Protection Agency if we do not reduce these emissions?"

The most obvious clues to the roles of your readers are the titles of the jobs they hold: for example, systems analyst, laboratory technician, bank cashier, or manager of public relations. However, job titles are often vague, not indicating precisely the responsibilities of the reader. Furthermore, you need to know not oniy the general but also the specific roles the readers will play while reading your communication. Why, exactly, will each person be reading your communication, and what kinds of information will he or she be looking for? As you try to answer those questions, you might find it helpful to think about the three major kinds of roles that readers play.

- *Decision-makers.* The decision-makers' role is to say how the organization, or some part of it, will act when confronted with a particular choice.
- *Advisers.* Advisers provide information and advice that decision-makers will consider when deciding what the organization should do.
- *Implementers.* Implementers carry out the decisions that have been made.

Of course, these general descriptions of the three basic roles are greatly simplified. For example, advisers usually are specialists who advise only on certain things: an accountant usually is an adviser only on financial matters, a mechanical engineer is an adviser only on certain other kinds of things, and so on. Therefore, the specific questions that any adviser—or decision-maker or implementer—will ask will be determined in part by the specific role he or she has been asked to play in the organization, a role that is determined by the individual's specialty.

Also, any particular individual can play more than one role. Often, for example, those who are asked to advise about a particular decision will also be asked to carry it out. For instance, the shop foreman asked by the manager of a plastics factory about the wisdom of installing a new type of injection mold will probably be required to retrain his department in the use of that new mold, if it is purchased.

Furthermore, although the roles of decision-maker, adviser, and implementer may seem to apply only to employees within large organizations, it is equally useful to consider these roles when thinking of people outside of organizations. For example, when deciding what brand of stereo or personal computer to buy, a consumer may be considered a decision-maker, perhaps one who will consult the "advisers" who review these products for magazines and journals such as *Stereo Review* and *Consumer Reports*. Once the consumer has decided what to purchase, he or she becomes an implementer, faced with the problems of buying, transporting, unpacking, and setting up the equipment and then using it.

## Find Out How Familiar Your Readers Are with Your Specialty

Another characteristic of your readers that affects the questions they will ask while reading your communications is the degree of familiarity they have with your specialty. Consider, for example, the reader of an instruction manual that says that the next step in operating a drill press is to "zero the tool along the Z-axis." If unfamiliar with this kind of tool, the reader might ask, "What is zeroing? What is the Z-axis?" If the instructions do not answer such questions, the reader would have to ask someone else for help, which would defeat the purpose of the instructions.

Of course, the writer of those instructions would not want to answer those same questions about zeroing and the Z-axis if all of his or her readers already knew the answers. Coming upon such explanations, the reader who is familiar with the terms would ask, "Why is the author of these instructions making me read about things I already know?"

Knowing the degree of your readers' familiarity with your specialty will help you predict some of the questions the readers will ask of your communication, particularly questions that you would answer by providing background and explanatory information.

## Find Out Your Readers' Reasons for Holding Their Present Attitudes

In the preceding chapter, you learned that one step in defining a communication's purpose is to tell what your readers' attitudes are now—before reading your communication—and what you want their attitudes to be after they read it. This guideline suggests that when you are gathering information about your readers you should find out why they hold their present attitudes.

To see why it is important for you to gather information about your readers' reasons for holding their present attitudes, suppose that you were assigned the task of writing the instructions that will accompany a small computer to be sold to housewives, construction workers, and other people who have no previous experience with computers. In describing your readers' attitudes toward the computer, you could say that almost all of them would be at least slightly apprehensive about operating the machine. Further, if you knew something about the source of the readers' anxieties, you would be in a good position to know how to allay them. For example, suppose you found out that many of your readers were afraid to operate a computer because they thought they might do something wrong that would break it. You could then include, at the beginning of your instructions, some appropriate reassurances.

Similarly, suppose you found that your readers' fears arose in part from an apprehension that the computer would make them feel stupid. You could then incorporate assurances that the computer is easy to operate and that it does nothing more than what operators ask it to do.

Knowing your readers' reasons for holding their present attitudes is equally important when you are writing communications to persuade people. For instance, suppose that you want to recommend that one of the parts of an electrical generator manufactured by your firm be made out of brass rather than stainless steel. It would be important to you to know whether or not your reader, the head of the Generator Department, knows of the problems created by using stainless steel. If she doesn't, you will have to persuade her first that there are problems with using stainless steel and then that using brass would solve these problems. On the other hand, if your reader does know of the problems, you will want to know why she hasn't changed to some other material already. If she didn't change because she didn't know of any good replacement for stainless steel, then you would probably concentrate on describing the relevant properties of brass to her. If she didn't change because she thought that, despite its problems, stainless steel is better than any other alternative, then you would either have to show her that she overlooked some important information about brass or else that she should be evaluating the materials by a different set of criteria.

Besides having more or less logical reasons for maintaining their attitudes, readers often have personal, nonlogical ones. As you write, you need to take these into account as well. For instance, the head of the Generator Department may have made the original decision to use stainless steel, so that she now fears that changing the material would be tantamount to admitting that she had made a mistake earlier. In that case, you might want to ease her fear in your report, perhaps by acknowledging that stainless steel had been a good choice at the time of the earlier decision, in light of what was then known about the generator and steel.

As these examples show, knowing why your readers hold their attitudes can help you considerably as you try to design a communication that will induce your readers to change those attitudes in the way you desire.

## Find Out If Your Readers Have Any "Special" Characteristics

Guideline
7

This guideline is a catch-all. A knowledge of your readers' roles (Guideline 2) and of their familiarity with your specialty (Guideline 3) will be useful to you in every writing situation. However, you may encounter situations in which it will be equally important for you to know some other characteristics of a reader, characteristics that you would not normally need to consider.

For example, you might have a reader who prefers a certain way of organizing communications, a certain kind of sentence structure, certain words. You might have a reader who has an especially high or low reading level, so that you would have to choose your words and sentence structures accordingly. Or you might have a reader who is interested in certain kinds of information that you would not have to supply to most readers.

Most of your readers will probably not have any such characteristics. However, each time you address a reader you haven't addressed before, you should ask, "Is there anything special I should take into account when writing to this person?"

## Guideline 8 Find Out about the Setting in Which Your Readers Will Read Your Communication

By learning about the setting in which your readers will be reading, you can decide how best to handle many of the physical characteristics of the communication you are writing. This information will be especially useful when your readers will read outside the normal office environment.

For example, suppose that you are writing a repair manual for hydraulic pumps. You know that your reader will use it around water, oil, and dirt. Therefore, you might want to have your manual printed on waterproof paper, or at least on paper that is coated to resist moisture and oil. Additionally, you realize that your reader may also have trouble finding a spot close to a pump where he or she can set your manual down. Consequently, the reader will have to read your manual from a greater distance than he or she normally would. You may therefore want to use a large typeface for your printing.

In this and many similar situations, you can greatly increase the usefulness of your communication by thinking about the setting in which your reader will read it.

## Guideline 9 Aggressively Seek Information about Your Readers

Many writers are reluctant to gather information about their readers, as if it were bad manners to ask about them. You will write much more effectively if you overcome any such shyness and aggressively seek information about your readers.

Of course, there are times when you will already know so much about your readers that you will not need to seek any additional information about them. That will happen, for instance, when you write to someone with whom you work regularly. On the other hand, at one time or another you will almost surely be asked to write to readers you do not know well—or even at all.

### Sources of Information about Readers

The best source of information about such readers is the readers themselves. If your boss asks you to prepare a report that will be used by people in some other department, visit or call that department. Tell them that you want to be sure that your report meets their needs. Then ask how they will use your communication, what kinds of questions they hope it will answer, what form they want the answer to be presented in, and so on.

By doing this, you will be using the technique employed by marketing and advertising personnel: you will be asking your target market (your readers) what would make them satisfied with your product (your communication). You can also ask your employer to send you to meet readers outside your organization. Employers are becoming increasingly aware of the need for writers to know about their readers, and some employers will even send writers to distant places so that the writers can learn firsthand about the people to whom they will be writing.

Even if you are unable to meet or telephone your readers directly, you can almost certainly gain some information about them. One obvious source of information is the person who assigned you to write in the first place. You must not assume that in making the assignment the person told you all that he or she knows about the readers. After all, that person may have assumed that if you need to know anything more than you already do, you will ask.

## Importance of Checking Your Assumptions

It is also important to ask others about your readers so that you can check on any assumptions you are making about them. Sometimes, the things writers think they know about their readers are inaccurate. The alarming result of assuming that your information is complete and shared by others can be seen in the following account of a meeting of the members of the internal auditing department of an international corporation. This department's sole product is a continuous stream of reports. When an outside consultant asked the two dozen auditors in this department who they thought were the primary readers of their reports, every auditor thought that he or she knew. When they spoke, however, they didn't agree. Only two of the auditors named the reader that was identified by their manager.

The manager was astounded. He thought his auditors knew—just as surely as they themselves thought they knew. Furthermore, it was important for them to know who their readers were because that knowledge would provide a sound basis for making many decisions about how to write the department's reports, including the kinds of material to provide, the amount of background information needed, and even the tone to be used.

In this case, once the manager knew that he needed to describe the reader to his employees, he was able to do it quite helpfully. However, many managers are not able to do so. Therefore, you will need to use a great deal of skill and judgment when gathering information about readers from secondhand sources, even your boss. When they talk about readers, some people tend to talk about their own writing preferences, rather than about the readers' true needs. Sometimes, too, such people base their advice on misinformation. Unless you are extremely unfortunate, however, you will get good information from such efforts at least as often as you will get bad; and the value of such good information is so great that it is always worth the effort to gain it.

# WORKSHEET FOR DEFINING OBJECTIVES

Figure 5-4 shows a worksheet that will help you follow the guidelines given in this chapter and the preceding one. In fact, this worksheet is an extended version of Figure 4-10 (page 90). At the bottom of the worksheet you will find a series of questions about your reader.

## SUMMARY

To be able to create a reader-centered communication, you need to know as much as possible about the ways your readers will read it. This chapter has presented nine guidelines to help you learn the characteristics of your readers that will determine how they will read what you have written.

1. *Create imaginary portraits of your readers in the act of reading.* By creating a portrait of your readers reading, you will be able to guess well about the questions that will arise in your readers' minds at any time during their reading activity.

2. *Portray your readers as people who ask questions while they read.* Thinking of your readers in this way will help you imagine their reactions as they read your communications.

3. *Learn who your readers will be.* At times, it may be difficult to learn who your readers are—especially if you are writing for a phantom reader or a complex audience. Two ways to identify readers are (1) to ask the person who requested that the communication be written and (2) to use an organizational chart.

4. *Find out whether your readers will be decision-makers, advisers, or implementers.* Their roles will greatly influence the kinds of questions your readers will ask while reading your communications.

5. *Find out how familiar your readers are with your specialty.* This information will help you determine which technical terms you should explain and which you should avoid as well as what background information is needed and what is not.

6. *Find out your readers' reasons for holding their present attitudes.* A knowledge of your readers' reasons for holding their present attitudes will help you devise a strategy for modifying those attitudes.

7. *Find out if your readers have any "special" characteristics.* Some readers have special characteristics (such as an unusually high or low reading level) that you must take into account to write effectively to them.

8. *Find out about the setting in which your readers will read your communication.* A knowledge of this setting can help you decide about the physical characteristics of your communication.

## Worksheet
## DEFINING OBJECTIVES

PROJECT: _____ DUE DATE: _____

**READERS:**

**FINAL RESULT** What final result do you want to achieve?

**READERS' TASK** What will your readers try to do while reading?

_____**Compare Facts.** What criteria will your readers use?

_____**Locate Specific Facts.** What will these facts be?

_____**Understand.** What are the key points for readers to understand?

_____**Follow Step-by-Step Instructions.**

_____**Apply Information.** In what situations will your readers apply it?

_____**Other.** What are the important parts of this task?

**READERS' ATTITUDES** How do you want to change your readers' attitudes? (Tell what the reader's attitudes are now and what you want them to be.)

**About you:**

**About your subject:**

**About themselves:**

**READERS' CHARACTERISTICS**

**Roles (decision-makers, advisers, implementers):**

**Familiarity with your specialty:**

**Special characteristics you should keep in mind:**

**Important features of place where readers will read:**

**Figure 5-4.**   Worksheet for Defining Objectives.

9. *Aggressively seek information about your readers.* Don't be reluctant to ask for information about your readers. Two good sources of that information are the readers themselves and the person who gave you the writing assignment.

## NOTE

1.   Robert W. Kelton, "The Internal Report in Complex Organizations," in *Proceedings of the 30th International Technical Communication Conference* (Washington, D.C.: Society for Technical Communication, 1984), RET54-RET57.

## EXERCISES

1. A. Find an example of a communication that you might write in your career. After studying the communication, write a paragraph about the readers it seems to have been addressed to. You may write this in the form of an essay or in the form of an imaginative portrait of a typical reader.
   B. Explain several features of the communication by telling how they seem to be tailored to the reader you have described.

2. Think of a project you might like to do in your writing class. Describe the readers for it. Discuss ways you would write your project so that it would be suited to these particular readers.

3. Imagine that you are employed in the kind of job you want. Find an article written to specialists in your field that might be useful to you. Tell how you would *use* that information in your job and what additional information you might want about the subject if you were on the job.

# Finding Out
# What's Expected

## PREVIEW OF GUIDELINES

1. Ask your boss, readers, and other knowledgeable people what's expected.
2. Obtain and study any written regulations concerning your writing.
3. Look at similar communications written by other people.
4. Find out your deadline and make a schedule for meeting it.
5. Find out what length is expected.
6. Find out what your communication should look like.
7. Find out what style you should use.
8. Find out what you are expected to say—or not say.
9. Find out what policies apply to the communication you are preparing.
10. Find out whether and when you will need to have your drafts reviewed.

Jeff, a recent college graduate, has just returned from a very embarrassing meeting with his boss. Yesterday, he turned in a report on his first project, quite proud of his six-page effort. He had worked on the report for two days, even taking his drafts home after work. In the meeting today, however, Jeff's boss asked him to rewrite the report. She wants him to cut the report in half, reorganize it, write a new beginning, remake his tables, and spell out most of the technical terms that he abbreviated.

"If only I'd known *beforehand* what my boss wanted," Jeff thinks to himself.

Such incidents are quite common, especially for new employees. They might seem to be easy enough to avoid. Weeks ago, for example, Jeff could have asked his boss what she wanted the report to be like. However, like many new employees Jeff was reluctant to ask, reluctant to seem unsure of himself.

On the other hand, Jeff's boss could have prevented the situation. She could have told Jeff, in advance, about *all* her expectations. She did explain *some* of them to Jeff when she first asked for the report: she told Jeff when she expected the report to be turned in and what form of communication (a memo) she expected Jeff to use for his report. But she did not explain her many other expectations about such things as organization, length, and writing style. Perhaps she is herself so used to reading communications that fit one particular pattern that she has forgotten that some people use other patterns.

On the job, you should be sure to learn what expectations apply to your writing. By doing so, you will avoid the embarrassment that Jeff feels. And you will enable yourself to plan and write efficiently, so as to save both you and other people the labor that is required when communications have to be rewritten.

This chapter presents ten guidelines for finding out what's expected of you when you write at work. First, though, it briefly discusses the nature of the expectations that you are likely to encounter on the job.

## THE MAIN ADVICE OF THIS CHAPTER

Underlying the ten guidelines of this chapter are three general pieces of advice: don't assume that you know what is expected, don't be afraid to ask about expectations, and remember that you may have to satisfy expectations about any and all aspects of your writing.

### Don't Assume that You Know What Is Expected

Jeff *thought* he knew what his boss was looking for in his report. He did well on written projects in college. Why weren't his assumptions good? There are two reasons.

First, on the job people write for much different reasons than they do in school. As pointed out in the preceding chapter, in school you write largely to learn and to show your professor what you know. On the job, you write to make something happen.

Because of this difference in aims, the writing that succeeds in school will not necessarily succeed at work. For example, in school you often succeed by writing a longer project rather than a shorter one. A longer project gives you a fuller chance to demonstrate your knowledge. As Jeff found out, however, shorter communications are often preferred on the job. Shorter communications get to the point more quickly, and they contain only the information the readers really need.

A second reason to be wary of assuming that you know what is expected is that expectations about writing vary greatly from one job to the next. This book attempts to prepare you *generally* for the writing that you will do at work; your writing instructor in the course for which you are using this book is doing the same. But you can prepare only so much while in school. At one level, what's expected of writing at General Electric is very similar to what's expected at General Motors or General Foods or IBM or DuPont Chemical. Therefore, you can prepare in school for the kinds of expectations that prevail in business and in various technical and scientific fields. At another level, however, each organization has its own way of doing things. And the same goes for individual managers within those organizations. You cannot prepare in school for the customs and policies of the particular organization that will employ you or for the specific preferences of the particular person who will be your manager.

Accordingly, even if you learn everything that this book and your instructor have to teach you about writing on the job, you will still need to learn about the expectations that your particular boss and organization and readers have for your writing.

## Don't Be Afraid To Ask about Expectations

On the job, especially during your first weeks and months, you may feel reluctant to let anyone know that you are unsure about what to do. You should ask anyway. Keep in mind that most experienced employees, including most managers, realize that new employees have much to learn. And they expect and welcome questions from them.

## Remember that You May Have To Satisfy Expectations about Any and All Aspects of Your Writing

Don't assume that you adequately understand all expectations simply because you know what the deadline for your communication is and how long your communication should be. Your organization, boss, or readers may also have expectations about how you will organize your material, how you will lay it out on the page, and so on. They may even have expectations (and formal policies) about what you will say.

Some expectations (such as deadlines) arise from management decisions. Others, such as those governing the style and appearance of communications, may arise from carefully considered policies that are clearly laid out in print. Or these expectations may take the form of customary ways of doings things. These customs are passed along by word of mouth or by imitation.

Whatever the source of these expectations, you stand the best chance of writing effectively on the job if you know what they are.

## INTRODUCTION TO GUIDELINES 1 THROUGH 3

The first three guidelines in this chapter tell you how to find out what's expected.

Guideline

**1**

### Ask Your Boss, Readers, and Other Knowledgeable People What's Expected

The best way to find out what's expected is to ask the people who know. Your boss is certainly chief among these people. He or she has probably written many similar communications and also has helped other people do the same.

Your intended readers are also good people to ask about expectations—especially when you are writing in response to a request from one of them. Generally, it is much more efficient to ask this person what he or she expects than it is to guess. When you guess, you run the risk of guessing incorrectly, thereby wasting both your time and your readers'. (In many situations, of course, your boss and your intended reader will be the same person.)

Other people can also provide you with valuable advice about what's expected of you when you write. For instance, if you are new on the job, you might ask any experienced co-worker. If you are writing product descriptions or operations manuals for customers and clients, you might ask people in the marketing department about the expectations of these readers.

As you gain experience, you will learn about the expectations that apply to your writing. Consequently, you will need to inquire less and less frequently. When you are new on the job, however, you should be prepared to ask often. You may be embarrassed at having to inquire, but the embarrassment of asking will be much less than the embarrassment of having to rewrite your communication because you failed to ask.

Guideline

**2**

### Obtain and Study Any Written Regulations Concerning Your Writing

Employers sometimes write out their expectations concerning writing. On the job, you should be sure to obtain and study these regulations—if they exist.

These written regulations sometimes take the form of *style guides*. Style guides vary considerably in scope. Some cover most aspects of organization, appearance, and writing style. Others cover only some of these topics.

Many employers create their own style guides whereas others rely on guides put out by professional societies (such as the Council of Biological Editors or the American Psychological Association). Some organizations use general manuals, such as *The Chicago Manual of Style*.

Another kind of printed regulations are the instructions that some government agencies (such as the National Aeronautical and Space Administration) and some private foundations (such as the Rockefeller Foundation) send to people who want to obtain grants or contracts with them. These regulations specify deadlines, organization, contents, and many other aspects of writing that must be observed by anyone who submits proposals to them.

Guideline
### 3
## Look at Similar Communications Written by Other People

To a large extent, the way people will expect you to write will be determined by the writing they are used to seeing. That's partly because people like to deal with what's familiar to them. It's also because effective strategies for writing tend to be imitated. What's customary is often what has been found to work.

When looking at sample communications, however, be cautious about *assuming* that they worked. You should ask whether or not the communications you study worked well or poorly. (Of course, you can also learn from communications that didn't work because they can teach you what to avoid.)

## INTRODUCTION TO GUIDELINES 4 THROUGH 10

The first three guidelines explained where you can find out what's expected of you when you write at work. The remaining seven guidelines tell you what kinds of information you should seek from those sources.

Guideline
### 4
## Find Out Your Deadline and Make a Schedule for Meeting It

At work, deadlines are much more serious than deadlines at school. At school, you can often obtain an extension on a project. And even if you miss a deadline, the penalty is relatively mild, provided that you finally complete your assignment.

At work, however, there are no incompletes. A missed deadline can cost your employer thousands or even millions of dollars. Business people sometimes advise new employees that "It's better to turn in your project 80% complete than to turn it in 100% late." If your organization is submitting a proposal, the proposal must reach the client by the deadline, or it may not be considered at all—no matter how good it is. If your company is about to release a new product, the manual you are writing must be ready by the release date or the product cannot be shipped. And if it cannot be shipped, it cannot be sold.

So, when you write at work, find out when the deadline is. Then, make a schedule for meeting it. In your schedule, be sure to leave time for all the writing activities you will perform, including revising.

## Guideline 5 — Find Out What Length Is Expected

At work, as at school, find out how long your communication is expected to be. At work you will need to interpret expectations about length much differently than you do at school, however. At school, when an instructor tells you how long a project should be, he or she probably has in mind a *minimum* length. When someone at work tells you how long a communication should be, he or she probably has in mind a *maximum* length.

## Guideline 6 — Find Out What Your Communication Should Look Like

When finding out what your communication should look like, you need to be concerned with three things: form, organization, and appearance.

The form of your communication is its type: for example, whether it is to be a report, memo, or letter.

The organization of your communication is the arrangement of its contents. Some employers prefer certain ways of organizing various kinds of communications. For instance, some employers want writers to place their recommendations and conclusions at the beginning of their reports and proposals, whereas others want those contents at the end.

Some employers also have preferences (and even strict rules) about the appearance of the communications their employees write. They specify exactly how wide the margins are supposed to be, whether the headings should be centered or at the left-hand margin, and so on. It may be that you will have a secretary, typing pool, or word-processing center prepare finished copies of your communications. If so, you will not have to worry about these matters yourself. But if *you* prepare the finished copy, be sure to find out what it should look like.

As mentioned above, some organizations publish style guides that tell their employees how their communications should be organized and how they should look. Even if your employer has no such printed instructions, it is likely that there are customary practices in your organization that you will want to follow. You can learn about these by looking at communications others have written and by asking.

## Guideline 7 — Find Out What Style You Should Use

The word *style* refers to many aspects of writing. Here are some that your boss or readers may have preferences about:

- Use or avoidance of personal pronouns *(I, we)*.
- General tone—whether relatively formal or relatively informal.
- The way that tables are set up.
- The use of abbreviations.

Like the other expectations, these may be spelled out in a style guide. If not, you will have to find out about them by asking and by looking at what your co-workers have written.

Guideline
## 8
## Find Out What You Are Expected To Say—Or Not Say

At work, other people may have some fairly strong expectations about what you will say—and what you won't say—in a particular communication. These people may include your boss or some other person who assigned you to write. They may also include your readers.

Such expectations often exist at work because the communications an employee writes can have a considerable effect—for better or worse—on the employer's organization. The communications you write may affect the future development of your department or of other departments. They may spur the creation of new product lines or research activities. When you write to people outside the organization, your communications will be official statements of the organization, not just personal statements from you.

Accordingly, there are good reasons for other people to have opinions about what should or should not be said.

Guideline
## 9
## Find Out What Policies Apply to the Communication You Are Preparing

As mentioned in the discussion of the preceding guideline, the communications you write can have profound effects on your organization. For that reason, beyond any informal *expectations* about what you might or might not say in a particular communication, your organization may have strict *policies* about the ways that certain topics are treated.

These policies may have been created for a great variety of reasons. For instance, they may have been intended to protect the organization's business interests by ensuring that information about secret manufacturing processes or product designs are not divulged to competitors. Or, the policies may have been designed to ensure the safety of purchasers of your organization's products by requiring you to provide proper warnings about the use of those products.

When organizations have policies about what must or cannot be written, they also have procedures by which they review communications for compliance to those policies. If you are working for such an organization, your communications will need to be reviewed before they can be sent to your intended readers. From your point of view, these review procedures are helpful. They protect you from making costly mistakes. Even though your work will be reviewed for compliance to these policies, however, it is still important for you to learn about these policies before you begin to write. You will save both yourself and your reviewers considerable time by knowing and following these policies when you first draft your communications.

Guideline

## Find Out Whether and When You Will Need To Have Your Drafts Reviewed

## 10

At work, it is common for writers to have their drafts reviewed before the drafts are sent to their intended readers. The preceding section mentioned one of the several reasons for such reviews: to ensure that your communications comply with your organization's policies. Reviews also allow your boss and other people in your organization to see that your communication is satisfactory in many other respects, including its style, organization, content, and appearance.

As you prepare to write something, one of the first things you should do is find out whether or not it must be reviewed. If so, you will need to adjust your schedule for writing so that you can satisfy this requirement and still make your final deadline. You must have a complete draft ready for reviewers well ahead of the final deadline so that the reviewers will have time to do their work and so that you will have time to incorporate changes suggested or required by your reviewers.

In Chapter 23 you will find advice about how to gain the greatest benefit from reviews.

## A FINAL NOTE

As you gain on-the-job experience you will quickly learn a great deal about how your employer's organization, your boss, and your readers expect you to write. Consequently, you will need to inquire about these matters less and less frequently. But you will never outgrow your need to check your assumptions, particularly when you are writing to new readers or preparing a type of communication that you have never prepared before.

## SUMMARY

This chapter has emphasized that other people may have many expectations about your writing. Underlying the ten guidelines for finding out these expectations are the following three pieces of advice.

- Don't assume that you know what is expected.
- Don't be afraid to ask about expectations.
- Remember that you may have to satisfy expectations about any and all aspects of your writing.

The ten guidelines for finding out about expectations fall into two groups. The first group tells you *how* you can find out what is expected of your writing.

1. *Ask your boss, readers, and other knowledgeable people what's expected.* Your boss, your reader, and co-workers are all good sources of information.

2. *Obtain and study any written regulations concerning your writing.* These may take the form of style guides provided by your organization. If you are writing for readers outside your own organization, your readers may have their own style guides, which you must follow.

3. *Look at similar communications written by other people.* By examining sample communications, you will learn a great deal about what your readers are accustomed to seeing. Be sure to ask whether the sample communications worked well or poorly.

The remaining seven guidelines tell *what kinds of information* you should seek from those sources.

4. *Find out your deadline and make a schedule for meeting it.* Deadlines at work are much more serious than deadlines at school. At work, a missed deadline can cost your employer thousands or even millions of dollars.

5. *Find out what length is expected.* Remember that when someone at work tells you how long a communication should be, he or she probably has in mind a *maximum* length.

6. *Find out what your communication should look like.* You need to be concerned with three things: form (for example, report, memo, or letter), organization, and appearance.

7. *Find out what style you should use.* Your boss or readers may have preferences about such stylistic matters as use of personal pronouns, formality of tone, use of abbreviations, and so on.

8. *Find out what you are expected to say—or not say.* The communications you write at work will often function as official statements of your department or organization. They can have a considerable effect on your employer. Also, people at work may have fairly strong expectations about what you should or should not say in a particular communication.

9. *Find out what policies apply to the communication you are preparing.* Because the communication you write can have such a profound effect on your organization, the organization may have policies about the way certain topics are treated. Be sure to learn about and follow these policies as you write.

10. *Find out whether and when you will need to have your drafts reviewed.* Find out *before* you begin writing, so you can allow time in your schedule for the reviews and any revisions you might have to make.

# Planning

In Part II, you read three chapters about defining the objectives of your communications. In Part III, you will learn how to use your definition of objectives to *plan* your communications.

## AIMS OF PLANNING

When planning a communication, you decide what to include and how to organize and present that material. When you have completed your plan, you should know what the parts of your communication will be, what they will contain, and how they will fit together.

Your plans can help you in three very important ways:

- *Planning saves you time.* Consider, for instance, some of the difficulties encountered by Charlotte because she drafted a report without planning first. She spent more than an hour polishing three paragraphs that she later realized she didn't need. She also spent half an hour revising two pages of material when she found that the information they contained would be much more effective if moved to a different place. Charlotte could have saved time if she had first planned what she would say and how she would organize it.
- *Planning helps you increase the effectiveness of your writing.* When Charlotte's boss read her report, he pointed out several places in which she didn't present and support her major points as clearly and forcefully as she might have. How did that happen? When

Charlotte was writing, she concentrated on the smaller parts of her report—focusing on each sentence and paragraph as she wrote it. She didn't think much about her overall strategy. Planning would have focused her attention on important decisions about emphasis and structure.

- *Planning helps you get good, timely advice about your writing.* When Charlotte's boss read her report, he also pointed out that she could have improved her report by saying some things in a different way and by including information she had omitted. If Charlotte had planned before drafting, she could have discussed her plan with her boss. Then, before spending any time at drafting, she could have revised her plan in light of his advice.

You, too, will be able to write most efficiently and effectively if you take the time to plan.

## THE FORMS PLANS CAN TAKE

How elaborate do your plans need to be? Sometimes you can plan effectively by simply making some mental notes or jotting down a few words on a scrap of paper. At other times you will benefit from writing a detailed outline. Here are some things to consider as you decide what kind of plans to make:

- *Length.* Generally, the shorter the communication, the less detailed your plan will need to be.

- *Complexity.* Generally, the simpler your message, the less detailed your plan will need to be.
- *Familiarity.* Generally, the more times you have written similar communications, the less detailed your plan will need to be.

There is one other important consideration: if you going to show your plan to other people to obtain their advice or approval, make your plans as detailed as possible. The more detailed your plans, the better able other people will be to give you helpful reactions and suggestions.

## THE BOUNDARIES OF PLANNING

Because all of this book's advice about planning appears in a single, separate section, you might think that you should treat planning as a separate stage in your writing. That's not exactly the case. Although you should set aside time specifically for planning before you begin to draft, planning plays an important role in all phases of your writing. Even when you are defining your objectives, you should make preliminary plans about the contents and organization of your communication. Likewise, as you draft, review, and revise, you should fill out and refine your plans in light of the good ideas you gain through your continuing work on your communication.

## MAKE FLEXIBLE PLANS

Because planning is an on-going activity of writing, make your plans flexible. That may be hard to do. After you have taken the trouble to develop a plan, you may be reluctant to change it. Keep in mind, however, that one cause of weak writing is the writer's unwillingness to change good plans for better ones.

Why can't your plans be perfect when you first make them? Sometimes they *will* be perfect. That is most likely to happen when you are writing short, uncomplicated messages.

Often, however, you will get better ideas about how to write your communications as you work on them. After all, the more you work with your material and the more you think about your readers and your readers' response to your writing, the more insight you will gain into ways to convey your message effectively. Also, when you are first planning longer, more complicated messages, you may simply be unable to anticipate all of the relevant considerations that affect your communication's content and organization. Consequently, when you are drafting or evaluating your communication, you may find out that what *appeared* to be a good plan is not.

In sum, you will write best if you set aside some time to plan before you begin to draft. But you will write even better if you continue to plan throughout your subsequent work on your communication.

## CONVENTIONAL PATTERNS CAN HELP YOU PLAN

When you are planning, nothing could help you more than detailed advice from someone who has succeeded previously in accomplishing the very things you hope to accomplish in your communication. If you are fortunate enough to have such an adviser, you will be very lucky indeed. Even if you find no such person, however, you can still benefit from the experience of others— many others—by following the many helpful customs and conventions about the way written communications are constructed at work.

Particularly helpful are the conventions concerning the following three aspects of a communication: its format, superstructure, and segments.

## Conventional Formats

A *format* is the "package" you put your message into when you present it to your readers. When you study formats, you learn how to make a letter *look* like a letter, a memo *look* like a memo, and so on.

Being able to make your communications look like others of their kind is an important writing skill. If you don't make your communication look familiar, your readers might think that you are a person who doesn't understand how things are done in the working world.

Knowing the conventional formats will also help you in another way, one that is more directly related to your efforts at planning. Each format not only tells how a communication will look, but also specifies some of its contents. For example, the format for a memo includes a line at the top stating what the message is about. The format for a formal report includes a table of contents and a brief summary written for the busy executive who wants to understand the gist of the report without taking time to read all of it. By knowing the format, you can be sure that you include all the appropriate parts as you plan a communication.

In Chapter 26, you will find extensive information about the three formats most often used at work: letter, memo, and book. (The book format is usually used for formal reports and proposals, instruction manuals, and other long communications.)

## Superstructures

Formats function independently of the message they contain. You can use the let-ter, memo, or book format, for instance, to report on research, propose a project, or give instructions. There are, in addition to formats, conventional patterns for creating various kinds of messages. Like a blueprint for a building, these patterns tell writers how to build their messages: what parts to include, what arrangement to give them, and what strategies to use to develop each part and relate it to the others. Such conventional patterns are called *superstructures*.

There are superstructures for each type of message mentioned above (research report, project proposal, instructions) and for a multitude of others, including minutes of meetings, progress reports, and trip reports, to name just a few. To see some of the features of a superstructure, look at Figure III-1, which briefly describes a pattern writers often use when seeking approval to start a new project.

### Superstructures Are Successful Patterns for Communicating

Each superstructure is a pattern that *works*. Consider, for instance, the superstructure shown in Figure III-1. It helps *readers* because it tells writers to include the kinds of information that readers need when deciding whether or not to approve a proposed project—information about what the writer wants to do, why, and how. The superstructure tells writers how to organize and develop their information to help their readers readily understand and use it. In fact, even the mere familiarity of the pattern helps readers. Because they know the pattern in advance, readers can use it to mentally organize the information in the communication as they try to understand the communication's larger meaning.

Superstructures also help *writers*. Each one describes a pattern for writing in a helpful

CONVENTIONAL SUPERSTRUCTURE
FOR PROJECT PROPOSALS

Introduction

- Brief description of what is proposed
- Description of the need the proposed action will meet or the problem it will solve

Objectives of the proposed project: the link between the need or problem and the proposed project

Plan for the project

- Description of the outcome of the project: what will be made or accomplished and how it relates to the project objectives
- Method to be used in the project
- Facilities and equipment needed
- Schedule

Management plan for the project

- Management structure
- Personnel and their qualifications

Budget, including costs and anticipated savings as a result of the project

Conclusion

Appendixes, glossary, and so on

**Figure III-1.** Conventional Superstructure for Project Proposals.

and persuasive manner in some particular kind of situation. For example, if you were writing a proposal, you could use the superstructure shown in Figure III-1 to be sure that you had included the various kinds of information you needed to provide your readers to achieve your purpose. And if you also followed the details of that superstructure, which is discussed at greater length in Chapter 32, you could develop and present that material in a way that other writers have found to be persuasive in situations similar to yours.

## How To Use Superstructures When Planning

To use superstructures effectively when planning, you should look to them for *suggestions* about how to construct your messages. You should not try merely to imitate them. Although superstructures present reader-centered structural patterns for achieving certain purposes, the patterns are too general to apply to your particular purpose and readers without some careful thought and adaptation. You must use your good judgment, common sense, and creativity to determine which aspects of a superstructure will work for you and how best to modify them to suit your situation. Chapters 8 and 9 provide specific advice about using conventional superstructures effectively as you plan the communications you write at work.

## Conventional Patterns for Segments

As you develop the parts of your superstructures, other conventional patterns can help you: the conventional patterns for organizing segments. A *segment* is a group of related sentences, such as the sentences in a paragraph or group of paragraphs. Thus, segments are the building blocks of overall communications. They are patterns you can use to organize the parts of your superstructure. There are many conventional patterns for segments, including the following:

- Classification
- Partition (description of an object)
- Segmentation (description of a process)
- Comparison and contrast
- Persuasion
- Cause and effect
- Problem and solution

Of course, every segment you write won't fit neatly into one of these patterns. Some segments combine elements of many of these patterns, and some are developed according to other principles. Further, even when you are using one of these conventional patterns, you still should follow the general guidelines for writing segments (see Chapter 11).

Patterns for segments differ from superstructures in an important way: whereas superstructures are each suited to one particular purpose and situation, the patterns for segments are *general* ways of arranging material. In the particular segment you are planning, you must choose carefully the pattern that is most likely to achieve your purpose with your readers.

## SUMMARY

When you plan, you decide what your communication will contain, how that material will be arranged, and how the parts of your communication will relate to one another. Your plans may take many forms, such as mental notes, words jotted down on a scrap of paper, or a carefully drafted outline.

To benefit from the insights you gain as you continue to work on your communications, you should be willing to alter (and improve) your plans throughout your work at drafting and evaluating.

As you plan, you can enlist the aid of three types of conventional patterns for constructing communications.

- *Conventional formats* help you make your letters look like letters, your memos look like memos, and so on.
- *Superstructures* are general plans for the overall organization of communications.[1] They are used to achieve specific purposes with certain kinds of readers under certain circumstances. To use superstructures effectively, you must adapt them carefully to your particular situation.
- *Conventional patterns for segments* provide ideas about how to organize the paragraphs, sections, and other parts of your overall structure.

## CONTENTS OF PART III

**Chapter 7**
**Deciding What to Say**

**Chapter 8**
**Ten Techniques for Generating Ideas about What To Say**

**Chapter 9**
**Organizing**

**Chapter 10**
**Using Outlines**

## NOTE

1. For a brief explanation of the source of the term *superstructure*, see the note at the end of the introduction to Part IX, "Conventional Superstructures."

# Deciding
# What To Say

## PREVIEW OF GUIDELINES

1. List the questions your readers will ask, and answer them.
2. List the additional things your readers will need to know.
3. List the persuasive arguments you will have to make.
4. Generate ideas freely before you evaluate them.
5. Review and revise the contents of your communication throughout your work on it.

The most basic step in planning a communication is deciding what to say. To make good decisions about the content of communications you write at work, you must pursue two goals that pull in opposite directions. On the one hand, you must prepare communications that are *complete*—that say everything necessary to achieve your purpose. If you omit an important argument, you may fail to persuade your readers. If you leave out a crucial fact, your readers may have to take the extra time and trouble to ask you for it. Or, your readers may not know that it is missing—and may make a wrong decision as a result.

On the other hand, you must write communications that are as *brief* as possible. It's quite common for recent college graduates to have one of their first memos or reports returned to them with instructions to "shorten it," "cut it in half," or "say it all in one page." Readers want only information they can use, and they want to be able to obtain that information as quickly and easily as possible. If you say things they feel are unnecessary or irrelevant, they may object that you are wasting their time and energy. Furthermore, unnecessary information can obscure your key points by burying them in prose irrelevant to your readers' needs and desires.

In short, to achieve your purposes you must include everything that is needed while including nothing that isn't. The five guidelines in this chapter will help you decide what to say—and not say—in the communications you write on the job.

## YOUR TWO ACTIVITIES WHEN DECIDING WHAT TO SAY

To pursue the twin goals of saying everything necessary and saying nothing extra, you need to perform two distinct activities: generating a list of things you *might* include, and selecting from that list the things you *should* include.

### Generating a List of Possible Contents

When generating a list of things you might include, your aim is to be sure that you don't overlook anything that will help achieve your purpose. In general, your list should identify two kinds of information. First, there are the things you already know about your subject—the information and ideas you already have. Much of this information you may have stored in your memory while thinking over your subject and its significance to your readers. In some kinds of projects, you may also have much of this information in notebooks, photocopies of articles, computer files, and so on.

Second, your list of possible contents should identify things you still need to find out and think about. Therefore, you should begin thinking systematically about what you might say well before you begin to draft. Your early thoughts about what you might say can guide your efforts to gather information about your subject.

## Selecting the Items You Should Include

After you have generated a complete list of possible contents, you should review and evaluate the list, ruthlessly eliminating anything that won't contribute directly to your communication's success. You may be able to evaluate some items on your list only after you have gathered additional information about them. Others you will be able to evaluate immediately. The items that remain after you have completed your investigation and evaluation will be the contents you organize and present to your readers.

# INTRODUCTION TO GUIDELINES 1 THROUGH 3

Guidelines 1 through 3 will help you identify the specific pieces of information you should include in a particular communication. They suggest that as you create and review your list of contents, you ask the following questions.

- What will my readers want me to tell them? (Guideline 1)
- What else do they need to know? (Guideline 2)
- What arguments will persuade my readers to alter their attitudes in the way I want? (Guideline 3)

Each of these questions prompts you to take a different perspective on your subject. Because your objective is to create a *complete* list of things you might say, you should answer all three in the ways suggested in Guidelines 1 through 3.

Guideline
**1**

## List the Questions Your Readers Will Ask, and Answer Them

When deciding what to say, begin by figuring out what your readers will want to learn from you. You can do that by following a strategy that is a natural companion of the second guideline in Chapter 5: "Portray your readers as people who ask questions while they read." The strategy is to list the questions your readers will ask, together with your answers to them.

How can you predict what your readers' questions will be? Use the imaginary portrait of your readers that you developed while following the guidelines in Chapter 5. Four kinds of information you gathered to create that portrait will be especially helpful.

- Your readers' roles.
- Your readers' familiarity with your specialty.
- The specific features of the situation in which your readers will be reading your communication.
- Any special characteristics your readers have.

Although all four of these types of information will help you predict what questions your readers will ask, only the first two are discussed further here. The last two kinds of information are so particular to your readers and your communication situation that it's impossible to make helpful general comments about the way to use them when deciding what to say. To use them effectively, you will have to rely upon your common sense.

## Your Readers' Roles

Knowledge of the roles your readers will be playing while reading can help you as you determine what to say to them. As Chapter 5 indicates, there are three basic roles your readers can play: decision-maker, adviser, and implementer. Each role gives rise to a different set of questions.

### Decision-Makers

Decision-makers are responsible for determining what an organization will do in the future—next week, next month, next year. When you write to them, you will usually give reports or make recommendations to help them choose the best course of action. As they read, decision-makers are likely to ask the following questions:

- *What are your conclusions?* Decision-makers are much more interested in your conclusions than in the procedures you used or the results you obtained. Conclusions present the outcome of your work in a manner that can readily serve as the basis for a decision.
- *What do you recommend?* If you have any recommendations to make, decision-makers want to know what they are. Indeed, decision-makers often assign people the task of making written recommendations.
- *What will happen?* Besides wanting to know what you recommend, decision-makers usually want to know what will happen if your recommendations are followed—and if they aren't. How much money will be saved? How much will production increase? How will customers react to the change?

While they are usually interested in conclusions, recommendations, and projections, decision-makers rarely ask for detailed information about how you arrived at them. Information about your procedures or the raw data you obtained is used primarily to check on the validity of your conclusions. Because decision-makers usually presume that writers are experts in their own specialties, they seldom feel obliged to check on the quality of a writer's work. Furthermore, if they do want to check, decision-makers rarely perform the job themselves: they ask their advisers to check for them.

In your answers to their questions, decision-makers want two things: brevity and simplicity. The director of a government research center expressed the typical attitude of decision-makers toward the proper length of a communication: "If it can't be held together by a paper clip, it is too long for me to read." Very common in all kinds of organizations are guidelines that state that reports to management must be no more than one or two pages long, even if the reports represent the culmination of several months' work.

Decision-makers want you to answer their questions simply so that they can understand and use your information rapidly. Many know very little about the specialized fields of those who write to them. Even decision-makers who did, at one time, specialize in the same field as a writer may have lost touch with new developments in that field since being promoted to managerial positions. As a result, decision-makers usually prefer communications that explain things in plain words rather than in specialized terminology. They rely heavily upon background and explanatory information designed to help them understand technical content that absolutely cannot be avoided.

### Advisers

Advisers help decision-makers choose the best course of action. They may be asked to advise based upon their experience and expertise, or because the decision being made will affect them directly. Their role is to conduct a detailed analysis and evaluation of your general conclusions, recommendations, and projections. Thus, advisers are likely to ask questions that touch upon the thoroughness, reliability, and impact of your work. Their questions include:

- Did you use a reasonable method to obtain your results?
- Do your data really support your conclusions?
- Have you overlooked anything important?
- If your recommendation is followed, what will be the effect on other departments?
- What kinds of problems are likely to arise?

Advisers want detailed answers to their questions—much more detailed than decision-makers want. They need to know enough detail about how you arrived at your conclusions so that they can evaluate the soundness of your work. Yet, even in their detailed study of your information, advisers are not likely to want to know every last particular of the procedures you used or every last piece of data you collected.

Despite the differences between the kinds of answers that decision-makers and advisers want, you will often find yourself writing a single communication that must meet the needs of both groups. (In that case you would be addressing your communication to a complex audience, as discussed in Chapter 5.) Usually, such reports fall into two distinct parts: (1) a very brief summary—called an *executive summary* or *abstract*—at the beginning of the report, which is designed for the decision-maker, and (2) the body of the report, which is designed for advisers. Typically, the executive summary is only a page long, a few pages at most, while the body of the report may exceed 100 pages.

In a study conducted by James W. Souther, a large group of managers were asked how often they read each of the major parts of a long report. As Figure 7-1 indicates, they reported that they read the summary 100% of the time, whereas they read the body only 15% of the time. When these decision-makers want more information than the summary provides, they usually go to the introduction (which provides background information) and to the recommendation section (where the writers explain their suggestions more fully). The rest of the report is read by advisers.

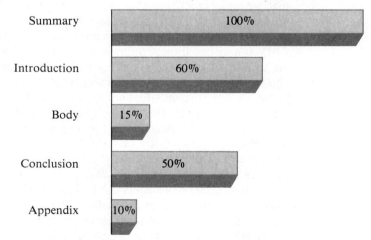

| SECTION OF REPORT | FREQUENCY OF READING |
|---|---|
| Summary | 100% |
| Introduction | 60% |
| Body | 15% |
| Conclusion | 50% |
| Appendix | 10% |

**Figure 7-1.**   How Often Decision-Makers Read the Various Parts of a Report.

From a study by James W. Souther that is described by Richard W. Dodge in "What to Report," *Westinghouse Engineer* 22:4–5 (1962):108–11.

## Implementers

Decisions, once made, have to be carried out. People at all levels within an organization are entrusted with doing that. To aid them, you will write a wide variety of communications. Three of the most common types are as follows.

- *Step-by-step instructions.* For example, you may write instructions that will teach a new employee to use your employer's lathes.
- *General instructions.* General instructions include such things as policy statements and general directives, which the readers must implement. You might, for instance, write up the company's new vacation policy, a policy that individual employees and managers will have to apply to their own situations.
- *Requests and orders.* You will write requests and orders when you are in a position to decide that something needs to be done and that your reader is the person who should do it—or help you do it. When you advance to a supervisory or managerial position, you will do a great deal of writing of this sort. However, you may also do it when you are new on the job—for instance when you request information or other help from another department that is supposed to work with you.

The most important question implementers will want you to answer is, "What do you want me to do?" To answer this question when you are writing step-by-step instructions, requests, and orders, you need to provide directions that are clear, exact, and easy-to-follow. Be sure to say what you want *explicitly.* All too often writers fail to get what they want from their readers for the simple reason that the readers do not know what the writers want.

You may find it more difficult to answer the question, "What do you want me to do?" when you are writing policy statements and general directives. You need to be sure that each of your readers can see how the policy applies to his or her specific situation. One excellent way to do that is by using concrete examples to illustrate otherwise abstract instructions.

Another important question asked by implementers is, "What is the purpose of the actions you are asking me to take?" Readers of policy statements and general directives, especially, often must know the intent of a policy if they are to produce satisfactory results. Imagine, for instance, the situation of the managers in a factory who have been directed to cut by 15% the amount of energy used in production. They will want to know whether they are trying to realize long-term energy savings or whether they are trying to compensate for a short-term energy shortage. If the latter, they may think of temporary actions, such as altering work hours and curtailing certain operations temporarily. However, if the reduction is to be long-term, they may think of long-term measures, such as purchasing new equipment and modifying the structure of the factory building. Similarly, the people to whom you send requests and orders can often help you best if they know why you are asking for their help and cooperation.

In many situations, implementers also need the answer to this question: "How much freedom have I to choose the way I will perform this action?" People who are following instructions often devise shortcuts or alternative ways of doing things. Where any of several ways is sufficiently efficient, effective, and safe, you may not care to specify one way over another. But where you have a preference, you should say so explicitly. For example, the managers in charge of reducing energy consumption might go about devising any number of plans, each distinctly different from the others. The managers need to know whether they must use some particular methods or obtain any particular results.

Finally, implementers often need the answer to the question, "When must I complete this task?" They need to schedule their work and can do so only if they know when you want it done.

## Your Readers' Familiarity with Your Specialty

When answering your readers' questions, you may use a word, concept, or symbol they don't understand. That gives rise to another sort of question: "What does that word (or concept or symbol) mean?"

The first step in answering this kind of question is to anticipate where your readers will ask it. For that reason, when you are deciding what you will include in a communication, you should look for places where you might use unfamiliar terms. Once you have identified them, you should ask yourself what kind of information you should provide to help your reader understand. You might, for instance, provide a definition:

> A headgate is a gate for controlling the amount of water flowing into an irrigation ditch.
>
> A rectifier is an electrical device for converting alternating to direct current.

Or, you might explain by means of analogy:

> Wood sorrel resembles clover.
>
> On the proposed extruders, we will use a feed system much like that found on a Banbury mixer.

You will find detailed advice about defining and explaining unfamiliar words and concepts in Chapter 18.

| Guideline | |
|---|---|
| 2 | **List the Additional Things Your Readers Will Need To Know** |

Your expertise and experience, your efforts to thoroughly study your subject, will lead you to realize that some information is indispensible to your readers even though they may not know they need it. Imagine, for instance, the situation of Daniel. He has been asked to investigate three computer programs that his employer might purchase to aid in designing its products. While gathering information about the capabilities, performance, and cost of the programs, Daniel learned that the company that makes one of them is having serious financial difficulties. If the company goes out of business, Daniel's employer will not be able to obtain the assistance and improved versions of the program that it would normally expect to receive over the coming years. Thus, even though Daniel's readers are not likely to ask, "Are any of these programs made by companies that appear to be on the verge of bankruptcy?" Daniel should include that information in his report. Similarly, when deciding what to say in your communications, you should be sure to identify all the information that is important to your readers, including information your readers won't know that they should ask for.

As you are considering items that your readers *should* be told, be careful not to include items that you want to say even though they will not be useful or persuasive to your reader.

<table>
<tr><td>Guideline<br>3</td><td>

## List the Persuasive Arguments You Will Have To Make

</td></tr>
</table>

When you are deciding what to say in your communications, remember to include the material needed to alter your readers' attitudes in the way you desire.

### Decide Upon the Claims You Will Make

The first step in identifying the material that is likely to alter your readers' attitudes is to decide upon the persuasive points you want to make. Each of these points is called a *claim*. Imagine, for instance, that you are writing a report requested by your plant manager, who has asked you to research two large incinerators that your company might buy for its factory. Further, imagine that your research has led you to believe that the first incinerator is preferable to the second. Among the claims that you might include in your report are the following:

> The first incinerator is safer for the environment than the second.
>
> Although the second furnace costs less to purchase, the first is so much more efficient that the additional profits it could generate will offset the additional expense in 18 months.

### Choose Claims that Will Appeal to Your Readers

As you list your claims, keep in mind that you must choose claims that your readers will find persuasive. These may not be the same claims that you find most convincing. To see the difference between a communication incorporating reader-centered claims and a less effective one incorporating writer-centered claims, consider an illustration used in Chapter 2: the two versions of the memo Mr. Prindle drafted to persuade his boss, Ms. Van Husen, to give him his own office (Figures 2-2 and 2-3).

In both memos, the basic claims are implicit yet obvious, as often happens in communications written at work. The implicit, writer-centered claim of the first memo is that, "There are serious disadvantages to me (Prindle) in not having my own office." The more reader-centered claim of the second version is that, "Our organization will benefit from giving me my own office."

You will be able to write much more effectively if you carefully choose claims that will appeal to your readers.

## List the Material You Will Include To Support Your Claims

In most cases, your readers will not accept your claims simply because you have made them. An effective method for determining what information you need to include to support each of your claims is to use a version of the strategy already discussed in Guideline 1: List the questions you imagine your readers would ask you after reading each of your claims.

For instance, you might imagine that, after hearing your claim that the first incinerator is better for the environment than the second, your readers would ask, "What evidence do you have to support that claim?" You might respond by saying that the first one emits fewer fumes, which might prompt your readers to ask, "How many fewer?" "What are the fumes it does emit?" and "How do you know?" The questions your readers ask will depend upon many things, particularly the roles the readers will be playing while reading (decision-maker, adviser, implementer) and their professional backgrounds (engineer, lawyer, accountant, and so on).

## Supporting Implicit Claims

Where your claims are implicit, you face a special challenge in anticipating your readers' questions. For instance, suppose that in your report on the incinerators you wished to persuade the plant manager not only that the first incinerator is preferable to the second, but also that you are a good, thorough researcher. Given the current etiquette in the working world, you could hardly make such a statement explicitly. Yet, to succeed in creating that impression, you would have to include material that would support such a claim.

You could begin to do that by identifying all the places in your communication where your reader might think that you have overlooked something. Then, at each of those places, you could include information that would forestall that thought. Thus, suppose that you compared twenty-five characteristics of the two incinerators, but felt that with respect to eighteen characteristics the two were so similar that these characteristics could not be used as a basis for preferring one incinerator over the other. You would be quite wise to discuss only the remaining seven in your text, for those would be the only ones of use to your reader. However, you would also want to mention that you had looked at the eighteen other factors and tell why you had decided not to discuss them. Unless you did, your reader might think that you had discussed only seven characteristics because you had considered only seven.

You will find additional, detailed advice about the content and structure of persuasive arguments in Chapter 13.

## INTRODUCTION TO GUIDELINES 4 AND 5

The three guidelines you have just read tell you *what* information to look for when deciding what to say in a particular communication. They suggest that you look for information that answers these three questions: "What will my readers want me to tell them?" "What else do they need to know?" and "What arguments will persuade my readers to alter their attitudes in the way I want?"

The next two guidelines tell you *how* to look for that information. They make suggestions that can greatly increase not only the effectiveness of your final decisions about what to say but also your overall efficiency as a writer.

Guideline
### 4
### Generate Ideas Freely before You Evaluate Them

When deciding what to say in a communication, you engage in two distinct kinds of activities. When you search for all the things you might say, you are acting *creatively*. When you evaluate each of those ideas to see if it will truly help achieve your purpose, you are acting *critically*. Guideline 4 suggests that you concentrate solely on your creative activity of generating ideas *before* you begin your critical activity of evaluating them.

If you try to generate ideas and evaluate them simultaneously you may encounter two difficulties. First, you may find it difficult to restart the creative flow of new ideas after repeatedly disrupting it to evaluate each idea as you think of it. Further, by eliminating ideas early, you may sometimes miss connections that those ideas have with other ideas you think of later. You will usually be able to generate much fuller lists of things to say if you postpone all your evaluation until your flow of new ideas has stopped by itself.

In Chapter 8, you will find detailed suggestions for ways to generate ideas about what to say.

Guideline
### 5
### Review and Revise the Contents of Your Communication throughout Your Work on It

Many writers focus their attention on the content of their communications when they begin work on them, but they don't do so again. That's a mistake. You should continue to review and revise the contents of your communications throughout your work on them.

Why? As you shape and reshape the sections, paragraphs, and sentences of your communications, you modify and refine your strategy for affecting your readers in the way you desire. If you do not constantly and consciously review your decisions about the information you will include, you may leave in information that was needed in the original plan but not in the revised and improved one. Or you may omit information irrelevant to your original plan, but necessary to the success of the revised one.

To keep alert to the need to refine your plans about what you will say, focus on the questions your readers will ask while reading. In particular, once you progress beyond your initial planning and begin to draft, focus on questions your readers might have about specific statements you make in your draft. Your ability to answer these specific questions will be just as important as your ability to answer the larger questions you anticipated when you first planned your communication.

## WORKSHEET FOR DECIDING WHAT TO SAY

As you think about what to include in communications you write at work, you may find it helpful to use the worksheet for deciding what to say shown in Figure 7-2.

## SUMMARY

This chapter has presented five reader-centered guidelines to help you decide what to say in communications you write on the job. To decide what to say, you must engage in two activities: searching for things you might say and selecting from among those possibilities. While searching, you need to assemble information you have already acquired, identify things you still need to find out, and identify issues on which you must still work out your thoughts. Your goal is to be complete but brief, to say everything needed to achieve your purpose, but nothing more.

The five guidelines are as follows:

1. *List the questions your readers will ask, and answer them.* Your readers' roles (decision-maker, adviser, implementer) will determine what many of their questions will be and what kinds of answers they will want. Your readers may have other questions because they are unfamiliar with the terms and concepts you use.

2. *List the additional things your readers will need to know.* Identify the things that, because of your expertise and experience, you realize are important even though your readers may not ask about them.

3. *List the persuasive arguments you will need to make.* Identify the arguments you must make to persuade your readers to modify their attitudes. Choose claims that your readers will find appealing. Then, identify the material you will use to support those claims.

4. *Generate ideas freely before you evaluate them.* If you try to search for and evaluate ideas simultaneously, you may stem the flow of your ideas and miss important and useful connections among them.

5. *Review and revise the contents of your communication throughout your work on it.* Many writers make the mistake of considering their contents only once, at the beginning of their work on a communication. Only by reviewing your contents throughout the writing process can you be sure to adjust for the refinements you make to your overall plan.

## EXERCISES

1. Imagine that you are now employed full-time. You have decided that you would like to take a course at a local college. Begin this exercise by naming a course that you might really like to take in this way. Next, use the worksheet for deciding what to say (Figure 7-2) to list the information you would include in a memo asking your employer to pay your tuition and permit you to leave work early two days a week to attend the class.

2. Think of some concept, procedure, or topic you have learned about in college that you might want to write about to a decision-maker while you are on the job. Using the worksheet for deciding what to say (Figure 7-2), create a list of things that you would want to tell that person about your subject. Be sure to include any things you still need to find out about your subject to make your report truly useful to the decision-maker.

3. Imagine that the English department at your college has outfitted one of its classrooms with computers, which it would like to use in a few sections of Freshman English. The department's objective is to determine whether it can teach writing better by using computers than it can without using them. Except for using computers, these experimental computer sections will be taught in the same way as all the remaining sections of Freshman English.

    It is now July, and the department has hired you to write a letter to all incoming freshmen in order to persuade 125 of them to volunteer to enter the experimental computer sections. By following the guidelines in this chapter, decide what you should say in that letter. Identify things you would want to investigate further before writing.

Worksheet
**DECIDING WHAT TO SAY**

Project: _____

1. What questions are my readers likely to ask while reading my communication? How will I answer them?

2. What additional things do I need to tell my readers to achieve my purpose?

3. What arguments do I need to make to persuade my readers to alter their attitudes in the way I desire? (Identify the claims you will make and the evidence you will use to support those claims.)

**Figure 7-2**   Worksheet for Deciding What To Say.

# Ten Techniques for Generating Ideas about What To Say

## PREVIEW OF TECHNIQUES

**Ask Your Readers**

**Talk with Someone Else**

**Look at Successful Communications Used in Similar Situations**

**Look at Conventional Superstructures**

**Brainstorm**

**Write a Throw-Away Draft**

**Make an Idea Tree**

**Make an Outline**

**Draw a Flow Chart**

**Make a Matrix**

**A**s explained in the preceding chapter, when you are deciding what to say in a communication you engage in two activities: generating a list of things you might say and selecting from among those possible items. This chapter describes ten techniques for generating a list of ideas. The first four are techniques for obtaining advice and assistance from other people, either directly or indirectly. The other six are techniques in which you work by yourself.

For many communications—especially long ones—you will benefit from using several of these techniques. For instance, you might use outlining to generate ideas about the broad categories of information you want to include and then brainstorm or make an idea tree to generate ideas about what to say within some of the sections and subsections. Because different techniques will help you in different stages of writing and in different parts of your communications, you can become a better writer if you learn to use many techniques, not just one or two.

## STRUCTURED VERSUS UNSTRUCTURED LISTS

Some of the techniques discussed in this chapter produce *unstructured* lists and some produce *structured* lists. For instance, if you *brainstorm,* you will get an unstructured list, one in which your ideas are listed without any particular order. You write down your thoughts on your subject as rapidly as they come into your mind. Relying on the spontaneous association of ideas, you let each thought suggest another. The resulting list will reflect the pattern of personal associations that led you from one recollection or thought to the next, but it will not have any other structure. In contrast, if you *make an outline,* you will produce your list of ideas in a highly structured pattern.

When using one of the techniques that provide a structured list, keep in mind that the structure is primarily to help you generate ideas. Once you have a list, even a structured one, you must still carefully organize your material to suit your purpose and readers (following the guidelines given in the next chapter). In doing so, you may need to devise a totally new structure, abandoning the one that helped you generate ideas.

## WHEN TO USE THESE TECHNIQUES

The techniques described in this chapter can help you at several stages in your work on a communication. Because the techniques can help you identify information you need to gather about your subject, you can often benefit from using them long before you are ready to write. For instance, if you are going to undertake any sort of investigation—whether in the laboratory or in the library—you can use these techniques as you plan your research. They will help you decide what kinds of information you need to gather in your investigation.

When you are ready to plan the organization of your communication, you can use these techniques to help you assemble the knowledge you have acquired so that you will know exactly what information you have to arrange. Using these techniques at that point will also help you determine whether or not you still need to obtain additional information that you had previously overlooked.

Even after you have begun to draft, you can use these techniques whenever you find yourself unsure about what to say in a particular part of your message.

## ASK YOUR READERS

The first of the ten techniques is to ask your readers what they would like you to say. When doing that you are simply following some of the guidelines from Chapter 6, "Finding Out What's Expected." The point is worth repeating here because so many writers overlook the simplest and most direct means of finding out what questions their readers want answered in their communications: ask them. Are you writing a report to your boss? Ask what he or she wants you to say. Are you writing instructions? Ask some of the people who will be using them what kinds of information they need.

Of course, there will be many times when you cannot ask. You may be writing to people in other cities or other organizations, so that contacting them is difficult. Or you may be writing to people in your organization who don't expect or desire inquiries of this sort.

On the other hand, some of your readers may be so eager to tell you what they would like that they will specify exactly the contents of your communication. For instance, organizations that regularly receive proposals from outside individuals or groups often publish instructions that tell writers exactly what to include in their proposals.

## TALK WITH SOMEONE ELSE

In addition to your readers, other people can help you generate ideas about what to say. On the job, you can benefit greatly from visiting with your boss or walking to the next desk or down the hall to find someone who is willing to talk with you about your communication. Simply by explaining your communication, its purpose, and its readers to someone else, you will gain a sense of what you need to include to make your message clear and complete. Beyond that, you can ask for advice about what to say. This advice will be especially valuable if the other person knows your readers well enough to help you understand how they will react to your communication.

The following suggestions will help you get good advice about what you should say:

1. Explain to the other person the purpose of your communication. If you haven't yet settled on your purpose or learned about your readers, ask for the other person's help with these matters as well.

2. Ask the other person, "What do you think I might say to these readers that would help me achieve my purpose?"

3. Keep asking, "What else could I say?" You will find that conversations about the contents of your communications move naturally toward the critical activity of evaluating the ideas that you and the other person have come up with. That evaluation will be very helpful to you, but (as explained in the preceding chapter) it can stem the flow of new ideas about what to say. By repeatedly asking "What else could I say?" you turn your attention back to the creative activity of generating additional ideas about what you can include.

4. Take notes, either during the conversation or immediately afterwards. Otherwise, you risk forgetting your good ideas.

5. When you and the other person seem to have run out of ideas, review the things you have said. You can do that by reading back through your notes. This review can suggest additional ideas.

Like the technique of asking your readers what you should say, that of talking with someone else enables you to obtain advice by conversing directly with other people. The next two techniques enable you to obtain other people's assistance *indirectly* by looking at what they wrote when facing situations similar to yours.

## LOOK AT SUCCESSFUL COMMUNICATIONS USED IN SIMILAR SITUATIONS

On the job, certain general situations arise repeatedly. Over and again, one person needs to tell someone else how to run a machine or a computer program. Over and again, an individual or group seeks approval or funding for a project or purchase. Consequently, one way to get ideas about what to say to your readers is to look at communications that other writers have written when facing a similar situation. If you are going to write to people in your own organization, it may be very easy for you to find such samples. The following advice will help you make good use of communications when you are deciding what to say:

1. Review your purpose and your understanding of your readers.

2. Find sample communications written in situations that resemble yours as closely as possible.

3. Carefully define the purpose and readers for each sample communication. If you aren't sure of its purpose and readers, make a guess after reading it carefully.

4. Evaluate each sample communication. Find out how well it worked at achieving its purpose. You would be foolish to rely for advice on a communication that didn't succeed. You may be able to ask other people who were familiar with the communication about how it worked. If not, your common sense and careful thought will allow you to make a good guess about its level of success—or lack of it.

5. Compare the purpose and readers of the sample communication with the purpose and readers for your communication.

6. Based upon your understanding of the differences in purposes and readers between the sample communication and your communication, decide how to change the content of the sample to fit your situation.

When you are looking at sample communications, it is crucial for you to remember that you are using them to give you ideas. You should not try merely to imitate them. Thus, Number 6 in the above list is absolutely essential: adapt the ideas you get from the samples to your own situation.

## LOOK AT CONVENTIONAL SUPERSTRUCTURES

Another way to obtain ideas based upon the experiences of other writers is to look at conventional superstructures. As explained in the introduction to Part III, superstructures are overall patterns for writing in commonly occurring situations. There are conventional superstructures for giving instructions, proposing projects and purchases, reporting on progress and experimental research, and so on.

To use conventional superstructures to generate ideas about what you might say, follow these steps:

1. Review your purpose and your understanding of your readers.

2. Look for a conventional superstructure that is adaptable to your situation. Part IX of this book describes several common superstructures.

3. Adapt the conventional superstructure to your particular situation.

The last step is crucial. To use conventional superstructures successfully, you need to adapt them to your situation.

Looking at conventional superstructures is the last of the four techniques for obtaining advice from others, either directly or indirectly, about what you should say. The remaining six techniques all help you draw upon your own knowledge and creativity.

## BRAINSTORM

As explained at the beginning of this chapter, when you brainstorm you generate ideas as rapidly as you can through the spontaneous association of ideas. Brainstorming is especially helpful when you have very little under-standing at first about what you need to tell your readers, or when you have been thinking a great deal about a subject and now want to begin to shape your thoughts into a communication. Brainstorming is helpful whether you are deciding what to say in a whole communication or in part of one. It also works very well when you are working on a group-writing project. All the group members can brainstorm aloud together. The ideas offered by one person often suggest additional ideas to other group members.

The usual procedure for brainstorming is as follows:

1. Review your purpose and your understanding of your readers.
2. Ask yourself, "What do I know about my subject that might help me achieve my purpose?"
3. As the ideas come, write them down as quickly as you can, using single words or short phrases. As soon as you have noted one idea, move on to the next.
4. When your stream of ideas runs dry, read back through your list to see if your previous entries suggest any new ones.
5. When you no longer have any new ideas, gather related items in your list into groups to see if this activity prompts new thoughts. This grouping of related items can be the first step towards organizing your material, which is the subject of the next chapter. While brainstorming, however, don't focus so much on the final structure as on looking at relationships among ideas that might suggest further ideas.

The keys to brainstorming are to write quickly and avoid evaluating or rejecting ideas that come to mind. Jot down *everything*. If you shift your task from generating ideas to evaluating them, you will disrupt the flow of free association that brainstorming relies upon.

Figure 8-1 shows part of the list of ideas generated through brainstorming by Kevin, who works for Meditech, a company that makes machines that keep surgery patients alive during organ transplants. Kevin is planning to recommend that Meditech use new procedures for ensuring that every one of its machines is in perfect working order when delivered to a hospital. After running dry of ideas while making the list shown in Figure 8-1, Kevin began to group items in the list, as suggested in Step 5, above. Part of the results he obtained are shown in Figure 8-2.

## WRITE A THROW-AWAY DRAFT

Writing a throw-away draft is very much like brainstorming. You sit down and write out your ideas as they come to you. As you write, you will often refine your sense of what to say. New ideas and new lines of thought will occur to you. You write them down without trying to make them fit into the material you have already written.

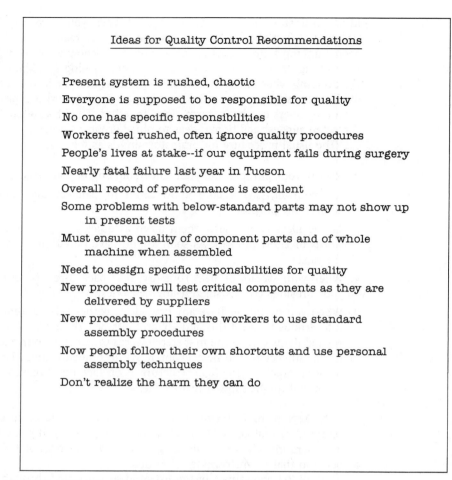

Figure 8-1. A List of Topics Produced by Brainstorming.

The resulting draft will be chaotic, but that is fine. You are trying to follow and develop your thoughts, not to produce a draft you can revise directly into a finished communication. Once you have run out of ideas, you can read back through your material to find the ideas that are worth including in your communication. The rest you throw away.

Throw-away drafts can be especially helpful when you are trying to develop your main points. They are not as good at helping you identify minor points. You can probably use these drafts best when you are writing brief communications or the parts of a long communication where you have yet to decide how to interpret your data or state your recommendations. It would not be efficient for you to write a throw-away draft for an entire long communication.

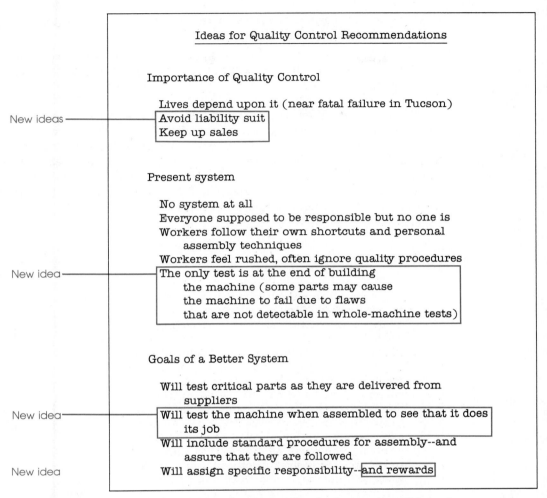

Figure 8-2.   Grouping of Ideas Produced by Brainstorming. (Note the new ideas suggested to the writer by the grouping.)

The procedure for writing a throw-away draft is very much like that for brainstorming.

1. Review your purpose and your understanding of your readers.

2. Ask yourself, "What do I know about my subject that might help me achieve my purpose?"

3. As the ideas come, write them down as sentences. Follow each line of thought until you come to the end of it, and immediately pick up the next line of thought that suggests itself.

4. Write without making corrections or refining your prose. If you think of a better way to say something, start the sentence anew.

5. Don't stop for gaps in your knowledge. If you find that you need some information you don't have, note the place you would use it and then keep on writing.

Figure 8-3 is a throw-away draft written by Miguel, an employee of a company that makes precision instruments. Miguel has spent two weeks investigating technologies that might be used to design instruments that will be placed aboard airplanes for detecting microbursts and wind shear, two dangerous atmospheric conditions that have been blamed for several crashes. Miguel wrote the throw-away draft shown in Figure 8-3 when trying to decide what to say in the opening paragraph of his report.

Wind shear and microbursts have been blamed for several recent crashes. (Find out which ones--Dallas??) People are studying several technologies for detecting these conditions. Then pilots can fly around them. There is also much interest in using these or other technologies for instruments to be used on-board planes. These would be much more helpful to pilots. Four technologies are currently being studied. The equipment would need approval of the Federal Aviation Administration, which is eager for such devices to be developed. Airlines would be eager to buy a reliable system as soon as it became available. The key to all devices is to allow pilots to detect in advance the special conditions that give rise to wind shear and microbursts, both very dangerous conditions that can overwhelm a plane that is landing or taking off. A key point is: we can make a lot of money if we can develop the right instrument first. We need to pick the best one and develop it. Criteria are: effectiveness, readiness for commercial development, and likely competition.

**Figure 8-3.**   Throw-Away Draft.

Miguel's throw-away paragraph is a jumble—but that is not a problem. The paragraph is a place for him to work out his ideas; it is not a draft he would try to develop into a finished product. Figure 8-4 shows the first paragraph that Miguel subsequently wrote. Note that it fills out some ideas found in the throw-away draft and that it omits others. It represents a fresh start—but one built on the thinking Miguel did while writing the throw-away draft.

> We have a substantial opportunity to develop and successfully market instruments that can be placed on-board airplanes to detect dangerous wind conditions called wind shear and microbursts. These conditions have been blamed for several recent air crashes, including one of a Lockheed L-1011 that killed 133 people. Because of the increasing awareness of the danger of these wind conditions, the Federal Aviation Administration is supporting considerable research into a variety of technologies for detecting them. Most of this research concerns systems that are placed at airports, but both the FAA and the airlines agree that systems that could be placed on-board airplanes would be even more helpful.
>
> In this report, I will review the four major technologies that are now being studied. About each, I will answer four questions:
>
> How does it work?
>
> Will it result in a product that will perform well enough to attract substantial sales?
>
> How long until the technology will be refined enough to make an effective instrument?
>
> If we develop such an instrument, what will our competition be like?

**Figure 8-4**    Introduction Written from Ideas Generated in the Throw-Away Draft Shown in Figure 8-3.

## MAKE AN IDEA TREE

An idea tree is a sketch of the various topics and subtopics you might discuss in your communication. It provides you with a structured list of things you might say in your communication. The following diagram shows what an idea tree looks like:

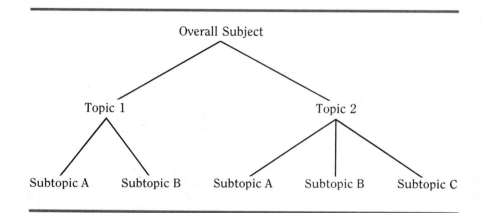

To create an idea tree, use the following procedure:

1. Review your purpose and your understanding of your readers.
2. At the top of a sheet of paper, write the overall subject of your communication. (Some people prefer to draw their trees from side to side, beginning at the lefthand margin.)
3. Ask yourself, "What are the major topics I might take up in my communication?"
4. List those topics across the page below the subject. Don't worry about the order of the topics.
5. Draw a line from each topic to the overall subject.
6. For each of the topics, ask yourself, "What are the major subtopics I might take up when I discuss this topic in my communication?"
7. List the subtopics in a row below the topics. Leave as much space as you can between the subtopics.
8. Draw a line from each subtopic to its major topic.
9. Repeat this process until you have run out of ideas. Don't worry if some of your branches have a different number of levels than do others. Take each branch as far as you can, whether to the topic level, subtopic level, or some lower level.

Figure 8-5 (page 156) shows an idea tree written by Carol, who works for an engineering firm. Carol leads a team that is helping a small city look for places it can drill new wells for its municipal water supply. The tree represents her thoughts about what she might say in her report to the city. Notice how easily Carol can add details to the tree as she thinks of additional things to say.

# MAKE AN OUTLINE

Most people view outlines as tools for organizing information that they have already decided to include in their communications. However, you can also use outlines to help you decide what to say.

The procedure for making an outline is similar to that for making an idea tree.

1. Review your purpose and your understanding of your readers.
2. At the top center of a sheet of paper, write the subject of your communication.
3. Ask yourself, "What are the major topics I might take up in my communication and in what order might I want to present them?"
4. List those topics in order along the left-hand margin. Leave a large amount of blank space between the topics. For the purpose of generating ideas about your communication, you don't need to label your topics with numbers.
5. Ask yourself, "What are the major subtopics I might take up in my communication and in what order might I want to present them?"
6. List each subtopic in order, slightly indented underneath its major topic.
7. Repeat this process until you have run out of ideas.

To enjoy the greatest help from outlining, remain flexible. If you find that you have overlooked a major topic, add it. If you find that you need to change the order of topics or subtopics, draw arrows to rearrange them. If you think of a better overall structure, begin again. If you think of ideas but can't find a place for them, jot them down along the margin so that you won't forget them. Remember that your major purpose is to generate ideas, so don't worry about the formalities of outlining (such as using roman and arabic numerals to label topics and subtopics).

Figure 8-6 (page 157) shows what Carol's notes for her report on the well drilling would have looked like if she had used an outline instead of an idea tree. You can see where she made marginal notes about the information she would include in some of the sections.

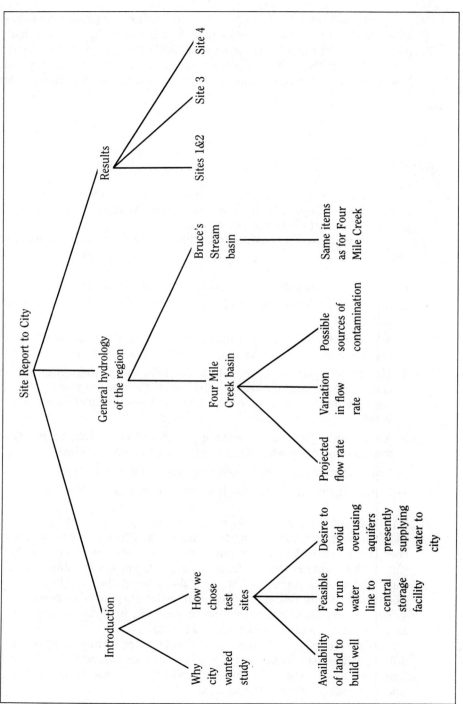

**Figure 8-5.** Idea Tree.

SITE REPORT TO CITY

Introduction

Why City Wanted the Study

How We Chose the Sites
Land available to build well
Feasible to run line to city's central water supply
Sites that would not overuse the aquifers that
currently supply water to the city. ← *Get data about capacity and use of existing aquifers.*

General Hydrology of the Region

Four Mile Creek Basin
Projected flow rate
Projected variation in flow
Possible sources of contamination ← *Mention that Delta-T plant does not pose a contamination threat.*

Bruce's Stream Basin
Projected flow rate
Projected variation in flow
Possible sources of contamination

Results of Drilling
Sites 1 and 2  ← *Say that both have too little flow to justify detailed discussion.*

Site 3

Site 4

**Figure 8-6.** Outline Used to Generate Ideas (same content as shown in Figure 8-5).

## DRAW A FLOW CHART

Flow charts can help you whenever you are trying to decide what you might say about a succession of events. That might happen, for instance, when you are describing a process, instructing your readers how to perform a series of steps, or proposing a sequence of actions. To generate ideas about what you might say, you can draw a flow chart in the following simple form:

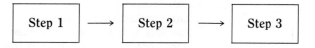

Here are some suggestions about using flow charts to generate ideas about your subject:

1. Review your purpose and your understanding of your readers.
2. List (perhaps mentally) the steps or activities in the process or procedure.
3. Draw the flow chart, leaving lots of blank space around each box, especially above and below.
4. Brainstorm about the things you might want to say about each activity or step. Write your ideas above or below the boxes.
5. When you no longer have additional ideas about a step, review what you have written about the others. Often a piece of information you might provide about one step or activity will suggest a parallel piece that you could provide about others.

Figure 8-7 shows a flow chart that Kevin used to figure out what to say about the new quality-control procedures that he wanted to propose to his employer, a manufacturer of medical machinery.

## MAKE A MATRIX

A matrix is a special kind of table that writers can use to decide what to say when they are comparing two or more things or are treating them in a parallel fashion. Figure 8-8 (page 160) shows the basic form of a matrix.

Here are the steps in using a matrix:

1. Review your purpose and your understanding of your readers.
2. List (perhaps mentally) the topics you will discuss and the general items or issues you want to treat when discussing each topic.

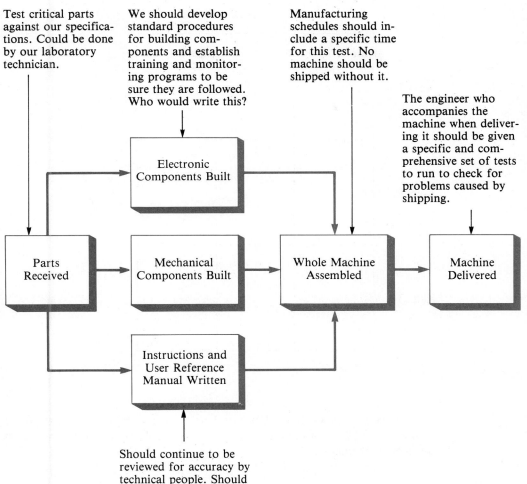

Test critical parts against our specifications. Could be done by our laboratory technician.

We should develop standard procedures for building components and establish training and monitoring programs to be sure they are followed. Who would write this?

Manufacturing schedules should include a specific time for this test. No machine should be shipped without it.

The engineer who accompanies the machine when delivering it should be given a specific and comprehensive set of tests to run to check for problems caused by shipping.

Electronic Components Built

Parts Received

Mechanical Components Built

Whole Machine Assembled

Machine Delivered

Instructions and User Reference Manual Written

Should continue to be reviewed for accuracy by technical people. Should also be tested by *users*.

**Figure 8-7.**   Flow Chart Used to Generate Ideas.

3. Draw a matrix, listing the items down the left-hand side and the issues across the top. Make as much room as possible in the matrix to write down your ideas.

4. Brainstorm to determine what you could say in each of the boxes of the matrix. Write your ideas in the appropriate box.

5. When you no longer have additional ideas for a box, review what you have written in the others. Often a comment that you make in one box will suggest a parallel piece of information that you might enter in others.

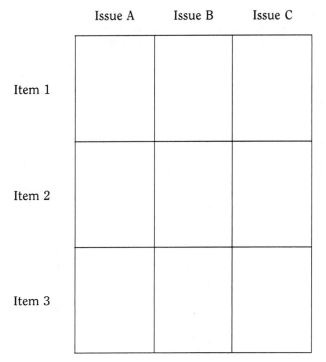

**Figure 8-8.** Basic Form of a Matrix.

Figure 8-9 shows the matrix that Miguel used to figure out what points to make about each of the technologies he wanted to discuss in his report on airplane instruments.

## EXERCISES

1. Choose a concept, process, or procedure that is important in your field. Imagine that one of your instructors has asked you to explain it to freshmen in your major.
   A. Use brainstorming to generate a list of things you might say in your talk.
   B. Use an idea tree, outline, flow chart, or matrix to generate a list of things you might say.
   C. Compare the two lists. What conclusions can you draw about the strengths and limitations of each technique?
2. This exercise will give you experience in using brainstorming in a group. Imagine that you and two or three of your classmates have been asked by the admissions office at your college to write a brief essay telling high

| SYSTEM \ TOPIC | How It Works | Technical Limitations | Readiness for Commercial Development | Existing or Anticipated Competition | Recommendation |
|---|---|---|---|---|---|
| **Air Speed Sensor Systems** | Sensors measure air speed at various places on the plane (nose, wing, tail, etc.). Large differences indicate wind shear. | Works only when plane is already in wind shear. | Now being used on some planes. | Sperry Delco Electronics. | Only if we can improve on presently marketed designs. |
| **Infrared Detectors** | Often wind shear raises temperature slightly. Infrared devices could detect rising temperatures. | Sometimes temperature may not rise. | Tests by Federal Aviation Administration have been suspended. | None expected. | Not worth pursuing. |
| **Doppler Radar** | New radar techniques can detect rapidly rotating masses of air, like those that accompany wind shear. | Technology still being researched. | Perhaps two more years of technical development needed. | None known. | Very promising. Should start preliminary design immediately. |
| **Laser Sensors** | Sudden wind shifts affect reflectivity of air. Lasers can detect that. | Limited to a 20-second warning for jet traveling at typical speed. | Ready. | Perhaps General Electric. | Reasonable to develop if we can do so quickly. |

Figure 8-9.   Matrix Used To Generate Ideas.

school juniors what they should do to prepare themselves to do well in their college studies. As a group, take ten minutes to brainstorm a list of things you might say in that essay. Then, as a group, generate a list of observations about how brainstorming works for groups. In what kinds of situations at work do you think it would be useful? When would it not be effective?

3. Imagine that a friend of yours wants to purchase some item about which you are knowledgeable (for example, a radio, motorcycle, sewing machine, or drafting set). This friend is unsure about what brand to buy and has asked you to write a letter giving advice. Identify two or three brands your friend might choose from and list three or four criteria your friend might use in choosing among them. Then design a matrix that you might use to decide what to say in your letter. Fill in the matrix as well as you can. The boxes you can't fill identify information you would have to find before you could write a thorough letter.

4. Imagine that you have been hired by your major department to write a brochure that will present the department in a favorable light to high school students thinking of entering your college and to students at your college who are thinking of transferring out of other majors. Using two techniques described in this chapter, generate ideas about what you might say in that brochure. As you work, keep in mind the guidelines for deciding what to say that are presented in Chapter 7.

# Organizing

## PREVIEW OF GUIDELINES

1. Arrange your material into groups meaningful to your readers.
2. Build a hierarchy by combining your groups into clusters meaningful to your readers.
3. Arrange the parts of your hierarchy into a reader-centered sequence.
4. Plan your visual aids.
5. Organize all the parts of your communication, small as well as large.
6. Adapt organizational patterns used successfully by other writers.

magine that yesterday you received a call from the dean. The school's computer has broken down and the registrar's office is too busy to compile a list that the dean wants. One of your teachers has recommended you to help the dean by making that list, which is to contain the names, addresses, majors, and class standings (freshman, sophomore, junior, senior) of all the students at your school who are taking an advanced writing course this term.

The instructors of these writing courses have agreed to help you by circulating sign-up sheets to their students. Now, with the completed sheets in front of you, you are ready to combine all the information into a single list for the dean.

How are you going to organize your list?

## VARIOUS APPROACHES TO ORGANIZING

Even in a communication as simple as a list, you have many choices. You might, for instance, take a "writer-centered" approach to organizing. That is, you might organize the material in the way that is easiest for you. In this case, you would simply copy the names in the order that they appear on the sign-up sheets.

Or you might take a "subject-centered" approach. You might say that there is one best or correct way of organizing this type of material. For instance, you might say that address lists should always be alphabetized.

Finally, you might take a "reader-centered" approach. That is, you might organize in light of the *purpose* of your communication. Imagine, for example, that the dean will use the list to send out a series of letters, one to all the seniors who are taking writing classes, another to all the juniors, and a third to all the sophomores. You would then organize the list according to the class standings of the students. On the other hand, if you knew that the dean wanted to know which departments on campus are sending the most students to advanced writing classes, you would arrange the list according to the students' majors.

As this example suggests, the reader-centered approach is the most likely of the three to produce a useful organization. You can sometimes hit upon an effective organization by choosing the one that is easiest for you or that seems always to be used with your particular subject matter. But you will create an effective organization more often by thinking about your communication's purpose and readers.

## IMPORTANCE OF THINKING ABOUT YOUR PURPOSE AND READERS

The observation that you should think about your purpose and readers when organizing your communications may seem redundant to you. After all, the most general advice offered in this book is to think about the ways you want your communications to affect your readers and then write accordingly. Thus, every chapter could begin with a statement like these: "When writing sentences, think about your purpose and readers." "When creating your visual aids, think about your purpose and readers." And so on.

Nevertheless, this is the only chapter that begins with such advice. Why does this point deserve special emphasis here? Writers are more likely to forget their purpose and readers when they are organizing than when they are performing any of the other activities of writing. When asked what kind of organization they feel they should create, they will often reply, "A *logical* organization, of course."

## A Logical Organization Is Not Necessarily an Effective One

Unfortunately, a logical organization is not necessarily an effective one. To be effective, an organization must aid in achieving your purpose. That is, it must help your readers perform the tasks they will be trying to perform while reading and it must help persuade your readers to alter their attitudes in the way you desire.

### Organization and Your Readers' Tasks

In the example at the beginning of this chapter, all the ways of organizing the list of students in advanced writing classes are logical in one sense or another. Only one, however, would be very useful to a person who wants to find the addresses of all the juniors.

In Chapter 4, you read another example that illustrates how your decisions about organization can affect your readers' ability to perform their reading tasks. In that example, Lorraine needed to write a report comparing two blast furnaces that a steel mill might build. She discovered that she might choose either of two organizational patterns shown in Figure 9-1.

In the first pattern, Lorraine would provide all the information about the first furnace and then all the information about the second furnace. In the second pattern, she would talk about the cost of the two furnaces, then the efficiency of the two furnaces, and so on.

These two organizations are equally logical. Yet the second one is much better at enabling the reader to compare the two blast furnaces on a point-by-point basis. Thus, by thinking about her reader's task, Lorraine could choose the most effective organization in a situation where logic alone was no help.

### Organization and Your Readers' Attitudes

Your communication's organization also affects its ability to change your readers' attitudes in the way you want. Consider, for instance, the situation

| Divided Pattern | Alternating Pattern |
|---|---|
| Furnace A | Cost |
|    Cost |    Furnace A |
|    Efficiency |    Furnace B |
|    Installation Time | |
|    Air Pollution | Efficiency |
|    Etc. |    Furnace A |
| |    Furnace B |
| Furnace B | |
|    Cost | Installation Time |
|    Efficiency |    Furnace A |
|    Installation Time |    Furnace B |
|    Air Pollution | |
|    Etc. | Air Pollution |
| |    Furnace A |
| |    Furnace B |
| | |
| | Etc. |
| |    Furnace A |
| |    Furnace B |

**Figure 9-1.**    Two Organizational Patterns Lorraine Could Use.

of Scott, who works for a company that manufactures and supplies hundreds of parts used by companies that drill gas and oil wells. Scott has thought of a new system his employer might use for organizing and retrieving parts from several warehouses. Compared with the current system, it will save both time and money. As Scott sits down to write a memo recommending his system, his first instinct is to organize in the following way:

    I. State his general recommendation.
   II. Explain why his way of warehousing is superior to the present way.
  III. Explain in detail how to implement his system.

As you think about that organization for a minute, you will see that it is well suited to a situation in which Scott's readers already feel that there is a problem with the present warehousing system and have been looking for a good alternative. If Scott knows that his readers want a new system, he can begin by describing his alternative and then explain why it is good. He can concentrate on shaping his readers' attitude toward his proposed method, without worrying much about their attitude toward the present system.

Suppose, however, that Scott is writing to readers who believe the present system is fine. He cannot persuade his readers to accept his suggestion without first persuading them that the present system should be abandoned:

I. Discuss the shortcomings of the present system.
II. Discuss the benefits of changing systems.
III. Make his general recommendation.
IV. Explain in detail how to implement his recommendation.

Both ways of organizing Scott's memorandum are logical. However, one is better suited to some situations and the other is better suited to other situations. To choose between these ways of organizing, Scott must think about his purpose and readers. Similarly, to organize your communications in ways that are not only logical but also effective, you must think constantly about your purpose and readers.

## Arrange Your Material into Groups Meaningful to Your Readers

Guideline
**1**

The essential act of organizing is to group related things together. For example, you organize the dishes in your kitchen by grouping all the drinking vessels together, all the plates together, and so on. In much the same way, when you organize a communication, you group its contents. You place closely related pieces of information together, setting them apart from other, less closely related pieces.

### Grouping Helps Readers Find Key Information

From your readers' point of view, such grouping is very important. To see why, look at the memo shown in Figure 9-2. This memo was written by Benjamin Bradstreet, who works in the New York office of a company that builds factories in other countries. In the memo, Benjamin relays information given to him by Dick Saunders, who called from Manila, where he is in charge of the construction of a factory for making roofing shingles.

Benjamin decided to spend as little time as possible organizing his memo. Rather than organize the information given to him by Dick during their long, rambling telephone conversation, he simply included each piece of information in the order in which he'd received it. By avoiding the task of grouping related facts, Benjamin saved himself some work.

What makes easy work for a writer, however, often makes hard work for a reader. Imagine, for instance, that you are one of the readers of Benjamin's memo. Further, imagine that your task in reading is to find out what things Saunders needs in Manila so that you can send them to him. Take a minute now to read the memo to find that information.

Did you find it all?

INTERNATIONAL MANUFACTURING, INCORPORATED

Interoffice Correspondence

DATE:  7 July 19--

FROM:  S. Benjamin Bradstreet, Marketing Department

TO:  Tom Wiley, New Factory Development Department

Dick Saunders called us on Tuesday, 5 July, from Manila, requesting some product information, an old formula, and information about the status of shipments to him. He also told us his return schedule.

He plans to return to the States on 12 July, arriving in Chicago and then going directly to North Hampton, Massachusetts. He will stay at the Colonial-Hilton, (413) 586-1211, with Mr. Rossini of the Appliance Factories Group (13-15 July).

A Victory luncheon is planned in Manila for Friday, 8 July, based upon the anticipated success of our roofing factory there. General Tobias, the equivalent of our Secretary of HEW, is pleased with the factory.

Dick requested information to be sent by cable. This includes when and how the Osprey drawings and Quikmold release agent were shipped. He also wants to know the exact cost of the release agent because it will be paid for with a check that he will carry back with him.

Dick said that Snyder and Leigh, of the American consulate, have been replaced by Wilson F. Brady. In future correspondence, we can use his name as a contact.

Dick said they were having considerable success in making the shingles at the new roofing plant. The Filipinos will be able to take over full operation of the plant in a few days. These are the results that please General Tobias so.

Dick will return to New York on 18 July.

He said that the hard-rubber roofs installed last fall look good and that people are living in the houses. The only problem is some holes where nails were pounded through in the wrong place.

Dick is interested in a fire-retardant formula for the dry blend phenolic system, and specifications for the Slobent 37, so he can find a locally available substitute before he leaves. Please send him the answers to these questions by cable.

Figure 9-2.    Memorandum Organized in a Writer-Centered Way.

As you noticed, the author has placed the information you need in two locations, the fourth paragraph and the last paragraph. This organization could cause you two problems. First, if you read the memo hastily (as you might have to do on a busy day at work), you could overlook some of the needed information. Second, this organization makes it difficult for you to use the memo as a checklist—for instance, to be sure that you have assembled all the things Saunders needs.

To make this a more helpful memo, Benjamin would need to put in one place all the information about what Saunders wants sent. Similarly, he would need to group the other pieces of related information. If you do the first exercise described at the end of this chapter, you will reorganize this memo so that you can see how much improvement such grouping brings.

## Your Grouping Can Affect Your Readers' Attitudes

As you look at Benjamin's memo, notice how its organization might affect his readers' attitudes. Although Benjamin is not trying consciously to persuade his readers, he does shape the readers' impressions of him. Someone who actually had to use his memo would, most likely, become irritated at the writer who wrote in a way that is needlessly difficult to read.

Besides influencing the readers' attitudes toward the writer, poor organization may also influence the readers' attitudes toward the writer's subject. When reading a technical report or proposal, for instance, a reader might believe that an easy subject is difficult to understand because the writer has organized the communication in a way that makes understanding difficult.

## Choose Carefully Your Method of Grouping

As the example of Benjamin's memo indicates, you will write more effectively if you group your material than if you don't. But any body of information can be grouped in several different ways. To choose the best grouping for your communications, you must think about your reader in the act of reading.

For example, consider the reports written by Gordon, who works for a chain of department stores. His job is to inspect stores in a ten-state region. When he inspects a store, he visits each of its major departments, one at a time. In each department, he studies the way the department undertakes certain activities, such as advertising, displaying merchandise, and relating to customers. Then he prepares a report in which he tells people at corporate headquarters what he found.

Because Gordon conducts his inspections one department at a time, it might seem to make sense for him to group the information in his reports by department, rather than by activity. However, his readers consist of specialists in the various activities Gordon considers in his inspections. Each reader wants to understand the store's overall performance in his or her specialty. If Gordon were to organize his reports by department, these readers would have a difficult time performing their reading tasks. Each specialist would have to piece together information from many sections—the section on the appliances department, the section on the women's clothing department, and so on. To avoid creating such difficulties for his readers, Gordon groups his material by activity. As a result, his readers have a relatively easy time performing their reading tasks.

Similarly, the way you choose to group the material in your communications will make a great deal of difference to your readers as they attempt to use the communications you write.

Guideline

2

## Build a Hierarchy by Combining Your Groups into Clusters Meaningful to Your Readers

When you follow Guideline 1, you are like a person who walks into a very messy kitchen and gives it order by grouping the forks together, the glasses together, the breakfast cereals together, and so on. When you have done that, you might have dozens of little piles of items lying around on your countertops and tabletops. The next step is to make clusters, or "supergroups," of items—large groups composed of smaller groups. Thus, you might put all the eating utensils together (knives, forks, spoons), all the drinking vessels together (glasses, cups, mugs), all the dry foods together, and so on.

In much the same way, when you organize a communication you must combine your small groups of information into clusters to create a hierarchy. Consider the example shown in Figure 9-3.

The contents of this recommendation report are arranged into four major groups, designated by the Roman numerals. Further, the materials contained in two of these groups are themselves arranged into subgroups.

### Hierarchies Help Readers

Why is it important for you to organize your communications hierarchically? While reading, your readers use the information in your communications to create a mental structure that represents your meaning. They organize that structure hierarchically—even if you don't organize your communication that way.[1] By providing a hierarchical organization, you save your readers a great deal of work.

In addition, creating a hierarchy is one of the ways you can be sure that your readers devote special care to understanding and absorbing your main points. Research shows that readers are better able to recognize the main ideas of a communication that is organized hierarchically than of one that is not. Furthermore, they pay more attention to information placed high in a communication's hierarchy, they devote more time to reading it, and they recall it better. Thus, besides helping your readers read efficiently, your organizational hierarchy helps them pick out and remember your main points.[2]

PARKING RECOMMENDATION TO GREENWOOD AREA COUNCIL

I. Introduction
II. Problems
   A. Insufficient parking
      1. Shoppers complain about inconvenience
      2. Shop owners complain about lost sales
      3. Police report many parking violations
   B. Traffic congestion
      1. People circle through the area waiting for a parking space to open up
      2. Additional on-street parking impedes the free flow of traffic
      3. Though illegal, some double-parking occurs, clogging streets
III. Possible solutions
   A. Build a parking garage in the center of the shopping area
      1. Advantages
      2. Disadvantages
      3. Cost
   B. Build a parking lot on the edge of the shopping area and run shuttle buses to major stores
      1. Advantages
      2. Disadvantages
      3. Cost
IV. Recommendations

**Figure 9-3.**   Hierarchical Organization of a Report.

## Which Hierarchy Will Work Best?

You stand the best chance of creating a communication that will affect your readers in the way you desire if you place your main points at the higher levels of its organizational hierarchy and if you create its hierarchy in light of the tasks your readers will perform while reading.

Consider how Toni can take these considerations into account. She has been assigned to write the final report for a federal task force seeking to determine the effects of sulfur dioxide ($SO_2$) emissions from automobiles and factories. The report will summarize the findings of hundreds of research projects conducted over many years.

Toni might organize these summaries in many different ways. Suppose that the purpose of her report is to teach readers about the history of research in $SO_2$ emissions. Overall, her readers would want to understand the relationship over time of the various studies that Toni summarizes. To help them do that, she could organize her report chronologically, so that her chapters might have titles such as "Research in 1970–1975" and "Research in 1976–1980." Within the chapters, she might organize her summaries around areas of research, such as effects on human health, on the natural environment, and so on. Her main points—the ones she would place high in her hierarchy—would concern the contributions that various research projects made to human knowledge about the effects of $SO_2$.

On the other hand, suppose that her purpose is to help federal and state legislators evaluate the effects of $SO_2$ emissions so that they can pass appropriate laws regulating those emissions. Then, she should organize her report so that her readers could find all the information about each kind of effect in one place, regardless of when that effect was studied. And her main points would concern the potential dangers of $SO_2$ to human health and the environment. Her organization might look like this:

---

    I. Effects on Human Health
      A. Effects on skin
      B. Effects on internal organs
        1. Heart
        2. Lungs
        3. Etc.
   II. Effects on the Environment
      A. Effects on man-made objects
        1. Buildings
        2. Cars
      B. Effects on nature
        1. Lakes, rivers, and oceans
        2. Forests
        3. Wildlife
        4. Etc.

---

As the example of Toni's report shows, you should tailor the hierarchy of each communication you write to your specific purpose and readers.

## How To Create a Hierarchy

While reading this discussion of hierarchies, you may have asked, "How do I go about creating a hierarchy?" You can use either of two procedures.

In the first, you begin with the individual facts and gradually build up your hierarchy. You arrange the facts into small groups that are meaningful to your readers. Then you gather those groups into still larger groups, and so on.

In the second procedure, you begin with your overall subject and gradually break it down into a set of interrelated parts. You start by identifying the major parts of your topic. Then, you divide those parts into subparts, and so on. This procedure is much like the one used to develop idea trees (see Chapter 8).

Actually, most of the time you will use both procedures, working toward the middle. What's important isn't the method but the result: a reader-centered hierarchy of information.

You may be wondering whether or not you need to outline to create a hierarchical organization. No. A hierarchical organization can always be represented by an outline, but you do not need to make an outline to create a hierarchy. In fact, some writers find that outlines are very little help, at least when they are preparing short communications. Others find outlines to be helpful much of the time. Chapter 10 provides guidelines for using outlines in a way that will help you, not get in your way.

Guideline

3

## Arrange the Parts of Your Hierarchy into a Reader-Centered Sequence

When you are organizing, you must not only arrange your groups into a hierarchy, you must also decide which groups of material you will present to your reader first, which second, and so on. Your surest guide to ordering your groups of information is your understanding of the purpose of your communication.

Are you writing a training manual that has the purpose of enabling the reader to perform a certain sequence of steps? Then arrange your material chronologically.

Are you writing a recommendation for a new system for pricing the merchandise your employer sells? Then begin by asking what your readers' present attitudes are. If your readers are dissatisfied with the present system, you might begin with the recommendation. On the other hand, if your readers are satisfied with the present system, you might want to discuss the shortcomings of that system before offering your recommendation.

Keep in mind that sometimes readers do not want to read communications from front to back. Instead, they want to locate specific pieces of information. You can help them do that by organizing thoughtfully. For example, imagine that you are writing a computer reference manual whose purpose is to enable the reader to quickly locate information about specific commands. You might organize your material around commands and then arrange your discussions of those commands alphabetically.

Guideline

4

## Plan Your Visual Aids

At work, your visual aids will be an essential part of your communications. You will often use tables, graphs, drawings, photographs, and other visual materials to carry a significant portion of your message. Used well, they can help you communicate more clearly, usefully, and persuasively than you could with words alone. You should treat visual aids as an integral part of your organizational plans.

## Plan Your Visual Aids from the Beginning

To be sure you give full consideration to the ways you can use visual aids, you should begin planning these powerful communication tools when you are making your very first plans for writing. Doing that will help you write efficiently. You will be able to draft your prose and visuals so that they work together harmoniously from the start. That takes much less time than writing the prose, then replacing part of it with visual aids, and finally recasting some of the prose to fit the visual aids you have added.

By planning your visual aids from the beginning, you can also ensure that they will be ready when you need them. You may have to ask an artist or photographer to prepare some of them. These people will need plenty of time to do their work for you.

## Deciding Where To Use Visual Aids

A key step in planning visual aids is deciding where in your communication to use them. Chapter 19, "Using Visual Aids," provides detailed advice about looking for those places. Often, however, you won't need anything more than common sense to find them. The important point here is that your visual aids are part of your communication's organization—and you should begin planning how and where you will use them even as you are planning the other aspects of your organization.

How can you do that? First, as you are deciding upon the organization of your communication, set aside time specifically for finding places where you will want to use visual aids. Second, if you are making written notes about your organization, include notes about your plans for your visual aids. For example, after the appropriate entries in an outline, list in parentheses the visual aids you will use.

In some situations, you may actually want to focus your initial planning effort on your visual aids. Some companies do that, for instance, when preparing proposals. They begin planning by listing the visual aids they will use. These include such things as diagrams of their engineering designs, tables summarizing the organization's past successes, drawings that explain important design features, and charts showing the management structure, schedule, and budget for the project. After they have listed and even drafted these visual aids, the writers turn their attention to their prose, deciding what points the prose should make in support of the visual aids.

Even if you don't make visual aids the focus of your efforts to organize your communications, you should still be sure to include them in your plans from the very start.

## Organize All the Parts of Your Communication, Small as Well as Large

Guideline

5

To write well, you must apply the four guidelines you have just read to all parts of your communication, the small as well as the large. One of the most common causes of ineffective writing is that the writers organize only at the broadest level. As a result, although the overall arrangement of their chapters or sections looks good, the material within those chapters or sections is difficult to understand and use.

This is especially likely to happen when writers are supplied an organization by their employers or when they are using a conventional superstructure. A typical outline of this kind follows:

---

    I. Introduction

   II. Literature Survey

  III. Research Method

  IV. Results

   V. Discussion

  VI. Conclusion

---

Some writers imagine that when they begin with an outline like this, all of their organizational problems are over: the whole organization has been given to them. Nothing could be further from the truth. They still have to organize each of the sections of the communication if they are to succeed in writing a communication that will affect their readers in the way they desire.

As a general guideline, when you are organizing a short communication, you should plan the specific role of every paragraph in it. When you are organizing a longer communication, say three pages or more, you should plan the specific role for every group of two or three paragraphs.

That doesn't mean, however, that you must organize every part of your communication before you begin to draft *any* of it. Many writers find that they can write very efficiently if, after planning the overall structure, they then plan the details of only the one or two sections they will draft first. They then plan the details of the other sections when they are ready to write them.

## Adapt Organizational Patterns Used Successfully by Other Writers

Guideline

6

Although you must devise an organization that suits your specific purpose, you do not need to develop a completely novel organization for *every* communication you write. Sometimes you can adapt the organizational patterns used successfully by other writers in situations similar to yours.

There are several advantages to adapting existing organizational patterns. First, they provide shortcuts to organizing by making many organizational decisions for you. Furthermore, if your communication is at all complicated, you may find that the patterns you borrow have worked out difficulties you would have difficulty overcoming as effectively on your own. Finally, if you use an existing pattern you may be using one that your readers have seen before. Familiar patterns can help your readers find, understand, and use the information you convey.

## Finding Appropriate Patterns

Here are three ways you can locate organizational patterns you might use:

- *Look at the superstructures customarily used in situations similar to yours.* Superstructures are successful patterns for communicating in specific kinds of situations. Many of the superstructures that you will find most helpful on the job are described in Part X of this book. If you are writing instructions, a proposal, research report, progress report, or any of several other kinds of communications, look there for a description of a superstructure commonly used for it.

- *Look at conventional patterns for organizing segments.* Whereas superstructures are overall frameworks for communications, segments are the smaller units, such as paragraphs and sections, from which superstructures are built. There are many conventional patterns for organizing segments. Seven that you will find to be especially useful when writing on the job are described in Chapter 12.

- *Look at samples of communications written in similar situations.*

## The Need to Evaluate and Adapt

Whenever you find an organizational pattern that seems to be suited to your situation, you must evaluate and adapt it very thoughtfully. The conventional superstructures and conventional patterns for segments are *general* patterns that you must tailor to your *particular* purpose and readers. Even the sample communications you find that have been tailored to situations similar to yours will require some modification to address exactly your purpose and readers. As long as you make these alterations carefully, however, the organizational patterns used by other writers can help you immensely.

## SUMMARY

Good organization is essential to good writing. The same body of information can be organized in many different ways, many of them perfectly logical. However, a logical organization is not necessarily an effective one. Some logical ways of organizing will make your readers' work easy. Some will make it hard. In addition, some logical organizations are more likely than others to persuade readers to change their attitudes in the way you desire. Consequently, you should think constantly about your purpose and readers when following the six guidelines for organizing.

1. *Arrange your material into groups meaningful to your readers.* By grouping your material, you help your readers understand and use the information you provide. To decide which groupings will be most helpful, think about your purpose and readers.

2. *Build a hierarchy by combining your groups into clusters meaningful to your readers.* A hierarchy is an interrelated set of groups and subgroups of information. We think in hierarchies. Therefore, when you organize your material hierarchically, you are presenting your material in a way that readers will find to be both understandable and helpful.

3. *Arrange the parts of your hierarchy into a reader-centered sequence.* To determine the order in which you arrange your material, consider your reader in the act of reading. Choose the order that will help your readers most and that is most likely to affect your readers' attitudes in the way you desire.

4. *Plan your visual aids.* In the communications you write at work, visual aids will often carry a major portion of your message. By including visual aids in your plans from the beginning, you ensure that you will focus sufficient attention on using them effectively and that they are ready by the time you need them.

5. *Organize all the parts of your communication, small as well as large.* Even if you carefully organize the large parts of a communication, you will write ineffectively if you overlook the need to organize the small parts. Readers benefit from careful organization at every level.

6. *Adapt organizational patterns used successfully by other writers.* There is no sense in devising an original organization if you can readily find one developed by other writers that will convey your message effectively. Consider conventional superstructures that are suited to your situation, conventional patterns for organizing segments, and specific communications that were written in situations similar to yours. In all cases, you will have to adapt the patterns you find to suit your purpose and readers.

## NOTES

1. Teun A. van Dijk, *Macrostructures* (Hillsdale, New Jersey: Erlbaum, 1980).
2. Thomas N. Huckin, "A Cognitive Approach to Readability," in *New Essays in Technical and Scientific Communication: Research, Theory, Practice,* ed. by Paul V. Anderson, R. John Brockmann, and Carolyn R. Miller (Farmingdale, New York: Baywood, 1983), 90–108.

## EXERCISES

1. Following the guidelines in this chapter, reorganize the memo from Benjamin Bradstreet (Figure 9-2). What are the major groups of information? Do any of the groups fit within larger groups? What would be a good

sequence for this information? When you have finished reorganizing, compare your organization with those made by other students.

2. Imagine that you work for a large company with offices scattered throughout North America. To communicate with people at other offices, the company's executives rely heavily on telephone calls and travel.

At the request of your supervisor, you have been studying videoconferencing systems. Videoconferencing allows groups of people at different locations to see and hear one another talk—as if they were speaking on telephones that also had television pictures.

Your supervisor wants to know whether or not your company should invest in a videoconferencing system. While reading your report, she will be comparing videoconferencing with the company's present means of communication.

Using the guidelines in this chapter, organize the information below. You do not need to rewrite the sentences. Just designate them by their letters.

A. Videoconferencing reduces the need for travel. People in different locations can meet face-to-face without being in the same room or city.

B. On average, videoconferencing systems cost about $12,000 to install per location.

C. In most major cities, hotels have videoconference rooms. They rent these rooms by the hour to companies that occasionally need such facilities.

D. Videoconferencing may inspire unnecessary and unproductive meetings. That happens because meetings can be arranged and held quickly and easily.

E. Videoconferencing involves the use of conference rooms in each location that are equipped with cameras, viewing screens, and microphones.

F. Last year, your company spent $150,000 on travel expenses.

G. Companies have found that videoconferencing speeds up the decision-making process. Meetings can be held as soon as they are needed.

H. The success of videoconferencing depends upon the willingness of people in the company to use it.

I. Some companies report that videoconferencing systems encourage central management to exert too much control over branch locations.

J. Reduced travel means reduced expenses: airfare, car rental, hotel bills, and meals.

K. Videoconference systems sometimes cause morale to suffer. Executives miss the personal contact with personnel in other locations.

L. Videoconferencing allows many people to be involved in making decisions. This supports a more participative management style.

M. Some executives resist videoconferencing. They are accustomed to the prestige of large travel budgets and expense accounts.

N. Executives often return from business trips too tired to work efficiently.

O. One of your company's long-term goals is to implement more participative management practices.

P. When many people are involved in making decisions, the decisions are often better than those made by few people.

Q. One company reports that it spends at least $2500 per year on maintenance for each videoconference room.

R. Executives are relatively unproductive during the time they are traveling.

# Using Outlines

## PREVIEW OF GUIDELINES

1. Outline to find the best content and organization for your communication.
2. Outline to check the content and organization of your draft.
3. Outline to obtain approval of your writing plan.
4. Outline when you are on a writing team.
5. Use a sentence outline to sharpen the focus of your writing.
6. Change your outline when circumstances warrant.
7. Don't outline when outlining won't help you.

Sonia sits at her desk, worried. Having been on the job for four weeks, she is preparing to write her first monthly report. She wants to impress people in her company with this report, so she wants to do everything right. Now, she is trying to decide whether she should start her writing by creating an outline.

In school, some teachers told Sonia to start *every* writing assignment by making an outline. However, she never liked outlining, and she made outlines only when required to. Even though she usually received good grades anyway, she is now worried that her teachers may have been correct. Maybe she needs to outline in order to write this monthly report well. On the other hand, she can't spend very much time writing this report. If she outlines without really needing to, she will waste time that she should be spending on an important project she must complete as soon as possible.

Sonia's uncertainty about outlining is understandable. Outlining can help writers in some situations, but in other situations outlining is a waste of time.

This chapter presents seven guidelines to help you decide when to outline—and when not to. Further, it presents some advice about how to obtain the greatest benefit from the outlining you do.

## OUTLINING VERSUS ORGANIZING

Before reading the guidelines, you may find it helpful to think briefly about the relationship between outlining and organizing. Though the two are closely related, they are not the same thing.

*Organizing* is the activity through which you structure your writing. *Outlining* is but one *technique* you can use to organize. There are many others. You can, for instance, adapt the organization of similar communications, rely on your intuition, or draw idea trees or flow charts.

Some people who fail to distinguish outlining from organizing mistakenly think that when writers don't outline, they are consequently neglecting to organize. That may explain why some of Sonia's teachers told her that she should outline everything she writes.

A related error made by some people who don't distinguish outlining from organizing is to think that because they have created an outline they have necessarily created a good organization. Although it is true that outlining helps writers think productively about organization, outlining does not, by itself, ensure good organization. The guidelines for creating good organization are given in Chapter 9. You should follow those guidelines whether you use outlining or some other technique to structure your communications.

## AN OUTLINE IS A SKETCH

Although outlining and organizing aren't exactly the same thing, outlining is a powerful technique for helping you organize. When you outline, you draw a special sort of sketch in which you focus only on the organization of your communication. In this sketch, you portray organization by drawing the contents of the communication in two dimensions: *sequence* (the order in which the contents will be arranged) and *hierarchy* (the relationships among the parts and subparts). As Figure 10-1 illustrates, an outline shows sequence by listing topics in the order in which they appear in the communication. An outline shows hierarchy chiefly by indenting subparts farther than it indents the larger parts that the subparts belong to.

In longer outlines, writers often add notations that help indicate sequence and hierarchy. Figure 10-2 illustrates two commonly used notational systems.

## THE VALUE OF OUTLINING

The chief value of the sharply focused sketch you create when outlining is that it enables you to make decisions about content and organization without being distracted by any of the other aspects of writing. For example, you can organize something you are writing without worrying about the particular words you will use to communicate your message. Or, you can consider the effect upon your reader of various changes in your organization without needing to write your entire communication each alternative way.

Because outlining enables you to focus so sharply on organization, it is also a very good means of *learning* how to improve your skills at organizing. For that reason, your writing instructor may ask you to write many outlines, which he or she may also ask you to discuss in class or conferences. Through

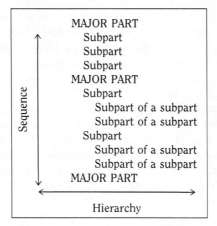

Figure 10-1. Two Dimensions of an Outline.

```
        ROMAN SYSTEM              DECIMAL SYSTEM
        I.  ....                  1.0 ...
           A.  ....                  1.1 ...
           B.  ....                  1.2 ...
           C.  ....                  1.3 ...
        II. ....                   2.0 ...
           A.  ....                  2.1 ...
              1. ....                   2.1.1 ...
              2. ....                   2.1.2 ...
           B.  ....                  2.2 ...
              1. ....                   2.2.1 ...
              2. ....                   2.2.2 ...
        III. ....                  3.0. ...
```

**Figure 10-2.**   Two Notational Systems for Outlines.

these activities and through studying the seven guidelines in this chapter, you will learn various ways to use outlining to improve your effectiveness and efficiency as a writer.

Guideline

## Outline To Find the Best Content and Organization for Your Communication

# 1

The most obvious use of outlining is to help you plan your organization before you begin to write. When an effective organization isn't immediately obvious to you, you can often save yourself a great deal of work by outlining. Consider the example of Ellen.

Ellen has just finished two days of writing. She has produced an eight-page draft in which she reports on a project she recently completed. Even as she was writing this draft, she realized that its organization wasn't working. While writing page 5, she found that she was saying something that would have helped her readers understand some of her earlier points—if only she had put it on page 2. She also realized that her writing digressed in several places.

Now that Ellen has written a full draft, she must reorganize it. In doing so, she will have to discard or substantially revise a great deal of material that she spent considerable time writing. If she had begun by outlining, rather than by writing, she could have briefly sketched out her original ideas for organizing, perhaps using fewer than one hundred words, and certainly in fewer than eight pages. Then, by analyzing her outline from her readers' point of view, she could have detected at least some of the weak spots in her original plan. And she could have revised that plan very economically by drawing a few arrows and by adding and deleting a small number of words. Instead, she must now rewrite and discard major portions of her work.

Like Ellen, you can often write more efficiently if you write an outline before you begin to draft your communications.

## You Can Outline Parts as Well as Whole Communications

When they think of outlining, many people think of outlining *entire* communications. However, you will probably find yourself in situations where it is unnecessary to outline all of a communication, because the best general organization either is obvious or is similar to that of other communications you have written recently. Nevertheless, it still may be helpful to outline some *part* that is giving you problems.

In sum, one important use of outlining is to help you plan a good organization *before* you invest time in drafting. You can use outlining in this way with whole communications or with parts of them.

## Outline To Check the Content and Organization of Your Draft

Guideline

2

Besides helping you plan your organization in advance, outlining can help you check the organization of material you are drafting. In this case, you make an outline that describes the organization of the already drafted material and then examine the outline from the point of view of the readers you are addressing.

When might you want to outline in order to check your organization? Anytime that you are having trouble with a section or feel uneasy about it, you should ask yourself whether your difficulty could possibly be related to organization. The following problems in particular are often rooted in poor organization:

- *Repetition.* Do you seem to be talking about the same thing in two places?
- *Digression.* Do you think that you may be wandering from the main point?
- *Poor emphasis.* Are you worried that you haven't presented your key points forcefully enough?
- *Poor flow.* Are you having difficulty making smooth transitions from one paragraph or section to another?

By outlining problematic passages or communications, you can step back from the details of your writing to look at its general structure. You cease to focus on the way you have phrased particular sentences and the way you have incorporated particular facts. You start to look at the general framework of the communication. That perspective can help you identify and solve problems rooted in weak organization.

<table>
<tr><td>Guideline</td><td></td></tr>
</table>

Guideline
3

## Outline To Obtain Approval of Your Writing Plan

At work, you may need to obtain approval of your writing plan before you begin to draft. For instance, when your boss asks you to write a particularly sensitive report, he or she might ask you to discuss your plans for the report's contents and organization before you begin writing. In this way, your boss can be sure you fully understand what you are supposed to do. Similarly, if you are given an assignment that you don't fully understand, you might want to discuss your writing plan with the person who gave you the assignment. You can also do that when you are unsure about what to include or how best to present your material.

By obtaining approval of your plans before you begin to write, you reduce the risk that you will invest a large amount of time in preparing a draft that you will have to discard. You will also save other people the annoyance and lost time of reading and discussing a draft that will have to be redone.

Outlining can help you considerably when you are seeking approval for a writing plan. By handing an outline to your boss and other advisers, you will give them a very specific idea of your intentions—especially if you prepare a sentence outline (discussed in Guideline 5). After reading your outline, those people will be able to give you detailed directions about the ways (if any) they want you to change your plan.

You will find it especially important to obtain early approval of your writing plans when you are new on the job. As you gain work experience, early approval will become less important to you because you will gain a good sense of what the people you work with need and want from the things you write. It is likely, however, that throughout your career you will at least occasionally need or want to secure approval for your plan before you begin to write. In those situations, describe your plan in an outline.

Guideline
4

## Outline When You Are on a Writing Team

At work, you may be asked to work on a team of people who must write something together. The team may assign each member to write one part of the communication so that all the parts can then be assembled into a complete communication. Or, one person may draft the entire communication, which the team can then review and revise.

In either case, your team can work most efficiently if it begins its work in the same way that an individual writer should: by planning.

Planning is much more difficult for a team than for an individual. When you write by yourself, you can create and keep part or all of your plan in your head. Teams, however, have many heads. Unless the team members make explicit agreements, each member may have a different plan in mind. That can lead to serious inefficiencies and hurt feelings. After investing considerable time and energy in drafting, a team member may find his or her work rejected because other team members had in mind different, incompatible plans. If that happens, the team member's time will have been wasted, and the person's desire to work cooperatively with the other people may be greatly reduced.

You can reduce the likelihood of such difficulties by urging your team to begin by planning. If several team members are drafting different parts, your joint planning should enable each member to know what to say and how the parts will fit together. Similarly, if your team is asking one member to do all the drafting, your joint planning will help that person understand what content and organization the team expects. Outlining also gives your team an opportunity to identify—and resolve—areas of disagreement before anyone undertakes the labor of drafting. It is easier and more economical to handle disagreements among team members when forging an outline than when reviewing someone's draft.

Guideline

5

## Use a Sentence Outline To Sharpen the Focus of Your Writing

There are two types of outlines: *topic outlines* and *sentence outlines*. A topic outline uses a word or phrase to tell what *kind of material* will be included in each section and subsection of a communication. A sentence outline describes the *main point* you want to make in each section and subsection. Figures 10-3 and 10-4 show topic and sentence outlines for the same communication.

How will you decide what main points you should make in your sentence outline? If you follow Guideline 1 in Chapter 7, you will have a list of your readers' questions. The answers to those questions can form the statements made in a sentence outline. The statements can also be some of the assertions (or claims) you want to make—and support—to persuade your readers to change their attitudes (see Guideline 3 in Chapter 7).

### Sentence Outlines Help You Choose and Emphasize Main Points

Because sentence outlining focuses on the points you will be making, it can help you sharpen the focus of your writing by helping you choose and emphasize your main points. It can help in this way even if you have already developed a list of the points you want to make before you begin to outline. After making such lists some writers fail to decide which points are their main ones. Because they don't have main points clearly in mind, their writing is unfocused.

When writing a sentence outline, however, you *must* decide what your point is. Once you have stated your main point in your outline, you can place that statement in a topic sentence or other prominent location in your draft.

---

PROPOSAL FOR FUNDS TO MAKE AN INSTRUCTIONAL
TELEVISION SERIES ABOUT ENERGY AND THE
ENVIRONMENT FOR HIGH SCHOOL STUDENTS

    I. Need
       A. Students' need
       B. Teachers' need

   II. Proposed project
       A. Television shows
          1. Contents
          2. Technical specifications
       B. Teacher's Guide

  III. Plan of action
       A. Planning and scripting
       B. Field testing
          1. Location
          2. Evaluation
       C. Production
       D. Promotion and distribution

  IV. Qualifications
       A. Institutional
       B. Personnel

   V. Budget

**Figure 10-3.** Topic Outline. (Compare with the sentence outline in Figure 10-4.)

## Sentence Outlines Help You Establish a Tight Fit among Parts

Sentence outlining can also help you sharpen your prose by helping you fit the parts of your communication tightly together. Imagine, for instance, that you are writing a proposal. Most likely, you will devote an early section to describing in detail the problem your proposed project will solve. Later in the proposal, you will describe the project itself, together with its costs and schedule.

All these parts must be carefully matched. Consider, for example, the relationship between your discussions of the problem and the proposed project. When discussing the problem, you will want it to appear to be a significant difficulty that requires attention. When discussing your proposed project,

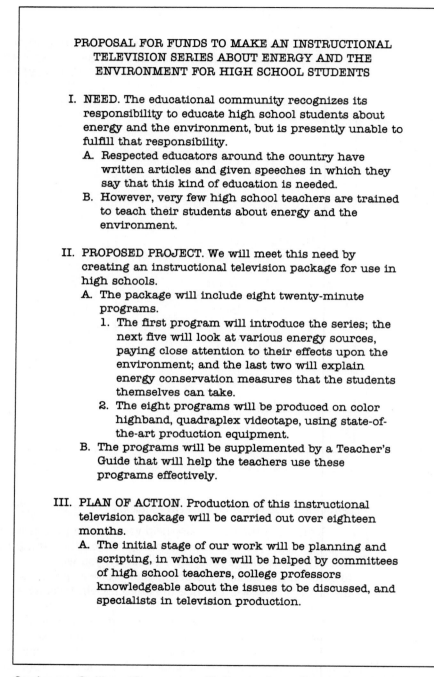

**PROPOSAL FOR FUNDS TO MAKE AN INSTRUCTIONAL TELEVISION SERIES ABOUT ENERGY AND THE ENVIRONMENT FOR HIGH SCHOOL STUDENTS**

I. NEED. The educational community recognizes its responsibility to educate high school students about energy and the environment, but is presently unable to fulfill that responsibility.
   A. Respected educators around the country have written articles and given speeches in which they say that this kind of education is needed.
   B. However, very few high school teachers are trained to teach their students about energy and the environment.

II. PROPOSED PROJECT. We will meet this need by creating an instructional television package for use in high schools.
   A. The package will include eight twenty-minute programs.
      1. The first program will introduce the series; the next five will look at various energy sources, paying close attention to their effects upon the environment; and the last two will explain energy conservation measures that the students themselves can take.
      2. The eight programs will be produced on color highband, quadraplex videotape, using state-of-the-art production equipment.
   B. The programs will be supplemented by a Teacher's Guide that will help the teachers use these programs effectively.

III. PLAN OF ACTION. Production of this instructional television package will be carried out over eighteen months.
   A. The initial stage of our work will be planning and scripting, in which we will be helped by committees of high school teachers, college professors knowledgeable about the issues to be discussed, and specialists in television production.

**Figure 10-4.**   Sentence Outline. (Compare with the topic outline in Figure 10-3.)

2

B. When scripting is completed, we will construct a
slide-tape mock-up of the series, which we will field
test.
   1. The field testing will occur in urban, suburban,
   and rural schools, to involve all the kinds of
   locations in which the series might be used.
   2. The test results will be evaluated with the
   assistance of an expert in educational research.
   3. The entire package will be produced at our own
   facilities.
   4. We will use established channels for distributing
   instructional television shows; these channels
   will make. . . .

**Figure 10-4.**   *Continued*

you will want to persuade your reader that your project is an effective way of solving the particular problem you have just described. If your discussions of the problem and the proposed project don't match, your proposal will lose its force. The reader will decide that, however important the problem might be, your project isn't a reasonable way to solve it.

Similarly, your discussion of your budget should be carefully matched to your other sections. It should, for instance, include all those items, and only those items, that are clearly related to the project you describe. Likewise, your schedule should include each of the activities necessary to carry out your project. In short, every part of your proposal must bear a clear and strong relationship to the other parts.

A *topic* outline will help you decide what kind of material to include on each important subject (problem, solution, budget, schedule, and so on). However, it may not help you write all the sections so that they fit together tightly. For instance, it will help you remember to include a budget and a schedule, but it will not help you relate the main points about the budget and schedule directly to the other sections in the proposal.

In contrast, a *sentence* outline will focus your attention on the statements you must make in each section if the section is to work harmoniously with the others. Once you have completed a thorough sentence outline, you will know precisely what major points you want to make in each section of your communication, and you will know exactly how those points contribute to your overall argument.

The topic and sentence outlines shown in Figures 10-3 and 10-4 are for the same proposal. You can judge for yourself how much more helpful the sentence outline would be for the writer.

## Guideline 6    Change Your Outline When Circumstances Warrant

An outline is a planning aid. It should help you, not tie you up. Avoid becoming bound by it: remain flexible. As you draft, evaluate, and revise your communication, it is only reasonable to expect that you will come up with ideas to improve its organization. After all, the more you think about something you are writing, the more likely you are to see ways to make it better. Also, as you draft you may discover unanticipated problems with the organization you planned. When you find that you need or have a better plan, abandon the one you had. Do not stick to a weak plan when you can make a stronger one.

## Guideline 7    Don't Outline When Outlining Won't Help You

For the most part, this chapter focuses upon situations in which outlining can help you write well. However, there are some situations in which outlining won't help you:

- When you are writing a short communication that does not require much thought about organization.
- When you are writing a kind of communication you have written several times before, so that you have already worked out the organizational issues involved with it.
- When you are confident that you can hit upon the best organization intuitively.
- When you are confident that you can organize effectively by adapting the organization of a similar communication that someone else has written.

Keep in mind that outlining may not help you if you are still unsure about the general nature of what you want to say. You may be a writer who needs to start drafting in order to figure out what you have to say—even to figure out what your purpose is. If that's the case, go ahead and draft first. *Then* outline. On the other hand, for some writers, outlining helps even when they are unsure of their general message because sitting down to outline forces them to think their message through.

There is one excuse for not outlining that you should never offer yourself. Never say that you are not going to outline because you don't have the time. When you fail to outline for that reason, you may end up *wasting* time. It can take far longer to revise a poorly organized draft than to outline the draft in advance.

## SUMMARY

When you outline, you draw a detailed sketch of your communication. In this sketch, you focus only on your communication's organization without needing to consider the other aspects of writing. By creating and evaluating this sketch, you can carefully plan an organization that will affect your reader in the way you desire.

The seven guidelines for outlining effectively are as follows:

1. *Outline to find the best content and organization for your communication.* If you begin writing without a good organizational plan, you may waste a lot of time creating a draft that you will have to revise considerably.

2. *Outline to check the content and structure of your draft.* Some writing problems have their roots in weak organization. These include repetition, digression, poor emphasis, and poor flow of the prose. When faced with such a problem, you may be able to identify and correct the underlying organizational problem very quickly by outlining.

3. *Outline to obtain approval of your writing plan.* At work, your boss or other people may want to know what you are planning to say before you

begin to draft. Also, there may be times when you want to have someone's approval of your writing plan before you invest your time and energy in writing. By presenting an outline for approval, you help others react to your plans in a specific, detailed manner.

4. *Outline when you are on a writing team.* Teams should begin their writing projects in the same way that individuals should: by planning. Outlining is an excellent tool for team planning because it allows the team to make detailed, specific plans that are agreed upon jointly.

5. *Use a sentence outline to sharpen the focus of your writing.* In a sentence outline you state the main point that you wish to make in each section and subsection. This effort can help you choose and emphasize your main points, and it can help you establish a tight fit among the parts of your communication.

6. *Change your outline when circumstances warrant.* Though advance planning is important, you should also take advantage of the insights and ideas you gain as you draft your communication. Don't be bound by your outline. Stay flexible.

7. *Don't outline when outlining won't help you.* It is a waste of time to outline when you are confident that you can hit immediately upon an organization that will need no revising. That can happen when you are writing short communications, communications of a kind you have often written before, or communications where you have an appropriate model to adapt.

## EXERCISES

1. After you have decided what to include in a project you are preparing for your writing class, make two outlines that show alternative ways of arranging this material. Write a paragraph comparing the two outlines from the point of view of the intended readers of the project.

2. Write a sentence outline for a project you are now preparing for your writing class. Be sure that sentences in your outline make clear how its parts fit together. (Look at Figure 10-4 for ideas about how you can write your sentences.)

3. Find a communication that is similar to one you are preparing to write for your class. Prepare a sentence outline of it. Write a brief description of the ways that the organization of that communication would be—and would not be—effective for the project you are going to write.

4. For this exercise, you are to work with the passage given below. From the following list of tasks, perform those assigned by your instructor:
   A. Write a topic outline for the passage.
   B. Write a brief analysis of the weaknesses of the passage that your outline reveals. (*Note:* your outline will have to include the various contents

of the longer paragraphs in the passage to uncover all the major weaknesses.)

C. Compare your outline with the outline prepared by one or more of your classmates.

D. Write an outline that shows how the material in the passage could be reorganized to better serve the needs of a reader who wants to reproduce the experiment.

E. Compare your outline with the outline prepared by one or more of your classmates.

The passage you are to work with is from a report prepared under contract from the U.S. Environmental Protection Agency (EPA). Many of the plastic parts used in new cars give off a gas, called vinyl chloride monomer (VCM), which is hazardous to human health in large doses. The EPA wanted to determine whether the level of VCM found in new cars was high enough to constitute a health hazard. The following passage describes an experiment in which the new cars were found to be safe. For the purpose of this exercise, suppose that the readers of this passage are researchers at other locations who desire to repeat this experiment using other cars.

To collect vinyl chloride monomer (VCM) from the interior of each test automobile, we used glass collection tubes purchased from SKC (Pittsburgh, PA). Each tube was filled with 100 mg of charcoal sorbent, through which we drew 8 liters of air at 50 milliliters per minute. The VCM present in the air was captured in the charcoal sorbent. We then capped the tubes, stored them in a freezer at minus 20 degrees centigrade, and shipped them to the laboratory, which performed a VCM analysis using an extraction/gas chromatographic technique.

For this procedure, we drew the air through the collection tubes using two electric Telmatic Air Samplers (Bendix Models 150 and C115), which we modified with remote controls so that we could turn them on and off from outside the test cars. We also placed the batteries used to power the pumps outside the cars. Each sampling pump was calibrated and adjusted for a constant flow rate of 50 milliliters per minute. We placed the collection tubes on the front seat next to the sampling pump. The collection tubes were attached to the air sampler by short connectors to minimize the exposure of the 8 liters of air to the connector materials before the air reached the carbon sorbent.

Each air sampler was equipped with a thermistor to measure the air temperature inside the test vehicle. An additional thermistor was used to measure the air temperature outside the car.

We placed the air sampler and thermistor on the front seat, either on the driver's side or on the passenger's side, wherever it seemed that the sun would heat the seat the most. We reasoned that sampling the hottest seat during the summertime would establish the worst-case condition for concentrations of the pollutant, VCM. Then we inserted the collection tubes on the connector to the air sampler and ran the control wires through the car window, which we quickly rolled up to the top position. The seal at the top of the window was such that the window could be tightly closed and still allow the wires to pass through.

After placing the sampling package in the cars, we waited at least 30 minutes before beginning to sample so that equilibrium could be reestablished inside the car. Because we were aware that disturbing the equilibrium would give us unreliable results, we had opened the car door carefully, slid the sampler in quickly, and closed the door with a minimum of air interchange. Further, we always used the downwind door, to minimize circulation of outside air into the car.

We activated the pumps until approximately 8 liters of air had been sampled. Then we turned the pumps off and prepared collection tubes for storage as previously described.

# Drafting
# Prose Elements

In the preceding four chapters, you learned how to plan your communications. That is, you learned how to decide what to include in your communications and how to organize that material. In the chapters that follow, you will find advice about how to turn your *plans* into full, effective *drafts*.

## THE PROCESS OF DRAFTING

When you are drafting, your ultimate objective is to prepare the final draft you will send to your intended readers. Sometimes your first draft will also be your final draft. You will sit down, write your draft, and send it off without making any changes at all. However, that will usually happen only when you are writing very short communications on routine matters. You will generally need to devote more effort to writing longer, less routine messages. In response to a survey, 122 professional and managerial personnel reported that when they prepare reports they spend an average of 3 hours per page.[1]

Why does writing at work require so much effort? At work, the standards for writing quality are high enough that even most experienced writers cannot meet them in a single draft. Consequently, you will be *expected* to write more than one draft. Sometimes, you will even be required to give your drafts to your boss and others so that they can make suggestions for improving your communication.

Similarly, the instructor in your writing class may require you to prepare more than one draft of your assignments. You could find this requirement frustrating if you are accustomed to writing your term papers and other assignments the night before they are due, so that you leave yourself no time to make a second draft. However, even if you have been earning good grades by preparing only one draft in school, it will be important for you to change your writing habits before you begin work.

## HOW TO DRAFT EFFICIENTLY

Because drafting takes so much time, you should strive to draft efficiently, so that you spend as little time and energy as possible reaching a satisfactory final draft.

### Prepare Carefully before Drafting

One way you can help yourself draft efficiently is to prepare carefully *before* you draft. If you begin by defining your objectives (as explained in Part II) and planning your communication (as explained in Part III), you will be able to create first drafts aimed specifically and accurately at achieving your aims.

### Draft in Stages

If you are like most people, you can also increase your efficiency by drafting in stages.

First, make a rough draft based upon the plans you made when following the guidelines in Part III. Concentrate on developing your general line of thought and your general strategy for expressing yourself. Do this without worrying about spelling, punctuation, grammar, or other fine points of writing. Eventually you must attend to these with the utmost care, but you will simply interrupt your creative momentum if you become too concerned about details when making a rough draft.

When you have completed your rough draft, *then* check it over, looking for ways to improve it. Based upon the insights you gain, make a second draft. Continue this cycle of drafting and evaluating until you have achieved the level of quality that you desire.

By separating your essentially constructive activities of drafting from your essentially critical ones of evaluating, you will be able to concentrate upon each more fully. Writers who suffer from writer's block or who get stuck in the middle of their drafts are often people who try to combine creativity and criticism—getting nowhere with either one.

### Learn the Guidelines for Drafting

Finally, you can increase your efficiency at drafting by studying and practicing the guidelines given in this book. With practice, you will gradually learn to apply the guidelines more or less automatically. The more skilled you become at applying them when creating your rough drafts, the less revising you will have to do before you can present your communications to your intended readers.

## ORGANIZATION OF THE CHAPTERS ON DRAFTING

When you draft a communication, you create both its *prose* and its *visual aids*. You will find advice about drafting visual elements later in this book (Part V). The eight chapters you are now about to read concern your prose. The chapters discuss in turn the largest parts of your communications (segments), smaller parts (sentences), and the smallest parts (words).

### Chapters on Segments

Chapters 11 through 15 deal with the largest parts of your communications: *segments* (which include paragraphs, sections, and chapters). Although segments vary considerably in size, the same basic guidelines apply to them all. In Chapter 11 you will learn what those basic guidelines are. In Chapter 12 you will learn how to create seven types of segments that you will find useful on the job, and in Chapters 13 and 14 you will learn about writing the beginning and ending of a communication—kinds of segments that also deserve your special attention. In Chapter 15 you will learn how to provide your readers with a "map" that shows how you have fit your segments together to build your overall communication.

### Chapters on Drafting Sentences

Chapters 16 and 17 deal with *sentences*. Chapter 16 will help you write strong, forceful sentences, and Chapter 17 explains ways to write sentences so your readers can readily follow your flow of thought from one sentence to the next.

### Chapter on Choosing Words

Chapter 18 provides advice about choosing *words* that will help you bring about the results you desire.

## CONTENTS OF PART IV

**Chapter 11**
**Writing Segments**

**Chapter 12**
**Seven Types of Segments**

**Chapter 13**
**Writing the Beginning of a Communication**

**Chapter 14**
**Writing the Ending of a Communication**

**Chapter 15**
**Mapping Your Communication**

**Chapter 16**
**Writing Sentences**

**Chapter 17**
**Guiding Your Readers**

**Chapter 18**
**Choosing Words**

## NOTE

1. Robert W. Kelton, "The Internal Report in Complex Organizations," *Proceedings of the 30th International Technical Communication Conference,* (Washington, D.C.: Society for Technical Communication, 1984), RET54-RET57.

# Writing Segments

## PREVIEW OF GUIDELINES

1. At the beginning of each segment, tell what it is about.
2. At the beginning of each segment, tell how it is organized.
3. At the beginning of each segment, generalize about its contents in a reader-centered way.
4. Put the most important information first.
5. Answer your readers' questions where they arise.
6. Provide transitional statements between segments.

In this chapter, you will learn how to write effective segments. Segments are groups of related sentences—such as paragraphs, sections, and chapters. This chapter treats all sizes of segments together because you can write any of them effectively by following the same basic guidelines.

As you will see when reading the discussions of this chapter's six guidelines, the way you structure your segments has a considerable effect, for better or worse, on the success of your writing. By following the guidelines provided here, you can make your segments much easier to understand than they would otherwise be. In addition, you can greatly increase your ability to shape your readers' attitudes in the way you desire.

## RECOGNIZING SEGMENTS

The first step in writing effective segments is to recognize them. You will need no help in spotting some segments such as paragraphs and chapters. However, others are harder to see. Some paragraphs contain more than one segment, and sometimes two or more paragraphs form a single segment that is not marked by a chapter title or section heading.

If you have prepared an outline for your communication, you can use it to help you identify your communication's segments. Look, for instance, at Figure 11-1, which shows the segments represented in an outline created by Jonathan. Jonathan works for an engineering consulting firm that advises amusement parks, museums, and similar businesses. His outline shows the hierarchy of segments he plans for a preliminary report to a zoo. In Figure 11-1, the overall report is called an A-level segment because it is a single, unified group of related sentences. Contained within the overall report are its major sections, which are B-level segments. Some of the B-level segments contain smaller, C-level segments, and so on.

When Jonathan writes his report, he may create additional segments that are not shown in his outline. When outlining long communications, writers seldom work their way down to the smallest groups of ideas. Consequently, when they draft their communications, they often break down the smallest segments identified in their outlines to several segments that are even smaller yet.

What do segments look like when written out in prose? Figure 11-2 shows the three levels of segments within a short passage from Jonathan's report.

It's important for you to be able to recognize all the segments in your communications so that you can apply this chapter's six guidelines to each of them.

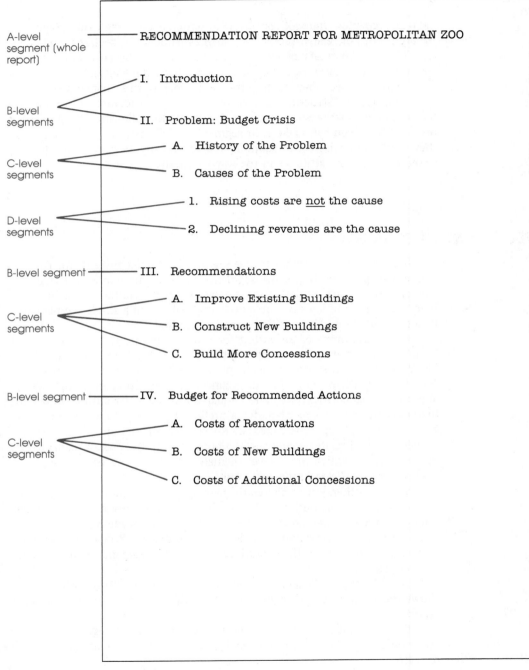

**Figure 11-1.**   The Segments in an Outline.

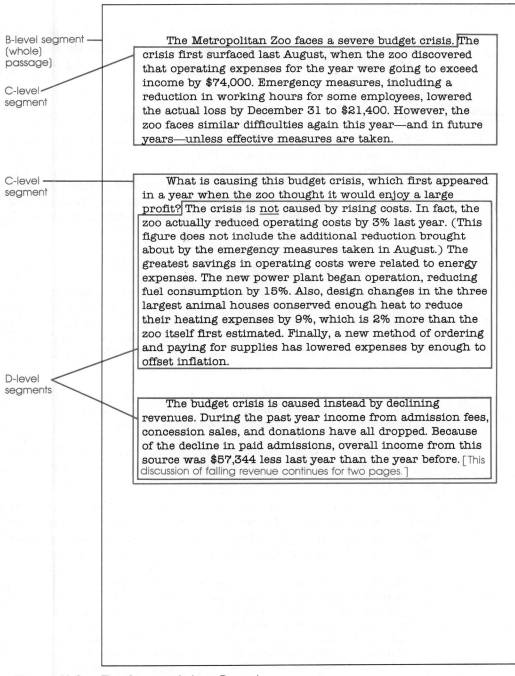

B-level segment
(whole)
passage)

C-level
segment

The Metropolitan Zoo faces a severe budget crisis. The crisis first surfaced last August, when the zoo discovered that operating expenses for the year were going to exceed income by $74,000. Emergency measures, including a reduction in working hours for some employees, lowered the actual loss by December 31 to $21,400. However, the zoo faces similar difficulties again this year—and in future years—unless effective measures are taken.

C-level
segment

What is causing this budget crisis, which first appeared in a year when the zoo thought it would enjoy a large profit? The crisis is <u>not</u> caused by rising costs. In fact, the zoo actually reduced operating costs by 3% last year. (This figure does not include the additional reduction brought about by the emergency measures taken in August.) The greatest savings in operating costs were related to energy expenses. The new power plant began operation, reducing fuel consumption by 15%. Also, design changes in the three largest animal houses conserved enough heat to reduce their heating expenses by 9%, which is 2% more than the zoo itself first estimated. Finally, a new method of ordering and paying for supplies has lowered expenses by enough to offset inflation.

D-level
segments

The budget crisis is caused instead by declining revenues. During the past year income from admission fees, concession sales, and donations have all dropped. Because of the decline in paid admissions, overall income from this source was $57,344 less last year than the year before. [This discussion of falling revenue continues for two pages.]

**Figure 11-2.**   The Segments in a Report.

## INTRODUCTION TO GUIDELINES 1 THROUGH 3

Guidelines 1 through 3 concern the way you begin your segments. Even though the beginning of a short segment may be only a single sentence, that sentence is very important. For longer segments, the beginning may be a paragraph or more.

Guideline

**1**

### At the Beginning of Each Segment, Tell What It Is About

You have undoubtedly heard that you should begin your paragraphs with topic sentences. This guideline extends that advice to *all* your segments, large and small: begin *every* segment by telling your readers what it is about. Figure 11-3 shows how Jonathan provided topic statements for three levels of segments in a passage from his report to the zoo. Notice how Jonathan's topic statements interlock to form a hierarchy that corresponds to the hierarchy he planned in Section II of his outline (Figure 11-1).

### How Topic Statements Help Readers

Why are topic statements so important? When reading a segment, readers look for a meaningful pattern to the various pieces of information conveyed within it. Sometimes that task of meaning-construction can be quite difficult. Consider, for instance, how hard it is to establish the overall meaning of the following group of sentences, which are from a technical report written by an engineer at a coal-fired electric plant.

> Companies that make cement and wallboard cannot use wet gypsum cakes. The cakes must be transformed into dry pellets, using a process called agglomeration. We could enter the agglomeration business ourselves or hire another company to agglomerate the gypsum for us. Also, the chloride content of the cakes is too high for use in wallboard. Our engineers can probably devise inexpensive cake-washing equipment that will reduce the chlorides to an acceptable level.

Can you figure out what that paragraph is about? Probably so, if you spend enough time thinking about it. Your job as a reader would be much easier, however, if the writer added a statement telling you what it is about. Try reading that paragraph again, this time with a topic statement added:

Topic
Statement

> Before we can sell the gypsum produced by our stack scrubbers, we will have to process the wet gypsum cakes they produce. Companies that make cement and wallboard cannot use wet gypsum cakes. The cakes must be transformed into dry pellets, using a process called agglomeration. We could enter the agglomeration business ourselves or hire another company to agglomerate the gypsum for us. Also, the chloride content of the cakes is too high for use in wallboard. Our engineers can probably devise inexpensive cake-washing equipment that will reduce the chloride to an acceptable level.

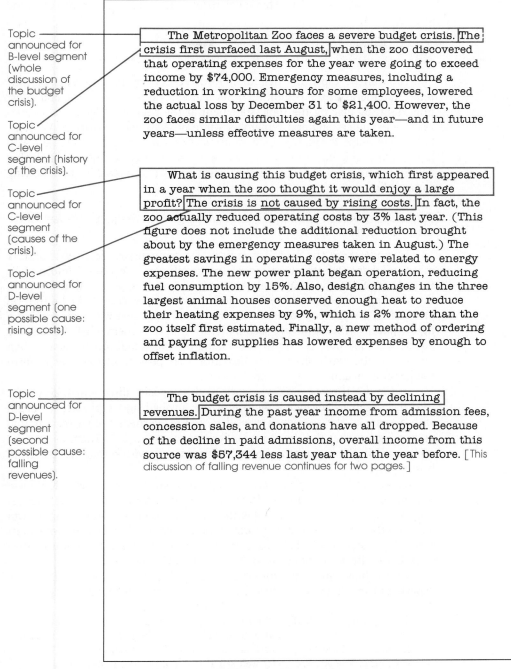

Topic announced for B-level segment (whole discussion of the budget crisis).

Topic announced for C-level segment (history of the crisis).

The Metropolitan Zoo faces a severe budget crisis. The crisis first surfaced last August, when the zoo discovered that operating expenses for the year were going to exceed income by $74,000. Emergency measures, including a reduction in working hours for some employees, lowered the actual loss by December 31 to $21,400. However, the zoo faces similar difficulties again this year—and in future years—unless effective measures are taken.

Topic announced for C-level segment (causes of the crisis).

Topic announced for D-level segment (one possible cause: rising costs).

What is causing this budget crisis, which first appeared in a year when the zoo thought it would enjoy a large profit? The crisis is not caused by rising costs. In fact, the zoo actually reduced operating costs by 3% last year. (This figure does not include the additional reduction brought about by the emergency measures taken in August.) The greatest savings in operating costs were related to energy expenses. The new power plant began operation, reducing fuel consumption by 15%. Also, design changes in the three largest animal houses conserved enough heat to reduce their heating expenses by 9%, which is 2% more than the zoo itself first estimated. Finally, a new method of ordering and paying for supplies has lowered expenses by enough to offset inflation.

Topic announced for D-level segment (second possible cause: falling revenues).

The budget crisis is caused instead by declining revenues. During the past year income from admission fees, concession sales, and donations have all dropped. Because of the decline in paid admissions, overall income from this source was $57,344 less last year than the year before. [This discussion of falling revenue continues for two pages.]

**Figure 11-3.**   Topic Statements for Three Levels of Segments.

As you can see, a topic statement helps readers understand the meaning of a segment.

## Topic Statements Help Most at the Beginning

You can be especially helpful to your readers by placing each of your topic statements at the beginning of its segment rather than at the end. Most important, by placing the topic statement at the beginning, you enable your readers to understand and remember more of what you've written. You also help them skim through your communication for particular facts, and you help them decide how to read the segment. These three advantages of placing a topic statement at the beginning of a segment are discussed in the following paragraphs.

### Topic Statements Placed at the Beginning Help Readers Understand and Remember More

To understand the overall meaning of a segment, readers must build a meaningful mental structure out of the small, separate bits of information they derive from the individual sentences. As readers, we do that by using two reading strategies: bottom-up and top-down processing. In bottom-up processing, we proceed in much the same way as people who are working a jigsaw puzzle without having seen a picture that tells them what the finished puzzle will look like. We try to understand the individual sentences and then guess how the small pieces of information they provide fit together to form the general meaning of the overall segment.

In top-down processing, we work in the opposite direction. Like people who have seen in advance a picture of the finished jigsaw puzzle, we begin with a sense of the segment's overall structure. Perhaps the writer has told us in advance what the structure is. Perhaps we guessed it. Then, as we read each detail in the segment, we place it immediately within that mental structure.

As we read, we engage continuously in both bottom-up and top-down processing. But the more a writer tells us in advance about the general frameworks he or she is working with, the more top-down processing we can do—and the more easily and accurately we can comprehend the writer's message. Topic statements help us perform top-down processing because they provide us with an initial clue to the meaning of a segment.

To see how much this initial clue can help, consider the results of an experiment in which people heard someone read the following passage.[1]

> The procedure is actually quite simple. First you arrange things into different groups. Of course, one pile may be sufficient depending on how much there is to do. If you have to go somewhere else due to lack of facilities that is the next step, otherwise you are pretty well set. It is important not to overdo things. That is, it is better to do too few things at once than too many. In the short

run this may not seem important but complications can easily arise. A mistake can be expensive as well. At first the whole procedure will seem complicated. Soon, however, it will become just another facet of life. . . . After the procedure is completed one arranges the materials into different groups again. Then they can be put into their appropriate places. Eventually they will be used once more and the whole cycle will then have to be repeated. However, that is part of life.

The researchers told one group the topic of this passage before the group heard it; they told the other group afterwards. To see how well each group understood the passage, the researchers asked the people to write down everything they remembered from it. People who were told the topic (washing clothes) before hearing the passage remembered much more than did people who were told afterwards.

Topic statements also help us read segments much larger than paragraphs. They do that by telling us in advance about the overall organization of the segment. Thus, a topic statement at the beginning of a major section of a communication helps us see how the smaller segments within it fit together to form a larger meaning.

Because topic statements are such a powerful aid to top-down processing, you should provide them at the beginning of all your segments, small and large.

### Topic Statements Placed at the Beginning Help Readers Find Particular Facts

By placing a topic statement at the beginning of a segment, you also help your readers locate particular pieces of information. Suppose, for instance, that someone asked you to locate the information in Figure 11-4 that explains the dangers of importing insects from other parts of the world. The topic sentence that opens the second paragraph would stand out like a large sign saying, "Reader, read this!"

### Topic Statements Placed at the Beginning Help Readers Decide How To Read Segments

Finally, by announcing the topic of a segment at the beginning, you help your readers decide *how* to read the segment. When readers come to a segment that contains information with which they are already familiar, they will know from the opening sentence that they should skim or skip it. Likewise, when they arrive at a segment that contains new information, they will know from the opening sentence that they should read carefully.

## How To Indicate the Topic

There are many ways to indicate the topic of a segment. The sentence you have just read, for instance, states the topic of the segment you are now

## Importing Insects

Countries sometimes benefit from importing insects from other parts of the world, but they also risk hurting themselves. By importing insects, Australia controlled the infestation of its continent by the prickly pear, a cactus native to North and South America. The problem began when this plant, which has an edible fruit, was brought to Australia by early explorers. Because the explorers did not also bring its natural enemies, the prickly pear grew uncontrolled, eventually rendering large areas useless as grazing land. The problem was solved when scientists in Argentina found a small moth, Cactoblastis cactorum, whose larvae feed upon the prickly pear. The moth was imported to Australia, where its larvae, by eating the cactus, reopened thousands of acres of land.

In contrast, the importation of another insect, the Africanized bee, could threaten the well-being of the United States. It once appeared that the importation of this insect might bring tremendous benefits to North and South America. The Africanized bee produces about twice as much honey as do the bees native to the Americas.

However, the U.S. Department of Agriculture now speculates that the introduction of the Africanized honey bee into the U.S. would create serious problems. The problems arise from the peculiar way the Africanized honey bee swarms. When the honey bees native to the United States swarm, about half the bees leave the hive with the queen, moving a small distance. The rest remain, choosing a new queen. In contrast, when the Africanized honey bee swarms, the entire colony moves, sometimes up to fifty miles. If Africanized bees intermix with the domestic bee population, they might introduce these swarming traits. Beekeepers could be abandoned by their bees, and large areas of cropland could be left without the services of this pollinating insect. Unfortunately, the Africanized honey bee is moving slowly northward to the United States from Sao Paulo, Brazil, where several years ago a researcher accidentally released 27 swarms of the bee from an experiment.

Thus, while the importation of insects can sometimes benefit a nation, imported insects can also alter the nation's ecological systems, thereby harming it.

**Figure 11-4.** Passage with Many Levels of Segments.

reading. Sometimes, you can indicate the topic through a single word. For example, the first word ("First") of the second sentence in the passage on washing clothes tells the reader, "You are now going to read a segment that explains the steps in the 'simple procedure' just mentioned." You can even indicate the topic of a segment by asking a question. For example, the question ("Why are topic statements so important?") that begins the second paragraph in the discussion of this guideline told you that you were about to read a segment explaining the importance of topic statements.

Whether you use a word or a sentence, a statement, or a question, remember to tell your readers at the *beginning* of each segment what that segment is about.

## At the Beginning of Each Segment, Tell How It Is Organized

Guideline

2

You can provide your readers with additional help in understanding each of your segments if you tell them in advance not only what the segment is about but also how you have organized it. By supplying a preview of your segment's organization, you will assist your readers in establishing in their own minds a meaningful pattern for the information you provide.

Consider the following statement, which is the opening sentence of a segment:

Our first topic is the trees found in the American Southwest.

The sentence certainly adheres to Guideline 1 ("At the beginning of each segment, tell what it is about"). However, a segment begun with that sentence could be organized in a wide variety of ways. It might first take up the evergreen trees and then deciduous ones. Or it might first take up healthy trees and then diseased ones. Or it might be arranged in any of various other patterns.

One way you can tell your readers about the organization of a segment is to supplement your topic statement with a separate forecasting statement:

Forecasting
Statement

Our first topic is the trees found in the American Southwest. Some of the trees are native, some imported.

Another way is to announce both the topic and the organization in a single sentence:

Our first topic is the trees—both native and imported—found in the American Southwest.

Your previews of your segments' organization may vary greatly in the amount of detail they provide. The sample sentences you have just read provide specific detail: the reader knows both the number and the names of the categories of trees to be discussed. Here is an example of a much more general preview:

> To solve this problem, the department must take the following actions.

This statement tells its readers to expect a list of actions, but doesn't tell what they will be or even how many there are.

When trying to decide how much detail to include in your preview of the organization of a segment, consider the following pieces of advice.

- *Provide enough detail so that your readers know something specific about the arrangement of the segment.* Usually, the more complex the relationship among the parts, the greater the amount of detail needed.

- *Do not provide more detail than your readers can easily remember.* The purpose of a forecasting statement is to help readers understand what is to come, not to test their memories. If you are introducing the three steps of a solution, you might want to name the three steps before explaining them. If the solution contains eight steps, however, you would probably be better off stating the number of steps without naming them in the forecasting statement.

- *Do not forecast more than one level at a time.* Writers are sometimes tempted to explain all the contents of a document at its beginning. By doing so, they would burden their readers with a confusing amount of detail. To work effectively, a forecasting statement tells only the major divisions of a particular segment. If those divisions are themselves divided, then each of the divisions will have its own forecasting statement.

Like topic statements, forecasting statements prepare readers for top-down processing. When you provide this help, your readers can more easily read and remember your message.

Guideline

3

## At the Beginning of Each Segment, Generalize about Its Contents in a Reader-Centered Way

Researchers say that an essential part of the process by which readers construct the meaning of a segment is generalizing.[2] When reading a segment, we not only arrange the information we find there into an orderly pattern but we also generalize about the information. We decide what we think the main point is. Consider, for instance, the following set of sentences:

> Richard moved the gas chromatograph to the adjacent lab.
> He also moved the balance.
> And he moved one of the XT computers.

When reading these sentences, we not only determine their topic (Richard's moving) but also make a generalization based upon them. For instance, we might realize that all of the things he moved are pieces of laboratory equipment, so that we would generalize in this way.

> Richard moved some laboratory equipment.

Furthermore, we use the generalizations we derive from small segments to build higher-level generalizations. Imagine, for instance, that the three sentences about Richard's moving are from a communication that also recounts other kinds of activities that Richard performed on a given day. It has a paragraph about his working on an experiment, another about his presiding over a meeting of people in his department, and still another about his completing a monthly report. From the generalizations we draw from each of these paragraphs, we might derive the higher-level generalization that "Richard had a busy day."

## Influencing the Generalizations Your Readers Make

From any segment readers can derive a wide variety of generalizations. Factors that influence what generalization they will derive include their professional roles and their purposes in reading. For instance, a member of the labor union in Richard's organization might generalize from the three sentences about Richard's moving that "Richard persisted in performing jobs that should have been performed by union employees, not by a research engineer." In contrast, the doctor treating Richard for a sore back might generalize that "Richard lifted several heavy objects."

As a writer, you should care what generalizations your readers draw from your segments. After all, the kinds of generalization your readers make will influence the kinds of action they take. For example, after reading about Richard's moving, the union member and the doctor will act differently from one another partly because of the different generalizations they draw.

Because some generalizations will affect your readers in the way you desire and others will not, you should do what you can to influence the generalizations that your readers draw from your segments. You can do that by suggesting generalizations yourself, rather than letting your readers draw their own conclusions. By stating generalizations yourself, you tell your readers what you want them to see as the main point of each segment.

However, to enjoy the full benefit of your generalizing statements, you must place them at the *beginning* of your segments. If you wait until later in the segments, your readers may already have formed their own generalizations before they read yours. Furthermore, you emphasize your generalizing statements by placing them at the beginning of your segments. Research has shown that a sentence placed first in a group of sentences is more likely to be seen by readers as the main point than if it is placed elsewhere.[3] Your overall generalizations merit more emphasis than the details that support them, and so should be placed at the beginning of your segments.

## Deciding How Much To Say in Generalizing Statements

There are two ways to decide what you should say in a generalizing statement. The first involves thinking of your readers as people who ask questions that you must answer (see Guideline 2 in Chapter 5). In your generalizing statement, summarize the answer. The second way is to decide what conclusion you most want your readers to draw from the information contained in the segment. State that conclusion in your generalizing statement.

Often the two approaches will lead you to make the same generalizing statement. For example, imagine that you are writing a segment in which you will include a great deal of information about the results of some laboratory tests you have conducted. Your readers will almost certainly want, more than anything else, the answer to the question, "What conclusion do you draw from the results?" Similarly, the point you most want your readers to derive from the results is probably best summarized in your conclusion. Thus, you might write:

> The test results indicate that the walls of the submersible room will not be strong enough to withstand the high pressures of a deep dive.

## Is It Illogical To Put Your Conclusion First?

Putting the conclusion (or any generalization) first seems illogical to some writers. They reason that, because they had to obtain the specific facts or data before they could draw any generalization about them, they should give those facts to their readers before telling them the conclusion.

Such a view is writer-centered. There is no necessary reason why readers must learn information in the particular way the writer obtained it. In almost every case, readers at work are much more interested in generalizations than in details. After all, the generalizations—the writer's conclusions—are what the readers will use as the basis of their own actions. From the point of view of readers, then, detailed facts are not the path to the generalization but rather are the support for it. The details enable readers to answer the question, "Is that generalization justified?"—a question they can ask only after they know what the generalization is. Stating your generalizations first is a reader-centered way of organizing a segment.

Imagine, for example, that you are a manager who finds the following sentences in a report.

> Using the sampling technique described above, we passed a gas sample containing 500 micrograms of VCM through the tube in a test chamber set at 25 degrees C. Afterwards, we divided the charcoal in the sampling tube into two

equal parts. The front half of the tube contained approximately ⅔ of the charcoal while the back half contained the rest. Analysis of the back half of the tube revealed no VCM; the front half contained the entire 500 micrograms.

As you read these facts, you might find yourself repeatedly asking the question, "So what?" Without any good answer to that question and without any general principle around which to organize the facts contained in the passage, you might not remember much from it. However, you would read the facts in a much different way if the writer had begun the segment with the following statement:

We have conducted a test that demonstrates the ability of our sampling tube to absorb the necessary amount of VCM under the conditions specified.

Knowing from the outset what conclusion the writer draws allows you, as a reader, to put the other information to good use: you can use it to determine whether or not the conclusion is valid. Of course, you may ultimately reject the writer's conclusion. However, even if you do, you will at least have gained an accurate understanding of the writer's message, something you might not have obtained if the writer had left you to form generalizations on your own.

# RELATIONSHIP AMONG GUIDELINES 1 THROUGH 3

Guidelines 1 through 3 concern the way you should begin each of your segments: by telling your readers what it's about, how it's organized, and what general point you are making in it. In some segments, you will be able to follow all three guidelines in a single sentence. Here, for instance, is the first sentence in a two-page discussion of a series of experimental results:

Our test showed three shortcomings with plastic resins that make it undesirable for us to use them to replace metal in the manufacture of the CV-200 housing.

After reading this sentence, people would expect that the segment is about the shortcomings of plastic resins (topic), that the segment will discuss three of these shortcomings in turn (organization), and that the writers believe the resins will not make good substitutes for metal (conclusion).

Similarly, you will often be able to begin your segments with a single sentence that follows all three of these guidelines, although in some situations you may need to take a paragraph (sometimes more) to do so. The key point is for you to remember the kinds of things that your readers will find helpful at the beginning of each of your segments—and then write accordingly.

# INTRODUCTION TO GUIDELINES 4 AND 5

While the three guidelines you have just read concern the beginnings of your segments, the next two focus on the way you present the rest of the information in them.

## Put the Most Important Information First

Whenever you are writing a segment that has several pieces or groups of information that can be ranked in importance, you should put the most important information first. In many of the segments you write at work, for example, you will present several parallel pieces of information, such as a list of five recommendations you have developed, or a list of three causes of a problem that faces your employer, or a list of four reasons for using a part design that you have created. Whether you present each of the items in a single sentence or with several paragraphs of discussion, you should usually present the most important item in your list first, arranging the rest in descending order of significance.

For instance, consider the segment (just mentioned) in which the writers will describe three shortcomings of plastic resins. This guideline suggests that the writers organize their two-page discussion so that they discuss first the most important or serious of those shortcomings, then the one that is next in importance, and so on.

You should also use this strategy of putting the most important information first when you are writing segments that do not include lists of parallel information. Suppose, for example, that you are writing a memo in which you request $23,000 to undertake a certain project. From the point of view of your readers, the most important content of that memo is the statement of your request—what you want and what you want to use it for. Thus you should announce your request at the beginning of the memo, *before* your detailed discussions of the problem your project will solve, the ways you will spend the money, and even the benefits to your employer of the project you want to undertake.

As you can see, this guideline complements Guideline 3, "At the beginning of each segment, generalize about its contents in a reader-centered way." To *begin* a segment, the most important information is the generalization you make about it. (Guideline 3). *Throughout* your segment, order your information from most important to least important. (Guideline 4).

### Advantages of Putting the Most Important Information First

Putting the most important information first has two advantages. First, it increases the likelihood that your readers will remember it, because material presented early in a communication is remembered better than information contained later in it.[4]

Second, putting the most important information first greatly increases the probability that your readers will read it at all. Reading is work. Readers who are forced to plow through relatively unimportant information may very well become so exasperated that they quit reading. If your readers don't even read your communication, or read only parts of it, your writing is not likely to affect them in the way you desire.

## How To Decide What Information Is the Most Important

The two strategies for deciding what information in a segment is the most important are very similar to the two strategies for formulating a generalizing statement. First, ask yourself what information your readers want. Second, ask what information has the best chance of affecting your readers in the way you wish. Thus, if you have five good reasons to offer in support of a recommendation, you should probably begin your discussion with the reason you believe your readers will find most powerful. Your readers will be grateful to find the key reason so easily and quickly, and they will be likely to read it carefully and remember it well.

Guideline

5

## Answer Your Readers' Questions Where They Arise

You have already read in Chapter 5 that you should think of your readers as people who ask questions that you try to answer in your communication. Your readers will already have some questions before they begin to read. Other questions will arise as the readers read. Usually you should answer questions that are likely to arise during reading close to the points at which they will arise. Exactly *how* close depends largely upon the type of question being asked.

## Three Types of Questions

Three types of questions are likely to arise in the readers' minds during reading:

- *"Can you explain?"* The readers cannot understand some word, concept, or statement because the material is unfamiliar.
- *"Can you tell me more?"* The readers understand your writing but want more information about some aspect of your subject.
- *"How can you say that?"* The readers question the validity of a statement.

Generally, you should define unfamiliar words and explain unfamiliar concepts immediately. The following sentence does that by answering the reader's question ("What is spalling?") in the very sentence in which it arises.

One excellent method for determining how long the metal wheels of an overhead crane will last is to measure the metal's tendency to spall, a process by which small chips split away from its surface.

Answer to readers' question

In contrast, your strategy for answering objections and requests for more information can depend upon the length of your answer. When your answers require a sentence or less, you can answer the questions exactly where they will arise. For example, the second sentence of the following passage answers the readers' question, "Why did you send your questionnaire to those particular people?"

We then mailed our questionnaire to each of the 47 members of the local chapter of the American Society for Training and Development (ASTD). They seemed more likely than anyone else to know about in-company training programs in first-aid. Then we called each of the 29 ASTD members who returned a questionnaire.

Answer to readers' question

However, if the answers will take longer than a sentence, they will become digressions that disrupt the organization of a segment into which they are inserted. Such extended answers can be made into segments of their own that *immediately* follow the segment in which the questions arise.

## An Example of an Extended Answer

Figure 11-5 shows a passage in which three researchers who were writing together provided an extended answer to an objection they expected their readers to raise. (You have read part of the passage earlier in this chapter.) The passage concerns tests performed to see whether a certain type of filter would capture enough of a substance called VCM that was to be used in a particular experiment. For practical reasons they knew their readers would understand, the researchers tested the filter at a different temperature than would be used in the actual experiment. However, the researchers also anticipated that readers of the test results might object, "But how do you know that the filter will behave in the same way in the experiment, where it will have to function at a different temperature?" Because the answer took several sentences, the researchers correctly decided to provide it in a segment of its own, and to place that segment immediately after the segment in which the question would arise in readers' minds.

Providing answers to your readers' objections shortly after they arise can greatly increase the effectiveness of your writing. If the readers' objections go unanswered, even for a short time, the readers may begin to form the following generalization: "This material is unsound." Once they state that generalization, your readers will look for other facts to support it.

Of course, to be able to predict what questions will arise in your readers' minds while they are reading, you need to study your readers closely and look at your communications from their point of view, not just your own.

The writers expect their readers to raise an objection when they read these words.

We have conducted a test that demonstrates the ability of our sampling tube to absorb the necessary amount of VCM under the conditions specified. Using the sampling technique described above, we passed a gas sample containing 500 micrograms of VCM through the tube in a test chamber set at 25 degrees C. Afterwards, we divided the charcoal in the sampling tube into two equal parts. The front half of the tube contained approximately 2/3 of the charcoal, while the back half contained the rest. Analysis of the back half of the tube revealed no VCM; the front half contained the entire 500 micrograms.

The writers explicitly state the question they expect their readers to ask.

The writers answer the question.

Because the proposed experiment will be conducted at temperatures significantly above the 25 degrees C used in our test, the reader might wonder how higher temperatures will affect the absorption efficiency of the charcoal. The sorption temperatures inside the actual experimental environment will range from 43 to 66 degrees C. To check the sorption efficiency of the carbon at the higher temperatures, we calculated the ratio of the equilibrium static absorptive capacity per unit weight of carbon at 65 degrees C as compared to that of 25 degrees C using the method recommended by Nelson and Harder. This calculation indicates that the equilibrium static sorptive capacity for carbon at 65 degrees C should be 18% less than it is at 25 degrees C. However, a full 33% of the charcoal in the tube we tested remained available to absorb additional VCM. Thus, there is definitely enough charcoal in the tube to perform the experiment.

**Figure 11-5.**   Passage in Which the Writers Answer a Question They Expect Their Readers to Raise.

## INTRODUCTION TO GUIDELINE 6

The first five guidelines in this chapter concern structure *within* segments. They provide advice about how to begin segments and how to organize the information that follows the beginning. The next guideline focuses on the links *between* segments.

## Provide Transitional Statements between Segments

As readers move from segment to segment, they can become confused about the relationships among the segments. If that happens, they will have difficulty figuring out how the segment they are just beginning to read fits into the overall communication. Even if this confusion is only momentary, it can interfere with their ability to understand your message. You can help your readers avoid such confusion by providing transitional statements between your segments.

Transitional statements indicate two things: what the upcoming segment is about and how that segment relates to the segment that is just ending. Transitional statements often serve also as topic sentences because, like topic sentences, they tell what is coming. To make a true transition, however, they must also indicate the relationship between what is coming and what has just concluded.

### How To Write a Transition

You can write transitions in a great variety of ways. For example, you can say almost literally, "We are now making a transition from one topic to the next." You might even take two full sentences to do that:

> In our guidelines on handling Exban, we now have completed our suggestions on guarding against accidental spills during transportation. We turn now to our guidelines for storing the product.

In more subtle transitions, you may devote only a word or phrase rather than an entire sentence to naming the discussion that has just been completed:

> After we developed our hypothesis, we were ready to design our experiment.
>
> Having described the proposed processing method, we will now discuss its advantages over the present one.

In these examples, the introductory phrase reminds readers of what was just described in the previous segment (the development of the hypothesis and the proposed processing method, respectively). The rest of the sentence tells what will be discussed in the next passage (the design of the experiment to test the hypothesis, and the advantages of the proposed method over the present one).

Some transitions achieve even greater subtlety by including in the main clause (not in an introductory clause or phrase) the reference to the segment just completed:

> We then designed an experiment to test our hypothesis.
>
> The proposed processing method has several advantages over the present one.

### Will You Bore Your Reader with Transitions?

Transitions of one kind or another are needed at all levels within a communication. However, some writers hesitate to provide transitions at all the points where readers need them. They fear they will bore their readers by saying what is obvious.

Of course, it is possible to include more transitional material than is required. However, relationships that are obvious to you as a writer may not be obvious to your readers. In general, it is wiser to risk boring your readers by including too much transitional material than to risk confusing them by leaving out something they may need.

## SUMMARY

A segment is a group of related sentences. Some segments are small—paragraphs or groups of paragraphs. Some are large—chapters and whole communications. When reading any segment, regardless of its size, readers try to build a meaningful pattern from the information supplied in the individual sentences. By following the six guidelines in this chapter you can help your readers build those patterns easily and you can increase the likelihood that they will understand and react to your material in the way you desire.

1. *At the beginning of each segment, tell what it is about.* By announcing the topic of each segment at the beginning, you can help your readers (1) understand and remember what you have written, (2) find particular pieces of information, and (3) decide how to read the segment.

2. *At the beginning of each segment, tell how it is organized.* You can help your readers read efficiently by telling them in advance not only the topic of each segment, but also how you have organized your discussion of that topic.

3. *At the beginning of each segment, generalize about its contents in a reader-centered way.* As they read, people try to generalize about the contents of segments. If you provide generalizations at the beginning of your segments, you will save your readers the work of trying to figure out how the facts in the segment fit together.

4. *Put the most important information first.* Doing so has two advantages: it increases the probability that your readers will read the important information, and it increases the likelihood that they will remember the important information. Generally, the most important information is either the information your readers most want or the information that has the best chance of affecting them in the way you desire.

5. *Answer your readers' questions where they arise.* As they read the various statements in your communication, your readers will have questions. If you fail to answer these questions quickly, the readers may be distracted or may even dismiss your points.

6. *Provide transitional statements between segments.* Transitional statements look both forward and backward: they indicate both what is about to be discussed and how that relates to what has just been discussed. As readers move from segment to segment, transitions help them understand how the segments fit together.

## NOTES

1. John D. Bransford and Marcia K. Johnson, "Contextual Prerequisites for Understanding: Some Investigations of Comprehension and Recall," *Journal of Verbal Learning and Verbal Behavior* 11(1972):717–26.
2. Teun A. van Dijk, "Semantic Macro-Structures and Knowledge Frames in Discourse Comprehension," *in Cognitive Processes in Comprehension,* ed. Marcil Adam Just and Patricia A. Carpenter (Hillsdale, New Jersey: Erlbaum, 1977), 3–32.
3. David E. Kieras, "Initial Mention as a Signal to Thematic Content in Technical Passages," *Memory and Cognition* 8(1980):345–53.
4. Kieras, 345–6.

## EXERCISES

1. Circle the various segments in the passage given in Figure 11-4 to show how larger segments contain smaller ones.

2. Identify the topic statements in Figure 11-4 by putting an asterisk before the first word of each sentence that indicates the topic of a segment.

3. To see how one author provided statements that preview the organization of a passage, circle all such forecasting statements in Figure 15-16 (page 317). Be sure to find the segments that do not have explicit forecasting statements and explain how the reader comes to know how they are organized.

4. Below are two versions of a passage from a proposal asking the federal government to sponsor a research project. In both versions the writer uses explicit generalizing statements. However, only in the second version is the generalizing reader-centered. Write a brief paragraph describing the ways that reader-centered generalization improves the passage. Why is the second version shorter?

   (Incidentally, this passage is *intended* to be on a subject unfamiliar to you, so that you can see all the more clearly how much reader-centered generalizations assist readers.)

   A. There are two types of magnetic susceptibility. Substances in which the molecules are arranged essentially at random (gases and liquids) are magnetically isotropic; that is, the magnetic susceptibility is the same in all directions. Solids too may be effectively isotropic if their components are randomly oriented. The magnetic susceptibility of these types of material is referred to as the average susceptibility. The magnetic susceptibilities along each axis (often unequal) are referred to as principal susceptibilities. Because the magnetic susceptibility of test dusts would be measured on bulk randomly oriented quantities, average susceptibility measurements would be appropriate.

There are many methods available for measuring magnetic susceptibility. Choice of method depends on the problem to be solved. There is no single method of universal applicability. One method of determining the average magnetic susceptibility of the test dusts is the Guoy method. The principle by which this method operates is shown in Figure 13 [not shown]. A cylindrical sample is suspended with one end in a uniform magnetic field and the other end in a region where the field is negligible. The force exerted on the sample by the magnetic field is measured by suspending the cylinder from the arm of a sensitive balance and finding the change in weight due to the field.

B.  In the proposed research project, we will examine the average magnetic susceptibility of the test dusts. Average susceptibility describes the magnetic behavior of substances that are magnetically isotropic, that is, whose magnetic susceptibility is the same in all directions. This behavior is exhibited by gases, liquids, and randomly oriented solids such as test dusts.

To determine the average magnetic susceptibility of the test dusts, we will use the Guoy methods, which is one of several suitable approaches. This method is illustrated in Figure 13 [not shown]. A cylindrical sample is vertically suspended with one end in a uniform magnetic field and the other end in a region where the field is negligible. To calculate average magnetic susceptibility, the force exerted on the sample by the magnetic field is measured. The force is measured by suspending the cylinder from the arm of a sensitive balance and finding the change in weight due to the field.

5. The following paragraphs form one segment from a long proposal submitted by an architectural firm to a company that owns a chain of fast-food restaurants. Write a persuasive opening sentence that emphasizes the significance to the readers of the facts reported in the segment.

In conventional restaurants, energy (in the form of heat) is pushed out of the building in several ways. For example, the heat generated by grills and ovens for cooking is captured in hoods and forced outside through vents and flues. Hot water is poured down drains. And the heat generated by refrigerators and air conditioners is also discharged outside the building.

The design we propose, however, will capture this heat so that you can use it in the normal operation of your restaurants. For instance, heat pumps capture the waste heat generated from each restaurant's ovens, grills, refrigeration units, and dishwashing machines. This heat would be used to raise the temperature of cold water coming into the restaurant. Once warmed, the water would be stored in two 3,000 gallon tanks, from which it can be drawn when needed for cleaning, dishwashing, and other purposes.

6. On the facing page is a memorandum whose contents have been scrambled. Each statement in it has been assigned a number.
   A. Write the numbers of the statements in the order in which the statements would appear if the memorandum were written in accordance with the guidelines in this chapter. Place an asterisk before each statement that would begin a segment. (When ordering the statements, ignore the particular phrasing of the sentences. Order them according to the information they provide the reader.)
   B. Using the list you have just made, rewrite the memorandum. In doing so, rephrase the sentences so that the finished memorandum conforms with all the guidelines in this chapter.

   As you work on this exercise, you may find it helpful to know that an extruder is a piece of manufacturing equipment that works like a toothpaste tube to make such things as long strands of copper wire or plastic filament. A large container is filled with the copper or plastic, which is pushed out through a small hole to make a continuous strand.

TO:        Jimmy Ru, Plant Engineer
FROM:      Chip Bachmaier, Polymer Production
SUBJECT:   NEW STUFFERS

1. When materials are stuffed into the extruder by hand, they cannot be stuffed in exactly the same way each time.

2. We have not been able to find a commercial source for an automatic stuffer.

3. A continuing problem in the utilization of our extruders in Building 10 is our inability to efficiently feed materials into the extruders.

4. An alternative to stuffing the materials by hand is to have them fed by an automatic stuffer.

5. If the materials are not stuffed into the extruder in exactly the same way each time, the filaments produced will vary from one to another.

6. I recommend that you approve the money to have the company shop design, build, and install automatic stuffers.

7. The company shop will charge $4,500 for the stuffers.

8. An automatic stuffer would feed material under constant pressure into the opening of the machine.

9. Presently, we are stuffing materials into the extruders by hand.

10. The shop has estimated the cost of designing, building, and installing automatic stuffers on our 3/4 inch, 1 inch, and 1-1/2 inch extruders.

11. It takes many man-hours to stuff extruders by hand.

12. There are no automatic stuffers available commercially because other companies, which use different processes, do not need them.

# Seven Types
# Of Segments

## PREVIEW OF TYPES

**Classification**
**Partition: Description of an Object**
**Segmentation: Description of a Process**
**Comparison**
**Persuasion**
**Cause and Effect**
**Problem and Solution**

n the preceding chapter, you learned six guidelines for writing segments. You can use those versatile guidelines regardless of the purpose, size, or subject of your segments. This chapter looks at segments in a different way. It concerns segments that have special purposes. Consider, for instance, the special purposes of some of the segments in a report Gene is writing to tell decision-makers in his organization about a new technique for applying coatings to the insides of television screens.

In one segment, Gene's purpose will be to describe the equipment used in this technique. In another, he will tell how this process works, and in still another he will compare this new technique with the one his employer currently uses. When writing each of these segments, as well as many others in his report, Gene can use one of the conventional patterns described in this chapter. These patterns include, for instance, one for describing an object (the equipment), another for describing a process (how the equipment works), and a third for making comparisons (how the new technique compares with the current one). This chapter also describes four other patterns that Gene might be able to use elsewhere in his report.

Similarly, you will have many opportunities to use the seven patterns described in this chapter. The particular patterns discussed here were chosen because of their special helpfulness in the kinds of writing you will do on the job.

## BENEFITS OF LEARNING AND USING CONVENTIONAL PATTERNS

What will you gain by learning and using these patterns? To begin, these patterns can help you become a more *effective* writer. When you use one of them, you are organizing your segment in a manner that other writers have found to be successful in situations similar to yours. Furthermore, these patterns are widely used and therefore very familiar to readers. When your readers see one of the patterns, they immediately know what kind of mental hierarchy to use to organize the information you present. That helps them read your message quickly and comprehend it thoroughly.

In addition, these patterns can help you become a more *efficient* writer. When you use one, you don't have to create an organizational pattern for your segment from scratch. You will still need to decide how best to apply the conventional pattern to your segment, but that will take much less time and effort than would be required for you to create the whole pattern by yourself.

You can enjoy the benefits of these conventional patterns in two different stages of the writing process. When planning, you can use the patterns to help you design the overall structure of your communication. As you plan a section or subsection that will describe an object, for instance, you can adapt the pattern used for that purpose. By combining these patterns with one another, you can use them as your building blocks when planning your communication's organization.

Similarly, when you are drafting your communication, you can use these same conventional patterns to guide you as you write the segments you planned. The patterns suggest frameworks around which you can develop your paragraphs and sections.

## FOUR IMPORTANT POINTS BEFORE YOU READ FURTHER

Before studying any of the special types of segments described in this chapter, there are four important points for you to note.

First, you must remember to use these conventional patterns only as guides. To make them work in your communication, you must adapt them to suit your purpose, readers, and situation.

Second, although the seven types of segments described here are useful in the writing done at work, you usually cannot write entire communications with these patterns alone. When you aren't using one of these seven patterns, you can still create effective segments simply by following the guidelines in Chapter 11. Further, there are many other helpful conventional patterns that you learn by reading other people's writing, especially the writing of other people in your organization and field.

Third, although this chapter treats various types of segments separately, in actual communications the types are often mixed with one another and with segments using other patterns. That's what will happen, for instance, in Gene's report on the new technique for coating television screens. Like many technical reports, his will employ a problem and solution pattern for its overall structure. First, he will describe the problem that makes the old technique undesirable for his employer, and then he will describe the technique he has been investigating because it might offer a solution to that problem. Within that overall pattern, Gene will use many other patterns, including three mentioned above: the patterns for describing an object, describing a process, and comparing two or more things. Similarly, in the communications you write at work, you will weave together various types of segments, selecting each type because of the way it helps you achieve the particular purposes of some specific part of your communication.

The fourth point for you to keep in mind as you read the rest of this chapter is that when writing each of the types of segments described here, you should follow the general guidelines for writing segments that are given in Chapter 11. In fact, much of the information given below is really advice about how to apply the general guidelines of Chapter 11 to the special types of segments described here.

## THE ORGANIZATION OF THIS CHAPTER

In order to describe the seven patterns for segments in as useful a way as possible, this chapter is organized differently than are most of the others in this book. Instead of being built around guidelines, this chapter contains separate discussions of each pattern. Each discussion provides you with detailed advice about when and how to use the type of segment it describes. Because the chapter is organized in this way, you can use it as a reference source in which you can quickly find complete information about the particular type of segment that you want to study or use.

## CLASSIFICATION

On the job, you will encounter many situations in which you must provide sense and order to what is initially a miscellaneous list of topics. Maybe you'll work for a manufacturing firm and need to present an organized discussion of the various strategies that might be used to manufacture a new toy. Maybe you'll work for a hospital and be asked to write an organized brochure telling patients about various kinds of cancer. Maybe you'll work for a chemical company and be asked to write an advertising brochure that describes in an organized fashion the various adhesives your employer produces. Whenever you are writing a segment in which you want to present an organized discussion of a group of related subjects, you can *classify* the subjects by arranging them into subgroups that share common characteristics.

To see how classifying can help your readers, imagine that the chairperson of your department has asked you to write a booklet to tell your fellow students about the kinds of employers who hire graduates in your major. By visiting the campus placement center, you gather information about sixty employers who have hired graduates of your department over the past few years. How would you organize that information? You might discuss them in the order you found out about each, or you might talk about them in alphabetical order. But neither one of those organizations will really achieve your purpose of helping other students to see what kinds of employment opportunities they will have. If you organized the sixty employers into groups that shared important characteristics, such as location in the same region, you could create a much more useful and meaningful discussion.

### Classification Involves Building Hierarchies

When you classify, you create a hierarchy of your material. For instance, to classify the collection of sixty employers into subgroups based on location of head offices, you might build the hierarchy shown in Figure 12-1.

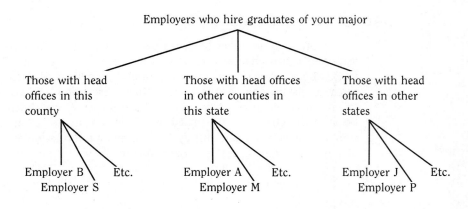

**Figure 12-1.** One Hierarchy To Classify Employers Recruiting on Campus.

Of course, you could create additional levels in your hierarchy. And you could use many other ways of creating the major groups and subgroups in your hierarchy. For instance, instead of organizing around the location of the head office, you might organize according to the number of employees, the kind of product or service offered, or the amount of money made last year. Figure 12-2 shows one of the many alternative hierarchies you might create.

## Two Kinds of Classification: Formal and Informal

The following paragraphs provide guidelines for two classification procedures, one formal and the other informal. When you use the *formal* procedure, you group the items in your list according to some objective, observable characteristic they possess. For example, you could find the address of each employer in the list of sixty in order to classify them by locations of their head offices, or you could count the number of employees to determine the size of the work force.

In contrast, you use the *informal* procedure when it is impossible or undesirable to classify according to observable characteristics. Suppose, for instance, that you wanted to classify the persuasive strategies used in advertisements that appear in the trade journals read by people in your field. Different people might group the advertisements in different ways because the very creation of the categories requires interpretation and judgment. Despite their differences, the two procedures share the same overall objectives:

- To arrange material in a way useful to your readers.
- To create groups (and subgroups) that include a place for every item you are trying to classify.
- To create a classification that has *only one* place for every item.

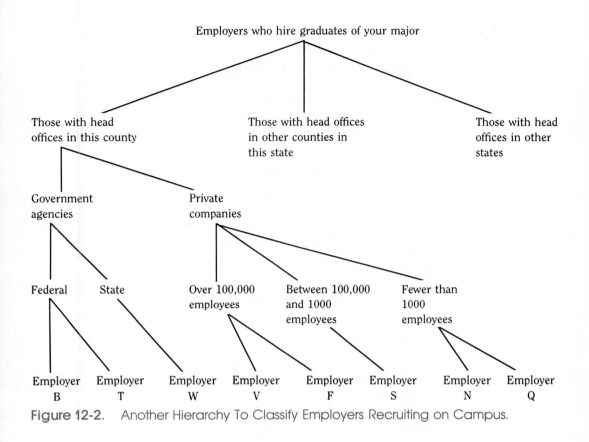

**Figure 12-2.**   Another Hierarchy To Classify Employers Recruiting on Campus.

## Guidelines for Formal Classification

1. *Choose principles of classification suited to your readers and purpose.* A principle of classification is the characteristic that you look at when you create your groupings and assign items to them. In the example of the employers, one principle of classification is the location of the head office. Employers with head offices in your county form one group, those with head offices in another county in your state form another group, and so on.

   Even though classification is largely a logical process, you must remember that your purpose in classifying your material is to make the information as useful as possible to your readers. Choose principles of classification that are suited to your readers and purpose. An initial classification of employers according to the location of the head office would probably be helpful because many students desire to work in certain regions of the country. A classification according to the size of the company also makes sense because many students start out by looking for employers of a certain size.

2. *Use only one principle of classification at a time.* As you classify a group of items, you should aim to provide one and only one place for each item at each level of the resulting hierarchy. If you use more than one principle of classification at a time, you will end up having two or more places for an item. For instance, suppose you classify the cars in a parking lot in the following way:

> Cars built in the United States
>
> Cars built in other countries

In this classification, you use only one principle of classification: the country in which the car was manufactured. Because each car was built in only one country, each would fit into only one group. On the other hand, suppose you grouped the cars in this way:

> Cars built in the United States
>
> Cars built in other countries
>
> Cars that are expensive

In this faulty classification, two principles are used at the same level: country of manufacture and cost. An expensive car built in the United States fits into two categories.

Of course, you can use different principles at different levels in your hierarchy:

> Cars built in the United States
>> Expensive ones
>> Inexpensive ones
>
> Cars built in other countries
>> Expensive ones
>> Inexpensive ones

In this case, an expensive car built in the United States would have only one place at each level of the hierarchy.

## Guidelines for Informal Classification

1. *Group your materials in ways suited to your readers and purpose.* Remember, even when you are not able to use the formal procedure, you are still classifying to achieve a particular purpose with respect to specific readers. Suppose that your employer wants you to classify advertisements as a way of thinking about advertising strategies it might use for its own

products. You might therefore classify the ads according to their focus: on the features of the product, on the general image of the product, or on the image of the manufacturer. Your readers could then judge which strategy might be most effective in ads for your employer's products.

2. *Use parallel groups at the same level.* For instance, if you are classifying advertisements, don't create a list of categories like this:

> Focus on price
>
> Focus on established reputation
>
> Focus on advantages over a competitor's product
>
> Focus on one of the product's key features
>
> Focus on several of the product's key features

The last two categories are at a lower level of detail than are the other three. To make the categories parallel, you could combine them in the following way:

> Focus on price
>
> Focus on established reputation
>
> Focus on advantages over a competitor's product
>
> Focus on the product's key features
> > Focus on one key feature
> > Focus on several key features

3. *Avoid overlap among groups.* Even when you cannot use strict logic to classify your material, strive to provide one and only one place for each item. To do that, you must avoid overlap among categories that would give an item two places. For example, in the following list, the last item overlaps the others because photographs can be used in any of the other types of advertisements listed.

> Focus on price
>
> Focus on established reputation
>
> Focus on advantages over a competitor's product
>
> Focus on the product's key features
>
> Use photographs

### Sample Classification Segments

Figure 12-3 shows the outline of a research report that the writers organized using informal classification. In the report, they describe the ways that investigators should go about identifying water plants that might be cultivated, harvested, and dried to make fuel for power generators.

IDENTIFYING EMERGENT AQUATIC PLANTS THAT MIGHT BE
USED AS FUEL FOR BIOMAS ENERGY SYSTEMS


INTRODUCTION

BOTANICAL CONSIDERATIONS
  Growth Habitat
  Morphology
  Genetics

PHYSIOLOGICAL CONSIDERATIONS
  Carbon Utilization
  Water Utilization
  Nutrient Absorption
  Environmental Factors Influencing Growth

CHEMICAL CONSIDERATIONS
  Carbohydrate Composition
  Crude Protein Content
  Crude Lipid Content
  Inorganic Content

AGRONOMIC CONSIDERATIONS
  Current Emergent Aquatic Systems
    Eleocharis dulcis
    Ipomoea aquatica
    Zizania palustris
    Oryza sativa
  Mechanized Harvesting, Collection, Densification, and
    Transportation of Biomass
  Crop Improvement
  Propagule Availability

ECOLOGICAL CONSIDERATIONS
  Water Quality
  Habitat Disruption and Development
  Coastal Wetlands
  Processing and Conversion

ECONOMIC CONSIDERATIONS
  Prior Research Efforts
    Phragmites communis
    Arundo donax
    Other Research
  Production Costs for Candidate Species
  Planting and Crop Management
  Harvesting
  Drying and Densification
  Total Costs
  End Products and Potential Competition

SELECTION OF CANDIDATE SPECIES

**Figure 12-3.** Outline of a Report Organized Using Informal Classification.

From S. Kresovich et al., *The Utilization of Emergent Aquatic Plants for Biomass Energy Systems Development,* (Golden, Colorado: Solar Energy Research Institute, 1982).

Figure 12-4 shows a group of paragraphs that the writers organized using formal classification. The paragraphs talk about two kinds of spruce forests. Both kinds are scattered throughout a large area of Alaska that was being studied for an environmental impact statement.

# PARTITION: DESCRIPTION OF AN OBJECT

At work, you will often need to describe a physical object to your readers. If you write about an experiment, for example, you may need to describe your equipment. If you write instructions, you may need to describe the machine your readers will use. If you propose a new purchase, you may need to describe the thing you want to buy.

To organize these descriptions, you can use a procedure called *partitioning*. In partitioning, you divide your subject into its major components and (if appropriate) divide those into their subcomponents. Partitioning is really a special kind of classification you use when you want to describe an object.

## How Partitioning Works

When you use partitioning to organize a description of an object, you follow the same basic procedure used in classification. You think of the object as a collection of parts and you use some principle to identify groups of related parts. Most often, that principle is either function or location. Consider, for instance, the ways that these two principles might be used to organize a discussion of the parts of a car.

If you organized your discussion of the car by location, you might talk about the interior, the exterior, and the underside. The interior would be those areas under the hood, in the passenger compartment, and in the trunk. The exterior would be the front, back, sides, and top, while the underside would include the parts under the body, like the wheels and transmission.

However, if you were to organize your discussion in terms of function, you might focus on parts that provide power and parts that guide the car. The parts that provide power are in several places (the gas pedal and gear shift are in the passenger compartment, the engine is under the hood, and the transmission, axle, and wheels are on the underside of the body), but you would discuss them together because they are related by function.

Often, these two commonly used principles of classification—location and function—coincide. For instance, if you were to partition the parts of a stereo system, you could identify the following major components: the turntable (whose function is to create electrical signals from the impressions made on the surface of a record), the amplifier (whose function is to increase the amplitude of those signals), and the speakers (whose function is to convert those signals into sound).

## SPRUCE FORESTS IN PROPOSED NOATOAK
## NATIONAL ARCTIC RANGE

Main topic and
subgroups are
announced
(principle of
classification is
elevation)

Spruce forest can be subdivided into two formations: one occurring on the well-drained uplands and in the upper portions of
tributaries of the Noatoak River and on a few isolated slopes in
the lower river valley, as at Napakutuktuk Mountains; the other
confined to river flats along the main channels of the Noatoak
and the lower portions of its major tributaries in the lower river
region.

First subgroup is
discussed.
Topics are
location,
species, stand
density, forest
floor, border,
and
reproduction

Upland spruce forests are found in areas of gentle to moderate
slope, from slightly above sea level near the Noatoak Delta to
over 1,200 feet near the headwaters of such rivers as the Eli.
Upland spruce forests are characterized by pure, usually rather
dense stands of white spruce with relatively few understory
shrubs. The forest floor consists largely of sphagnum moss,
reindeer lichen and related species, and dwarf ericaceous
shrubs, such as alpine blueberry, Labrador tea, and alpine bear-
berry. Upland spruce forest is usually surrounded by a belt of
shrubby vegetation, mainly willow, but it may interdigitate directly with shrub/tussock tundra. Scattered spruce trees often
occur in surrounding tundra areas, probably indicating the continuing colonization of tundra by spruce forest.

Reproduction is often high in and near upland spruce forest.
Upland spruce forest appears to cover 1 to 2 percent of the total
Noatoak Drainage, with isolated patches of spruce reaching
slightly beyond the upper end of the upper Noatoak Canyon. A
few isolated individual spruce trees are found along the upper
reaches of small creeks near the watershed between the Noatoak
and Kobuk drainages in the middle and upper portions of the
Noatoak Basin.

Second
subgroup is
discussed.
Topics are
same as for first
subgroup

Lowland spruce forests are found on islands and low-lying river
banks in and near the main channel of the Noatoak River,
mainly between the lower canyon and the lower reaches of the
upper canyon. The dominant species here is also white spruce,
but the understory and other associated vegetation differ
markedly from the upland white spruce forest. The lowland

**Figure 12-4.**   Segment Organized Using Formal Classification.

From U.S. Department of the Interior, *Proposed Noatoak National Arctic Range, Alaska: Final
Environmental Impact Statement* (Washington, D.C.: U.S. Department of the Interior, 1974), 51–52.

2

Discussion
of second
subgroup
continues

forest is usually rather open and consists mainly of mature trees, usually 15 to 20 meters tall and with a diameter at breast height of 25 to 40 cm. Ring counts on cut logs from these forests indicated that the average age of the mature spruce trees was 150 to 300 years. Reproduction is limited or nonexistent in undisturbed stands, yet mortality from damage by porcupines and bark beetles of several species is considerable.

Much of the understory of lowland spruce forest is tall willow brush. These species occasionally form pure stands within the forest, particularly along the shores of oxbow lakes and in abandoned river channels. Forest floor vegetation consists largely of typical subarctic forest species, such as one-sided wintergreen, and small northern bog orchis, as well as a number of forest–tundra species such as large-flowered wintergreen. Sphagnum moss is rare, but other mosses are common, and often form a significant portion of the forest floor vegetation. Lowland spruce is normally bordered by wet tundra, riparian willow thickets, and shoreline vegetation.

**Figure 12-4.**   *Continued*

Of course, other principles of classification are possible. You could organize your description of the parts of a car according to the material they are made from, a classification that could be useful to someone looking for ways to decrease the cost of materials in a car. You could also partition the parts of a car according to the country in which they were manufactured, a classification that could be useful to an economist studying the effects of import tariffs or quotas on American manufacturers.

## The Aims of Partitioning

Like classification, partitioning yields a hierarchy that might have one level or many levels. When you create that hierarchy, you pursue the following aims:

- To arrange your description in a way useful to your readers.
- To create groups (and perhaps subgroups) that include a place for every part you want to discuss. A rigorously thorough partitioning would identify every part of the object being described. In practice, however, partitioning is usually restricted to those parts relevant to the purpose and readers of the communication.
- To create an organization that has *only one* place for every item.

## Guidelines for Describing an Object

1. *Choose a basis for partitioning suited to your readers and purpose.* For instance, if you are trying to describe a car for new owners who want to learn about what they have purchased, you might organize according to location, so they will learn about the comforts and conveniences they will enjoy when sitting in the passenger compartment, when using the trunk, and so on. In contrast, if you are describing a car for mechanics who will have to diagnose and correct problems, it would make the most sense for you to organize your material around functions. Thus, when the mechanics must correct problems with the steering, they will understand the entire set of parts that make up the steering system.

2. *Use only one basis for partitioning at a time.* In the hierarchy of your description, you want one place and only one place for each part you describe. For that reason, you must use only one basis for partitioning at a time, just as you use only one basis for classifying at a time.

   In some communications, you may need to describe the same object from more than one point of view. For instance, in the introduction to a brochure advertising a new car, you might need to describe the parts of a car from the point of view of their comfort and visual appeal to the owner. Then you could organize by location. Later in the same brochure, you might describe the car from the point of view of performance, in which case you could organize by function. Each part of the car would appear in both sections of the brochure but would have only one place in each section.

3. *Arrange the parts of your discussion in a way your readers will find useful.* For instance, if you are partitioning by location, you might move systematically from left to right, front to back, or outside to inside. If you are describing things functionally, you might treat the parts in the order in which they are involved in an activity of interest to your readers. For instance, if you are describing the parts that provide power to a car, you might discuss them in the order in which the power flows through them: from engine to transmission, to drive shaft, to axle, and so on. If you are describing the sections of a food processing factory, you might describe the sections in the order in which the food flows through them: from the railroad car in which the raw materials arrive to the loading dock where the finished product is shipped to customers. There will also be times when you can arrange material from most important to least important. For instance, if you are writing a repair manual, you might treat first the parts that most often cause problems.

### Sample Description of an Object

Figure 12-5 shows an outline of part of a manual for owners of a personal word-processing system. In the manual, the writers have organized their description by systematically partitioning the system into its components and subcomponents.

## SEGMENTATION: DESCRIPTION OF A PROCESS

A description of a process explains the relationship of events over time. You may have either of two purposes for describing a process to your readers. First, you may want to describe a process so your readers can *perform* it. For example, you may be writing instructions so that your readers can analyze the chemicals present in a sample of liver tissue, make a photovoltaic cell, apply for a loan, or run a computer program.

Second, you may describe a process so your readers can *understand* it. For example, you might want your readers to understand any of the following kinds of things:

- *How something is done.* For instance, how coal is transformed into man-made diamonds.
- *How something works.* For instance, how a compact disk player uses laser beams to create music.

---

**IDENTIFYING THE COMPONENTS OF YOUR WANG
PROFESSIONAL COMPUTER**

INTRODUCTION

KEYBOARD
    Keys
    LED Lights
    Cable
    Speaker

ELECTRONICS UNIT

    System Card
    Disk Drives
      Diskette Drive
      Winchester Disk Drive
    Expansion Slots

DISPLAY MONITOR
    Screen
    Controls
    Adjustable Neck

---

**Figure 12-5.**   Segment Organized by Partitioning an Object.

Outline of eight pages from *The Wang Professional Computer: Introductory Guide* (Lowell, Massachusetts: Wang Laboratories, 1983), 1–2 through 1–10.

- *How something happened.* For instance, how the United States developed the space programs that eventually landed a man on the moon.

These two purposes of describing a process so your readers can perform it and so that they can understand it are much different from one another. In fact, Chapter 33 is devoted solely to the subject of writing instructions to enable readers to perform tasks. However, from the point of view of organizing the segments in a communication, these two ways of describing a process are so similar that one set of guidelines applies to them both.

To organize a description of a process, you *segment* the process. Like partitioning, segmenting is similar to classification. You begin with a long list of steps or events that all pertain to a single process. Then you separate that long list into smaller groups of related steps or events. Throughout, you treat the steps and events in sequence.

Although you begin segmenting by making a sequential list of the steps or events in the process you are describing, you should keep in mind that segmenting is much more than mere listing. The chief aim of segmenting is to show your readers the *structure* of those steps. Thus, if you were writing instructions for building a cabinet, you might group the several hundred steps into a hierarchy like this:

> Obtaining Your Materials
>
> Building the Cabinets
> > Cutting the Wood
> > Routing the Wood
> > Assembling the Parts
> > > Making the Frame
> > > Mounting the Doors
>
> Finishing the Cabinets
> > Sanding
> > Applying the Stain
> > Applying the Protective Sealant

## How Segmenting Works

To separate the long list of steps or events in a process into the smaller lists of a hierarchy, you use a principle of segmenting much as you use a principle of classification to classify or a principle of partitioning to partition. Commonly used principles of segmenting include the time when the steps are performed (first day, second day; spring, summer, fall); the purpose of the steps (to prepare the equipment, to use the equipment, to examine the results); and the tools that are used to perform the steps (for example, steps that use the function keys on a word processor, steps that use the numeric pad, and so on).

## Guidelines for Segmenting

1. *Choose a principle for segmenting suited to your readers and purpose.* For example, imagine that you are writing a history of the process by which the United States placed a man on the moon. If your readers are legislators or government historians, you might segment your account according to the passage of various laws and appropriation bills, or according to the terms of the various administrators who headed the National Aeronautics and Space Administration. If you were writing to scientists

and engineers, your segments might focus on the efforts to overcome various technical problems.

2. *Arrange your groups in an order your readers will find helpful.* When you are segmenting a process that has one long sequence of steps that must be performed in a certain order, you will have no difficulty deciding what order to use. Simply follow the order that your readers must follow.

   In many process descriptions, however, the steps may be arranged in any of several sequences. You should select a sequence that will help readers find and learn the steps efficiently.

3. *Make your smallest groupings of a manageable size.* One of the most important things for you to do when describing a process is to make your smallest groupings manageable. If you include too many or too few steps in them, your readers will lose the benefit of your segmenting. They will not see the process as a structured hierarchy of activities or events, but merely as a long, unstructured list of steps, first one, then the next, and so on. If you are writing instructions, the lack of structure will make it harder for your readers to learn the task, and if you are describing a process you want your readers to understand, the lack of structure will make the process more difficult for them to comprehend.

4. *Make clear the relationships among the steps.* Finally, keep in mind that your segment should help your readers understand and remember the entire process. To do that, your readers will probably need to understand the relationships among the events and steps that make it up. Sometimes, the relationship can be stated as simply as, "Here are the steps you perform to make your cake." In other instances, the relationships are much more complex.

   There are many ways you can make clear these relationships among steps and events in a process. Where the relationships are obvious, you may be able to do that simply by providing informative headings. At other times, you will need to explain the relationships in an overview at the beginning of the segment, weave additional explanations into your discussions of the steps themselves, and explain the relationships again in a summary at the end. The important thing is that your readers understand the relationships among the groups and subgroups in the hierarchy of your description.

## Sample Process Descriptions

Figure 12-6 shows a section of a research report. The report, which describes a special method for making solar cells, is organized by segmenting.

Figure 12-7 (page 242) shows a group of paragraphs in which the writer has described a general process—the way wood burns. Notice how the writer emphasizes how each phase of the process is significant to his particular readers, people who are concerned with the amount of energy that can be derived from burning wood.

## Section 2
## EXPERIMENTAL METHOD

General background is explained

The wafers for this study were processed simultaneously under identical conditions, and they lacked only a front contact. The 1-to-3 cm, boron-doped Czochralski-grown, single-crystal silicon wafers were texture-etched in NaOH. Then phosphorus, by means of phosphine gas, was diffused in to form a junction at a depth of 0.3 m. The cells also were etched in nitric and hydrofluoric acid (HF) mixture, and a 2% aluminum, silver-based ink was screen-printed and dried on their backs to form the electrical contact. The ink processing sequence included five major steps: printing, drying, firing, HF etching, and solder dipping.

Overview of the procedure is presented

### 2.1 PRINTING

First step is discussed. Topics are: procedure used, parameters adjusted, and data collected

A Presco (Model 435) automatic screen printer was used to print the thick-film inks. A grid pattern with lines 250 μm (10 mil) wide, a 2.5-mm square for adhesion testing, and two short 125-μm-wide lines were printed on each cell (Fig. 2-A). The adjustable parameters in the printing step included the screen mesh size, the snapback (distance of the screen above the substrate), and the squeegee pressure. After print-

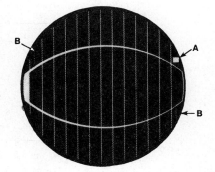

Figure 2–A.   Top View of Solar Cell with Screen-Printed Front Contact Showing 250-μm Grid Pattern with (A) Adhesion Test Pad, and (B) 125-μm Line.

**Figure 12-6.**   Segment Organized by Segmenting a Process.

From Steve Hogan and Kay Firor, *Evaluation of Thick-Film Inks for Solar Cell Grid Metallization* (Golden, Colorado: Solar Energy Research Institute, 1981), 3–5.

2

ing, data were obtained by observing how well the 125-μm line printed and how easily the ink cleaned off the screen after spraying it with trichloroethane solvent.

### 2.2 DRYING

After printing, the ink was allowed several minutes to settle. Visual inspection after this settling period determined how well the ink flow had smoothed out the screen impressions left on the printed surface of the film. The ink was then air-dried under an infrared (IR) lamp. Adjustable parameters during the drying stage included the time allowed for settling and the exposure time and temperature under the IR lamp.

### 2.3 FIRING

The cells were placed in a four-temperature-zone, Watkins-Johnson belt furnace for firing immediately after the ink dried. Figure 2-B shows a typical furnace profile, or plot, of

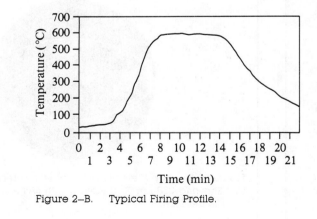

Figure 2–B.    Typical Firing Profile.

**Figure 12-6.**    *Continued*

3

temperature versus time. Firing took place in air, as recommended by all ink manufacturers. On several occasions, however, a nitrogen firing atmosphere was used to test for improved cell performance. Air flow rates, zone temperatures, and firing time were the adjustable parameters in this step. Immediately after the firing process, the first I-V curve was measured for each cell.

## 2.4 HF ETCH

Fourth step is discussed. Topics are: procedure used and parameters adjusted

The next processing step for most Ag-based inks was an HF etch. The cells were dipped in aqueous solution of either 2% or 5% (by weight) HF for 5 to 12 seconds. The cells then were rinsed in deionized water for 5 minutes, dried with air, and tested with another I-V curve. HF concentration and the etching time were the parameters adjusted during this step.

## 2.5 SOLDERING

Fifth step is discussed. Topics are: procedure used, data collected, and parameters adjusted

Each cell was solder-dipped to coat and somewhat protect the cell metallization from air moisture, which causes degradation of film adhesion. The cell metallization was fluxed with either Kester 1544 or Kester 1589 flux. The entire cell then was dipped into a bath of tin–lead solder (with 2% silver), typically heated to 210°C. A visual inspection then determined how well the solder coated the grid. An I-V curve was taken after soldering to determine the effects of this step on cell performance. Dip time, solder bath temperature, and the type of soldering flux were the adjustable parameters.

**Figure 12-6.**   *Continued*

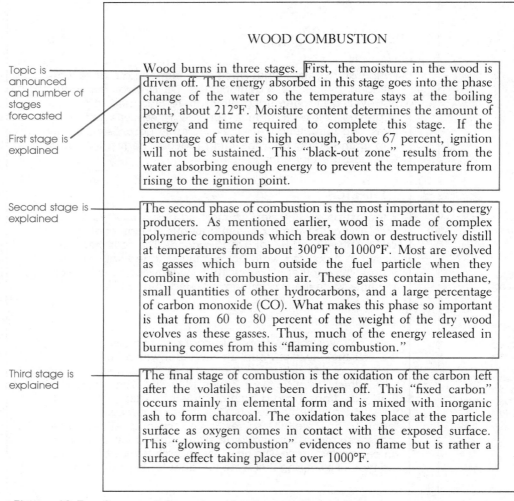

**Figure 12-7.**   Segment Organized by Segmenting a Process.

From N. Elliot, "Wood Combustion," in *Decision-Maker's Guide to Wood Fuel for Small Energy Users*, ed. Michael P. Levi and Michael J. O'Grady, (Golden, Colorado: Solar Energy Research Institute, 1980), 32.

## COMPARISON

At work, you will have many occasions to write segments that compare two or more things. Overall, these occasions fall into two categories.

● *When you want to help your readers make a decision.* The workplace is a world of choices. People are constantly choosing among available courses of action, competing products, alternative strategies. For that reason, you will often write communications in which you must compare and contrast two or more things.

- *When you want to help your readers understand something by means of an analogy.* One of the most effective ways to tell your readers about something new to them is to tell how it is like—and unlike—something they are familiar with.

In some ways, a comparison is like a classification. You begin with a large set of facts about the items that you will compare and then you create related groups of facts. In a comparison, you group your facts around points about the items that enable your readers to see how the items are like and unlike one another.

## How Comparisons Work

When you want to compare two things, you begin by choosing the points upon which you want to base the comparison. For instance, to compare two courses of action, you might select criteria such as cost and performance.

Then, you gather relevant facts about the items being compared. If the comparison is intended to help the readers make a decision, you can also draw general conclusions from the comparison.

There are two patterns for presenting comparisons. In the first, called the *alternating pattern,* you organize around your criteria:

Statement of Criteria

Overview of the Two Alternatives

Evaluation of the Alternatives
   Criterion 1
      Alternative A
      Alternative B
   Criterion 2
      Alternative A
      Alternative B
   Criterion 3
      Alternative A
      Alternative B

When you use this pattern, you enable your readers to make a point-by-point comparison without flipping back and forth through the communication. The pertinent information about both alternatives is presented together. As the outline given above suggests, when this pattern is used it is often necessary to precede the evaluation with an overview of the alternatives. Otherwise, readers may have to piece together an overall understanding of each alternative from the specific details you provide when discussing the alternatives point by point.

In the second pattern, called the *divided pattern,* you organize not around the criteria but around the alternatives themselves:

Statement of Criteria

Description of Alternatives
    Alternative A
       Overview
       Criterion 1
       Criterion 2
       Criterion 3
    Alternative B
       Overview
       Criterion 1
       Criterion 2
       Criterion 3

Comparison of Alternatives

The divided pattern is well suited to situations where both the general nature and the details of each alternative can be described in a short space, say one page or so. This pattern was used, for instance, by an employee of a restaurant who was asked to investigate the feasibility of buying a new turntable and amplifier for the restaurant's music system. He described each of three turntables and each of three amplifiers in about one page apiece. In his comparison, he briefly provided the analysis that enabled his readers to choose the best purchase.

## Guidelines for Writing Comparisons

1. *Choose points of comparison consistent with your purpose and readers.* When you are preparing comparisons for decision-makers, be sure to include both the criteria that those people consider important and any additional criteria that you—through your expert knowledge—realize are significant.

    When you are writing comparisons to create analogies, be sure to compare and contrast the things under consideration in terms of features that will help the readers understand the point you are trying to make. Avoid comparing extraneous details.

2. *If you are making complex comparisons, arrange the parts of your comparison hierarchically.* For example, group information on all the kinds of cost (purchase price, operating cost, maintenance cost, and so on) in one place, information on all the aspects of performance in another place, and so on.

3. *Arrange your groups in an order your readers will find helpful.* In comparisons made for the purpose of helping readers make decisions, you

should usually discuss at the outset the criteria that show the most telling difference between the things being compared.

In comparisons designed as analogies, you should usually discuss points of likeness first. In that way, you can begin with what your readers will find familiar, and then lead them to the less familiar.

### Sample Comparisons

Figure 12-8 shows a segment that compares two methods of studying the ways that mothers might affect their children if they take drugs during pregnancy.

# PERSUASION

Much of the writing you do at work will have the primary purpose of *persuading* your readers to adopt a certain point of view. Overall, these situations fall into two categories.

- *When you want to persuade your readers to take a certain course of action.* For example, you might want them to institute the policy you recommend, purchase the equipment you want, or approve the project you have proposed.

- *When you want to persuade your readers to accept a certain interpretation of some facts.* For example, when you report on an experiment or other research effort, you want to persuade your readers that your interpretation of the results is reasonable.

What can you do to construct persuasive arguments? The next several paragraphs answer that question by explaining the basic components of persuasion and then by offering advice about how to handle each of the components.

### Readers Are Skeptical

As you read the following discussion of persuasive segments—and as you write them—keep in mind that your readers will consider your arguments carefully. If your readers are decision-makers in your organization, they will be responsible for seeing that your organization's resources are spent wisely. For that reason, they will not simply accept your conclusions on your say-so, but they will want to know why they should agree with you. In the same way, specialists in any field are trained to accept positions only after carefully considering the arguments related to them. Scientists, for instance, read skeptically, looking for weak points.

## CLINICAL AND EPIDEMIOLOGICAL STUDIES OF THE EFFECTS OF PRENATAL EXPOSURE TO DRUGS

The need for epidemiological studies is explained

Since experimental administration of drugs to pregnant women is unethical, studies of the effects on their children of their using drugs is limited to clinical observation and epidemiological investigations. Although clinical reports can be of considerable importance in alerting physicians and health care providers to possible agents causing abnormal development, they are difficult to evaluate. For example, two early clinical reports of malformations in children born to marijuana users (Hecht et al. 1968; Carakushansky et al., 1969) were inconclusive since the mothers of these children used other drugs as well.

Comparison of two types of epidemiological studies is announced

When clinical reports are followed by epidemiological studies involving larger numbers of patients, a better appreciation of incidence and causation is possible. Such epidemiological studies can be divided into two types, retrospective and prospective, each of which has its own strengths and shortcomings.

First type is discussed: strength, then weaknesses

Second type is discussed: strength matched to weakness of first type.

In most retrospective studies, information from large numbers of cases is obtained from hospital records. However, such records are often inadequate or incomplete, thoroughness of reporting varies widely, and criteria for assessment of anomalies may also vary. In contrast to retrospective studies, prospective studies carefully establish criteria and protocols for maternal histories and examination of infants in prenatal health clinics. However, women who are usually most seriously at risk for giving birth to infants with drug-related anomalies may not attend prenatal health care facilities and, therefore, do not participate in prospective studies, resulting in underestimation of whatever problem is being investigated. Because prospective studies are so rigid in their design, they also are less flexible in allowing for changes to be incorporated as new information is obtained. Also, prospective studies cannot anticipate knowledge. For example, in the U. S. Collaborative Perinatal Project (Heinonen et al. 1977) which prospectively evaluated 55,000 consecutive births, no information was obtained with respect to maternal marijuana consumption because, at the time of the original protocol, marijuana was not a suspected teratogen.

**Figure 12-8.** Segment Organized by Comparison.

From Ernest L. Abel, "Effects of Prenatal Exposure to Cannabinoids," *Current Research on the Consequences of Maternal Drug Abuse,* ed. Theodore M. Pinkert (Washington, D.C.: U.S. Department of Health and Human Services, 1985), 20–21.

Their skepticism doesn't mean that your readers are hostile to you or your ideas. They are simply doing their job. Nevertheless, much of your success in writing persuasively will depend upon your ability to anticipate the specific questions your readers will ask as they read—so that you can supply convincing answers. The following discussion will help you do that.

## How Persuasion Works

In order to write persuasively, you must be sure that your readers understand all three of the following things.

- *Your claim.* The position you want your readers to agree with.
- *Your evidence.* The facts, observations, and other information you offer in support of your claim.
- *Your line of reasoning.* The line of thought that links your claim and your evidence. The reasons your readers should agree that your evidence proves your claim.

The following diagram illustrates the relationship among a claim, the evidence supporting it, and the line of reasoning that links the two:

Evidence ⟶ Claim

The following
facts, observations
and other evidence
support the claim.

Such and such should
be done.

Line of Reasoning

The evidence should be
accepted as adequate support
for the claim because of the
following reasons.

Imagine, for instance, that you work for a company that manufactures cloth. You have found out that one of your employer's competitors has increased its profits by installing computers to run its textile mill. If you were to write a recommendation urging your employer to buy similar computers, your basic argument would look like this:

Evidence ⟶ Claim

Our competitor has
increased profits
by using computers.

By using computers we
will increase our
profits.

Line of Reasoning

Experience has shown that
actions that increase the
profits of one company in an
industry will usually increase
the profits of other companies
in the same industry.

In this diagram, notice how important the line of reasoning is. Unless your readers agree that the experience of the other textile mill is relevant to the situation in your employer's mill, they will not be persuaded to accept your claim based upon evidence from the other mill. The following guidelines suggest ways that you can present your claims, evidence, and lines of reasoning as persuasively as possible when you write at work.

## Guidelines for Writing Persuasive Segments

1. *Begin your persuasive segment by stating your claim.* The reason for stating your claim explicitly should be obvious: you want your readers to consciously accept the position you hold, so you want to be sure that they know what it is. Further, as people read, they will want to assess the extent to which your evidence supports your claim. Without knowing what your claim is, they cannot do that.

   There is an important exception to this guideline. At times when you expect your readers to react immediately against your claim, you may be able to write more persuasively if you postpone introducing it. If your readers immediately react against your claim, they will begin to look for arguments against every bit of evidence you produce while they are reading it. By postponing your claim, you can present some of your evidence while they may be more open-minded.

2. *Present evidence that your readers will agree is sufficient and reliable.* Your whole argument on behalf of your claim rests upon your facts, observations, and other evidence. To persuade your readers that your point or position is a good one, you must convince them that your supporting evidence is strong. The two chief questions your readers are likely to ask about your evidence are:

- *Do you have a sufficient amount of evidence?* Readers may believe that your evidence is too skimpy to provide a sound basis for your claim. For instance, in the example of the textile mill they might feel that the experience of one single mill is not enough to base a decision upon. They might want to know what other mills have done. Or, they might feel that you haven't presented enough detail about that mill. Your readers would be unlikely to place much stock in your evidence if all you could produce was a vague report that the other mill had somehow used computers and had saved some money. Your readers would want to know how the mill had used computers, how much money they had saved, how those savings stacked up against the cost of the equipment they purchased, and so on.

- *Is your evidence reliable?* Your readers might feel that the evidence is unreliable because of the manner in which it was gathered. For instance, in the example of the textile mill, your readers would not put much stock in your evidence if it were based on rumors, but would put a great deal of stock in it if it were information published by the competitor's company in a trade journal.

The kinds of evidence that are considered reliable vary greatly from field to field. For instance, in the sciences and engineering fields, certain kinds of experimental procedures are widely accepted as reliable, whereas common wisdom and ordinary, unsystematic observation usually are not. Much of your professional education in your major may focus on learning to gather and analyze data in ways that people in your field will consider reliable.

If you expect that your readers are likely to question the adequacy of your evidence, there are several things you can do. The first is to add more evidence. If you think your readers will object that your evidence is too skimpy, you might try to provide more. If you expect them to object that you have overlooked some important piece of data, add a discussion of it. Alternatively, you might narrow your claim so that it corresponds more closely with the evidence you already have. Instead of trying to demonstrate that your claim applies to all people or situations, show why it applies especially well to the particular people or situations concerned in your communication. Finally, you can acknowledge the limitations and assumptions of your argument. For instance, if you are arguing that a certain savings will be enjoyed, acknowledge your assumption that prices will remain the same, or that loans will be granted at certain interest rates, or that demand will continue to increase.

Sometimes, the most effective way to address objections about your evidence is to state them directly—together with your refutations of them. To refute those objections, you must show the flaws or shortcomings in the evidence and lines of reasoning that your readers are relying upon—

or at least you must show why your evidence and lines of reasoning are more worthy of their agreement.

3. *Explicitly justify your line of reasoning if you believe your readers might question its validity.* In writing at work, as in everyday conversation, we often omit any explicit justification of the line of reasoning that links our evidence and our claims. That happens when we believe that our reasoning will be perfectly evident to our readers and when we also believe that they will accept that reasoning without question.

Imagine, for instance, that last week you randomly selected twenty 60-amp electrical motors from the hundred produced by your employer. Upon testing them, you found that two of the motors you selected had small but significant flaws. In a report based on these tests, you might claim that about 10% of all the 60-amp motors produced have some flaw. Your readers would most likely accept the link between your evidence and your claim without question—even without your mentioning it. That's because in engineering fields (as well as in the social and natural sciences) people generally agree that if data are gathered from a small group of randomly selected items, any major findings concerning that group also apply to the population overall. Similarly, in the construction industry, people generally agree that if you use the appropriate formulas to analyze the size and shape of a bridge you are building, the formulas will predict whether or not the bridge will be strong enough to support the loads it must carry. Because you are using a line of reasoning your readers will accept without question, you don't need to justify it.

When writing persuasive segments, you must look carefully for situations where your readers won't automatically accept your reasoning.

To find those places, determine where your readers might ask either of the following two questions:

- *Is your reasoning logically sound?* Some lines of reasoning simply aren't reliable. They involve fallacies of reasoning. One example is the hasty generalization, in which the writer tries to draw a broad conclusion based upon too few specific instances or to draw a conclusion based upon an untypical example.

- *Does your line of reasoning apply in this case?* At work, this question is asked often. Consider the example of the textile mill. Your line of reasoning in this case is that past experience predicts future performance. However, that reasoning is based upon the assumption that your mill and the competitor's mill are substantially the same. Someone might ask whether the two mills really are similar enough that such reasoning applies. Maybe the mills use different manufacturing processes, so that the computers may not work as well in your mill as in the other one. Or maybe the mills manufacture different products so that the other mill can enjoy greater savings than your mill could.

When you identify places where your readers might question your reasoning, state your line of thought explicitly and argue on behalf of it. For example, if you believe that your readers might suspect that you are drawing a hasty generalization, explain why you believe that you do have enough examples or that the examples you have chosen are in fact typical. If you think that your readers will object that your reasoning doesn't apply in this case, tell why you think it does.

## Structure of a Persuasive Segment

If you follow the guidelines for writing persuasive segments, you might mix the various elements of your argument together in many different ways. Here is one organizational pattern that you might find helpful:

Statement of claim

Evidence supporting the claim

Justification for the line of reasoning used (if need be)

Statement of anticipated objections (if any)

Refutation of objections

Restatement of claim (sometimes)

## Sample Persuasive Segment

Figure 12-9 shows a segment in which a researcher tries to persuade his readers that one method of measuring the yield of aquatic food and fiber crops is no good, and that another should be used instead.

# CAUSE AND EFFECT

You may use cause and effect segments in two very different ways. In one, your aim is *descriptive*. You help your readers understand the cause or consequences of some action or event, as when you answer a child's question, "What makes a rainbow?"

In other circumstances, your aim is *persuasive*. You try to persuade your readers that some event or action had the specific causes you believe it had or that some event or action will have the specific consequences you predict. For example, you might try to persuade your readers that the damage to a large turbine generator (effect) resulted from metal fatigue in a key part (cause) rather than from a failure to provide proper lubrication. Or you might try to persuade your readers that cutting the selling price of a product (cause) will increase sales enough to bring about a greater profit than the old price did (effect).

Because discussions of cause and effect deal with processes, they resemble discussions organized by segmenting a process. The key difference is that

## HOW TO ASSESS THE POTENTIAL YIELD
## OF AQUATIC PLANTS

*Major claim is stated: this method is best*

In order to assess the potential yield of such species there is no substitute to cultivating them throughout the year or growing season in either natural or experimental plots and harvesting and weighing the resulting crop. For species in mild climates that are able to grow continuously in a vegetative mode, such as all four of the examples shown in Fig. 1 [not shown], it is necessary to harvest the new growth frequently enough to maintain the density of plants at or near its optimum for maximum yield, if the full potential of the species for biomass production is to be determined.

*Major claim is restated: other methods are crude*

*One other method is explained*

Such experiments are difficult and time consuming and tend to be replaced by simpler but more crude yield estimates. One such approach has been to measure the growth rate of a given species experimentally, in the field or laboratory, and to apply such growth rates to a measured or estimated density of a natural stand of the plants to obtain annual yield values. In some cases short-term growth rates have been used to calculate annual yields, thereby ignoring seasonal effects. This general approach has been used to estimate the annual production of several kinds of aquatic plants including rockweeds and kelps in Nova Scotia, giant kelp off California, seagrasses, and water hyacinths.

*Subclaim is made: the other method exaggerates projections*

*Example of exaggerated projection is presented as evidence*

Reference to Fig. 1 clearly shows how the use of independent values of daily growth rate and density and extrapolation of the resulting daily to annual yields may result in greatly exaggerated projections. Using the maximum growth rates and densities for the seaweed *Gracilaria* and the freshwater *Eichhornia* as given in Fig. 1B and 1D, for example, would result in annual yield estimates in excess of 500 ash-free dry tons/ha.yr. Actual measured yields of small, experimental cultures of the two species maintained under the best possible conditions throughout the year in Central Florida were respectively 63 and 75 ash-free dry tons/ha.yr.

Figure 12-9.   Segment Organized for Persuasion.

From John H. Ryther, "Growth and Yield of Aquatic Plants," *in Cultivation of Macroscopic Marine Algae and Fresh Water Aquatic Weeds*, ed. John H. Ryther (Golden, Colorado: Solar Energy Research Institute, 1982), 91–93.

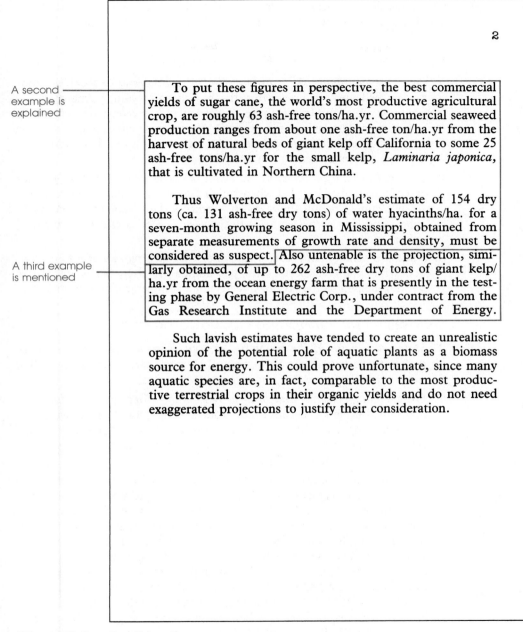

2

A second example is explained

> To put these figures in perspective, the best commercial yields of sugar cane, the world's most productive agricultural crop, are roughly 63 ash-free tons/ha.yr. Commercial seaweed production ranges from about one ash-free ton/ha.yr from the harvest of natural beds of giant kelp off California to some 25 ash-free tons/ha.yr for the small kelp, *Laminaria japonica,* that is cultivated in Northern China.
>
> Thus Wolverton and McDonald's estimate of 154 dry tons (ca. 131 ash-free dry tons) of water hyacinths/ha. for a seven-month growing season in Mississippi, obtained from separate measurements of growth rate and density, must be considered as suspect.

A third example is mentioned

> Also untenable is the projection, similarly obtained, of up to 262 ash-free dry tons of giant kelp/ha.yr from the ocean energy farm that is presently in the testing phase by General Electric Corp., under contract from the Gas Research Institute and the Department of Energy.

Such lavish estimates have tended to create an unrealistic opinion of the potential role of aquatic plants as a biomass source for energy. This could prove unfortunate, since many aquatic species are, in fact, comparable to the most productive terrestrial crops in their organic yields and do not need exaggerated projections to justify their consideration.

**Figure 12-9.**   *Continued*

in segmenting a process you focus on the steps of the process, and in discussing cause and effect you focus on the links between the steps.

## Guidelines for Describing Cause and Effect

1. *Begin by identifying the cause or effect you are going to describe.* Your readers will want to know from the beginning of your segment exactly what you are trying to explain so that they will know what they should be trying to understand as they read it. Sometimes you can describe the cause or effect in a single sentence. At other times a full description of it will take several sentences—or even longer.

2. *Carefully explain the links in the chain of cause and effect that you are describing.* Remember that when you are describing cause and effect, you are not simply listing the steps in a process. You want your readers to understand how each step leads to the next one or is caused by the preceding one.

3. *If you are dealing with several causes or effects, group them into categories.* When you create categories, you build a hierarchy that your readers will find helpful as they try to understand a complex chain of events.

## Guidelines for Persuading about Cause and Effect

1. *State your claim at the beginning of your segment.* Your claim will be that some particular effect was created by some particular cause, or that some particular cause will lead to a particular effect.

2. *Present your evidence and lines of reasoning.* Where possible, focus on undisputed evidence, because your readers' willingness to agree with your claim depends upon their willingness to accept your evidence. Use lines of reasoning your readers will agree are logically sound and appropriate to the situation you are discussing.

3. *Anticipate and respond to objections.* In cause and effect segments, as in any persuasive segment, your readers may object to your evidence or to your line of reasoning. The kind of reasoning that most often spurs objections by readers of cause and effect segments is one that contains the post hoc ergo propter hoc fallacy. In this form of faulty reasoning, a writer argues that because something happened after another event, it was caused by that event. You can see the flimsiness of this reasoning if you think back to the textile mill example. At the competitor's plant, profits may have risen after the introduction of the computers—but the rise in profits may have had many other causes: better designs, more sales of textiles generally throughout the industry, and so on. To prove that the computer caused the rise in profits, you would have to show by some direct chain of reasoning that the cause of the rise in profits was the introduction of the computers.

In responding to objections you expect your readers to raise, you can use any of the strategies discussed in the section on persuasive segments.

### Sample Cause and Effect Segment

Figure 12-10 shows a segment in which the author explains one theory about the cause of the extinction of the dinosaurs.

## PROBLEM AND SOLUTION

Like cause and effect segments, problem and solution segments fall into two categories: descriptive and persuasive. You can use problem and solution segments *descriptively* to explain things that have happened—how you and your co-workers solved the problem of the leaking reservoir, how you solved the problem of the large number of service calls on your employer's products, and so on. You can use problem and solution segments *persuasively* when you want your readers to agree with you that the particular actions you recommend will solve a problem that they would like to overcome.

### Guidelines for Describing Problems and Solutions

1. *Begin by identifying the problem that was solved.* Remember to make the problem seem significant to your readers. Emphasize the parts of the problem that were most directly addressed by the solution.

2. *Carefully explain the links between the problem and the solution.* Remember that you want your readers to understand *how* the solution overcame the problem.

3. *If the solution involves several actions, group the actions into categories.* When you create categories, you build a hierarchy that your readers will find helpful as they try to understand a complex solution.

### Guidelines for Persuading about Problems and Solutions

1. *Describe the problem in a way that makes it significant to your readers.* Remember that your overall aim is to persuade your readers to take the action you recommend because the action will solve some problem. Your readers will not be very interested in an action that solves a problem they don't care about.

2. *Present your evidence and lines of reasoning.* Be sure to use evidence that your readers will find sufficient and reliable, and lines of reasoning your readers will agree are logically sound and appropriate to the situation at hand.

3. *Anticipate and respond to objections.* In problem and solution segments, as in any persuasive segment, your readers may object to your evidence or

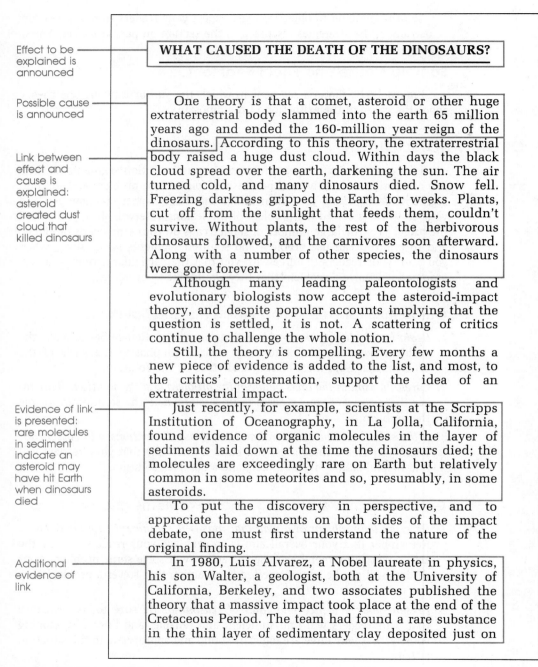

Effect to be explained is announced

Possible cause is announced

Link between effect and cause is explained: asteroid created dust cloud that killed dinosaurs

Evidence of link is presented: rare molecules in sediment indicate an asteroid may have hit Earth when dinosaurs died

Additional evidence of link

**WHAT CAUSED THE DEATH OF THE DINOSAURS?**

One theory is that a comet, asteroid or other huge extraterrestrial body slammed into the earth 65 million years ago and ended the 160-million year reign of the dinosaurs. According to this theory, the extraterrestrial body raised a huge dust cloud. Within days the black cloud spread over the earth, darkening the sun. The air turned cold, and many dinosaurs died. Snow fell. Freezing darkness gripped the Earth for weeks. Plants, cut off from the sunlight that feeds them, couldn't survive. Without plants, the rest of the herbivorous dinosaurs followed, and the carnivores soon afterward. Along with a number of other species, the dinosaurs were gone forever.

Although many leading paleontologists and evolutionary biologists now accept the asteroid-impact theory, and despite popular accounts implying that the question is settled, it is not. A scattering of critics continue to challenge the whole notion.

Still, the theory is compelling. Every few months a new piece of evidence is added to the list, and most, to the critics' consternation, support the idea of an extraterrestrial impact.

Just recently, for example, scientists at the Scripps Institution of Oceanography, in La Jolla, California, found evidence of organic molecules in the layer of sediments laid down at the time the dinosaurs died; the molecules are exceedingly rare on Earth but relatively common in some meteorites and so, presumably, in some asteroids.

To put the discovery in perspective, and to appreciate the arguments on both sides of the impact debate, one must first understand the nature of the original finding.

In 1980, Luis Alvarez, a Nobel laureate in physics, his son Walter, a geologist, both at the University of California, Berkeley, and two associates published the theory that a massive impact took place at the end of the Cretaceous Period. The team had found a rare substance in the thin layer of sedimentary clay deposited just on

**Figure 12-10.**    Segment Organized around Cause and Effect.

From Boyce Rensberger, "Death of Dinosaurs: A True Story?" *Science Digest* 94:5 (May 1986), 28–32.

2

Additional
evidence
continued

top of the highest, and therefore the most recent, stratum of rock contemporary with those bearing dinosaur fossils. It was the element iridium, which is almost nonexistent in the Earth's crust but 10,000 times more abundant in extraterrestrial rocks such as meteorites and asteroids. Deposits above and below the clay, which is the boundary layer separating the Cretaceous layer from the succeeding Tertiary, have very little iridium.

Because the same iridium anomaly appeared in two other parts of the world, in clay of exactly the same age, the Alvarez team proposed that the element had come from an asteroid that hit the Earth with enough force to vaporize, scattering iridium atoms in the atmosphere worldwide. When the iridium settled to the ground, it was incorporated in sediments laid down at the time.

Link is restated

More startling was the team's proposal that the impact blasted so much dust into the atmosphere that it blocked the sunlight and prevented photosynthesis (others suggested that a global freeze would also have resulted). They calculated that the object would have had to be about six miles in diameter.

Additional evidence of link

Since 1980, iridium anomalies have been found in more than 80 places around the world, including deep-sea cores, all in layers of sediment that formed at the same time.

Challenge to link is explained: molecules perhaps from volcano, not asteroid

One of the most serious challenges to the extraterrestrial theory came up very quickly. Critics said that the iridium could have come from volcanic eruptions, which are known to bring up iridium from deep within the Earth and feed it into the atmosphere. Traces of iridium have been detected in gases escaping from Hawaii's Kilauea volcano, for example.

Challenge is refuted by new evidence: other molecules couldn't have come from volcano

The new finding from Scripps appears to rule out that explanation, though, as a source for iridium in the Cretaceous–Tertiary (K–T) boundary layer. Chemists Jeffrey Bada and Nancy Lee have found that the same layer also contains a form of amino acid that is virtually nonexistent on Earth—certainly entirely absent from volcanoes—but abundant, along with many other organic compounds, in a type of meteor called a carbonaceous chrondrite.

**Figure 12-10.**    *Continued*

your lines of reasoning. In responding to objections you expect your readers to raise, you can use any of the strategies discussed in the section on persuasive segments.

### Sample Problem and Solution Segment

Figure 12-11 shows a memo organized into a discussion of a problem and its possible solution.

## EXERCISES

1. Some principles of classification you might use to group students include major and year in school. Think of five other principles you might use to classify students. Then identify a possible reader and purpose that would be served well by each. Are there any that seem to be useful to no one?

2. To choose the appropriate principle of classification for organizing a group of items, you need to consider your readers and purpose. Below you will find four topics for classification, each listed with two possible readers. First, identify a purpose that each reader might have for using a communication on that topic. Then identify a principle of classification that would be appropriate for each reader and purpose.

   Types of instrument or equipment used in your field
   Student majoring in your field
   Director of purchasing at your future employer's organization

   Types of communications you might write on the job
   You
   Your future typist

   Intramural sports
   Director of intramural sports at your college
   Student

   Flowers
   Florist
   Owner of a greenhouse that sells garden plants

3. Use classification to create a hierarchy having at least two levels. Some suggested topics are listed on page 260. After you have selected a topic, identify a reader and purpose for your classification. Depending upon your instructor's request, show your hierarchy in an outline or use it to write a brief discussion of your topic. In either case, state your principle or basis of classification. Have you created a hierarchy that, at each level, has one and only one place for every item?

MANUFACTURING PROCESSES INSTITUTE
Interoffice Memorandum                              June 21, 19--

To:      Cliff Leibowitz

From:  Candace Olin

RE:      Suggestion to Investigate Kohle Reduktion
          Process for Steelmaking

As we have often discussed, it may be worthwhile to set up a project investigating steelmaking processes that could help the American industry compete more effectively with the more modern foreign mills. I suggest we begin with an investigation of the Kohle Reduktion method, which I learned about in the April 1986 issue of High Technology.

Problem is identified — A major problem for American steelmakers is the process they use to make the molten iron ("hot metal") that is processed into steel.

Problem is explained — Relying on a technique developed on a commercial scale over 100 years ago by Sir Henry Bessemer, they make the hot metal by mixing iron ore, limestone, and coke in blast furnaces. To make the coke, they pyrolize coal in huge ovens in plants that cost over $100 million and create enormous amounts of air pollution.

Solution is announced — In the Kohle Reduktion method, developed by Korf Engineering in West Germany, the hot metal is made without coke.

Solution is explained — Coal, limestone, and oxygen are mixed in a gasification unit at 2500°. The gas rises in a shaft furnace above the gasification unit, chemically reducing the iron ore to "sponge iron." The sponge iron then drops into the gasification unit, where it is melted and the contaminants are removed by reaction of the limestone. Finally, the hot metal drains out of the bottom of the gasifier.

Link between problem and solution is explained — The Kohle method, if developed satisfactorily, will have several advantages. It will eliminate the air pollution problem of coke plants, it can be built (according to Korf estimates) for 25% less than conventional furnaces, and it may cut the cost of producing hot metal by 15%.

This technology appears to offer a dramatic solution to the problems with our nation's steel industry: I recommend that we investigate it further. If the method proves feasible and if we develop an expertise in it, we will surely attract many clients for our consulting services.

**Figure 12-11.**   Segment Organized around a Problem and Its Solution.

Boats

Cameras

Pets

Computers

Footwear

Physicians

The skills you will need on the job

Tools, instruments, or equipment you will use on the job

Some group of items used in your field (for example, rocks if you are a geologist, or power sources if you are an electrical engineer)

4. Partition an object in a way that would be helpful to someone who wants to use it. Some topics are suggested below. Whichever item you choose, describe one specific instance of it. For example, describe a certain brand and model of food processor, not a generic food processor. Be sure that your hierarchy has at least two levels, and state the basis of partitioning you use at each one. Depending upon your instructor's request, show your hierarchy in an outline or use it to write a brief discussion of your topic.

Food processor

Microwave oven

Theater

Bicycle

Some instrument or piece of equipment used in your field that has at least a dozen parts

5. Segment a procedure to create a hierarchy that you could use in a set of instructions. Give it at least two levels. Some topics are listed below. Show the resulting hierarchy in an outline. Be sure to identify your readers and purpose. If your instructor requests, use the outline to write a set of instructions.

Changing an automobile tire

Making a pizza from scratch

Making homemade yogurt

Starting an aquarium

Rigging a sailboat

Planting a garden

Developing a roll of film

Some procedure used in your field that involves at least a dozen steps.

Some other procedure of interest to you that includes at least a dozen steps

6. Segment a procedure to create a hierarchy that you could use in a general description of a process. Give it at least two levels. Some suggested topics are listed below. Show the resulting hierarchy in an outline. Be sure to identify your readers and purpose. If your instructor requests, use the outline to write a general description of the process addressed to someone who wants to understand it.

How a photocopy machine makes photocopies

How the human body takes oxygen from the air and delivers it to the parts of the body where it is used

How television signals from a program originating in New York or Los Angeles reach television sets in other parts of the country

How aluminum is made

Some process used in your field that involves at least a dozen steps

Some other process of interest to you that includes at least a dozen steps

7. In courses in your major field, you may be learning about alternative ways of doing things. For instance you may be learning how to keep patient records for a doctor's office on paper and on a computer. Write an outline comparing a pair of alternatives from the point of view of someone who has a practical need to choose between them. Use the alternating pattern of organization. If your instructor requests, use that outline to write a memo to the person you had in mind.

8. Imagine that one of your friends is thinking of making a major purchase. Some possible items are listed below. Help your friend by creating an outline having at least two levels that compares two or more good alternatives. If your instructor requests, use that outline to write your friend a letter.

Stereo

Binoculars

VCR

Personal computer

Bicycle

Some other type of product for which you can make a meaningful comparison on at least three important points

9. Think of some way that things might be done better in a club, business, or other organization you know of. Imagine that you are going to write

a letter to the person or people who actually have the power to bring about the change you desire. Write an outline having at least two levels that compares the way you think things should be done and the way they are done now.

10. Think of some way that things might be done better in a club, business, or other organization you know of. Imagine that you are going to write a letter to the person or people who actually have the power to bring about the change you desire. List the various specific claims you would make on behalf of the way you think things should be done. Identify the evidence you would offer and lines of reasoning you would use.

11. Imagine that a friend of yours has asked you to explain the causes of some occurrence or event. Some suggested topics are listed below. Write that person a brief letter explaining those causes.

> Sunspots
>
> Static on radios and televisions
>
> Freezer burn in foods
>
> Twin births
>
> Yellowing of paper
>
> Earthquakes

12. Think of a problem that you have some ideas about solving. The problem might be noise in your college's library, shoplifting from a particular store, or parking problems on campus. Then, briefly describe the problem and list actions you would take to solve it. Next, explain how each action will contribute to solving the problem. If your instructor requests, use your outline to write a brief memo explaining the problem and your proposed solution to a person who could actually take the actions you suggest.

13. Large segments that are structured according to one pattern often contain smaller segments structured according to another. The following table identifies the major pattern in each of four sample passages in this chapter. It also identifies a different minor pattern used in each passage. Reread the passages, locating the minor patterns listed. Can you find other minor patterns also?

| Figure Number | Overall Pattern | Smaller Pattern Contained in It |
|---|---|---|
| 12-8 | Comparison | Persuasion |
| 12-9 | Persuasion | Classification |
| 12-10 | Cause and Effect | Persuasion |
| 12-11 | Problem and Solution | Process |

# CHAPTER 13

# Writing the Beginning of a Communication

## PREVIEW OF GUIDELINES

1. Imagine your readers' thoughts, feelings, and expectations as they first pick up your communication.
2. Tell your readers how your communication is relevant to them.
3. Tell how your communication is organized.
4. Tell the scope of your communication
5. Tell your main points.
6. Provide the background information your readers will need to understand your communication.

I magine that you are attending the first day of a new course. You don't know anyone in the room. Afterwards, the person sitting next to you tries to start a conversation. How will you respond?

Your response will probably depend upon the other person's opening words. The subject this person talks about, the person's tone of voice, even the way the person phrases his or her comments may all determine whether you decide to chat or rush off to your next destination. If you stay, you will probably shape your own comments in light of the first things the other person says.

The opening words of the memos, reports, and other things you will write at work are very much like the opening words of a conversation. They may greatly influence the way your readers respond to what you have to say. They may determine whether or not your readers decide to read all of your message.

This chapter presents six guidelines for beginning your communications.

## THE TWO FUNCTIONS OF A BEGINNING

Before reading those guidelines, you may find it helpful to think about the two distinct functions that the beginning of a communication performs. First, the beginning introduces your readers to your message. You use it to persuade your readers to read attentively, and through it you try to shape your readers' response to what follows.

Second, the beginning of a communication serves as the beginning of a segment. As you will recall from reading Chapter 11, a segment is a group of related sentences. In any communication, the largest segment is the communication as a whole. For that reason, the things that Chapter 11 advised you to do at the beginning of any segment are things you should do when you write the beginning of the whole communication:

- Tell what your segment is about.
- Tell how it is organized.
- Generalize about its contents in a reader-centered way.

Those important pieces of advice are incorporated in the discussions of Guidelines 2, 3, and 5 of this chapter so that you can see how to apply them when you are writing the beginning of a communication.

Guideline

1

**Imagine Your Readers' Thoughts, Feelings, and Expectations as They First Pick Up Your Communication**

Following this guideline will help you write beginnings that successfully perform their first function, that of introducing your communication to your readers. Imagine three situations:

- You have written a manual for your company's computer system. A new employee sits down at a workstation for the first time and opens your manual.
- You have written a letter to a client to explain why you are behind schedule on a project she is paying your organization to conduct. She sits at her desk, opening your letter.
- You have written a memo to your boss. In it you suggest, for the first time ever, a substantial change in the way your department organizes its work. Your boss, who designed the current system, finds your memo on his desk.

If you were to approach these three individuals in person, you surely would begin your conversation with each in a different way. Similarly, if you were to write to the three, you would begin each communication differently. Your strategy for opening your communication to each individual should be based upon a thorough understanding of that particular person's thoughts, feelings, and expectations.

## How To Develop an Understanding of Your Readers' Initial Reactions

How can you develop a thorough understanding of your readers' mental state as they begin reading your communications? Sometimes you will know your readers well enough to anticipate their initial reactions without any special effort. At other times, you will need to think systematically about the many characteristics that will determine how your readers will respond to your message. These characteristics include the following:

- Readers' roles (decision-maker, adviser, implementer)
- Readers' jobs (for example, assembly-line worker, researcher, manager)
- Readers' technical or scientific specialties
- Readers' goals
- Readers' personalities
- Readers' attitudes as a result of previous relationship with you or your organization

Keep in mind, also, that your readers' expectations will be shaped by the conventions of the *type* of communication you are writing. We all expect that the beginning of a research article in a scientific journal will be much different from the beginning of a brief memo from a co-worker.

## Sometimes Telling a Story Can Help You

Although it is not necessary for you to do so, in some situations you may find that you can organize your knowledge of your readers in a particularly helpful way if you create a story about them. The central figure in your story should be your principal reader if you are writing primarily to one person or a typical reader if you are writing to a group of people. Begin the story a few minutes before this reader places a hand on your communication, and carry the story through the moment the reader begins to read your first words. This story can help you learn about your reader in a way that will assist you in deciding how best to introduce your communication. Keep in mind that your story's purpose is to help you decide how to introduce your communication to your readers. It is not something that you would actually include in your communication.

## Sample Story

Here is a sample story, written by a person preparing to write the beginning of an instruction manual for a word-processing system used at a newspaper:

> It's Sunday afternoon. Bob, the new, green reporter, sits down in front of the video display terminal (VDT) for the first time to type out his story, which is due in eighty minutes. Never having used computerized word-processing equipment before, he is confused and quite nervous. He knows that the equipment is expensive, and he does not want to damage it in any way.
>
> Bob has already been briefed on the use of the VDT, but that was last week, and he has forgotten much of what he was told. He will shy away from asking questions of the more experienced writers because (being new to the staff) he doesn't want to make a bad impression by asking dumb questions.
>
> Bob picks up the instruction manual I am writing for the VDT, hopeful that it will tell him quickly what he needs to know. He wants it to help him get his story done on time without breaking the machine and without asking embarrassing questions.

Through this story, the writer focused on several important facts about the reader—that the reader will be anxious, hurried, and (most likely) interested in accomplishing something immediately. All these insights helped the author write the opening to her manual. Here are the first few sentences:

> This manual tells you how to enter a story on the *Chronicle*'s VDT system. It covers the steps for opening a file for your story, establishing the format, typing and revising the copy, and sending it to an editor for editing.
>
> Be sure to follow the instructions carefully, so that you can avoid making time-consuming errors. At the same time, you should know that the *Chronicle*'s overall system is designed so that you cannot damage anyone else's file or the system itself.

Guideline

2

# Tell Your Readers How Your Communication Is Relevant to Them

The preceding discussion has emphasized that each individual reader comes to a communication with somewhat different thoughts, feelings, and expectations. There is one question, however, that almost all readers ask as they pick up any communication: "Why should I read this?" If you don't provide a persuasive answer quickly, your communications may go unread. That's because at work everyone is very busy with many responsibilities. To read something you have written, they may have to take time and energy away from their other duties. Consequently, you must persuade them from the outset that reading your communication is an important thing for them to do.

## Explaining Relevance versus Announcing the Topic

To persuade people to read a communication you have written, you must persuade them that you are going to say something related to their professional responsibilities, goals, desires, or interests. To do that you must, first, follow a guideline from Chapter 11: "At the beginning of each segment, tell what it is about." However, to persuade your readers that your communication is relevant to them, you must do more than merely announce your topic. You must also explain how your treatment of your topic will be useful to your readers. Compare the following sets of statements:

| Announcements of Topic (Avoid them) | Statements of Relevance (Use Them) |
|---|---|
| This memo tells about polymer coatings. | → This memo answers each of the five questions you asked me last week about polymer coatings. |
| This report discusses step-up pumps. | → Step-up pumps can save us money and increase our productivity. |
| This manual concerns the Cadmore Industrial Robot 2000. | → This manual tells you how to prepare the Cadmore Industrial Robot 2000 for routine welding tasks. |

## What a Statement of Relevance Says

The three sample statements of relevance that you have just read are all very short—only a single sentence each. Sometimes, you will need to write much longer statements of relevance, perhaps several pages long. Regardless of their length, however, all good statements of relevance ensure that the readers know the answers to all of the following three questions:

1. *What problem is this communication going to help us solve?* At work, people read to obtain information they need to fulfill their responsibilities as employees. The specific responsibilities they have will depend upon their particular job and professional role. In general, however, all have the same responsibility: to solve problems for their employers. These problems take many forms. Civil engineers are responsible for problems involved with roadways, bridges, and other structures their employers build. Systems analysts are responsible for problems involved with designing computer services that their employers want. Production managers are concerned with improving the productivity and reducing the costs of making the goods their employers manufacture. Although different readers work on different kinds of problems, all recognize the kind of problems they are responsible for—and they devote their time and energy to solving those types of problems but not other types. The important thing for you to remember is that, to persuade your readers that your communication is relevant to them, you must explain how it will help solve a problem that is significant to them. You must present yourself as your readers' partner, one who will assist them with a problem they are responsible for solving.

2. *What work has the writer done toward solving the problem?* To help your readers understand how your communication will help them solve the problem you have identified, you need to tell them what special activities you have undertaken toward solving it. These activities are the things you have done as a specialist in your field: developing a new formula for one of your employer's products, investigating the products offered by competitors, writing computer programs, or whatever. This description of your own activities is the part of your statement of relevance that will usually be easiest for you to write, but remember that your activities are significant to your readers only in terms of the problem you have identified.

3. *How will this communication help us with our efforts to solve the problem?* Your activities are significant to your readers only if you present the results in a way that will help your readers with their problem-solving efforts. For example, your report on your new formula for car wax is relevant only if you present your results in a way that enables your readers to develop a process for manufacturing the formula commercially. Similarly, your investigation of competitor's products is relevant only if you present your information in a way that helps your readers decide how to improve the appeal of your employer's product. Consequently, the third element of a complete statement of relevance is an indication of how your communication will help your readers in their own efforts to solve the problem.

## An Example

To see how to answer those questions, consider the situation of Carl, a junior systems analyst employed by a large hotel chain. Carl is writing a report based upon a trip to Houston, where he studied the billing system used by one of the hotels that his employer recently purchased. So far, he has drafted the body of his report, and now he must write the beginning, where he will explain the report's significance to his readers.

Carl starts by answering the first of the readers' questions given above, "What problem is this communication going to help us solve?" To answer, he simply retells the reason his boss asked him to go to Houston: the Houston hotel is making very little money and the home office suspects that a major reason is that the hotel is using a faulty billing system.

To answer the readers' second question ("What has the writer done toward solving the problem?"), Carl tells what he did in Houston and afterwards to evaluate the hotel's billing system and to recommend improvements in it.

To the readers' third question ("How will this communication help us with our efforts to solve the problem?"), Carl explains that his report provides the results of his analysis, together with his recommendations. Using this information, his readers can decide what action to take.

Here is the beginning Carl wrote for his report:

Overall problem Carl will help his readers solve

Over the past two years, our Houston hotel has shown a profit of only 4% even though it is almost always at least 78% filled. A preliminary examination of the hotel's operations suggests that its billing system may be inadequate: it may be too slow getting bills to customers, and it may be inefficient and needlessly unsuccessful in collecting overdue payments.

Carl's work toward solving that problem

Therefore, I have thoroughly examined the hotel's billing cycle and its collection procedures, and I have considered ways of improving them.

How Carl's report will help his readers with their efforts to solve the problem

In this report, I present the results of my analysis, together with my recommendations. To aid in the evaluation of my recommendations, I have included an analysis of the costs and benefits of each.

## Do You Always Need To Provide All This Information?

Statements of relevance can be much shorter than Carl's. Sometimes they need to be much longer. How can you decide how long your statement of relevance needs to be?

The crucial point is that your readers know the answer to each of the three questions listed above (What problem will your communication help your readers solve? What work have you done toward solving that problem? How will your communication help your readers solve the problem?).

In many situations, your readers will know the answer to the first and possibly the second question without being told explicitly. For example, if you are writing in response to your readers' request for information, you may not need to tell them what problem they were working on when they asked you to supply that information. It may be sufficient for you to remind your readers that you are providing the requested material. In such a situation, your activities toward solving the problem may likewise be obvious to your readers.

Here, for example, is the brief explanation of relevance that Carl might use if he were sure all his readers were familiar with the problem at the Houston hotel and with his activities there:

> In this report, I evaluate the billing system in our Houston hotel and recommend ways of improving that system.

Before using such an abbreviated statement of relevance, you should be sure that *all* the likely readers of your report will be able to understand *immediately* its relevance to their employer's goals without your stating its relevance more fully.

### Some Signs That You Should Provide a Full Explanation of Relevance

Here are some signs that you probably *should* provide a full explanation of relevance.

- *Your report will be read by more than two or three people.* The more readers your report will have, the less likely it is that all of them will be very familiar with the context of your report.

- *Your report will be longer than a page or two.* Long reports are often read by many people.

- *Your report will have a binding and a cover.* Reports prepared in this way are usually prepared for large groups of readers. They may also be filed so that people years in the future can use them. Unless you tell them, many readers of bound reports will have no idea of the organizational situation in which you wrote.

- *Your report will be used to make a decision that involves a significant amount of money.* Such decisions are often made at the upper levels of management. These managers usually need to be told of the organizational context of the reports they read. For example, although Carl's boss asked him for the report, the people who will make a decision based upon his recommendations are high-level managers responsible for the overall operation of 247 hotels. Carl cannot assume that they know in advance about the situation in Houston.

## Finding Out What the Problem Is

You may find it helpful to think further about the way that you answer the first of the readers' three questions about relevance: "What problem is this communication going to help us solve?"

Sometimes, you will have no difficulty answering that question. In the example given above, Carl could answer it quite easily because his boss told him what the problem was when he gave Carl his assignment.

In two other situations, however, you will have to devote a special effort to determining what problem your communication will help solve. The first will occur when (and if) your boss gives you an assignment without telling you why. In those cases you must *ask*. The question to ask is this, "What problem is the organization I work for trying to solve by asking me to perform this assignment?"

You will also have to make a special effort to determine the problem when you are writing *unsolicited* communications (communications you prepare on your own initiative, without having been asked to do so). Such situations may arise more often at work than you expect. According to a survey of college alumni, many write on their own initiative at least as often as they write on assignment.[1]

## Defining the Problem in Unsolicited Communications

When writing unsolicited communications, you can't ask someone else to tell you what the problem is. You must decide what it is on your own. Furthermore, you can't pick just any problem. You must identify one that your readers think important. Your efforts to solve a problem unimportant to your readers will seem irrelevant to them. And communications that are perceived to be irrelevant won't be read.

The key step in identifying a problem that is relevant to your readers is to forget (at least for the time being) your own reasons for writing. Instead, think about the situation you are writing about from the point of view of your readers.

Suppose, for instance, that you do not have a personal microcomputer to use at work and that you want to ask your boss to purchase you one. You may want this microcomputer primarily so that your job will be easier. In particular, you believe that it will enable you to complete much more rapidly a lengthy report that you must prepare at the end of each month. Currently, to complete that report on time each month you must work many overtime hours. At the same time, you know that your boss regularly works overtime, and that to your boss the problem of reducing the amount of overtime you work is not a top priority.

Accordingly, you must find another problem that the microcomputer will solve—a problem that will seem important to your boss. For example, you may know that your boss very much wants you to work on several projects that the two of you agree you do not have time to address. You might make that the problem you identify in your statement of relevance: with a microcomputer, you could finish all your work more rapidly, so that you would have some time to work on the projects that currently are neglected.

As this example makes clear, to explain the relevance of a communication, you must look at both the situation you are addressing and the communication itself from the point of view of your readers. In writing, there is no task for which taking a reader-centered approach is more critical than explaining the relevance of the things you have to say.

**Guideline**

**3**

## Tell How Your Communication Is Organized

This guideline echoes one given in Chapter 11: "At the beginning of each segment, tell how it is organized." The reasons for telling your readers about the organization of your overall communication are the same as for telling them about the organization of any other segment.

First, as readers try to understand your communication, they are helped by knowing how all the major parts of it fit together. Understanding the general framework of your message helps readers know what connections to draw among the various particular pieces of information you convey.

Second, as people read the beginning of a communication, they sometimes decide that they want to find certain pieces of information before (or instead of) reading the entire communication from front to back. If you tell how your communication is organized, you can help them find the information they want.

You can tell your readers about the organization of your communication in sentences:

> In this report, we state the objectives of the project, compare the three major technical alternatives, and present our recommendations. The final sections include a budget and proposed project schedule.

Or, you can describe your organization in a list:

> This manual includes the following sections:
> - Set-up
> - Calibration
> - Manual Operation
> - Automatic Operation
> - Troubleshooting

In some instances, your statement of the relevance of your communication also indicates its organization:

At our meeting on September 4, you expressed an interest in learning about the prospective market for each of the three major types of batteries that might be used in electrically powered cars in the next twenty years. In this memo, I describe each market and then compare them in terms of their likely profitability.

Guideline

4

## Tell the Scope of Your Communication

Readers often want to know, from the beginning, what a communication contains and what it doesn't contain. After all, even if you persuade them that you are writing on a subject that is relevant to them, your readers may still wonder whether you will discuss the specific aspects of the subject that they want to know about.

In many communications, you will have told your readers the scope of your communication if you follow the two preceding guidelines. For instance, if you explain the relevance of your communication (Guideline 2) by saying that you are providing the answers to the questions your readers asked you, then you are also telling them what the scope of the communication is. Similarly, when you tell how your communication is organized (Guideline 3), you indicate the topics you will treat in it.

However, there will be times when you will need to include additional information about the scope of your communication. That happens when you want to be sure that your readers understand that you are not treating your subject comprehensively or that you are treating your subject from a particular point of view. For instance, you may be writing a troubleshooting guide that a factory worker can use to solve certain kinds of problems with the manufacturing robots that he or she monitors. At the same time, there may be other problems that this factory worker should not try to solve. Such problems might require, for example, the assistance of a computer programmer or an electrical engineer. In that case, you should tell your readers explicitly about the scope of your troubleshooting manual:

> This manual treats problems that you can solve by using the tools and equipment normally available to you. It does not cover problems that require work by computer programmers or electrical engineers.

As another example, here is a statement of scope from a booklet about the B36-7 diesel-electric locomotive.

> This pamphlet is intended for use by maintenance personnel. It will enable them to become acquainted with the B36-7 locomotive in general and specifically with new electrical and mechanical details, many of which have not been previously used on any other locomotive.
>
> Detailed coverage of every one of the newer features in the B36-7 model and others in the Series-7 line is not intended. Instead, the focus is on the more significant departures from the usual equipment on General Electric locomotives, both in arrangement and function.

## Tell Your Main Points

This guideline is a special version of a guideline from Chapter 11: "At the beginning of each segment, generalize about its contents in a reader-centered way." The generalization that your readers want to find at the beginning of your communication is the one that expresses the main points you want to make.

To express their desire to learn the main points immediately when they read, people at work often urge writers to "put the bottom line first." The "bottom line," of course, is the bottom line in a financial statement. Literally, the writers are being told, "Before you enumerate the individual expenditures and the individual sources of income, tell your readers what they most want to know: whether we made a profit or took a loss."

However, this advice is applied to many kinds of communications prepared at work, not just to financial reports. Before people at work read the details of a report, proposal, or other decision-making document, they want to know the main points.

As a writer, you have three good reasons for trying to satisfy this desire.

First, the bottom line is usually the information that is *most important* to your readers. For example, if your boss has asked you to provide a recommendation, the thing your boss wants most from your memo or report is to learn what you have decided to recommend. Also, some decision-making readers don't need to know much more than the bottom line. They let their advisers work with the details. By putting the most important information first, you help your readers learn what they most want or need to know without requiring them to hunt extensively for it.

Second, by putting your most important points first, you increase the likelihood that your readers will *actually read* them. On the job many readers have so little time to read that they may put your communication aside if they can't find the main point quickly. Or, they may mistakenly assume that some minor point is the main one.

Third, even when your readers are going to read your entire communication from front to back, they will be able to *read more efficiently* if you state the most important points at the beginning. In most communications, the many details you give your readers derive their significance from their relationship to the main points. For example, if you are requesting $175,000 to renovate your employer's machine shop, the details you include when explaining the current problems in the shop take their significance in terms of your overall request. Similarly, the details of your test results are significant in terms of your recommendations for redesigning a key component of your company's major product. By giving your main point first, you let your readers see the significance of your more detailed information as they encounter it. If you don't give your main points first, you will require your readers to read details without knowing exactly what their significance is.

Guideline

6

## Provide the Background Information Your Readers Will Need To Understand Your Communication

As you are drafting the beginning of your communications, you should ask yourself whether you need to provide your readers with *general* background information that will enable them to understand what you have to say.

Here are examples of situations in which you may need to provide such background information at the beginning:

- *Your readers must learn some general principles to be able to understand your specific points.* For instance, your discussion of the feasibility of locating a new plant in a particular city may depend upon a particular analytical technique that you will need to explain to your readers.
- *Your readers are unfamiliar with some basic technical terms that you will use throughout your communication.*
- *You are writing instructions in which certain steps will be implicit.* For instance, in a manual on computer programming you might say, "Each time you type a command, you must follow it by pressing the RETURN key. To avoid unnecessary repetition, the instruction to press the RETURN key is *not* given in this manual."

There are some kinds of background information that don't belong at the beginning of your communication. Background explanations that pertain only to small, specific places in your communication should be provided at the beginning of those segments. In the beginning of your whole communication, include only the general explanations that will help the readers overall.

## APPLYING THESE GUIDELINES

When you apply the guidelines for writing the beginning of a communication, use your common sense. In some situations, a good, reader-centered beginning takes many paragraphs—even many pages. However, in some situations it takes only a sentence, or less.

### How To Decide How Long a Beginning To Write

How long does the beginning of a particular communication need to be? The answer depends upon how much information you need to provide your readers to be sure they know the following:

- Why they should read the communication (Guideline 2).
- How the communication is organized (Guideline 3).
- What the scope of the communication is (Guideline 4).
- What the main points of the communication are (Guideline 5).
- All the background information they need in order to understand and use the communication (Guideline 6).

If you are sure that your readers have each of these kinds of information, then you have written a complete beginning—regardless of how long or short it is.

## Examples of Brief Beginnings

Here is an opening prepared by a writer who followed all the guidelines in this chapter:

> In response to your memo dated November 17, I have called Goodyear, Goodrich, and Firestone for information about the ways they forecast their needs for synthetic rubber. The following paragraphs summarize each of those phone calls.

The following opening, from a two-paragraph memo, is even briefer. At first glance, this beginning may seem to have been written by a person who ignored all the guidelines given in this chapter:

> We are instituting a new policy for calculating the amount that employees are paid for overtime work.

This single sentence, however, *does* exemplify all the guidelines given in this chapter. From reading the topic of the memo (overtime pay), the particular people addressed by this writer would immediately understand the memo's relevance to them. This sentence also tells the main point of the memo: a new policy is being instituted. Furthermore, because the memo is only two paragraphs long, the scope of the memo is readily apparent without explicit comment about it. The brevity of this memo also suggests its organization, namely a brief explanation of the new policy—and nothing else. The writer correctly judged that the readers need no background information.

## Examples of Long Beginnings

Figure 13-1 shows a relatively long beginning from a recommendation report written by a consulting firm that was hired to recommend ways to improve the food service at a hospital. Figure 13-2 shows the long beginning of a 500-page service manual for the Detroit Diesel 53 engine, manufactured by General Motors.

INTRODUCTION

Problem

Wilton Hospital has added 200 patient beds through construction of the new West Wing. Since the wing was opened, the food service department has had difficulty meeting this extra demand. The director of the hospital has also reported the following additional problems.

Subpart of overall problem

1. Difficulties operating at full capacity. The equipment, some of it thirty years old, breaks down frequently. Absenteeism has risen dramatically.

Subpart of overall problem

2. Costs of operation that are well above average for the hospital industry nationally and in this region.

Subpart of overall problem

3. Frequent complaints about the quality of the food from both the patients and the hospital staff who eat in the cafeteria.

What the writers have done to help solve the problem

To study these problems, we have monitored the operation of the food service department, and interviewed patients, food service employees, and staff who eat in the cafeteria. In addition, we have compared all aspects of the department's facilities and operations with those at other hospitals of roughly the same size.

How this report will help the readers

How the report is organized

In this report, we discuss our findings concerning the food service department's kitchen facilities. We briefly describe the history and nature of these facilities, suggest two alternative ways of improving them, and provide a budget for each. In the final section of this report, we propose a renovation schedule and discuss ways of providing food service while the renovation work is being done. (Our recommendations about staffing and procedures will be presented in another report in thirty days.)

Scope

Main points

The first alternative costs about $730,000 and would take four months to accomplish. The second costs about $1,100,000 and would take five months. Both will meet the minimum needs of the hospital; the latter can also provide cooking for the proposed program of delivering hot meals to house-bound persons in the city.

**Figure 13-1.**   Beginning for a Recommendation Report.

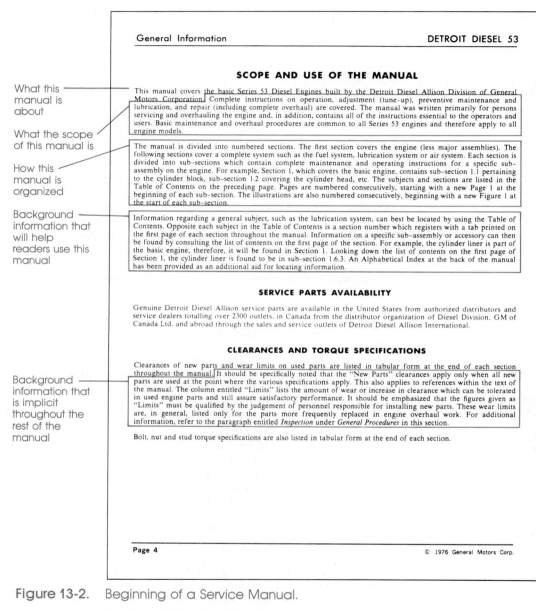

What this manual is about

What the scope of this manual is

How this manual is organized

Background information that will help readers use this manual

Background information that is implicit throughout the rest of the manual

General Information                                                                      DETROIT DIESEL 53

## SCOPE AND USE OF THE MANUAL

This manual covers the basic Series 53 Diesel Engines built by the Detroit Diesel Allison Division of General Motors Corporation. Complete instructions on operation, adjustment (tune-up), preventive maintenance and lubrication, and repair (including complete overhaul) are covered. The manual was written primarily for persons servicing and overhauling the engine and, in addition, contains all of the instructions essential to the operators and users. Basic maintenance and overhaul procedures are common to all Series 53 engines and therefore apply to all engine models.

The manual is divided into numbered sections. The first section covers the engine (less major assemblies). The following sections cover a complete system such as the fuel system, lubrication system or air system. Each section is divided into sub-sections which contain complete maintenance and operating instructions for a specific sub-assembly on the engine. For example, Section 1, which covers the basic engine, contains sub-section 1.1 pertaining to the cylinder block, sub-section 1.2 covering the cylinder head, etc. The subjects and sections are listed in the Table of Contents on the preceding page. Pages are numbered consecutively, starting with a new Page 1 at the beginning of each sub-section. The illustrations are also numbered consecutively, beginning with a new Figure 1 at the start of each sub-section.

Information regarding a general subject, such as the lubrication system, can best be located by using the Table of Contents. Opposite each subject in the Table of Contents is a section number which registers with a tab printed on the first page of each section throughout the manual. Information on a specific sub-assembly or accessory can then be found by consulting the list of contents on the first page of the section. For example, the cylinder liner is part of the basic engine, therefore, it will be found in Section 1. Looking down the list of contents on the first page of Section 1, the cylinder liner is found to be in sub-section 1.6.3. An Alphabetical Index at the back of the manual has been provided as an additional aid for locating information.

## SERVICE PARTS AVAILABILITY

Genuine Detroit Diesel Allison service parts are available in the United States from authorized distributors and service dealers totalling over 2300 outlets. in Canada from the distributor organization of Diesel Division. GM of Canada Ltd. and abroad through the sales and service outlets of Detroit Diesel Allison International.

## CLEARANCES AND TORQUE SPECIFICATIONS

Clearances of new parts and wear limits on used parts are listed in tabular form at the end of each section throughout the manual. It should be specifically noted that the "New Parts" clearances apply only when all new parts are used at the point where the various specifications apply. This also applies to references within the text of the manual. The column entitled "Limits" lists the amount of wear or increase in clearance which can be tolerated in used engine parts and still assure satisfactory performance. It should be emphasized that the figures given as "Limits" must be qualified by the judgement of personnel responsible for installing new parts. These wear limits are, in general, listed only for the parts more frequently replaced in engine overhaul work. For additional information, refer to the paragraph entitled *Inspection* under *General Procedures* in this section.

Bolt, nut and stud torque specifications are also listed in tabular form at the end of each section.

**Page 4**                                                                  © 1976 General Motors Corp.

**Figure 13-2.** Beginning of a Service Manual.

From General Motors, *Detroit Diesel Engines Series 53 Service Manual* (Detroit, Michigan: General Motors, 1980), 4. Courtesy of Detroit Diesel Allison Division, General Motors Corporation.

# SUMMARY

The opening of something you write will determine whether your readers continue to read, it will shape your readers' response to the rest of your message, and (if it is written well) it will help your readers read and use the rest of your message.

This chapter presents six guidelines for writing effective, reader-centered beginnings.

1. *Imagine your readers' thoughts, feelings, and expectations as they first pick up your communication.* As you do this, keep in mind your readers' roles, jobs, technical specialties, goals, personalities, and attitudes. In some situations, you may find it helpful to use your knowledge of your readers to tell a story about a key or typical reader picking up the communication you are writing.

2. *Tell your readers how your communication is relevant to them.* A good statement of relevance will answer three questions that readers ask: (1) What problem is this communication going to help us solve? (2) What work has the writer done toward solving the problem? and (3) How will this communication help us with our efforts to solve the problem?

3. *Tell how your communication is organized.* By doing so, you will help your readers understand the general framework of your message so that they will know what connections to draw among the various pieces of detailed information you convey. You will also help readers who desire to find certain pieces of information before (or instead of) reading the entire communication.

4. *Tell the scope of your communication.* Readers often want to know, from the beginning, what a communication contains and what it doesn't contain. This is especially important when you want to be sure that your readers understand that you are not treating your subject comprehensively or that you are treating your subject from a particular point of view.

5. *Tell your main points.* By telling your main points first, you provide, right away, the information that is most important to the readers—what they most want or need to know. You also increase the likelihood that your readers will actually read the main points. And you help your readers read efficiently because they will be able to see the significance of your more detailed information as they encounter it.

6. *Provide the background information your readers will need to understand your communication.* You may need to provide general background information if (1) your readers must learn some general principles to be able to understand your specific points, (2) your readers are unfamiliar with some basic technical terms that you will use in your communication, or (3) you are writing instructions where certain steps will be implicit.

## NOTE

1. Paul V. Anderson, "What Survey Research Tells Us about Writing at Work," in *Writing in Nonacademic Settings*, ed. Lee Odell and Dixie Goswami (New York: Guilford Press, 1985), 3–85.

## EXERCISES

1. The passages below are the openings to two memos. For each, identify the statements of relevance, organization, scope, main points, and background information.

   A. On August 3, 19——, Mordon Enviro-Chem asked Midlands Research to provide cost and engineering data for the recovery of elemental sulfur from three acid-gas waste streams of specified composition. In response, Midlands has negotiated estimates from several firms that design and build sulfur recovery facilities. This report summarizes information that management can use to compare the capital costs and operating data for the facility designed by Dodge, Hame, and Jefferson with similar data for facilities designed by the other firms.

   B. Plastic bottles are currently being used as containers for various soft drinks, and they are also being considered as containers for alcoholic beverages. For the latter use, one of the main concerns is that the alcohol may leach out some of the MHP polymer and make it a food additive subject to FDA scrutiny. Consequently, a gin sample was stored 6 months at 100 degrees F in a 12-ounce plastic bottle and then analyzed for MHP content. Those results were compared with the results of a UV analysis performed upon gin from the same distillery that was stored in glass under the same conditions. The tests indicated that while some MHP polymer is leached, the amount leached is below the minimum amount specified by the FDA.

2. Select a communication written to people in your field. This might be a letter, memo, manual, or report. (Do not choose a textbook.) Analyze the beginning of this communication in terms of the guidelines given in this chapter.

3. The instructions for many consumer products contain no beginnings at all. For instance, the instructions for some lawnmowers, cake mixes, and detergents simply provide a heading that says "Instructions" and then start right in. Find such a set of instructions and evaluate the writer's decision to omit a beginning in terms of the guidelines given in this chapter.

4. The following paragraphs are from the beginning of a long memo in which the manager of a forging department asks for better abrasives from a department that is providing unsatisfactory supplies. Does this seem to be an effective beginning for a memo that is essentially a complaint? Why or

why not? Analyze this beginning in terms of the guidelines given in this chapter.

I am sure that you have heard that the new forging process is working well. Our customers have expressed pleasure with our castings. Thanks, again, for all your help in making this new process possible.

We are having one problem, however, with which I have to ask once more for your assistance. During the seven weeks since we began using the new process, the production line has been idle 28% of the time. Also, many castings have had to be remade. Some of the evidence suggests that the problems are caused by the steel abrasive supplies we get from your department. If we can figure out how to improve the abrasive, we may be able to run the line at 100% of capacity.

I would be most grateful for help from you and your people in improving the abrasive. To help you devise ways of improving the abrasive, I have compiled this report, which describes the difficulties we have encountered and some of our thinking about possible remedies.

5. Write the beginning for a project that you are working on in your class.

# Writing the Ending of a Communication

## PREVIEW OF GUIDELINES

1. After you've made your last point, stop.
2. Repeat your main point.
3. Summarize your key points.
4. Focus on a key feeling.
5. Tell your readers how to get assistance or more information.
6. Tell your readers what to do next.
7. Identify any further study that is needed.
8. Refer to a goal stated earlier in your communication.
9. Follow the social conventions that apply to your situation.

I magine that you are talking with a friend. Suddenly, she turns around and leaves—without any warning at all. Or, imagine that at the time set for the end of your class, your instructor simply stops talking and walks out of the room, without making any concluding remarks. Situations like these would probably leave you a little unsettled, a little uncertain. Why didn't your friend say goodbye? Why didn't your instructor bring the class to a close? What do they want you to think? To do?

As these examples suggest, the end of a conversation is very special. We expect other people to end their conversations with us in certain ways, and they have the same expectations of us. Similar expectations apply to much of the writing you will do on the job. To maintain an effective relationship with your readers, you need to satisfy those expectations.

The ending of a communication is important for another reason as well. Endings create your readers' final impressions while reading your communications. They shape the last reaction your readers will have to your message before setting your pages aside. The nine guidelines in this chapter will help you write endings that satisfy your readers' expectations while also helping you achieve your overall objectives.

## THE THREE AIMS OF AN ENDING

In the communications you write at work, the endings will enable you to do the following important things.

- *Provide your readers with a sense of completion.* The most general aim of an ending is to provide your readers with the satisfying sense that something has been brought to an appropriate conclusion. In that way, you avoid making them feel the uneasiness and dissatisfaction that we all feel when something ends abruptly at what appears to be an arbitrary point.

- *Emphasize key material.* An ending is a place of emphasis. People are more likely to remember the points made at the end of a communication than they are to remember those made in the middle. Thus, the final impression is key.

- *Direct your readers' attention to future action.* An ending is a transition. While reading a communication, your readers are absorbed in it. The ending leads them out of the communication and into the larger stream of their activities. It provides an opportunity to direct the readers' next actions.

For these reasons, the endings of your communication deserve special care and attention.

## ABOUT THE GUIDELINES

The guidelines that follow describe a variety of ways to end a written communication at work. Sometimes, when you've said what you have to say, you can simply stop writing (Guideline 1). Other times, when you need to say something more, you can use the strategies suggested in Guidelines 2 through 9, either singly or in combination, to create an appropriate ending. How will you know which strategies to use in a particular communication? First, think about your readers in the act of reading. Determine what you want them to think, feel, and do as they lay down your communication. Then determine which strategy (or strategies) for ending is most likely to bring about that result. Second, look at what other writers in your organization and your field have done when writing in situations similar to yours. That will help you determine which kinds of endings your readers might be expecting—and which kinds other writers have found to be successful in similar circumstances.

Guideline
1

## After You've Made Your Last Point, Stop

Despite the importance of endings, you will often be able to end your communications without doing anything special at all. Much of the time when you write at work, you will be using patterns of organization that bring your communication to a natural stopping place. Here are some examples:

- *Proposals.* In proposals you will usually use an organizational pattern that ends with your detailed description of what you will do and how you will do it. Because that's where your readers expect proposals to end, they will enjoy a sense of completion if you simply stop after presenting your last recommendation. Furthermore, by ending after your recommendations, you will have given them the emphasis they require.

- *Formal Reports.* When you write formal reports (those with covers, title pages, and bindings), you will often use a conventional pattern that ends either with your conclusions or recommendations—both appropriate subjects for emphasis.

- *Instructions.* You can often end instructions after describing the last step. Your readers will certainly enjoy a sense of completion after performing that step.

- *Reference Manuals.* Reference manuals usually have no endings because there is no single place where readers finish their reading. Instead of reading reference manuals straight through, readers look selectively for particular sections. As soon as they have gotten the information they need, they close the manual. There would be no point for you to provide a special ending.

These are just some examples of communications that writers often—but not always—end directly after making their last point. If your analysis of your purpose, readers, and situation leads you to believe that you should say something additional after making your last point, look through the remaining guidelines in this chapter to see whether or not any of the strategies they suggest will meet your objectives.

Guideline

2

## Repeat Your Main Point

Because the end of a communication is a point of emphasis, you can use it to focus your readers' attention on one or more main points that you want to be foremost in their minds as they set your communication down.

Consider, for instance, the final paragraph of an article on "Preventing Wound Infections" that is directed toward family physicians.[1] The point made in this paragraph was stated in the abstract at the beginning of the article and then again in the fourth paragraph, where it was supported by a table. It was also referred to several other times in the article. Nevertheless, the writers considered it to be so important that in the final paragraph they stated it again:

> Perhaps the most important concept to be gleaned from a review of the principles of wound management is that good surgical technique strives to maintain the balance between the host and bacteria in favor of the host. The importance of understanding that infection is an absolute quantitative number of bacteria within the tissues cannot be overemphasized. Limiting, rather than eliminating, bacteria allows for normal wound healing.

You can use the same strategy in communications intended to help your readers make a decision or to persuade them to take a certain action. Here, for instance, is the final paragraph of a memo urging new safety measures:

> I cannot stress too much the need for immediate action. The exposed wires present a significant hazard to any employee who opens the control box to make routine adjustments.

Guideline

3

## Summarize Your Key Points

This guideline is closely related to the preceding one. The difference is that in repeating a main point, you are emphasizing something that you consider to be of paramount importance. In summarizing, you are concerned that your readers have understood the general thrust of your communication.

Here, for example, is the ending of a 115-page book entitled *Understanding Radioactive Waste*, which is addressed to the general public:[2]

It may be useful to the reader for us to now select some highlights, key ideas, and important conclusions for this discussion of nuclear wastes. The following list is not complete—the reader is encouraged to add items.

1. Radioactivity is both natural and manmade. The decay process gives radiations such as alpha particles, beta particles, and gamma rays. Natural background radiation comes mainly from cosmic rays and minerals in the ground.
2. Radiation can be harmful to the body and to genes, but the low-level radiation effect cannot be proved. Many methods of protection are available.
3. The fission process gives useful energy in the form of electricity from nuclear plants, but also produces wastes in the form of highly radioactive fission products.

This list continues for thirteen more items, but this sample should give you an idea of how this author used the strategy of ending with a summary of key points.

Be sure to note that summaries placed at the end of a communication differ significantly from the summaries placed at the beginning. A summary at the beginning summarizes the communication for someone who has not read it. Therefore, it must devote space to some things that are not of much concern at the end. For example, the summary at the beginning of a report will establish the background of the study, so that readers can figure out how it relates to them. In contrast, an ending that summarizes a report usually focuses strictly on conclusions and recommendations.

## Guideline 4    Focus on a Key Feeling

Sometimes, you may want to focus your readers' attention not on a fact but on a feeling. For instance, if you are writing instructions for a product manufactured by your employer, you may want to use your ending to encourage your readers' good will toward the product.

For example, here is the ending of an owner's manual for a clothes dryer.[3] Notice how the last sentence provides no additional information but does seek to shape the readers' attitude toward the company:

> The GE Answer Center™ consumer information service is open 24 hours a day, seven days a week.
> Our staff of experts stands ready to assist you anytime.

The following passage is the ending from a much different kind of communication, a booklet published by the National Cancer Institute that is addressed to people who have apparently been successfully treated for cancer but do not know how long the disease will remain in remission.[4] It, too, seeks to shape the readers' feelings.

Cancer is not something anyone forgets. Anxieties remain as active treatment ceases and the waiting stage begins. A cold or cramp may be cause for panic. As 6-month or annual check-ups approach, you swing between hope and anxiety. As you wait for the mystical 5-year or 10-year point, you might feel more anxious rather than more secure.

These are feelings we all share. No one expects you to forget that you have had cancer or that it might recur. Each must seek individual ways of coping with the underlying insecurity of not knowing the true state of his or her health. The best prescription seems to lie in a combination of one part challenging responsibilities that require a full range of skills, a dose of activities that seek to fill the needs of others, and a generous dash of frivolity and laughter.

You still might have moments when you feel as if you live perched on the edge of a cliff. They will sneak up unbidden. But they will be fewer and farther between if you have filled your mind with other thoughts than cancer.

Cancer might rob you of that blissful ignorance that once led you to believe that tomorrow stretched on forever. In exchange, you are granted the vision to see each today as precious, a gift to be used wisely and richly. No one can take that away.

## Guideline 5   Tell Your Readers How To Get Assistance or More Information

At work, one of the most common strategies for ending letters and memos is to tell the readers how to get assistance or additional information. Here are two examples:

> If you have any questions about this matter, call me at 523-5221.
>
> If you want any additional information about the proposed project, let me know. I'll answer your questions as best I can.

By ending in this way, you not only provide readers with useful information but encourage them to see you as a helpful, concerned individual.

## Guideline 6   Tell Your Readers What To Do Next

Another strategy for ending effectively is to tell your readers what you think should be done next. If more than one course of action is available, tell your readers how to follow up on each of them:

> To be able to buy this equipment at the reduced price, we will need to mail the purchase orders by Friday the 11th. If you have any qualms about this purchase, let's discuss it. If not, please forward the attached materials, together with your approval, to the Controller's Office as soon as possible.

## Identify Any Further Study That Is Needed

Much of the work that is done on the job is completed in stages. One study answers preliminary questions. If the answers look promising, additional study is undertaken. Consequently, one common way of ending is to tell the reader what needs to be found out next:

> This experiment indicates that we can use compound deposition to create microcircuits in the laboratory. We are now ready to explore the feasibility of using this technique to produce microcircuits in commercial quantities.

Such endings are often combined with summaries, as in the following example, which is from an article in *Brain Research*.[5] (Note that the highly technical language is appropriate because the authors are addressing other researchers in their field.)

> In summary, this study has demonstrated that there is a distinct population of reticular neurons that participate in the local circuitry of the interior olivary complex. These neurons can be distinguished from other medullary reticular neurons on the basis of their location, morphological characteristics and physiological responses. Our electron microscopic analysis has led to the hypotheses that portions of their dendrites may play a presynaptic role in olivary circuitry. The existence and functional importance of this interaction remains to be defined by further physiological and neurochemical analysis.

Identifies further study that is needed.

## Refer to a Goal Stated Earlier in Your Communication

Many communications begin by stating a goal and then describe or propose one or more ways of achieving it. If you end such a communication by referring to that goal, you remind your readers of the goal and emphasize that your communication has focused on matters related to it. The following two examples show both the beginning and ending of reports in which the ending refers back to the beginning.

This first example is from *Biotechnology,* a journal concerned with the synthesis of new organisms that are commercially useful.[6]

Opening

> Given the necessity of producing alcohol as an alternative fuel to gasoline, especially in countries like Brazil, where petroleum is scarce, it is important to have a yeast strain able to produce ethanol directly from starchy materials.

Ending

> We are convinced that the stable pESA transformants can be of technological value in assisting ethanol fermentation directly from starchy materials, and we

have described the first step towards this end. Continuing this work, genetic crosses with different *Saccharomyces distaticus* strains are presently being carried out to introduce maltase and glucoamylase genes into the stable transformants that secrete functional α-amylase.

The second example is from an article that concerns the development of chemicals that delay the flowering of fruit trees to protect them from late frosts.[7]

### Opening

Yield reduction resulting from spring frosts is a major problem in the fruit growing regions of the world. . . . [One] potential approach is to delay flowering. . . . Researchers have been looking for a chemical method to delay bloom.

### Ending

Ultimately, new cultivars which bloom late will be developed. Until then, chemical delay of bloom offers potential for avoiding some of the damage caused by spring frosts.

## Follow the Social Conventions That Apply to Your Situation

Guideline

9

All of the strategies for ending that you have read about so far in this chapter focus on the subject matter of your communications. It is also important for you to consider the social relationship between you and your readers. Be sure to observe all the social conventions that apply to each of the situations in which you communicate.

Some of the conventions that govern endings arise from the customary ways of writing particular kinds of communication. For example, letters usually end with some social gesture involving such things as an expression of thanks, or a statement that it has been enjoyable working with the reader, or an offer to be of further help if need be. In contrast, formal reports and proposals rarely end with such gestures.

Other conventions about endings belong to organizations. Consider, for example, some of the various conventions about ending memos. In some organizations, writers rarely add the kind of social ending often provided at the ends of letters. In other organizations, memos often end with such a gesture, and people who do not include such an ending risk seeming abrupt and cold. In still other organizations, ways of ending memos vary considerably from person to person.

There are also the very important conventions that apply to the personal relationships between you and your readers. Have they done you a favor?

Thank them. Are you going to see them soon? Let them know that you are looking forward to the meeting. Are you completing a project on which you worked together? Tell them that you have enjoyed the collaboration.

## SUMMARY

The endings of the communications you prepare at work can help you achieve three aims:

- Provide your readers with a sense of completion.
- Emphasize key material.
- Direct your readers' attention to future action.

This chapter has presented nine strategies for ending a communication.

1. *After you've made your last point, stop.* Sometimes you can achieve all of the above aims without writing anything special at all.
2. *Repeat your main point.* By repeating your main point at the end, you can place special emphasis on it.
3. *Summarize your key points.* A summary of your key points will help your readers understand the heart of your message.
4. *Focus on a key feeling.* Your ending provides a final opportunity to shape your readers' attitudes.
5. *Tell your readers how to get assistance or more information.* This kind of ending is especially common in letters, memos, and instructions for consumer products.
6. *Tell your readers what to do next.* You can help your readers by suggesting the action they should take after they finish your communication.
7. *Identify any further study that is needed.* Much of the work done on the job is completed in stages. You can use the ending of your communication to tell your readers what needs to be done to move to the next stage.
8. *Refer to a goal stated earlier in your communication.* Many communications begin by identifying a goal. You can end them by reminding your readers of that goal and of the information you have provided about achieving it.
9. *Follow the social conventions that apply to your situation.* These conventions might arise from many sources, including the customs associated with the format you are using (for example, letter or memo), the usual practices in the organization that employs you, and your professional and personal relationship with your readers.

## NOTES

1. Harold R. Mancusi-Ungaro, Jr., and Norman H. Rappaport, "Preventing Wound Infections," *American Family Physician* 33 (April 1986):152.
2. Raymond L. Murray, *Understanding Radioactive Waste* (Columbus, Ohio: Battelle Press, 1982), 100.
3. General Electric Company, *How To Get the Best from Your Dryer* (Louisville, Kentucky: General Electric Company, 1983), 9.
4. Office of Cancer Communications, National Cancer Institute, *Taking Time: Support for People with Cancer and the People Who Care about Them* (Bethesda, Maryland: National Cancer Institute, 1983), 53.
5. Georgia A. Bishop and James S. King, "Reticulo-Olivary Circuits: An Intracellular HRP Study in the Rat," *Brain Research* 371(1986):143.
6. Spartaco Astolfi Filho et al., "Stable Yeast Transformants that Secrete Function α-Amylase Encoded by Cloned Mouse Pancreatic cDNA," *Biotechnology* 4(April 1986):311–15.
7. C. D. Costan, "Delay Flowering for Frost Control," *American Fruit Grower* 106(April 1986):14–15.
8. K. M. Foreman, *Preliminary Design and Economic Investigations of Diffuser Augmented Wind Turbines (DAWT)* (Golden, Colorado: Solar Energy Research Institute, 1981), 23.

## EXERCISES

1. The following paragraphs are the ending of a report to the U.S. Department of Energy concerning the economic and technical feasibility of generating electric power with a special type of windmill.[8] The windmills are called diffuser augmented wind turbines (DAWT). In this ending, the writer has used several of the strategies described in this chapter. Identify each of them.

Section 6.0<br>
CONCLUDING REMARKS

We have provided a preliminary cost assessment for the DAWT approach to wind energy conversion in unit systems to 150 kW power rating. The results demonstrate economic viability of the DAWT with no further design and manufacturing know-how than already exists. Further economic benefits of this form of solar energy are likely through:

- Future refinements in product design and production techniques
- Economies of larger quantity production lots
- Special tax incentives.

Continued cost escalation on non-renewable energy sources and public concern for safeguarding the biosphere environment surely will make wind energy conversion by DAWT-like systems even more attractive to our society. Promotional actions by national policy makers and planners as well as industrialists and entrepreneurs can aid the emergence of the DAWT from its research phase to a practical and commercial product.

2. Describe the strategies for ending used in the following figures in this book:

| Figure | Page |
|--------|------|
| 2-1 | 16 |
| 2-2 | 23 |
| 2-3 | 24 |
| 2-4 | 28 |
| 3-3 | 68 |
| 5-2 | 104 |
| 9-2 | 168 |

3. Find examples of four of the nine types of endings described in this chapter by looking in magazines, textbooks, instruction manuals, and other publications. If you find an ending that uses more than one of these strategies, count it as more than one example.

   For each example, explain why the writer chose the ending strategy he or she used. Did the writer make a good choice? Explain why or why not.

# CHAPTER 15

# Mapping Your Communication

## PREVIEW OF GUIDELINES

1. Use headings often.
2. Phrase each heading to tell specifically what follows.
3. Write each heading independently, using parallel structure among headings where it is helpful.
4. Adjust the size and location of your headings to indicate organizational hierarchy.
5. Adjust the location of your prose to indicate organizational hierarchy.
6. Provide a table of contents if it is useful.
7. Provide an index if it is useful.

I magine that you have just moved to a new city. Standing at your front door, you look down the street—but you can't see very far. You'd like to see farther, to be able to picture your new town's geography, to know where the various neighborhoods are and how they fit together with the shopping areas, parks, and public offices. Also, you have some errands to run but don't know where the bank is, where to find the telephone company, or where you can purchase a screwdriver.

You need a map.

Readers need maps, too. To understand the communication fully, they must see how its segments fit together. Further, readers often want to obtain one small piece of information from a long communication, and so desire a map they can use to find that information quickly, without taking unnecessary sidetrips into sections that contain other material.

This chapter presents seven guidelines for providing your readers with a map of your communications.

## YOUR VIEW VERSUS YOUR READERS' VIEW

"But," you may be asking, "if I've organized my writing carefully, why will my readers need a map? Won't a good, logical organization be perfectly obvious to them?" Unfortunately, it may not be, because communications contain two distinct kinds of information:

- Information about their *subject matter*.
- Information about the way the writer has organized the discussion of that subject matter.

The relationship between these two kinds of information is illustrated in Figure 15-1.

When you read something you have written, *you* see both kinds of information, even if you included only subject matter information and no organizational information. That's because you created the organization, and you remember what you planned.

Your readers, however, must construct a mental map of your organization as they read. If you neglect to supplement your information about your subject matter with information about your organization, your readers' map may look very different from yours. As Figure 15-2 indicates, they may know the contents—but not the hierarchical organization—of what they have already read, and they will probably be unable to foresee either the content or the organization of the information that lies ahead.

You can demonstrate to yourself how difficult it is for readers to puzzle out the organization of a communication when the writer hasn't provided a map of it: try to revise the poor table of contents shown in Figure 15-3 (page

| INFORMATION ABOUT SUBJECT MATTER | INFORMATION ABOUT ORGANIZATION | FULL VIEW OF THE COMMUNICATION |
|---|---|---|
| Subject 1 | I. | I. Subject 1 |
| Subject 2 |   A. |   A. Subject 2 |
| Subject 3 |     1. |     1. Subject 3 |
| Subject 4 |     2. |     2. Subject 4 |
| Subject 5 |   B. |   B. Subject 5 |
| Subject 6 | II. | II. Subject 6 |
| Subject 7 |   A. |   A. Subject 7 |
| Subject 8 |   B. |   B. Subject 8 |
| Subject 9 | III. | III. Subject 9 |

(with a **+** between the first two columns and **=** between the last two columns)

**Figure 15-1.** Relationship of Information about Subject Matter and Information about Organization.

296) by converting it from a mere list into an outline. You will find that you *can* make the conversion, but you must work hard to do so.

In contrast, when a writer has provided a good map of his or her communication, readers can discern its organization without having any knowledge at all of the contents. Look, for example, at Figure 15-4 (page 297), which shows a page of a Chinese textbook on geology. Even if you cannot read the language, the writer has provided you with enough organizational information for you to outline the pages.

**Writer's View**

I. Part 1
  A. Part 2
    1. Part 3
    2. Part 4
  B. Part 5

Writer has read ──→ to here.

II. Part 6
  A. Part 7
  B. Part 8
III. Part 9

**Readers' View**

Part 1
Part 2
Part 3
Part 4
Part 5

??
??
??
??

Readers have ←─read to here, but have difficulty deducing organization or predicting content

**Figure 15-2.** Comparison of the Writer's View and the Readers' View of a Communication if the Writer Neglects To Include Organizational Information.

Pegasus Automatic Balance

1.  Introduction                                          1
2.  Right-Hand Application Knob                           2
3.  Left-Hand Application Knob                            2
4.  Micrometer Knob                                       3
5.  Zero-Point Adjustment Knob                            4
6.  Right-Hand and Left-Hand Weight Dials                 5
7.  Projection Weight Dial                                5
8.  General Features of the Weighing Components           6
9.  Special Features of the Weighing Pan                  6
10. Special Features of the Weighing Stirrup              7
11. Adjusting the Sensitivity                             8
12. Adjusting the Zero Point                             9
13. Determining Weight                                   9
14. Weighing to a Preselected Weight                     10

**Figure 15-3.**   Table of Contents from an Instructional Manual Whose Writer Has Failed To Provide Organizational Information.

## THE FOCUS OF THIS CHAPTER

You can use many strategies to provide your readers with a map of your communications. Some of these strategies involve the way you write the segments of your prose—your sentences, paragraphs, introductions, and so on. Guidelines for those strategies are presented in Chapter 11. The chapter you are now reading presents another kind of strategy: those you can use without changing a word of your prose. These strategies involve:

- Things you can *add* to your communications after you have written the prose.

- Ways you can *arrange* your prose on the page—without revising a syllable.

# 第十章　鲸鱼铬铁矿床

## 一、岩体地质概况

鲸鱼岩体位于新疆西部华力西褶皱带的一个复向斜轴部,与萨尔岩体毗邻。

岩体的围岩为上泥盆统石英砂岩、粉砂岩及中基性火山岩。岩体与围岩为侵入接触,外接触带具弱石墨化,内接触带蚀变为硅化碳酸盐化超镁铁岩。接触面产状同围岩产状斜交。

全部岩体被厚约 20 米的第四系所覆盖。经地面磁测及钻探圈定,岩体呈北东 60° 方向延伸,与区域构造线走向一致。岩体长约 2.4 公里,最宽处 840 米,面积 1 平方公里余。南侧与北侧接触面相向倾斜,成一漏斗状(图 77),北侧较南侧缓。岩体最大延深部位据物探推断在岩体东南部,深度大于 565 米,可能为岩浆通道。

## 二、岩相带划分

组成岩体的岩石主要是斜辉辉橄岩(占全岩体的 85%)和斜辉橄榄岩(占 13%)及少量纯橄岩(约占 2%)。

岩体中基性程度略高的岩相——斜辉辉橄岩(斜方辉石和绢石含量为 5—15%)杂有少量纯橄岩、斜辉橄榄岩异离体组合分布于岩体下部和边缘及中部的狭长地带;酸性程度略高的岩相——斜辉辉橄岩(绢石含量 15—25%)和斜辉橄榄岩组合分布于岩体的上部和中部。物探表明,边缘岩相为低磁带,上部及中部岩相为高磁带,矿群主要分布于具低磁高重力特点的中部狭长地带。

## 三、岩石学特征

### 1. 纯橄岩

以黑绿色为主。镜下观察已全部蛇纹石化,具网格结构,无橄榄石残晶。副矿物铬尖晶石含量一般可达 1—3% 或更多,半自形一他形,粒径 0.1—0.5 毫米,多蚀变为不透明。偶含少量绢石,成为含辉纯橄岩。

### 2. 斜辉辉橄岩和斜辉橄榄岩

呈黄绿色。镜下观察,岩石亦遭强烈的蛇纹石化,斜方辉石变为绢石,呈他形,偶见残晶。 $2V = (+)77°—80°$, $N_g = 1.675 \pm 0.02$, $N_p = 1.664 \pm 0.02$, Fs = 7—8,为顽火辉石。 橄榄石利蛇纹石化,具网格结构,可见少量残晶, $2V = (+)85°—88°$, $N_m = 1.662—1.665$, Fa = 7—8,属镁橄榄石。 偶见少量单斜辉石, $2V = (+)58°—59°$, $c \wedge N_g = 35°—44°$,为透辉石。

· 167 ·

**Figure 15-4.**   Page from a Chinese Book whose Writers Have Provided a Good Map for the Reader.

From Heng-sheng Wang, Wen-ji Bai, Bix-xi Wang, and Yao-chu Chai, *China's Chromium Iron Deposit and the Cause of Formation* (Beijing: The Science Press, 1983), 167.

## Use Headings Often

Headings are signposts that tell your readers what the parts of your communication are about.

At work, writers use headings often, not only in long documents, like reports and manuals, but also in short ones, such as letters and memos. Headings may be single words, phrases, or entire sentences. Generally, headings are simply inserted into the text, without requiring any change in the prose. To see how effectively the insertion of headings can reveal organization, look at Figures 15-5 and 15-6 (pages 300 and 301, respectively), which display two versions of the same memo, one without headings and one with them.

Headings are especially good at helping readers find information quickly. To see how much help they can be, use the headings in Figure 15-7 (page 302) to learn why you should ride your bicycle on a street with parked cars rather than on a street without parked cars. Could you find the answer quickly? Imagine how much longer it would have taken you to find that same information if you had not had the help of headings.

Headings also help readers see the organizational hierarchy of a communication. They do that by indicating the relationships of coordination and subordination among the segments. If you look at Figure 15-8 (page 303), you can see how clearly headings reveal those relationships to readers. For example, they tell you immediately that the paragraphs on "Role of the Yolk" and "Role of the Blastodisc" are at the same level in the organizational hierarchy and that both are subparts of the discussion of "Morphogenesis." In the discussion of the fourth guideline in this chapter, you will find advice about creating headings to reveal organizational hierarchy.

### Headings and Topic Sentences

Headings help readers largely by announcing the topic of the segments they label. In this way, they are very much like topic sentences.

Because headings and topic sentences serve this same purpose, you may be concerned that if you use both a heading and a topic sentence at the beginning of the same segment, you will make your writing redundant. According to conventional writing practices, however, you *should* use both when writing communications such as reports and proposals that develop ideas logically from front to back.

In those communications, topic sentences and headings supplement and reinforce each other. Headings specialize in bringing key words to the readers' attention so the readers can find particular parts of the communication quickly.

Topic sentences specialize in helping readers see exactly how each segment fits into the communication's overall structure. Consequently, in connected prose, writers usually repeat the key word of a heading in the first sentence that follows. For example, in a research report the first sentence under the heading "Research Method" might read, "To test the three hypotheses described in the preceding section, we used a method that has proven to be reliable in similar situations."

In contrast, writers sometimes use headings alone—without topic sentences—in communications that *catalog* information (that is, where readers treat each topic of a communication in isolation from the rest of what the communication has to say). Documents of this sort include warranties, contracts, troubleshooting guides, computer reference manuals, and fact sheets. Here, the headings serve mainly as markers that tell a reader where to find the answer to a particular question: How long does my warranty last? What if my lawnmower won't start? In such communications, topic sentences can become barriers between readers and the information that they seek.

Topic sentences are sometimes eliminated in other situations also, such as where the information following the heading is presented in a list, where the heading is a sentence that serves as a topic sentence, and where the heading states a question.

In the end, of course, conventional practices form only one source of advice about how to handle the relationship between headings and topic sentences. What is most important is that you consider the specific needs of the particular readers you are addressing.

## How Many Headings Should You Use?

The needs of your readers should also determine the number of headings you use in a particular communication. In communications to be read as connected prose, you can usually meet your readers' needs by using headings every few paragraphs. Major shifts in topic occur frequently in on-the-job writing, and headings help to alert readers to those shifts. Also, the headings help hurried readers find specific pieces of information and review parts of a discussion that they have already read.

On the other hand, in connected prose, writers rarely use headings for *every* paragraph. If they did, their headings would emphasize not the connectedness of the prose, but its disjointedness: every paragraph begins a major new topic.

When you are writing communications that catalog information, you may need to use headings for every paragraph in order to meet your readers' needs. In fact, writers sometimes use headings to label sentence fragments or brief bits of data. The effect, often, is to turn such a communication into something very like a table of facts. By looking at the advertising brochure shown in Figure 15-9 (page 304), you can see how such frequent headings help readers locate specific pieces of information. (Look especially at the right-hand column, which describes the camera's "specifications.")

Garibaldi Corporation
INTEROFFICE MEMORANDUM

June 15, 19--

TO:      Vice Presidents and Department Managers

FROM:   Davis M. Pritchard, President

RE:      PURCHASES OF MICROCOMPUTER AND
         WORD-PROCESSING EQUIPMENT

Three months ago I appointed a task force to develop
corporate-wide policies for the purchase of microcomputers
and word-processing equipment. Based upon the advice of
the task force, I am establishing the following policies.

The task force was to balance two possibly conflicting
objectives: (1) to ensure that each department purchase the
equipment that best serves its special needs, and (2) to
ensure compatibility among the equipment purchased so
the company can create an electronic network for all our
computer equipment.

I am designating one "preferred" vendor of microcomputers
and two "secondary" vendors.

The preferred vendor, YYY, is the vendor from which all
purchases should be made unless there is a compelling
reason for selecting other equipment. To encourage
purchases from the preferred vendor, a special corporate
fund will cover 30% of the purchase price so that individual
departments need fund only 70%.

Two other vendors, AAA and MMM, offer computers already
widely used in Garibaldi; both computers are compatible
with our plans to establish a computer network. Therefore,
the special corporate fund will support 10% of the purchase
price of these machines.

We will select one preferred vendor and no secondary
vendor for word-processing equipment. The Committee will
choose between two candidates: FFF and TTT. I will notify
you when the choice is made early next month.

**Figure 15-5.**   Memo without Headings (Compare with Figure 15-6).

Garibaldi Corporation
INTEROFFICE MEMORANDUM

June 15, 19--

TO:      Vice Presidents and Department Managers

FROM:    Davis M. Pritchard, President

RE:      PURCHASES OF MICROCOMPUTER AND
         WORD-PROCESSING EQUIPMENT

Three months ago I appointed a task force to develop
corporate-wide policies for the purchase of microcomputers
and word-processing equipment. Based upon the advice of
the task force, I am establishing the following policies.

Objectives of Policies

The task force was to balance two possibly conflicting
objectives: (1) to ensure that each department purchase the
equipment that best serves its special needs, and (2) to
ensure compatibility among the equipment purchased so
the company can create an electronic network for all our
computer equipment.

Microcomputer Purchases

I am designating one "preferred" vendor of microcomputers
and two "secondary" vendors.

> Preferred Vendor. The preferred vendor, YYY, is the
> vendor from which all purchases should be made unless
> there is a compelling reason for selecting other equipment.
> To encourage purchases from the preferred vendor,
> a special corporate fund will cover 30% of the purchase
> price so that individual departments need fund only 70%.

> Secondary Vendors. Two other vendors, AAA and MMM,
> offer computers already widely used in Garibaldi; both
> computers are compatible with our plans to establish a
> computer network. Therefore, the special corporate fund
> will support 10% of the purchase price of these
> machines.

Word-Processing Purchases

We will select one preferred vendor and no secondary
vendor for word-processing equipment. The Committee will
choose between two candidates: FFF and TTT. I will notify
you when the choice is made early next month.

**Figure 15-6.**   Memo with Headings (Compare with Figure 15-5).

## WELCOME TO THE WORLD OF CYCLING

Welcome to the world of bicycling! As a bike rider you're one of a fast-growing group of vigorous outdoors-loving people.

Bicycling in America has been growing at an amazing rate. Bicycles used to be sold to parents for their children. Now those same parents are buying them for themselves, as well as for their children. And grandma and grandpa and college and high school students are cycling too.

Many young executives ride bikes to work as an alternative to adding to the smog of smoky cities, and to fighting traffic jams. Young mothers are finding a way to do their shopping, complete with child carrier and shopping baskets, without having to drive hubby to work and fight for a parking place at the shopping center. College and high school students find bikes an economical alternative to cars or buses.

Unfortunately, the rise of ownership and riding of bicycles has seen a corresponding rise in traffic accidents involving bicyclists.

According to the National Safety Council, 1,000 U.S. bicyclists were killed and 60,000 injured in bicycle-motor vehicle accidents during a recent year. Over the same period, Ohio recorded 36 cyclist fatalities and 2,757 injuries.

This brochure is presented in an attempt to show how more people can happily enjoy the sport of biking without suffering needless tragedy.

## GUIDES FOR SAFE RIDING

In traffic situations, the bicyclist must remain as alert as the driver of a car or other motor vehicle. For adult cyclists, this means applying defensive driving techniques to cycling as well as auto driving. The adult cyclist may even find he is becoming a better driver because his experience on a more vulnerable bicycle makes him more aware of traffic hazards when he is in a car. For the child rider, safe cycling mastery will lead to safer adult driving.

### Ride with the Traffic

Riding in the same direction as the traffic on your side of the road minimizes the speed at which oncoming cars approach you. It unnerves drivers to see a cyclist heading toward them, yet this seems to be the most common mistake of young riders. One should always ride in the right hand lane, as close to the curb as possible. If you have to walk your bike, however, you become a pedestrian, and you should walk facing traffic, the same as any other pedestrian.

### Use Hand Signals

Signals should be given with your left hand, well before you get to the place where you want to stop or turn. Keep your hand out until you start to make your move, or until you need to put on hand brakes, if you have them. A straight-out arm indicates a left turn, a right turn is signaled by holding your arm up bent at the elbow, with your thumb pointing to the right. (Don't use your right hand to signal right, because drivers can't see it as well.) A stop is signaled by holding your left hand straight down and out from your side.

LEFT
STOP OR SLOW
RIGHT

### Ride One on a Bike

You may "look sweet upon the seat of a bicycle built for two" but unless you have a tandem, only one will do. Carrying passengers on handlebars or luggage carrier makes the bike unwieldy, hard to handle in traffic, and easy to spill. The same rule should apply to large packages. Put packages on the luggage carrier and don't tie up your hands with bulky objects.

A child can be carried safely in a child seat attached to the bicycle. Expert riders recommend rear child carriers because front seats make it difficult to steer and offer less protection for the child.

### Ride Single File

Always ride single file in congested areas. Double-file riding is only safe on deserted country roads or special bike paths where cars aren't allowed.

### Use Streets with Parked Cars

Expert cyclists say city cycling is safest on streets with parked cars. A lane about three-feet wide is left between the parked cars and the moving traffic, so if you can ride in a straight line, you should be able to maneuver easily between the parked cars and the traffic.

Watch for car doors opening, however. Always use extra caution when you see a person sitting in a parked car; he might just open his door and jump out without warning, throwing you into moving traffic.

Streets which do not allow parking are usually more heavily traveled and not safe for cyclists, say the experts. Naturally, you should never ride on freeways, toll roads or heavily traveled state and federal highways.

### Exercise Care at Intersections

The safest way to cross an intersection is to walk your bike across with the pedestrians. Left turns can be made from the center lane, as a car does, or by walking across one street and then the other. The latter procedure is safer for busy intersections and less experienced riders.

Figure 15-7.   Brochure with Headings.

Courtesy of State of Ohio, Department of Highway Safety.

# Cell Movements in Teleost Fish Development

Wilbur L. Long

The pattern of morphogenetic movements in teleost embryos is reviewed with special consideration of the formation of a bilaterally symmetrical embryo through epiboly and convergence of cells. A model is presented for the control of cell movements by contact guidance during epiboly and convergence. The yolk syncytial layer is considered to be the source of the control. *(Accepted for publication 21 September 1983)*

The Osteichthyes is the largest and most diverse of the seven classes of living vertebrates. It includes the fleshy-finned fishes (lungfish, crossopterygians) and four groups of living ray-finned fishes. These are the Chondrostei (sturgeons and paddlefishes), the Ginglymodi (gars), the Cladistia (bichirs and reed-fishes), and the Halecostomi. The Halecostomi consists of two major subgroups, the Halecomorphi (bowfin) and the Teleostei (most "ordinary" fish). Together the various living teleosts are the largest vertebrate group, both in terms of individuals and number of species (Orr 1982).

Whether a single pattern of early morphogenesis exists in this large and diverse group remains to be seen. However, the few cases where detailed information on cell movements in early embryos is available fit a common scheme, subject to some variation, which I will describe here. It differs radically from morphogenetic patterns in other classes of vertebrates. The differences suggest that selective mechanisms that were a revolutionary departure from previous schemes were active when and where teleosts first evolved.

### STRUCTURE OF THE TELEOST EGG

Cleavage in the teleostean fishes is meroblastic. The egg during late cleavage is divisible into three compartments; the blastodisc, the yolk, and the yolk cytoplasmic layer. These are illustrated in Figure 1.

The *blastodisc*, also called the *blastoderm*, is a mound of cells at the animal

Long is with the Department of Biology, Western Maryland College, Westminster, MD 21157. © 1984 American Institute of Biological Sciences. All rights reserved.

pole of the egg. It forms from cleavage divisions of a cytoplasmic cap in that region and has two recognized cell types; epithelial cells, which form a *cellular envelope* (CE) covering the disc, and *deep cells*, which are destined to form all but the epidermis of the embryo.

Teleost embryos store their yolk in a large membrane-bound vacuole. This contrasts with the condition in amphibia, where yolk is stored inside the blastomeres. The two primary functions of the yolk vacuole are to store material for later use by the embryo, and to provide a spherical base upon which the embryo is constructed.

The *yolk cytoplasmic layer* (YCL) consists of an inner membrane, which surrounds the yolk mass, an outer membrane, forming the external surface of the living egg not covered by the blastodisc, and cytoplasm sandwiched between the two (Lentz and Trinkaus 1967). The YCL thus forms a sac that encloses the yolk. The portion of the YCL that extends beneath the blastodisc is called the *yolk syncytial layer* (YSL). It has a function in the transportation of nutritional and other material from the major storage area to the embryo (Van der Ghinst 1935, Romanini et al. 1969). More interestingly, it "directs" the morphogenetic cell movements that produce the embryo (Betchaku and Trinkaus 1978, Long 1983). I concentrate further discussion on the YCL on this portion, since more is known about its activity than of the thinner region, which covers the external yolk surface.

### MORPHOGENESIS

The period of morphogenesis in teleosts is one in which cells move from one place to another, changing the overall appearance of the embryo from that il-

lustrated by the cross-section in Figure 1 to that of a little fish. The movement of each cell is coordinated with that of its neighbors, so that groups of cells move together in patterns that are characteristic for the development of each species. These patterns are called the *morphogenetic movements*. The three compartments of the egg, yolk, blastodisc, and YCL (or YSL) each play separate roles in producing or supporting the morphogenetic movements.

### Role of the Yolk

Little is known about the morphogenetic influence of the yolk except that it forms a necessary substratum for proper development of the other portions of the egg. Interference with this function by disruption of the yolk vacuole leads to serious interference with normal morphogenesis (Battle 1944, Long unpublished).

### Role of the Blastodisc

The cellular envelope (also known as the enveloping layer) and deep cells of the blastodisc form all of the teleost embryo, including the major ectodermal, mesodermal, and endodermal organs, with the possible exception of a portion of the liver (Williams 1939). Inward movement of cells from the surface of the blastodisc by invagination and involution, though once considered to be a

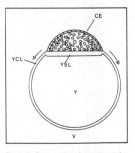

**Figure 1.** Cross section of a representative teleost egg during late cleavage. D = deep cells of the blastodisc, CE = cellular envelope of the blastodisc, YCL = yolk cytoplasmic layer, YSL = yolk syncytial layer, Y = yolk, V = vegetal pole of the egg. Epiboly carries both cells and the YSL margin toward the vegetal pole, as shown by the arrows.

**Figure 15-8.**   Headings That Indicate Organizational Hierarchy.

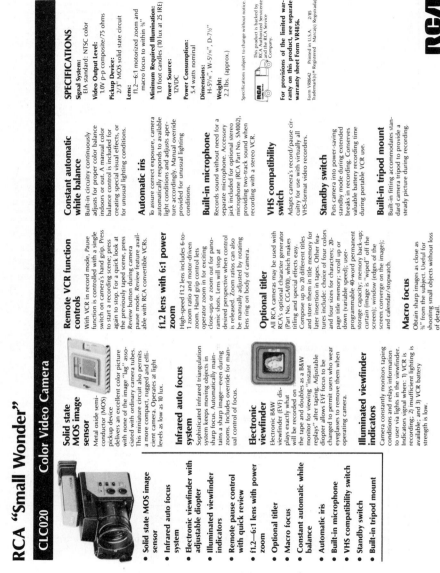

# RCA "Small Wonder"

## CLC020    Color Video Camera

- **Solid state MOS image sensor**
- **Infrared auto focus system**
- **Electronic viewfinder with adjustable diopter**
- **Illuminated viewfinder indicators**
- **Remote pause control with quick review**
- **f1.2—6:1 lens with power zoom**
- **Optional titler**
- **Macro focus**
- **Constant automatic white balance**
- **Automatic iris**
- **Built-in microphone**
- **VHS compatibility switch**
- **Standby switch**
- **Built-in tripod mount**

### Solid state MOS image sensor

Metal oxide semiconductor (MOS) pickup device delivers an excellent color picture with none of the image "lag" associated with ordinary camera tubes. This miniature circuit also permits a more compact, rugged and efficient camera. Operates at light levels as low as 10 lux.

### Infrared auto focus system

Sophisticated infrared triangulation system keeps moving objects in sharp focus. Automatically maintains a sharp image—even during zooms. Includes override for manual control of focus.

### Electronic viewfinder

Electronic B&W viewfinder (EVF) displays exactly what will be recorded on the tape and doubles as a B&W monitor for viewing "instant replays" after taping. Adjustable diopter allows EVF focus to be changed to permit users who wear eyeglasses to remove them when operating camera.

### Illuminated viewfinder indicators

Camera constantly monitors taping conditions and relays information to user via lights in the viewfinder. Indicators signal when: 1) VCR is recording; 2) insufficient lighting is available; and 3) VCR battery strength is low.

### Remote VCR function controls

With VCR in record mode, Pause function is controlled with a single switch on camera's hand grip. Press to start a recording scene; press again to stop. For a quick look at the previously taped scene, press Review button during camera pause mode. Review feature available with RCA convertible VCRs.

### f1.2 lens with 6:1 power zoom

High-speed f1.2 lens includes 6-to-1 zoom ratio and motor-driven zoom. Hand grip control lets operator zoom in for exciting close-ups or zoom out for panoramic shots. Lens will stop at desired perspective when control is released. Zoom ratios can also be manually adjusted by rotating lens ring on body of camera.

### Optional titler

All RCA cameras may be used with RCA's optional character generator (Part No. CGA010), which makes titling and special effects easy. Compose up to 20 different titles and store them in title memory for later insertion in tapes. Other features include: choice of four colors and four sizes for characters; 20-page title memory; scroll up or down (variable speed); user-programmable 40-word permanent storage capacity; memory back-up; curtain (image is "wiped" off the screen); window (edges of the screen converge on the image); and calendar/stopwatch.

### Macro focus

Obtain sharp images as close as ⅝" from the subject. Useful for shooting small objects without loss of detail.

### Constant automatic white balance

Built-in circuitry continuously adjusts for proper color balance indoors or out. A manual color balance control is included for creating special visual effects, or for unusual lighting conditions.

### Automatic iris

To assure correct exposure, camera automatically responds to available light conditions and adjusts aperture accordingly. Manual override provided for unusual lighting conditions.

### Built-in microphone

Records sound without need for a separate microphone. Accessory jack included for optional stereo microphone (RCA Part No. SM002), providing two-track sound when recording with a stereo VCR.

### VHS compatibility switch

Adapts camera's record/pause circuitry for use with virtually all VHS-format video recorders.

### Standby switch

Puts camera into power-saving standby mode during extended breaks in recording. Conserves valuable battery recording time during portable VCR use.

### Built-in tripod mount

Built-in fitting accommodates standard camera tripod to provide a steady picture during recording.

## SPECIFICATIONS

**Signal System:**
EIA standard: NTSC color

**Video Output Level:**
1.0V p-p composite/75 ohms

**Pickup Device:**
2/3" MOS solid state circuit

**Lens:**
f1.2—6:1 motorized zoom and macro focus to within ⅝"

**Minimum Required Illumination:**
1.0 foot candles (10 lux at 25 IRE)

**Power Source:**
12VDC

**Power Consumption:**
5.4 watts nominal

**Dimensions:**
H-5⅝", W-5⅜", D-7½"

**Weight:**
2.2 lbs. (approx.)

Specifications subject to change without notice.

This product is backed by RCA Authorized Servicenters and the RCA Service Company.

**For provisions of the limited warranty on this product, see separate warranty sheet Form VR485i.**

Form V8642   Printed in U.S.A.   2/85
Trademark(s)® Registered   Marca(s) Registrada(s)

**RCA**
©1985 RCA Corporation

**Figure 15-9.** Advertising Brochure with Headings for Every Paragraph.

By permission of RCA Corporation, Consumer Electronics Division.

<table>
<tr><td>Guideline<br>**2**</td><td>

## Phrase Each Heading To Tell Specifically What Follows

</td></tr>
</table>

Guideline

**2**

## Phrase Each Heading To Tell Specifically What Follows

To help your readers, your headings must tell clearly and specifically what kind of information is included in the segments they label. Vague headings are not useful. Researchers found that out when they tested the vague headings used in a set of government regulations. Readers could not accurately predict the content of many of the segments the headings labelled. Furthermore, when given these headings and sentences extracted from the segments, readers could not tell which sentences went with which headings.[1] You must write your headings so that your readers know what information they will find in the segments that follow.

Here are three strategies you can use to write helpful headings:

- *Write the question that the segment will answer for your readers.* Headings that say things like, "What happens if I miss a payment of my loan?" or "Can I pay off my loan early?" are especially helpful in communications that readers will use to guide their actions. Figure 15-10 shows an informational brochure about donating organs (eyes, hearts) that uses questions for its headings.
- *State the main idea of the segment.* This strategy was used by the person who wrote the brochure on bicycling safety shown in Figure 15-7. Its headings say such things as "Ride with the Traffic" and "Use Streets with Parked Cars." Headings of this sort are particularly effective at focusing your readers' attention on the key point of a passage. For that reason, in this book, each of the guidelines for writing is in the form of a heading.
- *Write a key word or phrase that indicates the topic to be discussed.* This strategy can be especially effective when you are writing a communication where a full question or statement would be unnecessarily wordy. Imagine, for instance, that you were writing a fifteen-page report on the feasibility of purchasing a new piece of equipment for a research laboratory. You could label the section that discusses prices with this title: "How Much Will This Equipment Cost?" However, in the context of the proposal, a heading that consists of the single word "Cost" would provide the same information much more succinctly.

Guideline

**3**

## Write Each Heading Independently, Using Parallel Structure among Headings Where It Is Helpful

As the discussion of Guideline 2 indicates, you can write headings in any of three ways (as questions, statements, or key words and phrases). Should you restrict yourself to using only one type of heading in a particular communication? Not necessarily. Each heading has its own individual function to perform: to announce the topic of its segment as clearly and usefully as possible. In many communications, different segments require different types of heading to achieve that goal.

## DONATION OF TISSUES AND ORGANS

Florida law allows you to indicate on your driver license or identification card if you want to donate your organs or tissues after your death.

Reasons for making tissue and/or organ donations:
1. Organs and tissues are given so others may live, or see or walk.
2. It is a humanitarian thing to do — it is a priceless gift.
3. The beneficiaries are unknown, but they could be anybody — a friend, a relative, a neighbor or a stranger.
4. Whoever receives your gift will be grateful for a gift of life.

**If you needed a cornea so you could see, or a kidney so you could live, or a bone so you could walk, or skin if you were burned — would you get it?**

You would if people are generous in giving permission to remove tissues and organs for transplantation after death.

Medical advances now make it possible to restore sight, to keep burn victims from dying, to restore normal life of many patients with kidney disease and to preserve limbs by transplanting bone and joints. Progress

is also being make with vital organs such as the liver, pancreas and heart. It requires the right organ — from people like you who care enough to give this priceless gift.

**Is there a need for tissue and organ donors?**

Yes there is! In Florida there are now more people who need organs and tissues than there are donors.

**How are tissues and organs obtained?**

They are donated by people like you.

**Will tissue and organ donation affect funeral arrangements?**

No. Removal of tissues or organs will not interfere with the customary funeral arrangements.

**Are there limits on who can donate tissues and organs?**

Yes. Tissues and organs can't be taken if there is a chance of transmitting disease to the person receiving the transplant. Vital organs such as kidneys are taken only when medical experts have determined that there is no brain activity, but the body is being kept functioning by artificial means. Corneas, skin, bone, etc. can be taken from most donors.

**Can you limit the tissues or organs you donate?**

Yes, by saying so on the registration form. You can say eyes only or kidneys only. If you like, you can also specify who you want to receive the organ or tissue.

**How can you make sure that doctors know you want to donate tissues or organs?**

Simply fill out the registration form on the back of this pamphlet and bring it with you to the Driver License Examination Station when you go to apply for a driver license or identification card. The examiner will arrange for your driver license or identification card to have the words "Organ Donor" appear on the license or card, and send the completed will to Tallahassee for filing.

Figure 15-10.    Headings that Ask Questions.

Courtesy of the Statewide Tissue and Organ Donor Program, Department of Health and Rehabilitative Services, Tallahassee, Florida.

On the other hand, you can often increase the clarity and usefulness of headings by writing them in a parallel fashion. This occurs especially when the headings label parallel segments, such as parallel parts of a process (turning on the machine, calibrating the instruments, and so on) or parallel items in a list (fruits, vegetables, grains, and so on).

Figure 15-11 shows the table of contents from a booklet entitled *Getting the Bugs Out.* Like most tables of contents, it is constructed by listing the headings used inside the document. Notice that all three types of headings are used. Notice also the places where parallel structures are used—and where they are not.

<table>
<tr><td>Guideline<br><br>4</td><td>

## Adjust the Size and Location of Your Headings To Indicate Organizational Hierarchy

</td></tr>
</table>

You can use headings not only to tell your readers what your segments are about, but also to indicate the organizational hierarchy of them. To use headings to indicate relationships of coordination and subordination, thoughtfully apply the following two principles:

- Make headings at the same level in the organizational hierarchy of the communication look the same.
- Make the headings for major segments more prominent than the headings for minor segments.

You can apply these principles both in texts that are prepared on a typewriter or in texts set in type by a printer. In either case, you may manipulate two variables: size and location.

### How To Adjust the Size of Your Headings

To make headings more prominent, you can make them larger. If you are preparing your communication on a phototypesetting machine (used by printers), you can easily vary the size of type used for your headings. However, you can also accomplish this same objective with documents prepared on typewriters.

One technique is to type the major headings with all uppercase letters and to type minor headings with only the initial letter of each major word uppercase. (It is customary to avoid headings that have no capital letters.) On a typewriter, you can also underline major headings to make them look larger. If you are using a word-processor, you can have it print major headings in boldface.

Another technique is to use "press-on" letters. Available at most bookstores, press-on letters come in sheets containing alphabets in various typefaces and sizes. By rubbing the sheets, you can transfer the letters to your typed sheet, upon which you would leave room for the headings.

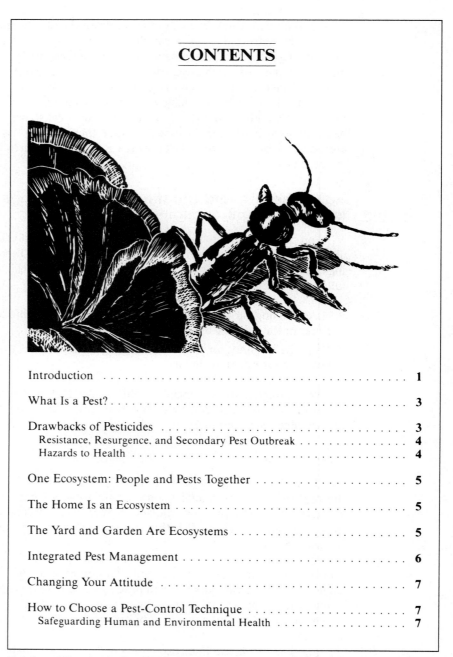

# CONTENTS

Introduction . . . . . . . . . . . . . . . . . . . . . . . . . . . . . . . . . . . . . . . . . . . . . . . . . . . . . . . **1**

What Is a Pest? . . . . . . . . . . . . . . . . . . . . . . . . . . . . . . . . . . . . . . . . . . . . . . . . . **3**

Drawbacks of Pesticides . . . . . . . . . . . . . . . . . . . . . . . . . . . . . . . . . . . . . . . . **3**
    Resistance, Resurgence, and Secondary Pest Outbreak . . . . . . . . . . . . . . **4**
    Hazards to Health . . . . . . . . . . . . . . . . . . . . . . . . . . . . . . . . . . . . . . . . . . . **4**

One Ecosystem: People and Pests Together . . . . . . . . . . . . . . . . . . . . . **5**

The Home Is an Ecosystem . . . . . . . . . . . . . . . . . . . . . . . . . . . . . . . . . . . . . **5**

The Yard and Garden Are Ecosystems . . . . . . . . . . . . . . . . . . . . . . . . . . **5**

Integrated Pest Management . . . . . . . . . . . . . . . . . . . . . . . . . . . . . . . . . . **6**

Changing Your Attitude . . . . . . . . . . . . . . . . . . . . . . . . . . . . . . . . . . . . . . . **7**

How to Choose a Pest-Control Technique . . . . . . . . . . . . . . . . . . . . . . **7**
    Safeguarding Human and Environmental Health . . . . . . . . . . . . . . . . . . **7**

**Figure 15-11.**    Table of Contents that Contains All Three Kinds of Headings (Question, Statement, Key Word or Phrase).

Physical Change of the Ecosystem . . . . . . . . . . . . . . . . . . . . . . . . . . . **7**
Selective Treatment . . . . . . . . . . . . . . . . . . . . . . . . . . . . . . . . . . . . . **7**

Eight Basic Strategies for Controlling Pests . . . . . . . . . . . . . . . . . . . **8**
Plant Selection . . . . . . . . . . . . . . . . . . . . . . . . . . . . . . . . . . . . . . . . **8**
Habitat Modification . . . . . . . . . . . . . . . . . . . . . . . . . . . . . . . . . . . . **8**
Physical Controls . . . . . . . . . . . . . . . . . . . . . . . . . . . . . . . . . . . . . . . **8**
Barriers . . . . . . . . . . . . . . . . . . . . . . . . . . . . . . . . . . . . . . . . . . . . . . **8**
Traps . . . . . . . . . . . . . . . . . . . . . . . . . . . . . . . . . . . . . . . . . . . . . . . . **8**
Cultural Controls . . . . . . . . . . . . . . . . . . . . . . . . . . . . . . . . . . . . . . . **8**
Biological Controls . . . . . . . . . . . . . . . . . . . . . . . . . . . . . . . . . . . . . **9**
Chemical Controls . . . . . . . . . . . . . . . . . . . . . . . . . . . . . . . . . . . . . **9**

Sanitation and Home Maintenance . . . . . . . . . . . . . . . . . . . . . . . . **9**

Pests in the House . . . . . . . . . . . . . . . . . . . . . . . . . . . . . . . . . . . . . . **10**
Cockroaches . . . . . . . . . . . . . . . . . . . . . . . . . . . . . . . . . . . . . . . . . **10**
Flies . . . . . . . . . . . . . . . . . . . . . . . . . . . . . . . . . . . . . . . . . . . . . . . . **11**
Rats and Mice . . . . . . . . . . . . . . . . . . . . . . . . . . . . . . . . . . . . . . . . **11**
Ants . . . . . . . . . . . . . . . . . . . . . . . . . . . . . . . . . . . . . . . . . . . . . . . . **12**
Termites . . . . . . . . . . . . . . . . . . . . . . . . . . . . . . . . . . . . . . . . . . . . . **13**
Fleas . . . . . . . . . . . . . . . . . . . . . . . . . . . . . . . . . . . . . . . . . . . . . . . . **14**
Mosquitoes . . . . . . . . . . . . . . . . . . . . . . . . . . . . . . . . . . . . . . . . . . **14**

Pests in the Garden . . . . . . . . . . . . . . . . . . . . . . . . . . . . . . . . . . . . **14**
Bees and Wasps . . . . . . . . . . . . . . . . . . . . . . . . . . . . . . . . . . . . . . **14**
Aphids, Mealybugs, Scales, Whiteflies . . . . . . . . . . . . . . . . . . . . . **15**
Caterpillars . . . . . . . . . . . . . . . . . . . . . . . . . . . . . . . . . . . . . . . . . . **15**
Japanese Beetles . . . . . . . . . . . . . . . . . . . . . . . . . . . . . . . . . . . . . . **16**
Mites . . . . . . . . . . . . . . . . . . . . . . . . . . . . . . . . . . . . . . . . . . . . . . . **16**
Weeds . . . . . . . . . . . . . . . . . . . . . . . . . . . . . . . . . . . . . . . . . . . . . . **16**

Criteria for Selecting Pesticides . . . . . . . . . . . . . . . . . . . . . . . . . . . **17**
The Most Selective . . . . . . . . . . . . . . . . . . . . . . . . . . . . . . . . . . . . **17**
The Smallest Area, the Shortest Application . . . . . . . . . . . . . . . . . **18**

Applying Pesticides . . . . . . . . . . . . . . . . . . . . . . . . . . . . . . . . . . . . **18**

Disposing of Unused Pesticides and Empty Containers . . . . . . . . . . . **18**

Hiring Professional Help . . . . . . . . . . . . . . . . . . . . . . . . . . . . . . . . . **19**

Glossary . . . . . . . . . . . . . . . . . . . . . . . . . . . . . . . . . . . . . . . . . . . . . **20**

Recommended Reading . . . . . . . . . . . . . . . . . . . . . . . . . . . . . . . . **21**

Other Resources . . . . . . . . . . . . . . . . . . . . . . . . . . . . . . . . . . . . . . . **22**

**Figure 15-11.**   *Continued*

A third technique is to use a machine that makes a photographic strip of letters, which you select one at a time to spell out your headings. Once you have the strip, you can paste it to your typed sheet. Headings prepared in this way will look just like headings prepared using press-on letters. Many schools have machines of this sort that students can use.

## How To Adjust the Location of Your Headings

Headings that are centered appear to be more prominent than the same headings would if they were tucked against the margin. Likewise, headings on a line of their own seem more prominent than they would if they were placed on the same line as the first sentence of the section they label.

### Examples

You can see how two students adjusted the size and location of their headings by looking at Figures 15-12 and 15-13. The student whose page is in Figure 15-12 (page 312) used a typewriter only, and the student whose page is in Figure 15-13 (page 314) used press-on letters.

As those sample pages suggest, you can use the variables of size and location to supplement each other. For example, in Figure 15-7, major headings (for example, "GUIDES FOR SAFE RIDING") are in all capital letters *and* are centered, while subheadings (for example, "Ride with the Traffic") have only the initial letters of each major word capitalized *and* are against the left-hand margin.

## How Many Levels of Headings Should You Use?

Theoretically, you could establish very elaborate visual hierarchies for your headings by manipulating their size and location. However, as the number of levels in a hierarchy increases, so too does the difficulty that readers have in comprehending that hierarchy. For most communications, two or three levels are about as many as readers find useful.

If additional levels are desirable, you can create them in either of two ways. The first is to make the most major sections into separate chapters. Because each chapter usually begins its own page and because the titles of chapters often use yet another, larger size of type than do headings, chapters can add one easily recognized level to the hierarchy of signposts in your communications. The second method is to reinforce the headings with the letters and numbers (or decimals) of an outlining system, which will enable readers to keep track of many levels of headings more easily than if they had to rely only on the cues provided by differences in typesize and location:

| Roman Outlining System | Decimal Outlining System |
|---|---|
| I. | 1. |
| A. | 1.1 |
| B. | 1.2 |
| 1. | 1.2.1 |
| 2. | 1.2.2 |
| 3. | 1.2.3 |
| C. | 1.3 |
| II. | 2. |

You will find an example of the use of the Roman outlining system in Figure 15-14 (page 315). Figure 15-15 (page 316) presents a table of contents that summarizes the headings in a research report that uses the decimal outlining system.

Once you have designed a hierarchy that works, you will be able to use it for many different documents that require the same number of levels of subordination. In fact, some employers require that one particular hierarchy be used in all long communications prepared by their employees.

As this and the three preceding guidelines make clear, the writing of headings requires judgment and skill. Like every other task of writing, it is something you will be able to carry out most successfully if you think constantly about your readers in the act of reading the communication you are writing.

## Adjust the Location of Your Prose To Indicate Organizational Hierarchy

Guideline
5

In addition to adjusting the size and location of your headings, you can adjust the location of your prose to tell your readers which parts of your text are parallel with one another and which are subordinate or superordinate. There are two variables you can manipulate: horizontal and vertical location.

You can adjust the horizontal location of your prose by establishing alternative left-hand margins, that is, by varying the left indentation of the prose. The wider blocks of prose, those you indent least, will appear more prominent than narrower blocks, those you indent more. Figure 15-16 (page 317) shows a page in which the writer has used indentation (and headings) to indicate relationships of coordination and subordination.

You can adjust the vertical location of blocks of prose by controlling the amount of white space between them, using relatively greater amounts of white space to indicate breaks between more major portions of the text. For

<table>
<tr><td>

Section
Heading

</td><td>

## INTRODUCTION

Southwestern Senior Services, Inc., is requesting support to develop an innovative behavior-management approach for nursing home residents who wander. The approach will be tested at Locust Knoll Village, a comprehensive philanthropic retirement community in the Chicago area.

</td></tr>
</table>

Section
Heading

### INTRODUCTION

Southwestern Senior Services, Inc., is requesting support to develop an innovative behavior-management approach for nursing home residents who wander. The approach will be tested at Locust Knoll Village, a comprehensive philanthropic retirement community in the Chicago area.

A-Level
Heading

### "Wandering" Defined

Wandering is one of several deviant behaviors associated with the aging process. The term refers to disoriented and aimless movement toward undefinable objectives and unattainable goals. Snyder et al. (1978) describe the wanderer as moving about and traversing locations 32.5% of waking hours, as opposed to 4.2% for nonwanderers. In testing, the wanderer also evidences greater involvement in nonsocial behaviors, such as sleeping, and disruptive behaviors, such as screaming and moaning. Wanderers differ significantly from nonwanderers on a number of psychological variables, and, overall, have more psychological problems, as assessed by the Human Development Inventory (Pyrek and Snyder, 1977). In nursing homes, wandering patients often violate the privacy rights of nonwandering patients by invading their rooms or areas of activity.

A-Level
Heading

### Causes

The cause of wandering has been studied, but research is limited. Research has focused on organic and psychosocial factors.

B-Level
Heading

**Organic Factors.** Early research typically attributed wandering behaviors in old age to damage to the brain as a result of disease, rather than to cognitive impairment. More recently geropsychologists and gerontological practitioners are beginning to believe that organic pathologies do not account for displays of functional disorders or explain the causes of disruptive behavior (Monsour and Robb, 1982). The main reason for this shift in thinking is the research

**Figure 15-12.**   Typewritten Headings That Indicate Organizational Hierarchy. (From a research proposal written by a student.)

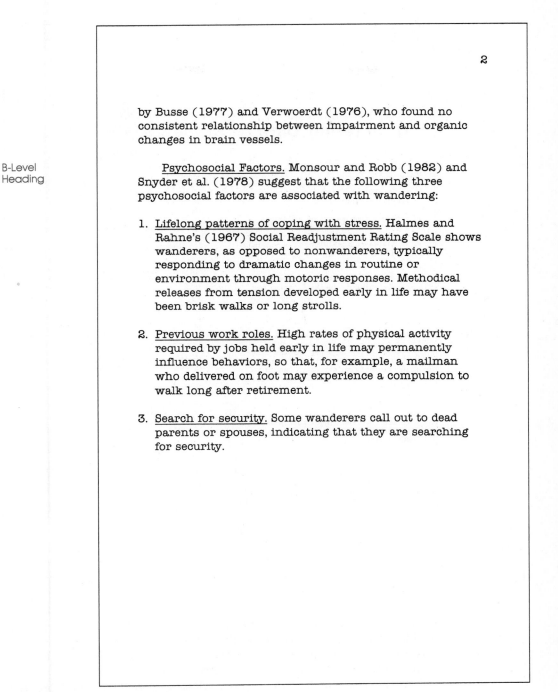

2

by Busse (1977) and Verwoerdt (1976), who found no consistent relationship between impairment and organic changes in brain vessels.

B-Level
Heading

<u>Psychosocial Factors.</u> Monsour and Robb (1982) and Snyder et al. (1978) suggest that the following three psychosocial factors are associated with wandering:

1. <u>Lifelong patterns of coping with stress.</u> Halmes and Rahne's (1967) Social Readjustment Rating Scale shows wanderers, as opposed to nonwanderers, typically responding to dramatic changes in routine or environment through motoric responses. Methodical releases from tension developed early in life may have been brisk walks or long strolls.

2. <u>Previous work roles.</u> High rates of physical activity required by jobs held early in life may permanently influence behaviors, so that, for example, a mailman who delivered on foot may experience a compulsion to walk long after retirement.

3. <u>Search for security.</u> Some wanderers call out to dead parents or spouses, indicating that they are searching for security.

**Figure 15-12.** *Continued*

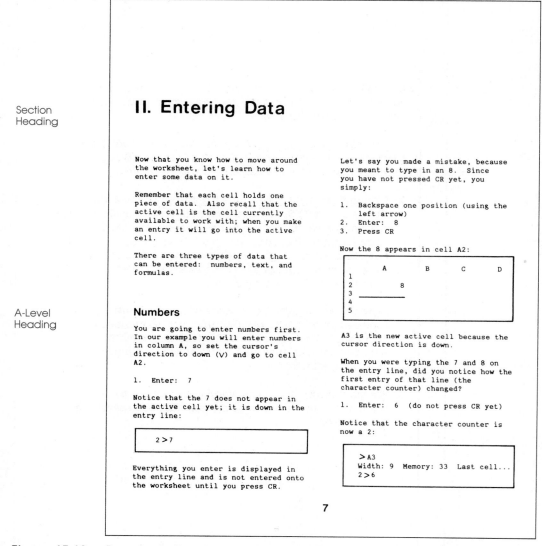

## II. Entering Data

Now that you know how to move around
the worksheet, let's learn how to
enter some data on it.

Remember that each cell holds one
piece of data.  Also recall that the
active cell is the cell currently
available to work with; when you make
an entry it will go into the active
cell.

There are three types of data that
can be entered:  numbers, text, and
formulas.

### Numbers

You are going to enter numbers first.
In our example you will enter numbers
in column A, so set the cursor's
direction to down (∨) and go to cell
A2.

1.  Enter:   7

Notice that the 7 does not appear in
the active cell yet; it is down in the
entry line:

    2 > 7

Everything you enter is displayed in
the entry line and is not entered onto
the worksheet until you press CR.

Let's say you made a mistake, because
you meant to type in an 8.  Since
you have not pressed CR yet, you
simply:

1.  Backspace one position (using the
    left arrow)
2.  Enter:  8
3.  Press CR

Now the 8 appears in cell A2:

|   | A | B | C | D |
|---|---|---|---|---|
| 1 |   |   |   |   |
| 2 | 8 |   |   |   |
| 3 |   |   |   |   |
| 4 |   |   |   |   |
| 5 |   |   |   |   |

A3 is the new active cell because the
cursor direction is down.

When you were typing the 7 and 8 on
the entry line, did you notice how the
first entry of that line (the
character counter) changed?

1.  Enter:   6  (do not press CR yet)

Notice that the character counter is
now a 2:

    > A3
    Width: 9  Memory: 33  Last cell...
    2 > 6

7

Figure 15-13.    Press-on Letters Used To Make Headings. (From an instruction manual
written by a student.)

example, in a single-spaced text prepared on a typewriter, you could place a
double-space before the heading at the top of each minor section, but place
a triple-space before the heading at the top of each major section. In long
reports, writers often signal the very largest divisions by beginning each of
the major sections on its own page, regardless of where the preceding section
ended on the previous page.

IEEE TRANSACTIONS ON SYSTEMS, MAN, AND CYBERNETICS, VOL. SMC-14, NO. 2, MARCH/APRIL 1984                257

# Anatomical and Physiological Correlates of Visual Computation from Striate to Infero-Temporal Cortex

ERIC L. SCHWARTZ, MEMBER, IEEE

*Abstract*—A review of the calculus of two-dimensional mappings is provided, and it is shown that cortical mappings are in general locally equivalent to the sum of a conformal mapping and a shear mapping. Currently available data suggests that a conformal (isotropic) mapping is sufficient to model the global topography of primate striate cortex, although the addition of a shear component to this mapping is outlined. A similar analysis is applied to the local map structure of striate cortex. Models based on a local reiteration of the complex logarithm and on Radon (and backprojection) mappings are presented. A variety of computational functions associated with the novel architectures for image processing suggested by striate cortex neuroanatomy are discussed. Applications to segmentation, perceptual invariances, shape analysis, and visual data compression are included. Finally, recent experimental results on shape analysis by neurons of infero-temporal cortex are presented.

## I.   Introduction

RECENT ADVANCES in understanding the anatomy and physiology of visual cortex have focused attention on the problem of visual data formatting in the central nervous system (CNS). It appears that the primate visual system may utilize a variety of novel architectures to represent visual data in the CNS. This possibility is reinforced by the use of the term "functional architecture" that Hubel and Wiesel [1] have used to describe the spatial patterns of columnar architecture in striate cortex. Here the term functional architecture will be expanded to include the global topographic structure of striate cortex, as well as the local columnar architecture represented by the orientation and ocular dominance column systems described by Hubel and Wiesel [1].

It will be shown that a conformal mapping provides a good model for global cortical topographic mapping, as well as a possible model for the local hypercolumn pattern of striate cortex. Furthermore, this mapping suggests that several integral transforms (Radon transform, backprojection) may provide workable models of the local architecture of striate cortex that are in agreement with current phenomenological descriptions.

Finally, a variety of computational applications of these anatomical models will be reviewed. These include applications to perceptual invariance, binocular segmentation, data compression, and shape analysis. Recent experimental work

Manuscript received August 8, 1982; revised June 1983. This work was supported in part by the AFOSR Image Understanding Program no. F49620-83-C-0108 and in part by the Systems Development Foundation.
The author is with Brain Research Laboratories, New York University Medical Center, 550 First Avenue, New York, NY 10016

(in monkey infero-temporal cortex), which is related to this analysis, will be reviewed. It will be suggested that the novel architectures of image representation suggested by primate visual cortex neuroanatomy may provide an example of "computational anatomy," which has relevance both to contemporary machine vision applications, as well as to current attempts at understanding the biological basis of visual computation.

## II.   Mappings of Two-Dimensional Surfaces into Two-Dimensional Surfaces

The concept of "topographic" mapping of the surface of the retina to the surface of striate cortex naturally invites an attempt at mathematical modeling. Which mapping fits the available data? In the following a complete characterization of planar mapping will be provided in an attempt to answer this question.

### A.   Whitney Mapping Theorem

A broad classification of two-dimensional mappings is provided by the Whitney mapping theorem, which states that topologically stable planar mappings are locally equivalent to either a "fold," a "cusp," or a "regular" mapping [2]. In the following, we will ignore the singularities represented by folds and cusps. Regular maps are characterized by having a Jacobian matrix of partial derivatives whose determinant is finite and nonzero. In the following section we will regroup the terms of this Jacobian in order to better understand the biological significance of the isotropic (ie., conformal) and nonisotropic maps that might occur in the nervous system.

### B.   Jacobian Matrix and Magnification Tensor

The (retinal) point $(x, y)$ is taken to a cortical point $(f, g)$ by the functions $f(x, y)$ and $g(x, y)$. The Jacobian of this map [2] is

$$J = \begin{pmatrix} \dfrac{\partial f}{\partial x} & \dfrac{\partial f}{\partial y} \\[2mm] \dfrac{\partial g}{\partial x} & \dfrac{\partial g}{\partial y} \end{pmatrix} = \begin{pmatrix} f_x & f_y \\ g_x & g_y \end{pmatrix}. \tag{1}$$

A common manipulation from the literature of continuum mechanics (and fluid mechanics) will be useful: the Jacobian is rewritten in terms of its symmetrical and

**Figure 15-14.**   Headings That Use the Roman Outlining System.

*IEEE Transactions on Systems, Man, and Cybernetics*, Vol. SMC-14, No. 2, © 1984 IEEE. Used with permission.

TABLE OF CONTENTS

| Section | | Page |
|---|---|---|
| 1.0 | SUMMARY | 1 |
| 2.0 | INTRODUCTION | 3 |
| 3.0 | DESIGN | 6 |
| | 3.1  Variable Cycle Engine (VCE) Description | 6 |
| | 3.2  Split Fan Description | 11 |
| | 3.3  Forward Variable Area Bypass Injector (VABI) | 12 |
| | 3.4  Variable Area Low Pressure Turbine Nozzle (VATN) | 14 |
| | 3.5  Rear Variable Area Bypass Injector for Forward VABI Test | 14 |
| | 3.6  Coannular Nozzle and Associated Rear VABI | 17 |
| | 3.7  Control System | 20 |
| |    3.7.1  Forward VABI Test Control System | 20 |
| |    3.7.2  Acoustic Nozzle Test Control System | 23 |
| 4.0 | FORWARD VABI TEST | 27 |
| | 4.1  Test Objectives | 27 |
| | 4.2  Test Description | 27 |
| | 4.3  Test Results | 30 |
| |    4.3.1  Summary of Results | 30 |
| |    4.3.2  Engine Performance | 32 |
| |       4.3.2.1  Overall Performance | 32 |
| |       4.3.2.2  Transitioning | 32 |
| |       4.3.2.3  Distortion | 41 |
| |    4.3.3  Forward VABI Aero Performance | 41 |
| 5.0 | ACOUSTIC TEST OF THE COANNULAR PLUG NOZZLE | 46 |
| | 5.1  Acoustic Nozzle Test Objectives | 46 |
| | 5.2  Test Results | 47 |
| |    5.2.1  Test Summary | 47 |
| |    5.2.2  Summary of Results | 53 |
| | 5.3  Test Facility Description and Data Reduction Procedures | 56 |
| |    5.3.1  Edwards Air Force Base Test Facility | 56 |
| |       5.3.1.1  Test Site | 56 |
| |       5.3.1.2  Sound Field | 57 |

v

**Figure 15-15.**  Table of Contents from a Research Report That Uses the Decimal Outlining System in its Headings.

From General Electric, *Aerodynamic/Acoustic Performance of YJ101/Double Bypass VCE with Coannular Plug Nozzle* (Cincinnati, Ohio: General Electric, 1981). NASA CR-159869.

### B. PRODUCT HANDLING

In our guidelines on processing, we now have completed our suggestions for guarding against microbial contamination. We turn now to our guidelines on how to handle the product. Here, we consider the receiving of the new raw material and the processing of it.

### 1. Receiving Raw Materials

By raw materials, we mean both the fish and any other raw materials used in processing.

**a. Fish.**—We consider first the fresh fish and then the frozen fishery products.

**(1) Fresh fish.**

Check fresh fish for sign of spoilage, off odors, and damage upon their arrival at your plant. Discard any spoiled fish.

Immediately move fresh fish under cover to prevent contamination by insects, sea gulls, other birds, and rodents. If the fish are to be scaled, scale them before you wash them.

Unload the fish immediately into a washing tank. Use potable, nonrecirculated water containing 20 parts per million of available chlorine and chill to 40°F or lower. Spray wash the fish with chlorinated water after taking them from the wash tank (Fig. 11).

If incoming fresh fish cannot be processed immediately, inspect them, cull out the spoiled fish, and re-ice the acceptable fish in clean boxes; then store them preferably in a cold room at 32° to 40°F or, at least, in an area protected from the sun and weather and from insects and vermin. Wash, rinse, and steam-clean carts, boxes, barrels, and trucks used to transport the fresh fish to the plant if any of these are to be used again. Reusable containers should be rinsed again with chlorinated or potable water just before use. *Note:* wooden boxes and barrels should not be reused. It is virtualy impossible to satisfactorily sanitize used wooden containers such as fish boxes and barrels. If disposable-type containers are used, rinse them off and store them in a screened area until you remove them from the premises.

**(2) Frozen fishery products.**

Use a loading zone that provides direct access to a refrigerated room.

32

**Figure 15-16.** Indentation of Text Used To Indicate Organizational Hierarchy. This page also shows a new section beginning at the top of a page.

From Perry Lane, "Sanitation Recommendations for Fresh and Frozen Fish Plants," *Fishery Facts* 8 (November 1974): 32. National Oceanic Atmospheric Administration, Department of Commerce.

Guideline

6

## Provide a Table of Contents If It Is Useful

Another way you can help your readers understand the organization of your communication is to include a table of contents. Through a table of contents, you provide a route through which your readers can quickly locate a particular part of your communication. Your readers need only find the appropriate title in your table of contents and then turn directly to the page indicated.

By including a table of contents, you can also help readers who will study your communication by starting to read on page 1 and then proceeding sequentially through your text to the final page. A table of contents enables such readers to see very quickly the general scope and arrangement of the material you cover, so they can begin to build the mental framework in which to organize the various pieces of information they will gain from the communication. Because such frameworks are so helpful to readers, many books that provide advice about how to read effectively suggest that readers read the table of contents before they look at anything else in a communication. Figure 15-15 shows a table of contents prepared on a typewriter and Figure 15-11 shows one typeset by a printer.

When writing your table of contents, think of it as an outline of your communication. Construct it from headings you have included in the text. However, you don't need to include *all* levels of headings. In deciding which headings to include, think of your readers. If your table of contents becomes too long (over a page or two), it may be difficult for your readers to find the organizational pattern in it. On the other hand, you need to include enough information to allow your readers to create a general picture of the organization of the communication and also to find the specific pieces of information they are likely to seek.

One way to determine which levels of headings to include in a table of contents is to ask which segments of the report might be sought by readers who want to find only *one* part of the communication. The segments you identify in this way should be included in the table of contents, along with all the other elements at that same level within the organizational hierarchy. Also, you should keep in mind that readers almost always find it useful to have more than one level of organization included in the table of contents.

Tables of contents can be just as useful to the readers of short communications as they are to the readers of long ones. Although some writers don't consider including tables of contents in communications shorter than 20 or 30 pages, readers often welcome them in communications as brief as 10 pages.

Guideline

7

## Provide an Index If It Is Useful

You can also provide readers with a quick route to specific pieces of information by including an index. In fact, in long documents an index can be much more useful than a table of contents. In an index, readers can locate items alphabetically but in a table of contents they must scan through many entries, which can be a time-consuming chore if the table of contents is very long.

The procedure for determining which items to include in your index is as follows:

1. Identify all the specific kinds of information your readers might seek when using your communication as a reference document.

2. List the words your readers might use while searching for those pieces of information. Be sure to include words they might use even if you have not used those words in your text. Where readers might be using any of a variety of words in their search for a particular kind of information, include *all* the most likely ones with cross-references to the main entry. For example, if you have a section on trees and expect that some readers may look under the word *Evergreen* in the index for information you've indexed under *Conifers,* then include an entry for *Evergreen* that directs readers to the *Conifer* entry.

   Some word-processing programs help you create an index by generating a list of the words used in your communication. From this list, you can select the words that will help your readers find the information they desire.

3. Where possible, use headwords to gather related terms (for example, *Conifers* as a headword for *Fir, Cedar, Redwood,* and so on). Be careful, however, to use cross-references if an entry will be located under a headword where some readers may not think to look for it. One cookbook makes it difficult to find the recipe for pancakes because in the index the writer listed *Pancakes* not under *P* but as a subgroup of *Breads*.

Figure 15-17 (page 320) shows a table of contents that uses headwords.

## SUMMARY

By providing maps to your communications, you will help your readers understand your text and find particular pieces of information quickly. This chapter has presented seven guidelines for mapping the communications you write at work.

1. *Use headings often.* Headings serve as signposts that tell what the parts of your communication are all about.

2. *Phrase each heading to tell specifically what follows.* To be useful, headings must tell readers precisely what the segment is about. They can do that by stating a question that is answered by the section, by telling the

```
                                  INDEX

Character Codes ......................................................  D-1
Character Names .....................................................  D-2
Control Code Listing ...........................................  E-1, E-2
Control Codes ...............................................4-1 to 4-24
       Definition of Terms .......................................  4-1
       Control Codes in the Text Mode ...........................  4-2
       Print Action Codes ........................................  4-3
       Paper Formatting Codes ....................................  4-4
       Character Designation Codes .............................  4-11
       Control Codes in the Bit-Image Mode .....................  4-18

Indicators .........................................................  3-2
       Power ......................................................  3-2
       Ready ......................................................  3-2
       Paper Out ..................................................  3-2
       On Line ....................................................  3-2
Installation ............................................... 2-1 to 2-6

Loading Cut Sheet Paper ........................... 3-11 to 3-15
Loading Fanfold Paper ................................... 3-7. 3-8

Paper Controls .............................................. 3-2, 3-3
       Paper Release Lever .......................................  3-2
       Gap Adjustment ............................................  3-3
Paper End Detector .................................................  3-4
Paper Separator ....................................................  2-5
Power Cable ........................................................  2-5
Print Head Replacement .............................................  B-1
Printer Initialization .............................................  C-2
Printer Lid ................................................... 2-3, 2-4

Relationship Between Data and Dot Wires ...........................  4-20
Removing Fanfold Paper .............................................  3-9
Ribbon Cartridge Removal and Installation ................... 3-4 to 3-6
       Removal ............................................... 3-4, 3-5
       Installation .......................................... 3-5, 3-6

Self Test ..........................................................  3-4
Signal Cable .......................................................  2-6
Specifications .............................................A-1 to A-2
Switches ...................................................3-1 to 3-2
       Power ......................................................  3-1
       On Line ....................................................  3-2
       Form Feed ..................................................  3-2
       Line Feed ..................................................  3-2

Top of Form Position ..............................................  3-10
Tractor Unit .......................................................  3-16


                             INDEX-1
```

**Figure 15-17.** Index That Uses Headwords.

From *Matrix Printer Manual,* Wang Laboratories. Used with permission.

main idea of the section, or by giving a key word or phrase that indicates precisely what is covered in the section.

3. *Write each heading independently, using parallel structure among headings where it is helpful.* Each heading has its own individual function to perform: to announce the topic of its section as clearly as possible. In many communications, different sections require different types of head-

ings to achieve that goal. However, you can often increase the helpfulness of headings by writing them in a parallel fashion.

4. *Adjust the size and location of your headings to indicate organizational hierarchy.* Change the size and location of your headings to (1) make headings that are at the same level in the organizational hierarchy look parallel and (2) make headings for major segments more prominent than headings for minor segments.

5. *Adjust the location of your prose to indicate organizational hierarchy.* Blocks of prose that you extend all the way to the left margin will appear more prominent than blocks of prose you indent. Before major segments, use more blank lines than you use before subsegments.

6. *Provide a table of contents if it is useful.* Think of the table of contents as an outline of your communication, and construct it from headings you have included in the text.

7. *Provide an index if it is useful.* Like a table of contents, an index provides readers with a quick route to specific pieces of information.

## NOTE

1. Heidi Swarts, Linda S. Flower, and John R. Hayes, *How Headings in Documents Can Mislead Readers* (Pittsburgh: Carnegie–Mellon University, 1980).

## EXERCISES

1. For this exercise, you are to find headings that display the various features discussed in this chapter. You may use the same heading to illustrate more than one feature.

   A. In communications you find outside of class, locate examples of at least four of the following ways of *writing the text* of a heading.

      _____ Heading is a keyword or phrase that is repeated in the topic sentence.

      _____ Heading is a keyword or phrase that is not repeated in the topic sentence.

      _____ Heading is a statement.

      _____ Heading is a question.

      _____ Heading includes letter or number from the Roman outlining system.

      _____ Heading includes number from the decimal outlining system.

   B. In communications you find outside of class, find headings that use each of the following techniques for *indicating relationships of coordination and subordination.* (These examples may be from the same communications you used in Part A of this exercise.)

_____ *Size.* Headings at the same level in the organizational hierarchy are the same size and use capital letters in the same way. Headings at higher levels are larger, use a more prominent typeface (such as boldface), or use more capital letters than headings at lower levels.

_____ *Location.* Headings at higher levels are placed on the page so that they are more prominent than headings at lower levels.

2. Find a communication that has headings you could improve by following the guidelines in this chapter. Rewrite four of those headings. For each, explain (from the point of view of the reader) why your version is better than the original.

3. A. Find a good table of contents (but not in a textbook or magazine). Taking the point of view of the reader, explain why you believe it is good.

   B. Find a table of contents that you could improve by following the advice provided in this chapter. Taking the point of view of the reader, explain why you believe your revisions improve the table of contents.

4. A. Find an index that uses headwords and cross-references in ways that readers would find helpful. Justify your choice by explaining specifically how the readers are aided.

   B. Find an index that you could improve by following the advice provided in this chapter. Taking the point of view of the readers, explain why you feel that your revisions improve the index.

5. Refer back to Figure 15-3, which shows a table of contents from an instruction manual for an electronic balance, a device used to weigh samples in a laboratory. The author has done a good job of telling what the communication includes, but has done a poor job of telling about the relationships of coordination and subordination among those contents.

   A. Using common sense, determine the general outline of the manual, making an outline of three levels (roman numerals, capital letters, arabic numerals). Do *not* change the order of the items in the table of contents. (Note that after the "Introduction," the manual falls into two major parts, each with subparts.)

   B. By following the guidelines given in this chapter, make a revised table of contents that provides an overview of the organization of the manual.

6. The following text is from a pamphlet that police departments around the country distribute to homeowners to help them protect themselves from burglary.

   A. Indicate each place in the pamphlet where you think a heading would help the reader.

   B. Write the heading for each place you think appropriate.

The crime rate goes ever upward—and burglary is the fastest-growing felony. A prowler enters someone's home every 15 seconds! Next time, it could be your home. Of the $400 million worth of goods stolen each year, only 5% is recovered. Serious crimes against the person also result from the failure of people to protect their homes and families against illegal entry.

Despite all the publicity about complex and expensive electronic burglary-prevention systems, the rules for protecting your home and family are really quite simple and basic. The best precautionary measures you can take involve old-fashioned common sense and good, modern locks. Let's cover the common-sense rules first. These are day-and-night, year-around precautions that will work effectively—if you use them.

You can't make a house absolutely burglar-proof. But you can make entry so difficult that the prowler will go elsewhere in search of an easier victim. Burglars don't like delay, risk or noise.

Make entry difficult with:

- Substantial door and window locks.
- Exterior lights at night when you're home, plus inside lights when you're not.
- A good watch dog (but be sure the "doggy door," if you install one, is not the way in for a burglar as well as the dog).

Make entry simple and you're easy game.

One of the most important ways of protecting your home is to know your neighbors.

- Make them fully aware of your family's living habits—who comes and goes to your house and when. Get to know theirs equally well.
- When you are familiar with regular visitors, strangers in the neighborhood are easy to spot.
- If you should notice a stranger who appears to be doing something out of the ordinary, call the police immediately.

The home with a "lived-in" or "at-home" look is a deterrent to burglars. Follow these simples rules, even when you're leaving the house for "just a couple of minutes":

- Lock all outside doors and windows.
- At night, leave one or more lights on in locations not visible from windows. For extended absences, there are inexpensive plug-in timers that will turn lights on and off.
- Leave a radio playing, preferably tuned to a talk show with volume low. This will create the impression of a conversation.
- Do not leave notes indicating your absence.
- Shut and lock the garage door.

Take extra care to guard against intruders when you're in the house. Your personal safety is also at stake!

- Keep outside doors locked.
- Never admit strangers under any pretext.
- Install a chain lock or have caller identify himself.
- Report any solicitor or salesman without proper credentials to the police.
- Don't keep valuables out in the open.

If a burglar enters your home at night, call police quietly on the phone, if possible—and remain calm! Try to stay on the phone until police arrive.

Don't try to shoot at a suspected prowler unless there is a real danger of harm to you or your family. You may injure an innocent person (possibly a child sleepwalking!) or your gun may be taken away and used against you. In most cases, it is against the law to use a gun if the suspected burglar is merely on your property, not actually in your home. Even with a prowler in your home, police officials say you should be in fear of your life to justify shooting.

If, when you return from a long trip—or indeed, even a trip to the store—there are obvious signs of a burglary, DO NOT ENTER your home. Call the police from a neighbor's house and wait until they arrive.

You can increase the safety of your home easily and relatively inexpensively. The many new, improved locks that are now available for your home can delay a prowler to the point that he will go elsewhere. Locks more than five years old are probably outmoded and should be replaced. With most locksets, replacement is surprisingly easy.

One of the most important lockset security features is a pin-tumbler mechanism, by far the safest type for keyed locksets. Locksets using solid-brass pin-tumbler construction allow far more keying combinations, thus reducing the chance of other keys operating your lock. Pin-tumbler construction is also more precise than disc or wafer types.

Locksets with deadlatches provide inexpensive additional security. The deadlatch is a little piece of hard brass that rides on the face of the latch and prevents jimmying with a celluloid strip or plastic card. Today, most quality locksets include deadlatches as standard features at no additional cost.

In these days of soaring crime, every home should have extra-security cylinder deadlocks, with full 1" bolts, on all entry doors. They should be installed on front, back and side doors in addition to existing locksets. Most have a key lock on the outside and a bolt-retracting thumb key on the inside. *Insist* on all steel and brass construction; a heavy-duty 1" solid-steel deadbolt; a solid-brass cylinder guard that will turn free with any attempt to remove it with a pipe wrench or similar tool; and solid-brass pin-tumbler cylinders and plugs. Don't buy any deadlocks that don't have *all* these features because you're not getting the most protection for your money.

You could spend thousands of dollars on an elaborate electronic security system, go shopping for a couple of hours—and return to find your home ransacked. If someone wants to break into your home, he can do it. But, if you follow the advice and precautions in this booklet, the odds are greatly in your favor that a would-be prowler will bypass your home and move on to an easier target.

Reprinted by courtesy of Kwikset, a Division of Emhart Corporation.

# Writing Sentences

## PREVIEW OF GUIDELINES

1. Eliminate unnecessary words.
2. Keep related words together.
3. Use the active voice unless there is good reason to use the passive voice.
4. Put main actions in verbs.

As a speaker of English, you have incredible power over the language. And the language itself, tremendously flexible, awaits your command. To see just how much power you have, and just how flexible the language is, consider the following simple sentence:

Dogs eat red meat.

The information communicated by that sentence could also be conveyed in sentences with much different structures:

Dogs eat meat that is red.
Red meat is eaten by dogs.
The meat that dogs eat is red.
Red meat, that's what dogs eat.
Dogs are eaters of red meat.

Most of the sentences you write communicate much more information than does the short sentence, "Dogs eat red meat." Just think how many different variations you could write for each of them.

Each version of a sentence is, by definition, at least slightly different from the others. Each will affect a reader in a slightly different way. That fact, however, is ignored by many writers. They think all versions of a sentence are equally effective as long as they are grammatically correct. As a result, those writers search for a grammatically correct version that is easy for them to write. In doing so, they are taking a writer-centered approach to writing.

To take a reader-centered approach to writing, you must keep your readers in mind while writing your sentences. In particular, you should pursue two reader-centered goals. The first is to make your readers' work at reading as easy as possible. The second is to structure your sentences so that they will affect your readers in the particular way that you desire. The guidelines presented in this chapter will help you achieve both those interrelated goals.

Guideline

1
## Eliminate Unnecessary Words

Reading sentences is hard work. Every extra word makes that work harder.

Consider all that your readers must do to read even a single word. First they need to attend to the word, focusing their eyes and attention upon it; then they must figure out what the word is; then they must determine which of the possible meanings of that word is the one that applies in the context of the words that surround it. Therefore, if you can convey essentially the same meaning in a sentence of thirteen words (for example) and also a sentence of eight words, you are helping your readers considerably if you use only eight words.

Compare the following versions of the same sentence. (Note that in the rest of this chapter the sample sentences are numbered for easy reference.)

The physical size of the workroom is too small to accommodate this equipment. (1)

The workroom is too small for this equipment. (2)

Sentences 1 and 2 communicate essentially the same information, but sentence 2 does so in 38% fewer words. Therefore, from your readers' point of view, sentence 2 is preferable to sentence 1.

## Efficient Sentences Increase the Effectiveness of Your Writing

Eliminating unnecessary words from your sentences is more than a considerate act performed by one person (the writer) on behalf of others (the readers). By increasing the efficiency with which your readers can comprehend your sentences, you also increase your chances of affecting them in the way you desire. Sentences that are needlessly long are fatiguing and difficult to concentrate upon. Your readers may feel exasperated by having to work so hard to understand what you are saying.

If your readers tire or become frustrated, they may miss important parts of your message. Perhaps they will become inattentive. Perhaps they will switch to what might seem to be a more efficient form of reading—skimming—in which they stand a much greater chance of overlooking important points than when engaged in close reading. Perhaps your readers will become so frustrated that they simply put your communication down, refusing to read any more at all.

## Why Do People Write Wordy Sentences?

Despite the obvious case against unnecessary words in your sentences, you will probably find yourself tempted toward wordiness at work. There are three primary sources of this temptation.

The first is the desire to sound very precise. The result is often a ponderous *over*precision. Instead of simply saying that the workroom is too small, the writer of sentence 1 succumbed to the temptation to specify the precise way in which the room is too small: in terms of its physical size. But the writer did not need to include the words *physical size:* when a person says that a room is too small for a piece of equipment, what feature of the room could he or she have in mind except its size?

The second cause of wordy writing is the use of long phrases that have insinuated themselves into our spoken language, where they do not create so much difficulty as they do in written communications. Thus, we are not troubled when we *hear* someone tell us,

> Because of the fact that the price of oil rose, Gulf Consolidated Services       (3)
> received many new orders for its fittings for drilling rigs.

When reading, however, we would prefer to have the words *of the fact that* stricken:

> Because the price of oil rose, Gulf Consolidated Services received many       (4)
> new orders for its fittings for drilling rigs.

A third cause of wordiness in writing is laziness: instead of finding the word that will communicate their meaning exactly, writers sometimes use an entire string of other words. For example, consider the following sentence:

> They do not pay any attention to our complaints.       (5)

With a little more effort, the writer could have found the one word that contains the meaning of six that he used:

> They *ignore* our complaints.       (6)

Eliminating unnecessary words requires extra work. However, by performing that work (and thereby reducing the amount of work your readers must do), you also increase the chances that your communication will affect your readers in the way you want.

<table>
<tr><td>Guideline<br>2</td><td>

## Keep Related Words Together

There are two good reasons for keeping related words together, both based solidly upon what is known about how people read.
</td></tr>
</table>

### Word Order Indicates Meaning

The first reason to keep related words together is that when we read English, we depend upon the order of the words in a sentence to indicate the relationships that the words have to one another. For example, even though both of the following sentences have exactly the same words, they do not mean the same thing:

> The woman in the seat hit the man.       (7)
> The woman hit the man in the seat.       (8)

The placement of *in the seat* in sentence 7 tells where the woman was, while the placement of the same words in sentence 8 tells where the man was—or else where the woman hit him.

In some sentences, the words are arranged so that readers cannot know which of two possible meanings the sentence is supposed to have. Here is an example.

We sent a note about the proposal to Chester.                     (9)

Unless we have some independent information about the situation, we cannot tell from this sentence whether the note went to Chester or whether the note went to someone else and it concerned a proposal made to Chester.

In most cases, however, a reader can figure out the meaning of the sentence even when a modifier (like *in the seat*) is separated from the word it is intended to modify. For example, most readers quickly figure out the meaning of the following sentence:

A large number of undeposited checks were found in the file cabinets,   (10)
which were worth over $41,000.

According to the way our language works, that sentence tells us that the file cabinets were worth over $41,000. Yet as we read we quickly reject that meaning. After all, it scarcely seems probable that the file cabinets were worth that much money, whereas it is very plausible that a large number of checks might be. Thus, the problem you would create by failing to keep related words together isn't so much that your readers will misunderstand the sentence as that they will have to pause in their reading to figure out your meaning. They can do that only by performing extra work that they will feel very strongly you should have saved them.

## Limitations of Short-Term Memory

The second reason you should keep related words together has to do with the limitations of short-term memory. As you recall from reading Chapter 2, the short-term memory performs the work of determining the meaning of sentences. It has a very limited capacity, being able to hold only about five to nine bits of information at a time. If you write a sentence in which related words are placed too far apart, your readers will forget the earlier words by the time they reach the later words. Consider, for example, the following sentence:

A new factory to produce chemicals for the OPAS system, which enables   (11)
large manufacturers of business forms to make carbonless copy paper
as part of their own manufacturing process, began to operate late last year.

If you are like most readers, you had to read that sentence twice to understand it. By the time you got to the verb of the sentence *(began),* you could not remember what the subject was. Therefore, you had to move your eyes to the left and up the page to remind yourself that the subject is *factory.* You could not remember the subject because the large number of words following the subject pushed *factory* from your short-term memory before you reached *began.*

You should be careful not to misinterpret this explanation of the difficulty in sentence 11. You can place interjections between words. In fact, placing a statement between the subject and verb of a sentence is one way to emphasize it:

> We managed, despite the recession, to make record profits that year.     (12)

The point is this: be sure that your interjections are brief enough not to interfere with the work being done by your readers' short-term memories.

## Use the Active Voice Unless There Is Good Reason To Use the Passive Voice

Guideline
3

To follow this guideline, you must first know what the active and passive voices are. In many of our sentences, we report an action in which someone or something does something to someone or something else. The person or thing doing the action is called the *actor* and the person or thing receiving the action is the *patient.*

| Actor | Action | Patient | |
|-------|--------|---------|---|
| Jack | hit | the ball. | (13) |
| The ball | broke | the window. | (14) |
| The owner | forced | Jack (to pay). | (15) |

A sentence is in the *active* voice if the *actor* is its subject; it is in the *passive* voice if the *patient* is the subject. For example, sentences 13, 14, and 15 are all in the active voice because the actors—Jack, the ball, and the owner—are the subjects of the sentences.

The same information that is contained in those three sentences can readily be expressed in the passive voice. Sentences 16, 17, and 18 are all in the passive voice because the patients—the ball, the window, and Jack—are the subjects of the sentences.

| Patient | Action | Actor | |
|---------|--------|-------|---|
| The ball | was hit | by Jack. | (16) |
| The window | was broken | by the ball. | (17) |
| Jack | was forced (to pay) | by the owner. | (18) |

## Why Use the Active Voice?

The first reason for using the active voice instead of the passive should become clear as soon as you compare the active and passive versions of the statement about Jack's batting the ball: the active version (sentence 13) communicates the message in four words; the passive version (sentence 16) in six —an increase of 50%. By using the active voice you are also often following Guideline 1: eliminate unnecessary words.

A second reason for using the active voice is that readers understand sentences in the active voice more quickly than sentences in the passive voice—even when the sentences have the same number of words.[1] Reading passive sentences requires more work than does reading active sentences, quite apart from the extra work of reading the additional words.

A third reason for using the active voice is to avoid the vagueness that the passive permits but the active doesn't. In the active voice, you must always tell who the actor is; in the passive you don't need to. "The ball was hit" is a grammatically correct sentence, even though it doesn't tell who or what hit the ball. The writer of that sentence knows who did the hitting, but the readers don't.

At work, you will encounter many situations in which you will want to be sure that your readers know who the actor is:

> The bushings must be adjusted weekly to ensure that the motor is not (19) damaged.

Will the supervisor on the third shift know that *he* is the person responsible for adjusting the bushings? Sentence 19 certainly allows him to think that someone else is, perhaps the supervisor of one of the other shifts.

## When To Use the Passive Voice

Although you should generally use the active voice, there are some situations in which the passive voice is preferable. That is why Guideline 3 reads, "Use the active voice *unless* there is good reason to use the passive."

One good reason is that you don't want to identify the agent:

> I have been told that you are using the company telephone for an excessive (20) number of personal calls.

Perhaps the person who told the writer about that breach of the corporate telephone policy did so in confidence.

> The lights on the third floor have been left on all night for the past week, (21) despite the efforts of most employees to help us reduce our energy bills.

Perhaps sentence 21 is from a memorandum in which the writer wants to urge *all* employees to work harder at saving energy, but at the same time does not want to point the finger at any particular person who might improve his or her efforts in this area.

A second good reason for using the passive voice is to focus your readers' attention on the patient, not the actor—something you can do by placing the patient in the subject position of the sentence. You will learn more about this use of the passive voice in Guideline 1 of the next chapter.

### Converting Passive Sentences to Active Ones

When you want to convert a passive sentence to an active one, you can use the following four-step procedure.

1. Identify the main action of the sentence. In sentence 19, for instance, that action is "adjusting the bushings."
2. Identify the actor, the person, or thing that performs the main action. For sentence 19, the person who must adjust the bushings is "the foreman of the third shift." (In this example, the actor must be supplied from outside the sentence because the sentence does not identify the actor.)
3. Start to say a sentence that begins with the actor and then immediately tells the action: "The foreman of the third shift must adjust the bushings. . . ." In most cases, the rest of the words will follow automatically.
4. Once you have re-created the sentence by performing the preceding steps, revise it, if necessary, in accordance with the other guidelines for writing sentences.

Guideline

**4**

## Put Main Actions in Verbs

This guideline contains a traditional piece of advice, given to writers again and again over the years. And for good reason: putting actions in verbs, not in nouns or modifiers, leads to forceful writing. For example, consider the following sentence:

> Our department accomplished the conversion to the new machinery in     (22)
> two months.

It could be improved by putting the action ("converting") into the verb:

> Our department converted to the new machinery in two months.     (23)

Not only is the new version briefer, it is also more emphatic and lively.

### Spotting Sentences without the Main Action in the Verb

When trying to put action in your verbs, you need to be able to identify sentences that have their main action somewhere else. Two kinds of sentences, both easily spotted, often displace the action from the verb. Therefore, a good way to apply Guideline 4 is to search your writing for these kinds of sentences.

The most common is the sentence in which the verb is some form of the verb *to be.*

> This is a process that anneals the metal twice as rapidly.                    (24)

The major action in this sentence is "annealing," but the main verb is *is.* The action can be put in the verb by rewriting the sentence in this way:

> This process *anneals* the metal twice as rapidly.                    (25)

The second kind of sentence that often has the action displaced from the verb is one in which an important word ends in one of the following suffixes: *-tion, -ment, -ing, -ion, -ance.*

> Consequently, I would like to make a recommendation that the           (26)
> department hire two additional programmers.

The main action in sentence 26 is "recommending," but the verb is *would like to make.* To put the main action in the verb, the writer could revise the sentence in the following way.

> Consequently, I *recommend* that the department hire two additional          (27)
> programmers.

### Placing Action in the Verb

Once you have identified a sentence that does not have its main action in the verb, you can follow the same four-step procedure given above for converting a sentence from the passive to the active voice.

1. Identify the main action of the sentence. Ask yourself, "What is happening in this sentence that is really crucial?" In sentence 24, that action is "annealing," even though the verb is *is.*

2. Identify the actor, the person, or thing that performs that main action. In sentence 24, the actor is the person or thing that does the annealing: "this process."

3. Start to say a sentence that begins with the actor and then immediately tells the action: "This process anneals. . . ." In most cases, the rest of the words will follow automatically.

4. Once you have re-created the sentence by performing the preceding steps, revise it, if necessary, in accordance with the other guidelines for writing sentences.

As you can see by comparing sentence 25 with sentence 24 or sentence 27 with sentence 26, putting the action into the verb often enables you to eliminate unnecessary words as well (Guideline 1). Even when you don't eliminate any words, however, you still increase the effectiveness of your writing by putting the actions of your sentences into verbs, not into nouns or modifiers.

## SUMMARY

As a speaker of the English language, you have tremendous power over its vast resources for expressing ideas in sentences. This chapter has presented four guidelines that will help you use that power to write clear, forceful sentences. The four guidelines are:

1. *Eliminate unnecessary words.* Every word creates work for your readers. By eliminating unnecessary words, you lessen the chance that your readers will become tired or frustrated, and you increase your chances of affecting them in the way you desire.

2. *Keep related words together.* For example, try to place modifiers close to the words they modify, and try not to separate the subject and verb of a sentence with an overly long insertion. If related words are not close together, your readers will have to pause to understand your meaning.

3. *Use the active voice unless there is good reason to use the passive voice.* A sentence is in the active voice if the actor (the person or thing doing the action) is the subject of the sentence. A sentence is in the passive voice if the patient (the person or thing receiving the action) is the subject.

4. *Put main actions in verbs.* By putting action in your verbs, not in nouns or modifiers, you will make your sentences shorter, more emphatic, and more lively. Action is often displaced from the verb in sentences that use some form of the verb "to be" and in sentences where an important word ends in one of the following suffixes: *-tion, -ment, -ing, -ion, -ance.*

## NOTE

1. David R. Olson and Nikola Filby, "On the Comprehension of Active and Passive Sentences," *Cognitive Psychology* 3(1972): 361-81.

## EXERCISES

1. Without altering the meaning of the following sentences, reduce the number of words in them.

   A. Those who plan federal and state programs for the elderly should take into account the changing demographic characteristics in terms of size and average income of the composition of the elderly population.

   B. After having completed work on the compression problem, we turned our thinking toward our next task, which was the reentry project.

   C. Would you please figure out what we should do and advise us?

   D. The result of this study will be to make total white-water recycling an economical strategy for meeting federal regulations.

2. Rewrite the following sentences in a way that will keep the related words together.

   A. This stamping machine, if you fail to clean it twice per shift and add oil of the proper weight, will cease to operate efficiently.

   B. The plant manager said that he hopes all employees would seek ways to cut waste at the supervisory meeting yesterday.

   C. About 80% of our pulp, to be made into linerboard and corrugated cardboard (much of it used for beverage containers), supplies our plants for manufacturing packages.

   D. Once they wilt, most garden sprays are unable to save vegetable plants from complete collapse.

3. Rewrite the following sentences in the active voice. Follow the four-step procedure explained in the discussion of Guideline 3.

   A. Periodically, the shipping log should be reconciled with the daily billings by the Accounting Department.

   B. Fast, accurate data from each operating area in the foundry should be given to us by the new computerized system.

   C. Since his own accident, safety regulations have been enforced much more conscientiously by the shop foreman.

   D. No one has been designated by the manager to make emergency decisions when she is gone.

4. Rewrite the following sentences to put the action in the verb. Follow the four-step procedure explained in the discussion of Guideline 4.

   A. The experience itself will be an inspirational factor leading the participants to a greater dedication to productivity.

   B. The system realizes important savings in time for newspaper reporters.

   C. The implementation of the work plan will be the responsibility of a team of three biophysicists experienced in these procedures.

D. Both pulp and lumber were in strong demand, even though rising interest rates caused the drying up of funds for housing.

5. For each of the sentences given below, do the following:

A. Read the sentence.

B. In the space provided, write in the number(s) of the revision(s) that would most improve it. The possible revisions are:
   1. Eliminate unnecessary words.
   2. Keep related words together.
   3. Use the active voice.
   4. Put action in verbs.

C. Revise the sentence.

_____ a.   Marex Advanced Systems is engaged in the development of ink-jet printing equipment and other high technology products.

_____ b.   Productivity will be increased greatly by the installation of robots on the assembly line.

_____ c.   She wanted to decide the question as to whether the Baltimore plant should try the process.

_____ d.   Our purpose is to achieve the identification of inefficiencies so that management can bring about their elimination.

_____ e.   During the past six months, 180 payroll advance checks have been written by Jane.

_____ f.   Provisions should be included in the plans for the new office building for the handicapped.

_____ g.   Our Canadian affiliate will accomplish a doubling of software sales in the next 18 months.

_____ h.   This is a machine that applies the welds twice as rapidly.

_____ i.   The parking lot, when the dayshift leaves, especially on hot afternoons, becomes a dangerous place.

_____ j.   That year marked the beginning of a sharp decline in the nation's attitude toward business, which sunk to its lowest level since 1954.

_____ k.   Our office used three weeks for the preparation of the proposal.

_____ l.   The circular chart recorder will aid the Process Control Department in controlling the process by providing a means by which its personnel can check to see that the speed has been properly set.

_____ m.   The new billing procedure will be welcomed by our department because the accountants' workload will be reduced by it.

_____ n.   In the event that you cannot attend the meeting, please call my secretary.

_____ o.   I recommend that the electric stencil machine be moved to the scheduling office, where decisions on production runs are made, for the following reasons.

_____ p.   Recently that client gave him an invitation to make a presentation of his ideas for reorganizing their production line.

_____ q.   Please draw your attention to the fact that your quarterly report is three months overdue.

_____ r.   Vacation schedules are coordinated by the Assistant to the Plant Manager.

# Guiding
# Your Readers

## PREVIEW OF GUIDELINES

1. Avoid needless shifts in topic from one sentence to the next.
2. Use transitional words and phrases at the beginning of your sentences.
3. Use echo words at the beginning of your sentences.
4. Use paragraph breaks to signal major shifts in topic.

Good writers are good guides. Think of your readers as novice mountain climbers, making their way across a snowfield high in the Rocky Mountains. Hidden from view are numerous crevasses, deep fissures that are covered over with snow. With any step, the climbers might place their weight on that thin covering and plunge through. You are their guide, who must lead them across this field, responsible for seeing that their every step is safe.

## RELATING NEW INFORMATION TO OLD

In a way, the communications you write are like fields of mountain snow. Each sentence is a safe place for your readers to stand, and the space between that sentence and the next is a crevasse into which your readers might tumble. Why? As you will recall, we read by deriving meaning from small chunks of text (a sentence or less) and then constructing those small bits of meaning into the larger meaning of the overall communication. As we progress through a communication sentence by sentence, we continually try to figure out how the "new" information we are now reading relates to the "old" information from the preceding sentence.

If we cannot immediately discern that relationship, we interrupt the forward progress of our reading. Instead of moving our eyes from left to right and from top to bottom across the page, we either pause or go backwards—rethinking those two sentences, trying to identify their relationship to one another. That pause, that break in our forward progress, is the crevasse into which we will fall if our writer does not guide us carefully from one sentence to the next.

## YOUR READERS' VIEW VERSUS YOUR VIEW

Too few writers are aware of the dangers that present themselves to readers at each and every break between one sentence and the next. As the writer of a communication, you already know how the new information in one of your sentences relates to the old information in the preceding one. And you may be tempted to assume that this relationship is equally obvious to your readers. However, you the writer gained that knowledge when deciding what you wanted to say. Your readers, in contrast, must rely solely upon the words you place on the page for clues about the relationship of one sentence to the next. If you view your sentences only from your own perspective—not from the readers'—you are taking a writer-centered approach to writing.

In contrast, if you work carefully at guiding your readers from sentence to sentence so that they can read efficiently, you are taking a reader-centered approach to writing. This chapter describes four guidelines that will help you take a reader-centered approach.

## Avoid Needless Shifts in Topic from One Sentence to the Next

The simplest relationship between the information in two adjacent sentences is this: the first sentence says something about a particular topic, and the second says something more about that same topic. We often write two or more adjacent sentences that provide information about the same thing:

> The links of the drive chain must fit together firmly. They are too loose if you can easily wiggle two links from side to side more than ten degrees.

Often a writer writes about the same thing in two adjacent sentences but the readers cannot see that both sentences are about the same thing, at least not immediately. How can you help your readers rapidly detect that sort of relationship between sentences when it occurs? To answer this question, you must first understand how people determine what a sentence is about.

### How Readers Determine What a Sentence Is About

People usually perceive the topic of a sentence to be whatever is the grammatical subject of the sentence. They interpret everything else in the sentence as a comment upon that topic:

| TOPIC (Subject) | COMMENT | |
|---|---|---|
| The engineer | was rewarded by the board of directors for his outstanding contributions to the corporation. | (1) |

The information in this (or any sentence) can be rewritten in a variety of ways to have different topics:

| TOPIC (Subject) | COMMENT | |
|---|---|---|
| The board of directors | rewarded the engineer for his outstanding contributions to the corporation. | (2) |
| A reward for his outstanding contributions to the corporation | was given to the engineer by the board of directors. | (3) |

Thus, by choosing the word that you will make the grammatical subject of your sentence, you control what the reader will perceive to be the topic of the sentence. Using this control, you can help your readers see when adjacent sentences are about the same topic. For example, imagine that you had just written this sentence:

| TOPIC | COMMENT | |
|---|---|---|
| (Subject) | | |
| Our company's new | permits tighter | (4) |
| accounting procedure | control over cash flow. | |

And suppose also that in your next sentence you wanted to communicate information that could be expressed either in this way:

| TOPIC | COMMENT | |
|---|---|---|
| (Subject) | | |
| Thousands of | have been saved by the | (5) |
| dollars | procedure this year alone. | |

or in this way:

| TOPIC | COMMENT | |
|---|---|---|
| (Subject) | | |
| The procedure | has saved thousands of dollars this | (6) |
| | year alone. | |

Would sentence 5 or sentence 6 be the best one to follow sentence 4? Both contain essentially the same information. But sentence 6 has the same topic—"procedure"—as does sentence 4, and sentence 5 does not. Therefore, your readers would find it easier to relate the "old" information in sentence 4 to the "new" information in sentence 6 than to relate it to the same "new" information in sentence 5:

> Our company's new accounting procedure permits tighter control over cash flow. The procedure has saved thousands of dollars this year alone.

The preceding example involves a pair of sentences in which you could aid your readers by using the same topic ("procedure") in the subject position of the adjacent sentences. You will also encounter many situations in which you can avoid needless shifts in topic by keeping the same *kind of thing* in the subject position. Consider, for example, the following paragraph.

> The *materials* used to construct and furnish this experimental office are designed to store energy from the sunlight that pours through the office's large windows. The *special floor covering* stores energy more efficiently than wood. The *heavy fabrics* used to upholster the chairs and sofas also capture the sun's energy. Similarly, the *darkly colored paneling* holds the sun's energy rather than reflecting it as lightly colored walls would.

In that paragraph, the subject of the first sentence is *materials*. Although the same word is not the subject of the sentences that follow, the subjects of all those sentences are kinds of "materials," namely the *special floor covering, heavy fabrics,* and *darkly colored paneling.* Thus, although the word placed in the subject position of the various sentences changes, the topic, in a sense, does not.

## How To Avoid Making Your Sentences Monotonous

You may wonder whether your writing will become monotonous if you strive, where possible, to keep the same topic in the subject position of adjacent sentences. Your writing *can* get monotonous if you follow this guideline without imagination and without sensitivity. There are two ways you can avoid that monotony.

First, use pronouns. Words like *he, she,* and *it* enable you to keep the same topic in the subject position of your sentences without keeping the same word there:

> *A new paper-making machine* is on order for the Chillicothe plant. *It* will produce carbonless copy paper, as well as copy machine and duplicator papers.

A second way to avoid monotonous writing is to vary your sentence structure. The grammatical subject of a sentence does not *have* to be the first word in the sentence. In fact, if it did, the English language would lose much of its power to emphasize more important information and to deemphasize less important information. Here is a passage in which the topic (italicized) remains the same from sentence to sentence, but the sentence structure varies:

> Through hard work and chance, *the present secretary* of our organization has assumed a dangerously wide variety of responsibilities. *He* takes and distributes the minutes of our meetings, as secretaries customarily do in organizations like ours. *He* (not the president) writes the agenda and *he* (not the treasurer) manages our bank account. Furthermore, just before last year's elections, *he* became the nominating committee, selecting the pool of people from which our new officers would be elected. In short, *the secretary* has become the most powerful person in the organization, one who has transformed our democratically run organization into a monarchy.

## Using the Passive Voice To Maintain Consistent Focus

There is one important implication of this guideline that you should note. At times you will be able to follow it only if you use the passive voice. In the preceding chapter, you learned that it is generally desirable to use the active voice, not the passive. But you also learned that there are times when the passive is appropriate, even preferable to the active. One of those times is when the passive voice enables you to avoid a needless shift in the topic of two adjacent sentences. Consider the following paragraph from an accident report:

*Tom* was injured on his way back from the cafeteria. *He* finished lunch late, so he ran through the plant toward his work station. As he rounded the corner into the shop, *a forklift truck* hit him. *He* missed 17 days of work.

The topic of most of the sentences is *Tom* or *he*. However, the third sentence *(a forklift truck hit him)* shifts the topic from Tom to the forklift truck. Furthermore, because the third sentence shifts, the fourth must also shift to bring the focus back to Tom. The writer could avoid these two shifts by rewriting the third sentence so that it is about Tom, not about the forklift truck. That means making Tom the grammatical subject of the sentence, and, therefore, making the verb passive:

As he rounded the corner into the shop, he was hit by a forklift truck.

In summary, then, one way to guide your readers from one sentence to the next is to notify them that adjacent sentences are about the same thing. You can provide this notice by placing that thing in the subject position of both sentences.

## Use Transitional Words and Phrases at the Beginning of Your Sentences

Guideline

2

Guideline 2 (as well as Guidelines 3 and 4 in this chapter) pertain to situations not covered by Guideline 1. Guideline 1 applies when two adjacent sentences are about the same topic. That situation arises often. However, if you look at any page in almost any communication, you will see that most of the sentences shift the topic. Even so, every one of the sentences is related in some way to the sentence that precedes it—unless the communication is simply incoherent. One of your major tasks as a writer is to guide your readers from one sentence to the next by indicating how the topic of the succeeding sentence relates to the *different* topic of the preceding sentence.

One way to do that is to include transitional words and phrases that explicitly state the relationship. Some of the most commonly used are:

| | |
|---|---|
| Links in time | After, before, during, until, while |
| Links in space | Above, below, inside |
| Links of cause and effect | As a result, because, since |
| Links of similarity | As, furthermore, likewise, similarly |
| Links of contrast | Although, however, nevertheless, on the other hand |

Such words and phrases will help your readers most if you place them at or near the beginning of your sentences. That's because, even while they are reading the beginning of a sentence, your readers already want to know what relationship that sentence has to the preceding one. Consider, for example, the following pair of sentences:

> The Jefferson County Courts, which hear approximately 400 cases per month, use videotape machines to record court proceedings. Hamilton County Courts, which hear about 1350 cases per month, rely on specially trained legal stenographers.

How does the information in the second sentence relate to that in the first? Does the author intend that we compare the two county court systems? If so, what is the point of the comparison: are the two systems essentially similar or are they fundamentally different? A transitional word would tell us. Furthermore, if the transitional word were at the beginning of the second sentence, the entire sentence would be easier to read because we would know from the start how it relates to the preceding one:

> The Jefferson County Courts, which hear approximately 400 cases per month, use videotape machines to record court proceedings. *In contrast,* Hamilton County Courts, which hear about 1350 cases per month, rely on specially trained legal stenographers.

Thus, the second guideline for guiding your readers from one sentence to the next is to use transitional words and phrases at the beginning of your sentences.

Guideline<br>3

## Use Echo Words at the Beginning of Your Sentences

Most sentences contain some word or phrase that recalls to the readers' minds some information they've already read. Such words and phrases might be called *echo* words, because they echo something the readers already know. Consider the following example:

> Inflation can be *cured.* The *cure* appears to require that consumers change their basic attitudes toward consumption.

In this example, the noun *cure* at the beginning of the second sentence echoes the verb *cured* in the first. It tells the readers that what follows will discuss the "curing" that they have already heard about.

### Kinds of Echo Words

In the example about curing inflation, the echo word is another version of the word being echoed *(cured:cure).* In other cases, the echo word can be a pronoun:

We had to return the *copier*. *Its* frequent breakdowns were disrupting our work.

And sometimes an echo word is another word from the same "word family" as the word being echoed:

I went to my locker to get my *lab equipment*. My *oscilloscope* was missing.

In this example, *oscilloscope* in the second sentence echoes *lab equipment* in the first, because people know that an oscilloscope is a piece of lab equipment.

Finally, an echo word can be a word or set of words that recall some idea or theme expressed but not explicitly stated in the preceding sentence:

The company also purchased and retired 17,399 shares of its $2.90 convertible, preferred stock at $5.70 a share. *These transactions* reduce the number of outstanding convertible shares to 635,250.

In this example, the words *these transactions* tell readers that what will follow in the sentence concerns the "purchasing and retiring" that were mentioned in the preceding sentence.

You may have noticed that in each of the examples given so far, the echo word is in the main clause of the sentence. Echo words can also be used effectively in subordinate clauses:

The primary objective of the proposed campaign is to attract new members. For *this campaign* to be effective, the club's executive committee must carefully assess the willingness of the club's current members to undertake the projects that are suggested.

Furthermore, a sentence may contain more than one echo word:

Because the park does not own enough horses, the stable manager is required to meet the demand for riding tours by making the horses work as much as seven hours a day, rather than a more reasonable four or five. The *horses, laboring* under *this heavy workload,* grow tired and uncooperative by the end of the day, so that beginning riders have trouble getting their horses to perform.

## Importance of Placing Echo Words at the Beginning of Sentences

Like transitional words and phrases (Guideline 2), echo words should be placed at or near the beginning of a sentence. The reason is the same. As soon as people begin reading a new sentence, they are already trying to understand its relationship to the preceding sentence.

To understand the kind of difficulties you can create for your readers by postponing the appearance of your echo words until the end of a sentence, imagine that you are reading a report in which you have just completed this sentence:

> The metal is then coated by the Barnhardt process.                    (7)

After completing that sentence, you begin the next. The first word you read is:

> Sales

What has the new information contained in the word *Sales* to do with the old information in sentence 7? You can't tell. The next words are:

> have increased threefold

Do you know the relationship between the two sentences yet? You may have a guess, but you can't be sure. The next words are:

> in the year since

which are followed by:

> our engineers

Your suspicions about the relationship between the two sentences may now be very strong, but you still have only a guess. You know for sure only after you read the last three words of the sentence:

> modified this process.

Think how much easier your reading would have been if the second sentence had been written in the following way, so that the echo words appeared at the beginning, not the end, of the sentence:

> The metal is then coated by the Barnhardt process. Since *this process* was modified by our engineers, sales have increased threefold.

In sum, this guideline concerns situations in which you change the topic from one sentence to the next by using one thing as the grammatical subject of the first sentence and something else as the grammatical subject of the second. The guideline suggests that at the beginning of the second sentence you include some word or phrase that echoes the meaning of some part of the preceding sentence. By doing so, you will help your readers quickly and easily discern the relationship that you intend those two sentences to have.

## Use Paragraph Breaks To Signal Major Shifts in Topic

Guidelines 2 and 3 concern situations in which you desire to shift from one topic to another as you move from one sentence to the next. Following them will help you guide your readers across the gap that divides the sentences from each other. Your readers will receive additional help if you also tell them whether the gap is a narrow one or a wide one, whether the transition from one topic to another is a minor transition or a major one. One method you can use to tell your readers how large a gap to expect is to use paragraph breaks judiciously.

Many people have been led to believe that there are unambiguously "correct" places to break a text into paragraphs, and unambiguously "incorrect" ones. They have been told, for instance, that a paragraph is a group of sentences that all concern the same subject. However, a chapter is usually about a single subject—and so are most books—yet we rarely encounter either chapters or books that are written all in one long paragraph. Furthermore, many paragraphs address more than one aspect of a particular subject and could "correctly" be divided into smaller paragraphs. Consequently, as a writer you have a great deal of freedom in deciding where to begin each new paragraph.

### Deciding Where To Put Paragraph Breaks

To decide where to make paragraph breaks in your communications, you must do two things. The first is to identify places you *might* place them. You can do that by looking at the organizational hierarchy of your communication. You could put a paragraph break before the first sentence of each segment and subsegment in that hierarchy because at each of those locations you shift your topic.

Next, you must decide which of the potential spots for paragraph breaks should really have them. Certainly, you should place a paragraph break before each major section in your communication. You may even want to accompany those breaks with headings, especially if your communication is more than one page long. To decide which of the smaller segments should be marked by a paragraph break, you need to consider two factors: emphasis and custom.

When you make a paragraph break, you emphasize the sentences placed immediately before and after the break. The sentence just before the break is emphasized because readers are likely to look to it for a summary of the paragraph just being brought to a close. The sentence just after the paragraph break is emphasized because the first sentence of a paragraph typically announces the most important thing the paragraph has to say. Single-sentence paragraphs, although abominated by some teachers of writing, are especially useful for creating strong emphasis if used sparingly.

A second consideration is the customary length of paragraphs in the kind of communication you are writing. Some forms of communication customarily have paragraphs that are relatively long or relatively short. Newspapers, for instance, usually have relatively short paragraphs, sometimes averaging less than two sentences apiece. The paragraphs in letters and memos are often shorter than those in reports and books. On the other hand, communications written at work rarely have paragraphs that are an entire page long, even in documents that are typed double-spaced.

Figures 17-1 and 17-2 show two ways of paragraphing the same set of sentences.

---

Time is a precious resource for a handicapped student. A disability can increase the amount of time required to perform many simple physical tasks such as showering, dressing, and eating. For the disabled student, this means time taken away from studying, doing research, and (in some cases) attending classes. In order for a disabled student to make the best use of his or her time, a reliable, efficient transportation service must exist on campus.

To provide transportation for the handicapped, the university three years ago made available a two-door sedan. Since the inception of this service, however, the quality of the service it provides to the handicapped student has deteriorated, largely because the demand for the service is greater than can be handled by the single-passenger sedan the university has provided. Three years ago, the service made 7,675 student pickups; two years ago it made 10,825 pickups; and last year it made 12,623 pickups. For last year, that is an average of 68 passenger pickups per academic day. The single-passenger sedan used exclusively for this service currently travels around campus an average of 75 to 80 miles per day.

Unfortunately, as demand for the service has increased, so too has the amount of time disabled students must spend simply waiting for the sedan to arrive. In fact, more than 80% of the handicapped students responding to a questionnaire indicated that they have been late to class because they have had to wait for the sedan to arrive. In addition, one-fourth of the students indicated that they have missed class completely due to a lack of available transportation. The net effect of these time-related problems is unfortunate from the standpoint of the students and the university.

---

**Figure 17-1.**   One Way of Paragraphing the Passage Also Shown in Figure 17-2.

Time is a precious resource for a handicapped student.

A disability can increase the amount of time required to perform many simple physical tasks such as showering, dressing, and eating. For the disabled student, this means time taken away from studying, doing research, and (in some cases) attending classes.

In order for a disabled student to make the best use of his or her time, a reliable, efficient transportation service must exist on campus. To provide transportation for the handicapped, the university three years ago made available a two-door sedan.

Since the inception of this service, however, the quality of the service it provides to the handicapped student has deteriorated, largely because the demand for the service is greater than can be handled by the single-passenger sedan the university has provided. Three years ago, the service made 7,675 student pickups; two years ago it made 10,825 pickups; and last year it made 12,623 pickups. For last year, that is an average of 68 passenger pickups per academic day. The single-passenger sedan used exclusively for this service currently travels around campus an average of 75 to 80 miles per day.

Unfortunately, as demand for the service has increased, so too has the amount of time disabled students must spend simply waiting for the sedan to arrive.

In fact, more than 80% of the handicapped students responding to a questionnaire indicated that they have been late to class because they have had to wait for the sedan to arrive. In addition, one-fourth of the students indicated that they have missed class completely due to a lack of available transportation.

The net effect of these time-related problems is unfortunate from the standpoint of the students and the university.

**Figure 17-2.**   One Way of Paragraphing the Passage Also Shown in Figure 17-1.

## SUMMARY

Your readers need your help to see how the "new" information contained in one sentence relates to the "old" information they gathered from the preceding one. When readers cannot discern immediately the relationship between adjacent sentences, they stop their forward progress to puzzle out the relationship. Such efforts are annoying and tiring to readers and detract from the effectiveness of your communication.

The four guidelines for guiding your readers from sentence to sentence are:

1. *Avoid needless shifts in topic from one sentence to the next.* Readers can most easily see the relationship between adjacent sentences if both sentences are about the same thing. You can tell your readers that you are continuing to talk about the same topic by putting that topic in the subject position of both sentences.

2. *Use transitional words and phrases at the beginning of your sentences.* When you must shift topics between sentences, tell your readers right away how the two sentences relate to one another. You can do that by using transitional words and phrases such as *however, nevertheless, because, until,* and the like.

3. *Use echo words at the beginning of your sentences.* Echo words are words or phrases that recall to the readers' minds some old information they've already read. Echo words can be: (1) another version of the word being echoed (for example, *cured:cure*), (2) a pronoun that refers to the word (for example, *copier:it*), or (3) another word from the same family (for example, *lab equipment:oscilloscope*).

4. *Use paragraph breaks to signal major shifts in topic.* Your readers will be helped by this visual sign of a major shift. Put paragraph breaks before the beginning of all major segments. When deciding which subsegments to mark by paragraph breaks, consider the amount of emphasis you want to give them and the customary length of paragraphs in the kind of communication you are writing.

## EXERCISES

1. In each of the following pairs of sentences, underline the subject of each sentence. For pairs in which the first and second sentences have different topics, rewrite one or the other so that their topics are the same.

   A. To fluoridate the drinking water, a dilute form of hydrofluorisilic acid is added directly to the municipal water supply at the main pump. An automatic control continuously meters exactly the right amount of the acid into the water.

B. "Grab" samplers collect material from the floor of the ocean. Rock, sediment, and benthic animals can be gathered by these samplers at rates as high as 8,000 tons per hour.

C. Fourteen variables were used in these calculations. The first seven concern the volume of business generated by each sales division each week.

D. The city's low-income citizens suffer most from the high prices and limited selection of food products offered by commercial grocers. Furthermore, information concerning nutrition is difficult for many low-income citizens to find.

2. As you read the memo in Figure 17-3 (page 352), circle the various devices used by the writer, Mr. Bradshaw, to help his readers move from one sentence to the next. In the margin, explain which of the guidelines in this chapter have been followed by the writer in each case.

3. The purpose of this exercise is to help you see how your decisions about where to start new paragraphs can affect the emphasis of your writing.

A. Compare the two versions of the passage shown in Figures 17-1 and 17-2. What is the chief difference in the effects these two versions create? Write a paragraph describing that difference and explaining its sources.

B. Write a paragraph explaining which passage—that shown in Figure 17-1 or that in Figure 17-2—would be most appropriate for an essay in which handicapped students wanted to win sympathy for their complaints against the university's special transportation service. Which would be most appropriate for a proposal in which the university seeks money from the federal government to purchase a van especially designed to transport the handicapped? Why?

4. The sentences listed below form a single passage. Without changing the order of the sentences, rewrite them in a way that will guide readers smoothly from sentence to sentence. As you revise the sentences, make notes in the margin that tell which of the guidelines given in this chapter you have used. You may rework the sentences freely, but try not to leave out any ideas. Some sentences may not have to be rewritten. To start your work on each sentence, decide what you want the topic of each sentence to be.

A. Catalytic converters are required equipment on almost all new cars sold in the United States.

B. Removing undesirable emissions, particularly sulfates, from the cars' exhaust gases is the purpose of these converters.

C. More efficient operation is demonstrated by some catalytic converters than by others.

TO:        Wally Nugent
           Don McNeal

FROM:      Kent Bradshaw

SUBJECT:   Response to Your Request for Names of
           Courses in Management Fundamentals

I apologize for the delay in following up our conversation;
however, it turns out that there are fewer courses available
in management fundamentals than I expected, so it took
some digging. The digging uncovered only two courses that
I would feel comfortable recommending.

First choice is the BASIC MANAGEMENT FOR THE
NEWLY APPOINTED MANAGER seminar offered by the
University of Michigan. This is a three-day program
designed as an introduction and a brush-up. I have
attached a description of it for your review.

The second choice is a public session of THE
EXCEPTIONAL MANAGER seminar offered by the Forum
Corporation. The next offering is in Chicago on September
19 through 21. One of the outcomes of this seminar is a
personal action plan, which may not be useful to you if
your primary purpose in taking a course is to show your
superiors that you have had formal training in
management theory; nevertheless, this course is a good
one, and you should consider it.

In addition, there are a few supervisory training
programs offered at local universities; however, I believe
these programs focus on too low a level of management to
meet your needs. They tend to be directed to production
supervisors, particularly foremen.

None of the courses I found has the emphasis on the
financial aspects of management that you want. If you wish
to look at some basic finance courses, please let me know.
I will be glad to help you enroll in either of the courses
listed above, or to help you get further information if you
need it.

**Figure 17-3.**   Memo for Use in Exercise 2.

D. The Environmental Protection Agency (EPA) has recently compared two converter designs.

E. The catalyst is in pellets (like mothballs) in the first design, called the pelleted converter.

F. The second design, called a monolithic converter, has its catalyst in a single cylinder, like an ancient Egyptian monolith.

G. Cars equipped with each type of converter were driven by the EPA over a standardized route in order to compare the two designs.

H. Sulfates were stored by both converters when used at the low speeds of city driving, whereas both emitted sulfates when used at the higher speeds of highway driving.

I. The pelleted converter exhibited a greater sulfate storage capacity.

J. At low speeds, only 23% of the sulfur in the gasoline was released by the pelleted converter.

K. The monolithic converter released 92% at low speeds.

L. The pelleted converter released a much greater proportion of its stored sulfur when used at higher speeds.

M. A characteristic desired by the EPA is possessed by the pelleted converter: it is able to store sulfates in the more densely populated urban areas, where people drive slowly, and then release those sulfates in the less densely populated suburban and rural areas, where people drive faster.

# Choosing Words

## PREVIEW OF GUIDELINES

1. Use concrete, specific words.
2. Use words accurately.
3. Choose plain words over fancy ones.
4. Avoid unnecessary variation of terms.
5. Use technical terms when your readers will understand and expect them—but not otherwise.
6. Explain technical terms that are not familiar to your readers.

# CHAPTER 18

D avid is reading over the first draft of the first report he has written on the job since he graduated. He keeps wondering about the words he has chosen. Whenever he comes across a simple word (like *use*), he wonders whether he should substitute a longer one (like *utilize*) so he will seem more knowledgeable. But every time he comes to a long or technical word, he wonders whether he will seem to be showing off by using it instead of a simple word. "Will my readers think I don't know my subject if I use simple words?" he asks. "Or, if I use lots of technical terms, will I confuse and annoy my readers?"

David's concerns are quite natural. When choosing the words you will use at work, you will need to pursue three goals which sometimes conflict with one another. The first is to encourage your readers to view you favorably: the words you use will affect the impression you make on others. The second is to make your meaning clear: you must use words that communicate your meaning as precisely as possible. The third is to choose words that your readers understand. These three goals can come into conflict, for instance, when the word that most precisely expresses your meaning is a technical term your readers won't understand and may think you used just to show off.

The six guidelines in this chapter will help you achieve the sometimes very difficult objective of choosing words that will shape your readers' attitudes in the ways you desire while still communicating precisely and understandably.

## THE MAIN ADVICE OF THIS CHAPTER

Before reading those guidelines, you might find it helpful to consider one general piece of advice that underlies all of them: choose words suited to your purpose and readers.

Your purposes for writing at work are different from those you may encounter in some of your courses at school—and your way of thinking about your vocabulary should be different, too. In courses in literature and composition, for instance, you may learn how to write to increase your readers' sensitivity to various aspects of human experience. Accordingly, you sometimes choose colorful, out-of-the-ordinary words that will create new impressions on your readers' imaginations.

At work, however, your purposes for writing are much more practical. As explained in Chapter 4, you write to make things happen. Your writing is an instrument, a tool your readers use as they engage in some practical activity: making a decision, performing a procedure, and so on. When writing on the job, you should choose words that will enable your readers to understand your message as quickly, easily, and accurately as possible. Thus (as you will read below), you should usually pick plain words, not fancy ones, and you should select words your readers already know, not ones that will strain your readers' vocabularies and imaginations. That doesn't mean that the advice you are learning in other courses is incorrect, just that it usually doesn't apply in situations encountered on the job.

## INTRODUCTION TO GUIDELINES 1 AND 2

One of your chief objectives when writing on the job will be to enable your readers to understand exactly what you mean. Guidelines 1 and 2 will help you achieve that objective.

Guideline
1

### Use Concrete, Specific Words.

Words may be ranked according to their degree of abstraction. When a group of related words is ranked in this way, the words create a hierarchy. At the top of such hierarchies are very general words, such as *clothes, skill,* and *natural phenomena.* At the bottom of the hierarchies are very specific terms, such as *t-shirt, ability to return volleys in tennis,* and *the eruption of Mount St. Helens in 1980.*

The hierarchy shown in Figure 18-1 presents words that might be used when discussing the various pieces of equipment owned by ACR Corporation. At the top of the hierarchy is an abstract word (*equipment*) that refers to a general category of things. At the bottom of the hierarchy are concrete words, which refer to specific items we can perceive with our senses (such as IBM XT computer with serial number 0014327).

In some hierarchies the words at the bottom are not concrete but they are still more specific than those at the top. Figure 18-2 provides an illustration. In it, *zoology, botany, chemistry,* and *physics* are abstractions, but they are also much more specific than *major field of college study.*

Different levels of abstraction and generality serve different purposes. Sometimes you will want to refer to a general category (*equipment*) or an abstract quality (*efficiency*). Then abstract and general words are perfectly appropriate. At other times, however, you will be able to convey your meaning much more precisely by choosing more concrete, specific words.

To see how your purpose affects the generality of the words you should use, imagine that you have developed, at your boss's request, a plan for moving your department from its current building to a new one. When describing your overall plan, you might write the following:

Abstract term        I have developed special provisions for moving the department's equipment.

When describing the details of those provisions, you might include these sentences:

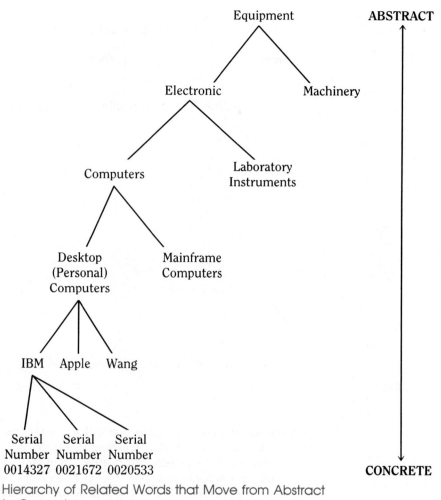

**Figure 18-1**   Hierarchy of Related Words that Move from Abstract to Concrete.

More
concrete
terms

> The four mainframe computers will be moved two weeks before anything else. The desktop computers will be moved much later, with the furniture.

And, in a memo to the company's security department, you might include these sentences:

Most
concrete
term

> We request that on June 18 you provide a security officer to accompany Dr. Wozniak's desktop computer (IBM serial number 0014327) from its present location in room 233 of the old building to room 1608 in the new building. This computer must be transported under such protection because of the highly confidential files contained in its memory.

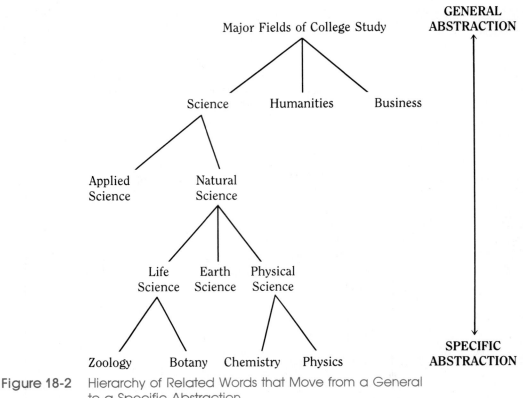

**Figure 18-2**    Hierarchy of Related Words that Move from a General to a Specific Abstraction.

As these sample sentences show, you will sometimes need to use general and abstract terms to convey your meaning. However, you should never select words that are more general and abstract than your meaning requires. The following pairs of sentences show how you can make your writing more meaningful to your readers by replacing general and abstract terms with concrete and specific ones.

- The costs of one material have risen recently.

  The costs of the bonding agent have tripled in the past six months.

- Some of the planned construction may harm animals.

  Building a fence along the west side of the highway may interrupt the natural migration of caribou, causing many to starve.

- Workers were ordered to work under unsafe conditions.

  Miners in Carlson's crew were ordered to place blasting caps in an area where the ventilation equipment had not been completely installed.

Guideline

2

## Use Words Accurately

Each word means certain things. When writing, you should be sure to use the words that say accurately what you mean. If you don't, your readers will notice. That creates two problems, even if your readers can still understand what you mean. First, inaccurate use of words distracts your readers from your message by drawing their attention instead to your problems with diction. Second, by showing your readers that you are not skillful or precise with one of the basic tools of thought—language—you might lead them to believe that you are not skillful or precise with your other tools—such as laboratory techniques or analytical skills.

One difficulty you should guard against is using the wrong word from pairs of words that are often confused. Here are some examples of such pairs. These words are drawn from the general vocabulary of all speakers of our language. It will be equally important for you to use accurately the technical terms of your special field.

| | |
|---|---|
| adapt, adopt | *Adapt* means "to adjust or make suitable for a certain use": |
| | "When you travel overseas, you will need to bring a special attachment to adapt your personal computer to the voltages used in European homes and offices." |
| | *Adopt* means "to select or take up as one's own": |
| | "The corporation has adopted a policy concerning the protection of company secrets." |
| affect, effect | *Affect* is a verb meaning "to influence": |
| | "Temperature will affect the seals." |
| | *Effect* is a noun meaning "result" or a verb meaning "to cause" or "to bring about": |
| | "They studied the effect of the acid on tin." (noun) |
| | "They effected many improvements." (verb; could be simplified to "They made many improvements.") |
| already, all ready | *Already* means "before the time specified": |
| | "The engineers had already completed the design." |
| | *All ready* means "completely set or prepared": |
| | "The components are all ready for final assembly." |
| altogether, all together | *Altogether* means "entirely" or "thoroughly": |
| | "The control panel is altogether too small." |

*All together* means "in a single group":

"The elk were all together at the bend in the river."

among, between

Usually, *among* is used with three or more things:

"KL-Grip is the strongest among the several glues available for this use."

Usually, *between* is used with two things:

"The managers had difficulty choosing between these two courses of action."

complement, compliment

*Complement* means "to fill out," "to complete," or "to mutually supply each other's lacks":

"The two systems complement one another effectively."

*Compliment* means "to praise":

"We complimented him on his plan."

continual, continuous

*Continual* means "happening again and again":

"The engine requires continual maintenance."

*Continuous* means "occurring without interruption":

"The continuous action of the river for two million years has worn a deep canyon in the rock."

eminent, imminent

*Eminent* means "distinguished":

"Dr. Korf is an eminent researcher."

*Imminent* means "about to take place":

"Feeling that a breakdown was imminent, he recommended that extra safety precautions be taken."

explicit, implicit

*Explicit* means "stated directly and unambiguously":

"She gave us explicit orders to shut down the line at 9 P.M."

*Implicit* means "implied" or "expressed indirectly":

"Though the commission didn't name the company that was suspected of illegally discharging the waste, the company's identity was implicit in the commission's report."

foreword, forward

*Foreword* is a noun meaning "a brief introductory section at the beginning of a report, manual, or similar communication":

"In the foreword of their report, the design team identified the three sources of funding for their work."

*Forward* is an adjective meaning "at or near the front" or an adverb meaning "toward what is in front":

"To engage the gears, push the handle to the forward position." (adjective)

"The conveyor belt carries the chassis forward to the painting booth." (adverb)

| | |
|---|---|
| imply, infer | *Imply* means "to suggest" or "to hint": |
| | "He implied that the operator had been careless." |
| | *Infer* means "to draw a conclusion based on certain evidence": |
| | "We infer from your report that you do not expect to meet the deadline." |
| its, it's | *Its* is the possessive form of *it*: |
| | "That is its control panel." |
| | *It's* is a contraction of *it is*: |
| | "It's a significant difference." |
| principal, principle | *Principal* means "primary" or "main": |
| | "Metal fatigue was the principal cause of the break." |
| | *Principle* means "basic law, rule, or assumption": |
| | "In dealing with clients, we observe three principles." |
| stationary, stationery | *Stationary* means "fixed in one position": |
| | "This pole is supposed to remain stationary." |
| | *Stationery* means "paper and envelopes for writing and sending letters": |
| | "The company had a graphic designer create new stationery for its official correspondence." |

---

# INTRODUCTION TO GUIDELINES 3 AND 4

Guidelines 3 and 4 will help you choose words that create a plain, direct writing style. Even though other, fancier styles are appropriate in other situations, at work the plain style works best—for reasons explained in the discussions of the two guidelines.

Guideline
**3**

## Choose Plain Words over Fancy Ones

At work, writers are often tempted to use fancy words rather than plain ones. Why? Perhaps they think that the fancy words seem more official, or that they will seem to be more knowledgeable if they use fancy words. Some of the fancy words used most often at work are included in the following list. The list includes only verbs, but might include nouns and adjectives as well. It is not intended to be exhaustive but to help you see the kinds of words you should try to avoid.

| Fancy Verbs | Common Verbs |
| --- | --- |
| ascertain | find out |
| commence | begin |
| compensate | pay |
| constitute | make up |
| endeavor | try |
| expend | spend |
| fabricate | build |
| facilitate | make easier |
| initiate | begin |
| prioritize | rank |
| proceed | go |
| transmit | send |
| utilize | use |

There are two important reasons for substituting plain words for these and any other fancy words that you might be tempted to use on the job: plain words are often more familiar than their fancy counterparts, and they are less likely to create a bad impression.

## Plain Words Are More Familiar

Abundant research confirms what common sense tells us: readers can read familiar words more easily than they can read unfamiliar ones.[1] People are quicker both to recognize and to understand words that are more familiar to them than words that are less familiar. Plain words are often more familiar than fancy words. We all recognize and understand the word *liar*, but most of us are slower to recognize and understand its synonym *prevaricator*—if we can recognize it at all. When readers don't immediately see the meaning of a word you use, they have to stop their forward progress to guess at its meaning from its context, or else they must consult a dictionary. In literary essays and other writing in the humanities there is often no problem in using words that will send readers to a dictionary. At work, however, readers want to understand as readily as possible without needing to consult outside sources.

### Fancy Words Can Create a Bad Impression

If you use fancy words instead of plain ones, you risk creating a bad impression. Part of your responsibility on the job is to make your knowledge useful to other people. If you use words they don't understand, you make your work useless—or at least difficult to use. Readers will think that you are not carrying out your responsibilities properly. In addition, they may become annoyed at you because you weren't considerate enough to write in a manner they could understand easily.

Even if they have no trouble understanding the fancy words you use, your readers may draw the conclusion that you are pompous, or showing off, or trying to hide a lack of ideas and information behind a fog of fancy words. Consider the effect, for instance, of the following sentence, which one writer included in a letter of application for a job:

> I am transmitting the enclosed resume to facilitate your efforts to determine the pertinence of my work experience to your opening.

### A Caution

From reading this guideline, you may have gotten the impression that you should use only simple language when writing at work. That is not necessarily the case. When you are writing to readers with vocabularies comparable to your own, you should use all the words at your command—provided that you use them accurately and appropriately. This guideline does not advise you to limit your vocabulary unnecessarily whenever you write at work. Instead, it cautions you against committing these two errors: using needlessly inflated words and using words that (although perfectly understandable to some people) will not be understood by your particular readers.

## Guideline 4   Avoid Unnecessary Variation of Terms

Most people have heard at one time or another that their writing will become monotonous if they use the same word over and over again. As a result, they use a variety of terms for the same thing. What they call a *dog* in one sentence, they call a *hound* or *cur* in the next. There are, however, at least two good reasons for calling a dog a dog throughout the communications you write at work.

The first reason is ease of recognition. Readers recognize and understand a term on its second appearance more quickly than they do an entirely new word for the same thing. If you've been discussing *structures* in a report but abruptly begin referring to *buildings*, your readers may have to pause to decipher whether you intend for *structures* and *buildings* to mean the same thing, or for one to be a subcategory of the other, or for the terms to refer to related but separate things. Or, your readers may simply fail to recognize that there might be any connection at all between the two terms.

The second reason to avoid unnecessary variation of terms is to avoid ambiguous reference, in which it is unclear which item a term refers to. Consider the following example:

> We have a new typewriter, for which we bought a special ribbon that lifts errors off the page. This item was very reasonably priced.

It is not entirely clear in the second sentence whether the *item* is the typewriter or the ribbon. In this case, variation in the terms used to designate something has led to an ambiguity that is not settled by the sentences themselves.

A common kind of variation of terms that can contribute to the problem of ambiguous reference is the careless use of pronouns as substitutes for nouns. For example:

> This job has much slack time, which the operator spends cleaning up. It takes much getting used to.

In the second sentence, *it* could refer to the job, the slack time, the cleaning up, or even to the whole idea that the job entails slack time that is spent in cleaning up.

Nevertheless, writing can become monotonous if the same word is used repeatedly throughout a sentence or adjacent sentences. One way to avoid that monotony is to use pronouns carefully. Compare the following sentences:

> If the operators encounter a problem, the operators should report the problem directly to the shift supervisor.

Pronoun

> If the operators encounter a problem, they should report it directly to the supervisor.

The second sentence is easier than the first to read, and the meaning of the pronoun *they* is unambiguous.

However, when your choice is between monotony and ambiguity, you should choose monotony. It is better to use the same word repeatedly to keep your meaning clear than to use a lively array of words that makes your meaning difficult to decipher.


## INTRODUCTION TO GUIDELINES 5 AND 6

The final two guidelines in this chapter provide advice about using the specialized vocabulary of your field.

<table>
<tr><td>

Guideline

5
</td><td>

## Use Technical Terms When Your Readers Will Understand and Expect Them—but Not Otherwise

</td></tr>
</table>

Guideline

5

## Use Technical Terms When Your Readers Will Understand and Expect Them—but Not Otherwise

One of the most pressing questions you will face when deciding what words to use will involve the technical terms of your special field. In some situations, technical terms help you communicate effectively because they convey precise meanings very economically. Many have no exact equivalent in everyday speech. Others would take many sentences—even many paragraphs—to explain to someone who doesn't know them. Besides helping you communicate effectively with your fellow specialists, technical terms can help to establish your credibility. Much of your study in your major involves mastering its special vocabulary. When you use that vocabulary accurately on the job, you show your readers that you are adept in your field.

Consequently, you should use the specialized terms of your field when your readers will understand those terms and expect you to use them. If you fail to do so, you will almost surely write ineffectively.

On the other hand, if you use technical terms when writing to people who are unfamiliar with them, you will make your communications very difficult for your readers to understand. Consider the following sentence:

> The major benefits of this method are smaller in-gate connections, reduced breakage, and minimum knock-out—all leading to great savings.

Although this sentence would be perfectly clear to anyone who works in a foundry that manufactures parts for automobile engines, it would be unintelligible to most other people because of the use of the technical terms *in-gate connections*, and *knock-out*. Similarly, if you use the technical terms of your field when writing to people who are unfamiliar with those terms, your readers won't be able to understand you.

### Identifying Technical Terms Your Readers May Not Know

To be able to avoid using technical terms inappropriately, you must identify ones your readers won't know. Remember that many of the terms that have become ordinary to you through your college studies will not be familiar to many of your readers. Many recent college graduates are so used to talking with their instructors and with other students in their majors that they think that the specialized terms used in their fields are more widely understood than they actually are. When you begin work, remind yourself that much of your writing will go to readers who are *not* in your field. Even when addressing fellow employees in your employer's organization, you may be writing to people who will not understand the technical terms that seem ordinary to you.

Your task of identifying technical terms your reader won't know can be complicated by that fact that in many fields—perhaps including yours—some of the technical terms are widely known but others are not. For instance, most people are familiar enough with chemistry to know what an acid is, and many have some sense what a base is. But many fewer know what a polymer is. Similarly, with respect to computers, many know what disk drives and programs are, but few know what an erasable, programmable, read-only memory is. Thus, when writing to people who are not familiar with your specialty, you must distinguish between those technical terms your readers know and those they don't. To do that, you must have a fairly specific knowledge of your readers, especially of their level of familiarity with your special field.

## Explain Technical Terms That Are Not Familiar to Your Readers

If you follow Guideline 5, you will generally avoid using technical terms when writing to readers who do not understand them. However, there will be times when you will need to use specialized terms unfamiliar to your readers. That might happen, for example, when you have a large audience composed of some people in your field and some people outside your field, or when you are explaining an entirely new subject not familiar to any of your readers.

When you must use unfamiliar words, you should explain them to your readers. Sometimes you will be able to do that simply by describing them in ordinary terms:

> On a boat, a *line* is a rope or cord.
>
> The *exit gate* consists of two arms that hold a jug while it is painted, then let it proceed down the production line.

At other times, you will need to use somewhat more elaborate procedures. Two that are especially useful in writing on the job are to make a classical definition and to use an analogy. The essential strategy of both of these forms of explanation is to help your readers understand the unfamiliar by describing it in terms of things they know well.

### Classical Definition

To write a classical definition, you need to identify three things:

- The thing being defined.
- Some familiar group of things to which the thing being defined belongs.
- The key distinction between the thing being defined and the other things in the group.

The following list provides examples of classical definition:

| Word | Group | Distinguishing Characteristic |
|------|-------|-------------------------------|
| burrow | hole in the ground | dug by an animal for shelter and habitation |
| crystal | solid | in which the atoms or molecules are arranged in a regularly repeated pattern |
| modem | electronic device | that permits computers to talk with one another over the telephone |
| vine | plant | that has a stem too flexible or weak to support itself |

Of course, a definition can work only if the reader is already familiar with the group to which the item being defined belongs and with the terms that are used to distinguish the thing being defined from the other things in the group. For example, a reader would need to know what a slurry is to understand this definition: *"Pulp* is a slurry made of fiber, usually from wood."

That does not mean, however, that you should try always to pick ordinary rather than technical terms for your definitions. When defining something for an audience familiar with your specialty, you should employ the specialized terms they already know to define those they don't. Thus, it would be appropriate for you to say that "wood sorrel is a trifoliate leguminous plant" when you are addressing a botanist or other person who would understand the terms *trifoliate* (having three leaves) and *leguminous plant* ("one of the dicotyledonous plants having a superior ovary and usually dehiscing into two valves with the seeds attached to the ventral suture").

## Analogy

Another way to help your readers understand unfamiliar words is to use analogy. In an analogy, you explain the unfamiliar thing by comparing it with something your readers know well:

An *atom* is like a miniature solar system in which the nucleus is the sun and the electrons are the planets that revolve around it.

A *computer memory* is like a filing cabinet in which information is stored in various drawers.

To explain complex ideas, you can extend your analogies for a paragraph or more. In lengthier analogies, it is sometimes important to tell your readers how the thing you are defining differs from the thing they are familiar with—

so that they do not draw incorrect inferences from the analogy. Nevertheless, in an analogy, the emphasis is upon similarities:

Unfamiliar thing being explained

Familiar thing it is being compared to

Key similarities

Key differences

> The suspension system on the monorail is very similar to that used on a passenger automobile. At the front and rear of each car, an automotive-type axle is attached to the body of the cars by a swivel-bar assembly that permits vertical and lateral movement of the car with respect to the axle. Shock absorbers like those used on a car damp vertical and lateral movement. The pneumatic tires are identical to those used on some busses.
>
> This suspension system differs from that of an automobile in one important respect. On each axle, two guide wheels are mounted horizontally between the tires. These guide wheels press against an I-beam that is laid along the middle of the track. By following the curves of the I-beam, the guide wheels steer the cars along their route.

## SUMMARY

1. *Use concrete, specific words.* Concrete, specific words are much easier than abstract, general ones for readers to understand.
2. *Use words accurately.* If you use words inaccurately, your readers may conclude that you are also inaccurate in the rest of your work.
3. *Choose plain words over fancy ones.* Fancy words can lead your readers to feel you are pompous, showing off, or hiding something.
4. *Avoid unnecessary variation of terms.* Using two words for one thing can confuse readers.
5. *Use technical terms when your readers will understand and expect them—but not otherwise.* Technical terms provide a concise, precise way of communicating to your fellow specialists, but they can also make your writing unintelligible to other people.
6. *Explain technical terms that are not familiar to your readers.* You can describe unfamiliar terms in ordinary words, or use classical definitions or analogies.

## NOTE

1. George R. Klare, "The Role of Word Frequency in Readability," *Elementary English* 45(1968): 12–22.

## EXERCISES

1. In the memo shown in Figure 18-3, identify places where the writer has ignored the guidelines given in this chapter. You may find it helpful to use a dictionary. Then write an improved version of the memo by following the guidelines in this chapter and in Chapter 16 ("Writing Sentences").

July 8, 19--

To:      Gavin MacIntyre, Vice President, Midwest Region

From:   Nat Willard, Branch Manager, Milwaukee Area
        Offices

The ensuing memo is in reference to provisions for the cleaning of the six offices and two workrooms in the High Street building in Milwaukee. This morning, I absolved Thomas's Janitor Company of its responsibility for cleansing the subject premises when I discovered that two of Thomas's employees have surreptitiously been making unauthorized long-distance calls on our telephones.

Because of your concern with the costs of running the Milwaukee area offices, I want your imprimatur before proceeding further in making a determination about procuring cleaning services for this building. One possibility is to assign the janitor from the Greenwood Boulevard building to clean the High Street building also. However, this alternative is judged impractical because it cannot be implemented without circumventing the reality of time constraints. While the Greenwood janitor could perform routine cleaning operations at the High Street establishment in one hour, it would take him another ninety minutes to drive to and fro between the two sites. That is more time than he could spare and still be able to fulfill his responsibilities at the High Street Building.

Another alternative would be to hire a full-time or part-time employee precisely for the High Street building. However, that building can be cleaned so expeditiously, it would be irrational to do so.

The third alternative is to search for another janitorial service. I have now released two of these enterprises from our employ in Milwaukee. However, our experiences with such services should be viewed as bad luck and not effect our decision, except to make us more aware that making the optimal selection among companies will require great care. Furthermore, there seems to be no reasonable alternative to hiring another janitorial service.

Accordingly, I recommend that we hire another janitorial service. If you agree, I can commence searching for this service as soon as I receive a missive from you. In the meantime, I have asked the employees who work in the High Street building to do some tidying up themselves and to be patient.

**Figure 18-3**   Memo To Be Used in Exercise 1.

2. List three words used in your field that are not familiar to people in other fields. These may include words that people in other fields have heard of but cannot define precisely in the way specialists in your field do. For each definition, underline the word you are defining. Then, circle and label the part of your definition that describes the familiar group of things that the word you are defining belongs to. Finally, circle and label the part of your definition that identifies the key distinction between the thing you are defining and the other things in the group.

3. Select two words or concepts that are familiar to people in your field, but unfamiliar to people outside it. For each term, create an analogy that will explain the term to people outside your field.

# Drafting
# Visual Elements

When you write at work, you will be able to call upon several resources for achieving your communication objectives. The first is prose: your words, sentences, paragraphs, and so on. The eight chapters in Part IV provide you with extensive advice about how to write your prose. The three chapters here in Part V focus instead on two additional communication resources that you have at your disposal:

- *Visual Aids.* These include tables, graphs, diagrams, photographs, and many similar devices. Using them, you can communicate more clearly, briefly, and forcefully than you could through prose alone.

- *Page Design.* To design your pages, you plan the size and placement of their headings, prose, visual aids, and other elements. Thoughtful page designs not only look attractive but also emphasize your main points and help your readers read efficiently.

When preparing visual aids and designing pages, you should use the same reader-centered approach that you use when writing your prose. Think constantly about the reading tasks you want your visual aids and page designs to help your readers perform and about the ways you want them to alter your readers' attitudes. Think also about your readers' moment-by-moment reactions while reading.

From your readers' point of view, your prose, visual aids, and page design form a single unit. All must work together harmo-niously. Consequently, you should plan, draft, evaluate, and revise the visual elements of your communications at the same times that you perform those activities for your prose. In that way, you can ensure that your visual aids, page design, and prose support and reinforce each other.

The next three chapters will help you create communications that are as effective visually as they are verbally.

## CONTENTS OF THIS SECTION

**Chapter 19**
**Using Visual Aids**

**Chapter 20**
**Twelve Types of Visual Aids**

**Chapter 21**
**Designing Pages**

# Using Visual Aids

## PREVIEW OF GUIDELINES

1. While planning a communication, look for places to use visual aids.
2. Choose visual aids appropriate to your purpose.
3. Choose visual aids your readers know how to read.
4. Make your visual aids easy to use.
5. Make your visual aids simple.
6. Label the important content clearly.
7. Provide informative titles.
8. Introduce your visual aids in your prose.
9. State the conclusions you want your readers to draw from your visual aids.
10. Put your visual aids where your readers can find them easily.

During her first two weeks on the job, Sandra has been assigned to learn as much as she can about her department's work. She's having fun. She is visiting each of her co-workers, one by one, so that they can tell her in detail about their specialties. And she is reading lots of things those people have written, including proposals, reports, and even a few published articles.

When she began this reading, she was struck by the abundance of tables, graphs, drawings, and other visual aids that she found. Her co-workers use these devices much more often than she and her classmates did in college. Furthermore, as a reader, she likes the visual aids a great deal. They make writing more informative, easier to use, and more persuasive than it would otherwise be.

As she read more, Sandra discovered something else: two of her co-workers, Elizabeth and Ravi, use visual aids much more effectively than do the others:

- Elizabeth and Ravi use visual aids frequently to help their readers understand and use the information in their communications. Others in Sandra's department occasionally write long, tedious prose passages that could have been replaced by a single, informative, helpful visual aid.
- Elizabeth and Ravi create visual aids that are clear and easy to use. Others occasionally include confusing visual aids.
- Elizabeth and Ravi weave their visual aids smoothly into their prose. At times, others seem to paste visual aids in thoughtlessly.

From her reading, Sandra concluded that people who use visual aids skillfully write better than people who don't. Now, she wants to know what she must do to use visual aids effectively in her own writing. The ten guidelines in this chapter tell her. Like Sandra, you can apply these *general* guidelines to all types of visual aids, whether tables, graphs, drawings, flow charts, budget statements, or any of the many others used on the job. In the next chapter, you will find *specific* advice about employing and constructing twelve kinds of visual aids that you are likely to find useful at work.

## MORE THAN JUST AIDS

Before reading the guidelines in this chapter, you need to know one important fact about the name *visual aids*. It's misleading. It suggests that in the writing you do at work, your *words* are primary and the visual *aids* are merely assistants.

Nothing could be further from the truth. When used creatively and effectively, visual aids are an integral part of communications. They can carry some parts of a message more effectively than prose can. In fact, in some situations, visual aids can carry the *entire* message. No words are needed at all.

For instance, if you've ever flown, you may recall reaching into the pocket on the back of the seat ahead of you to pull out a sheet of instructions for leaving the plane in an emergency. Many airlines use instruction sheets that are wordless. Figure 19-1 shows an example. Figure 19-2 shows the instructions that one company provides for unpacking and setting up an electric typewriter. These, too, communicate without words.

Although you may never create a communication that relies solely on visual aids, you should remember that visual aids are powerful communication tools, not mere decorations or supplements.

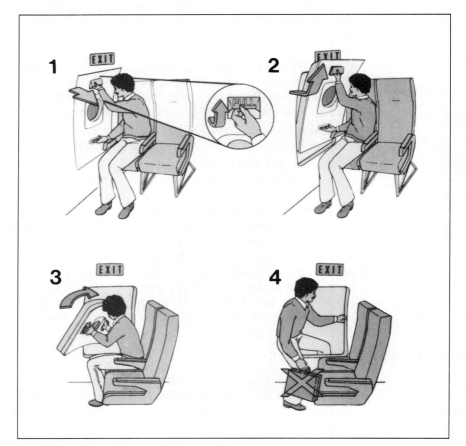

**Figure 19-1.**   Wordless Instructions for Leaving a Plane.
Courtesy of USAir.

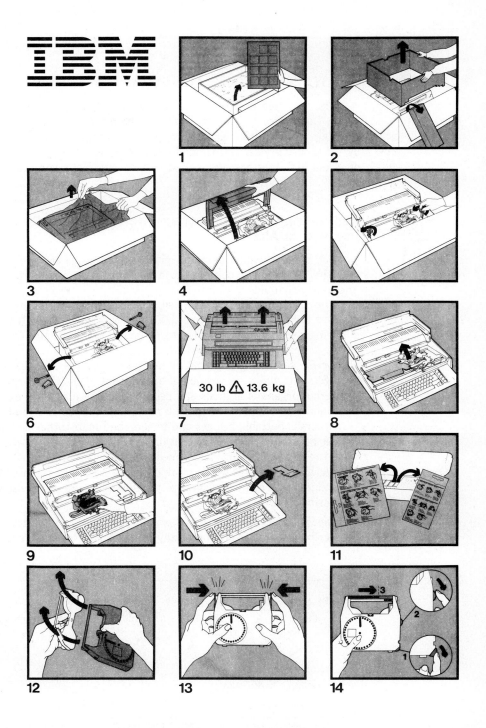

**Figure 19-2.** Wordless Instructions for Setting Up a Typewriter.
Courtesy of IBM.

## INTRODUCTION TO GUIDELINES 1 THROUGH 3

The first step in using visual aids effectively is to make careful plans for them. Guidelines 1 through 3 will help you make those plans. Guideline 1 will help you identify places where you can strengthen your communication by using visual aids, and Guidelines 2 and 3 will help you choose the types of visual aids that are best suited to achieving your communication objectives.

<u>Guideline</u>

**1**

### While Planning Your Communication, Look for Places To Use Visual Aids

You should begin looking for places to use visual aids when making your very first plans for a communication. Think from the start about ways to integrate visual aids fully into it.

The situations in which visual aids can make a communication more effective are numerous and varied. To give you an idea of what they are, the following paragraphs describe some of the things visual aids can help you do.

- *Describe something physical.* In every field, people describe physical objects: machines, experimental apparatuses, geological formations, internal organs of animals and people, to name just a few examples. For instance, an engineer designed a new hinge that can be used in spacecraft, portable bridges and towers, and many other applications. Next he wanted to describe that design and its virtues to other engineers who might want to use it. By means of the diagrams shown in Figure 19-3, he described it much more clearly than would be possible in words. Similarly, you can use diagrams, photographs, and other visual aids to describe physical objects.

- *Describe something that is not physical.* At work, you will need to describe the organization of many things that are not physical, such as a

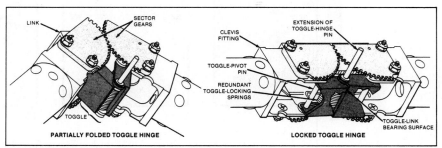

Figure 2. **Meshing Sector Gears** brace a strut as it deploys. The toggle unfolds and finally snaps and locks in place. An extension on the toggle-hinge pin allows it to be grasped by a toggle-opening tool for retraction.

**Figure 19-3.**    Drawing that Shows the Structure of Something Physical.

From "Toggle Hinge for Deployable Struts," *NASA Tech Briefs* 9 (Fall 1985):132.

plan for completing a project, the management structure of a company, a manufacturing process, or the overall design of a large computer software system. In all those situations, visual aids can help you.

For example, a group of computer researchers integrated a series of computer programs into a single system for helping engineers design manufacturing processes. By using the diagram shown in Figure 19-4, they were able to provide their readers with an excellent understanding of the relationships among these programs. A great variety of visual aids can be used to describe the organization of nonphysical things.

- *Explain the relationships among data.* Often, you will want your readers to understand the relationship among various pieces of data. These may be data you obtain through laboratory research, survey research, or other forms of information gathering.

For example, three scientists studying the causes and prevention of cancer conducted an experiment to find out what would happen if they treated a certain kind of cell with Interleukin-3, a peptide that promotes growth. In their experiment, they measured three variables at each of five concentrations of Interleukin-3. Although they could have presented their data in prose, they used instead the graph shown in Figure 19-5, which is much more effective than prose at showing the relationships they found.

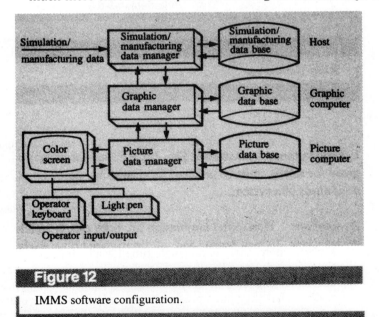

**Figure 12**

IMMS software configuration.

**Figure 19-4.**   Drawing that Shows the Organization of Something that Is Not Physical.

From H. Engelke et al., "Integrated Manufacturing Modeling System," *IBM Journal of Research and Development* 29 (July 1985):350. Copyright 1985 by International Business Machines Corporation.

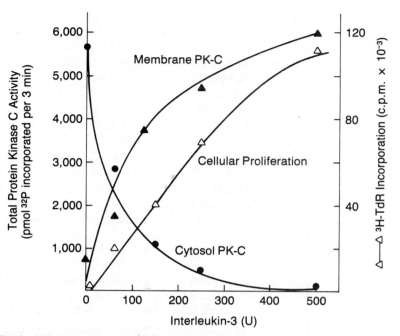

**Fig. 3** Effect of treatment of FDC-P1 cells with increasing concentrations of IL-3 on the subcellular redistribution of protein kinase C activity and on DNA synthesis.

**Figure 19-5.** Graph that Explains the Relationships among Separate Pieces of Data.

From William L. Farrar, Thomas P. Thomas, and Wayne B. Anderson, "Altered Cytosol/Membrane Enzyme Redistribution on Interleukin-3 Activation of Protein Kinase C," *Nature* 315 (1985):237. Reprinted by permission. Copyright 1985 by Macmillan Journals Limited. Courtesy of Dr . William L. Farrar, National Cancer Institute.

You too can use visual aids to help your readers see the relationships between various pieces of data.

- *Make detailed information easy to find.* For some tasks, visual aids are much easier than prose for readers to use. For example, a manufacturer of photographic film wanted to tell its customers how to develop their own negatives. One crucial piece of information these customers need is the length of time they should leave the film in the developer solution. Unfortunately, there is no single time for all situations. The developing time depends upon the type of developer used, the temperature of the developer, and the size of the developing tank. In addition, the writers realized that they must present two warnings to their readers.

     The writers decided to present the information in the table shown in Figure 19-6. The table enables readers to quickly find the correct developing time for the particular developer, temperature, and tank they are using. Readers would have much more difficulty finding the correct time if they had to rely on Figure 19-7 (page 380), which presents only

| KODAK Developer | Developing Time (in Minutes) | | | | | | | | | |
|---|---|---|---|---|---|---|---|---|---|---|
| | SMALL TANK (Agitation at 30-Second Intervals) | | | | | LARGE TANK (Agitation at 1-Minute Intervals) | | | | |
| | 65°F (18°C) | **68°F (20°C)** | 70°F (21°C) | 72°F (22°C) | 75°F (24°C) | 65°F (18°C) | **68°F (20°C)** | 70°F (21°C) | 72°F (22°C) | 75°F (24°C) |
| HC-110 (Dil B) | 8½ | **7½** | 6½ | 6 | 5 | 9½ | **8½** | 8 | 7½ | 6½ |
| D-76 | 9 | **8** | 7½ | 6½ | 5½ | 10 | **9** | 8 | 7 | 6 |
| D-76 (1:1) | 11 | **10** | 9½ | 9 | 8 | 13 | **12** | 11 | 10 | 9 |
| MICRODOL-X | 11 | **10** | 9½ | 9 | 8 | 13 | **12** | 11 | 10 | 9 |
| MICRODOL-X (1:3)* | — | **—** | 15 | 14 | 13 | — | **—** | 17 | 16 | 15 |
| DK-50 (1:1) | 7 | **6** | 5½ | 5 | 4½† | 7½ | **6½** | 6 | 5½ | 5 |
| HC-110 (Dil A) | 4½† | **3¾†** | 3¼† | 3† | 2½† | 4¾† | **4¼†** | 4† | 3¾† | 3¼† |

*Gives greater sharpness than other developers shown in table.

†Avoid development times of less than 5 minutes if possible, because poor uniformity may result.

**Note:** Do not use developers containing silver halide solvents.

**Figure 19-6.**   Table that Makes Information Easy to Find.
Courtesy of Eastman Kodak Company.

*some* of the same information in prose. To use the prose, they would have to do a lot of searching—and they might miss the opening warnings and background information altogether.

To spot places where visual aids can make your communication easier to use, look for places where your readers will want to find detailed information without scanning through line after line of prose.

● *Support your arguments.* Visual aids are often an excellent means of presenting information in support of your persuasive points. For example, the manufacturer of a plastic insulating material wanted to persuade greenhouse owners that they could greatly reduce their heating bills by applying sheets of the plastic to their greenhouses. Although the manufacturer could have explained the test results that support that claim, it decided instead to use the bar graph shown in Figure 19-8.

To spot places where bar graphs and other visual aids can increase the visual impact of your communication, look for places where you make persuasive points that could be represented in visual images.

## Guideline 2   Choose Visual Aids Appropriate to Your Purpose

Once you have decided where to use visual aids, you must decide which type to use. There are many types of visual aids, and most information can be presented in more than one type. For example, numerical data can often be presented in a table, bar graph, line graph, or pie chart. The components of an electronic instrument can be represented in a photograph, sketch, block diagram, or electronic schematic.

How can you determine which type of visual aid will most effectively communicate a certain point to your particular readers? First, consider the purpose of your visual aid. You can define the purpose of an individual visual

Developing Times for Kodak Tri-X Pan Film, Size 135

To develop 35-mm rolls of Kodak Tri-X Pan Film with one of Kodak's packaged developers at the recommended dilutions, use the developing times given below. Note that development times shorter than five minutes may result in unsatisfactory uniformity. In the following paragraphs, developing tanks that are one-quart size or smaller are called "small" tanks. They should be agitated at 30-second intervals. Bigger tanks, called "large" tanks below, should be agitated at 1-minute intervals. Do not use developers containing silver halide solvents.

With HC-110 (Dil B) in a small tank, develop for 8½ minutes if your developer is at 65°F (18°C), 7½ minutes at 68°F (20°C), 6½ minutes at 70°F (21°C), 6 minutes at 72°F (22°C), and 5 minutes at 75°F (24°C). In a large tank, develop for 9½ minutes at 65°F (18°C), 8½ minutes at 68°F (20°C), 8 minutes at 70°F (21°C), 7½ minutes at 72°F (22°C), and 6½ minutes at 75°F (24°C).

With D-76 in a small tank, develop for 9 minutes if your developer is at 65°F (18°C), 8 minutes at 68°F (20°C), 7½ minutes at 70°F (21°C), 6½ minutes at 72°F (22°C), and 5½ minutes at 75°F (24°C). In a large tank, develop for 10 minutes at 65°F (18°C), 9 minutes at 68°F (20°C), 8 minutes at 70°F (21°C), 7 minutes at 72°F (22°C), and 6 minutes at 75°F (24°C).

With D-76 (1:1) in a small tank, develop for 11 minutes if your developer is at 65°F (18°C), 10 minutes at 68°F (20°C), 9½ minutes at 70°F (21°C), 9 minutes at 72°F (22°C), and 8 minutes at 75°F (24°C). In a large tank, develop for 13 minutes at 65°F (18°C), 12 minutes at 68°F (20°C), 11 minutes at 70°F (21°C), 10 minutes at 72°F (22°C), and 9 minutes at 75°F (24°C).

With Microdol-X in a small tank, develop for 11 minutes if your developer is at 65°F (18°C), 10 minutes at 68°F (20°C), 9½ minutes at 70°F (21°C), 9 minutes at 72°F (22°C), and 8 minutes at 75°F (24°C). In a large tank, develop for 13 minutes at 65°F (18°C), 12 minutes at 68°F (20°C), 11 minutes at 70°F (21°C), 10 minutes at 72°F (22°C), and 9 minutes at 75°F (24°C).

With Microdol-X (1:3), do not develop at 65°F (18°C) or 68°F (20°C). In a small tank develop for 15 minutes at 70°F (21°C), 14 minutes at 72°F (22°C), and 13 minutes at 75°F (24°C). In a large tank, develop for 17 minutes at 70°F (21°C), 16 minutes at 72°F (22°C), and 15 minutes at 75°F (24°C).

With DK-50 (1:1) in a small tank, develop for 7 minutes if your developer is at 65°F (18°C), 6 minutes at 68°F (20°C), 5½ minutes at 70°F (21°C), 5 minutes at 72°F (22°C), and 4½ minutes at 75°F (24°C). Note, however, that you should avoid development times of less than 5 minutes if possible, because poor uniformity may result. In a large tank, develop for 7½ minutes at 65°F (18°C), 6½ minutes at 68°F (20°C), 6 minutes at 70°F (21°C), 5½ minutes at 72°F (22°C), and 5 minutes at 75°F (24°C).

**Figure 19-7.**   Prose Presentation of Part of the Information Given in Figure 19-6.

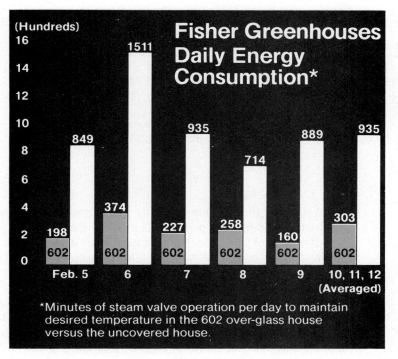

**Figure 19-8.**   Bar Graph Used To Support an Advertising Claim for Plastic Insulation.
Courtesy of Monsanto Company.

aid in the same way that you define the purpose of an entire communication: by identifying the task you want the visual aid to enable your readers to perform while reading the communication and the way you want it to affect your readers' attitudes.

## Your Readers' Tasks

Different types of visual aids are suited to different reading tasks. Imagine, for instance, that you have just surveyed people who graduated over the past five years from three departments at your college. Now, you want to tell those same people what you learned from them about their average starting salaries. You could do that in a table, bar graph, or line graph. Which choice would be best? The answer depends upon the way you want your readers to be able to use that information.

If your purpose is to enable alumni of the departments to find out the average starting salary for people who graduated in their year from their department, you could use a table like that shown in Figure 19-9. (The mar-

Average salary for reader's department, my year.

| Department | Year of Graduation | | | | |
|:---:|:---:|:---:|:---:|:---:|:---:|
| | 1983 | 1984 | 1985 | 1986 | 1987 |
| A | 17,350 | 17,725 | 18,250 | 18,600 | 19,225 |
| B | 21,650 | 22,875 | 23,300 | 24,000 | 25,250 |
| C | 19,250 | 19,975 | 22,100 | 23,225 | 24,625 |

**Figure 19-9.** Table of Average Starting Salaries by Department and Year of Graduation.

ginal notes in this and the following two figures show where one particular graduate would find the pertinent facts about his class.)

If you want to enable the alumni of each department to see how their average salary compared with the average salaries received by people who graduated in that same year from the other departments, you could use a bar graph like that shown in Figure 19-10.

And if you want to enable the alumni to see how the average starting salary in their department has changed over the years, and to compare that change with the changes in salaries for the other departments, you could use a line graph like that shown in Figure 19-11 (page 384).

In Chapter 20, you will find considerable information about the kinds of reading tasks that are served by twelve types of visual aids: tables, bar graphs, pictographs, line graphs, pie charts, photographs, drawings, diagrams, flow charts, organizational charts, schedule charts, and budget statements.

## Your Readers' Attitudes

When you want to use visual aids to affect your readers' attitudes, pick the type that communicates most quickly and directly the evidence that supports your persuasive point. Suppose, for instance, that you want to show that because of a design change you recommended, your company has had to make many fewer service calls on one of its products. Your evidence is a tally of the number of calls made in each of the six months before the design change and each of the six months after the change. As Figure 19-12 (page 385) shows, if you present that data in a table, your readers will have to do a lot of subtracting to appreciate the extent of the decrease. If you present the data in a line graph, however, your readers will be able to see the extent of the decrease at a glance.

In a similar fashion, whenever you desire to make a persuasive point in your communication, choose the type of visual aid that will enable your readers to grasp most quickly and easily the significance of your evidence.

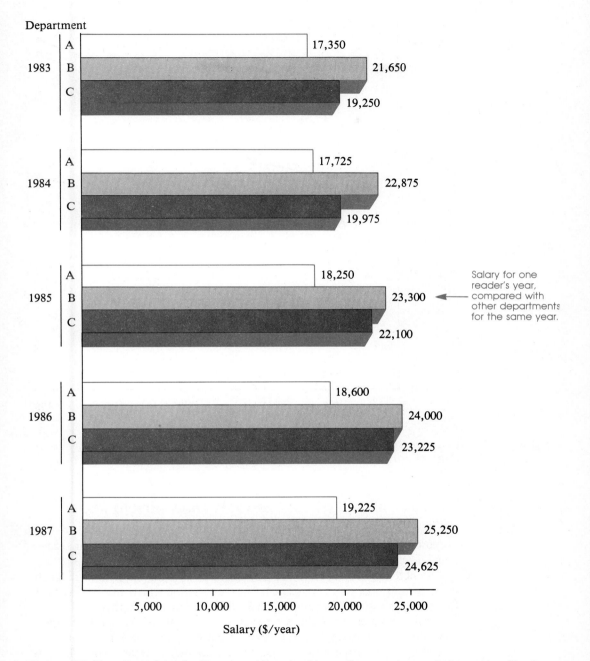

**Figure 19-10.**   Bar Graph Showing Year-by-Year Comparison of Average Starting Salaries for Three Departments.

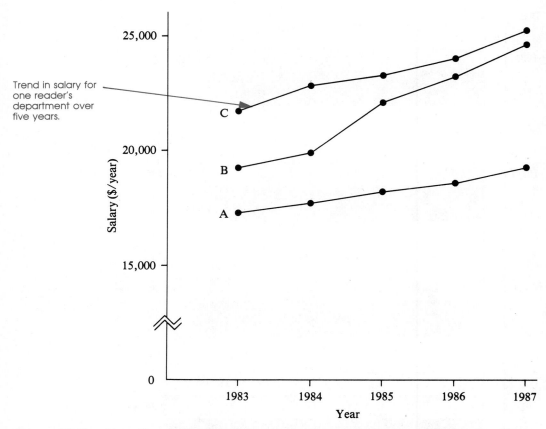

Trend in salary for one reader's department over five years.

**Figure 19-11.**   Line Graph Showing Trends in Average Starting Salaries for Three Departments.

Guideline
3

## Choose Visual Aids Your Readers Know How To Read

The types of visual aids you choose should be suited not only to your purpose but also to your readers. Be sure to use visual aids that your readers know how to interpret.

Some types of visual aids are familiar to us all—the table, the bar graph, the organizational chart, and so on. They are used so widely in so many fields that you can assume that any reader will know how to read them.

Other types of visual aids are much more specialized. Figure 19-13 (page 386) shows four examples. You may be learning to create these or other special types of visual aids in courses in your major department. They are informative to people who know how to read them. But they will only baffle readers who do not understand the symbols or other conventions used in them.

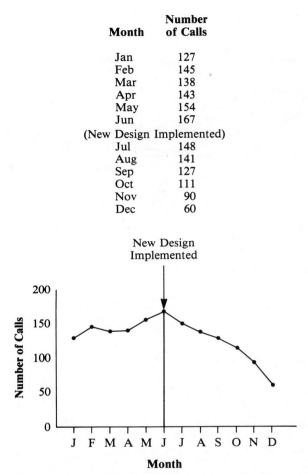

| Month | Number of Calls |
|-------|-----------------|
| Jan | 127 |
| Feb | 145 |
| Mar | 138 |
| Apr | 143 |
| May | 154 |
| Jun | 167 |
| (New Design Implemented) | |
| Jul | 148 |
| Aug | 141 |
| Sep | 127 |
| Oct | 111 |
| Nov | 90 |
| Dec | 60 |

**Figure 19-12.**   Comparison of Data Presented in a Table and in a Line Graph.

For that reason, you must use specialized visuals only when you are sure that your readers will be able to read them. Be especially careful to avoid the mistake of assuming that everyone in your organization knows how to read the specialized figures commonly used in your own field or department. Specialists in other departments might not be familiar with them. Also, be cautious in situations where you expect that your communication will be read by people from several different fields. To meet the needs of all these readers, you may need to include both basic and specialized visual aids.

One important implication of Guidelines 2 and 3 is that you should resist the temptation to choose a particular type of visual aid simply because you already have one of that type available or because you are accustomed to using such visual aids in your work. If you were to do either of those things,

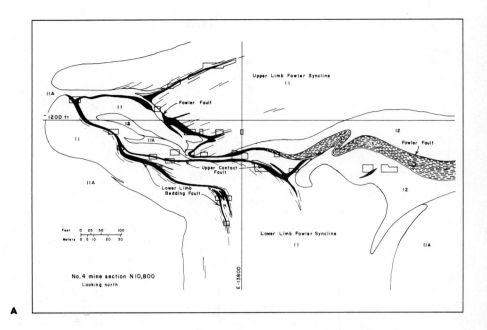

**A**

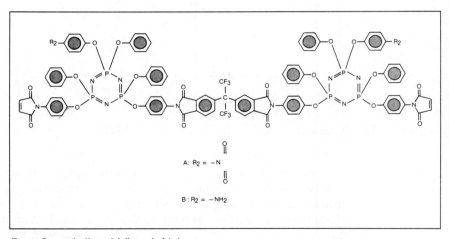

**B**

**Figure 19-13.** Four Specialized Visual Aids.

*Diagram A,* which displays plans for a mine, is shown by the courtesy of St. Joe Mineral Company. *Diagram B,* which shows chemical structure, is from "Imide Cyclotriphosphanzene/ Hexafluoroisopropylidene Polymers," *NASA Tech Briefs,* (Fall 1985):88. *Diagram C,* a computer-generated model, is from J.J. Sojka et al., "Diurnal Variation of the Dayside, Ionospheric, Mid-Latitude Trough in the Southern Hemisphere at 800 km: Model and Measurement Comparison," *Planetary and Space Science* 1985 (December):1378. *Diagram D,* a schematic used in electronics, is from Frank Montalbano III, et al., "A Solar-Powered Time-Lapse Camera to Record Wildlife Activity," *Wildlife Society Bulletin* 1985:179.

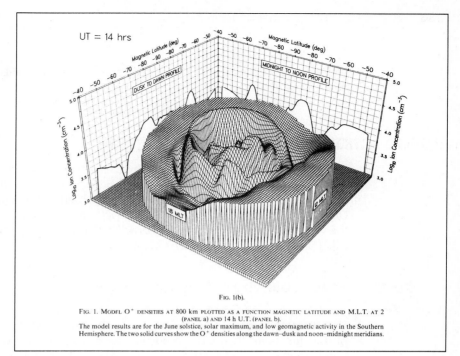

Fig. 1(b).

Fig. 1. Model O$^+$ densities at 800 km plotted as a function magnetic latitude and M.L.T. at 2 (panel a) and 14 h U.T. (panel b).

The model results are for the June solstice, solar maximum, and low geomagnetic activity in the Southern Hemisphere. The two solid curves show the O$^+$ densities along the dawn–dusk and noon–midnight meridians.

**C**

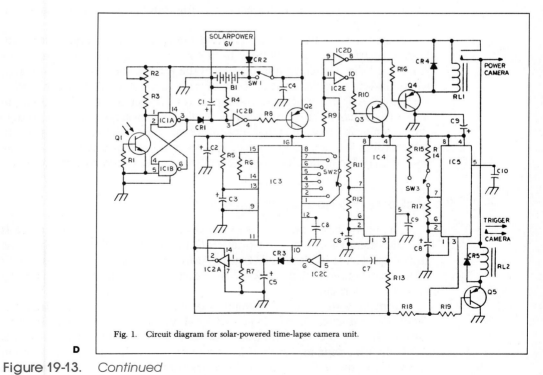

Fig. 1. Circuit diagram for solar-powered time-lapse camera unit.

**D**

# Figure 19-13. *Continued*

you would be taking a writer-centered approach to visual aids, one in which you choose the type that is easiest for *you* to prepare, not the one that is most likely to be useful or persuasive to your *readers.*

## INTRODUCTION TO GUIDELINES 4 THROUGH 7

The three guidelines you have just read will help you identify places where visual aids will increase the effectiveness of your communication, and they will help you decide what types of visual aids to use in those places. Once you have made those preliminary decisions, you must go on to design the visual aids themselves. Guidelines 4 through 7 will help you design visual aids that communicate directly and forcefully.

Guideline
## 4

### Make Your Visual Aids Easy To Use

Like your prose, your visual aids should be easy for your readers to use. Unfortunately, there are so many types of visual aids and such a great variety of uses for them, that it is not possible to identify one or two *rules* you can follow to make all your visual aids easy to use. There is, however, one *strategy* you can use in every instance. That strategy will sound familiar to you: think about your readers in the act of reading your visual aid and then design it accordingly.

For example, when you are creating drawings or photographs for step-by-step instructions, show things from the same angle of vision that your readers will have when trying to perform the actions you describe. Similarly, when making a table, arrange the columns and rows in an order that is compatible with the procedure you expect your readers to use when trying to find a particular piece of information. Maybe that means you should arrange the columns and rows in alphabetical order, or according to a logical pattern, or according to some other system. The key point is that you should arrange your material in a way that your readers will find easy and efficient to use.

Guideline
## 5

### Make Your Visual Aids Simple

One of the best ways to make your visuals persuasive and easy to use is to make them simple and uncluttered. Unnecessary detail creates the same sorts of problems in visual aids that unnecessary words create in prose. Both make extra, unproductive work for the readers, and both obscure the really important information.

To make your visual aids simple, examine each of them for extraneous detail. For example, Figure 19-14 shows two versions of a graph. In the version on the left, the long grid lines obscure the shape of the curves, which is the truly important information in this particular visual aid. In the version on the right, the shape of the curves emerges much more clearly because the

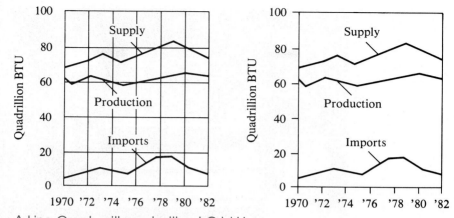

**Figure 19-14.**   A Line Graph with and without Grid Lines.

Based on a graph without grid lines that appeared in U.S. Bureau of the Census, *Statistical Abstract of the United States,* 106th ed. (Washington, D.C.: U.S. Government Printing Office, 1986), 552.

long grid lines are replaced by short tick marks along the horizontal and vertical axes.

As another example, Figure 19-15 shows two photographs of the same experimental apparatus. The main parts of the apparatus are obscured in the photograph on the top because that photograph includes much irrelevant material in the background. That material is removed from the photograph on the bottom.

As these two examples suggest, from situation to situation you will find great variation in the kinds of details that you need to eliminate. The general procedure, however, remains the same: look for and eliminate details that your readers don't need in order to understand and use your visual aid.

Guideline

6

## Label the Important Content Clearly

By clearly labeling the important content in your visual aids, you help your readers in two ways. First, you help them find the things they seek. For example, people using your instructions to repair the transmission on a tractor might need to locate a particular bolt to perform one of the steps. Before locating that bolt on the tractor, the readers will need to locate it in your visual aid. By providing a clear label, you help them do that.

Second, by providing labels, you also help readers when they want to know what they are seeing when they read a visual aid. For instance, readers looking at a drawing of a new product, or an aerial photograph of a manufacturing plant, may want to identify the various objects they see.

The first step in creating labels is to determine what parts need labeling. In some cases, that's easy to do. In a table, for instance, every row and column usually needs a heading that labels it. In other visual aids, you should

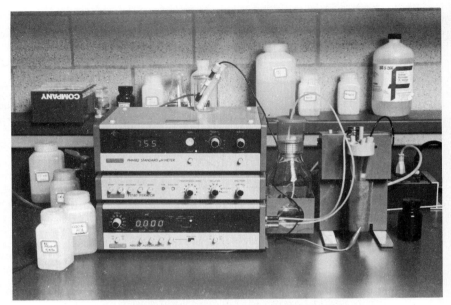

**LESS EFFECTIVE PHOTOGRAPH**

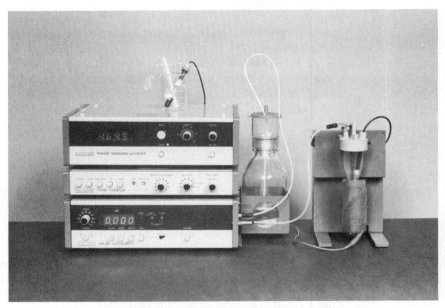

**MORE EFFECTIVE PHOTOGRAPH**

Figure 19-15.   Two Photographs of the Same Experimental Apparatus.

Photographs courtesy of Professor Alan Mills.

label every part that the readers are likely to want to find when they look at that particular visual aid. Of course, you should avoid labeling parts the readers won't be looking for, because unnecessary labels clutter the visual aid, making it difficult for readers to understand and use.

Sometimes, you may use figures that *don't* need labels. That may happen, for instance, when the title for your figure makes clear what the important parts of the figure are (for example, "Figure 3-2. Dents Caused by Hail"). At other times, the prose your readers will read before looking at the figure will serve the same purpose—although you should not rely on the prose to communicate that information if it seems likely that your readers will flip through your communication, hoping to be able to use the information provided in the figures without reading your prose.

After you've decided that a certain part needs a label, you must label that part clearly. Usually, that involves choosing the appropriate word or words for the label, placing the label in a position that makes it easy to see, and (if necessary) drawing a line from the word or words to the part of the visual aid they identify. Figures 19-16, 19-17, and 19-18 show visual aids that are

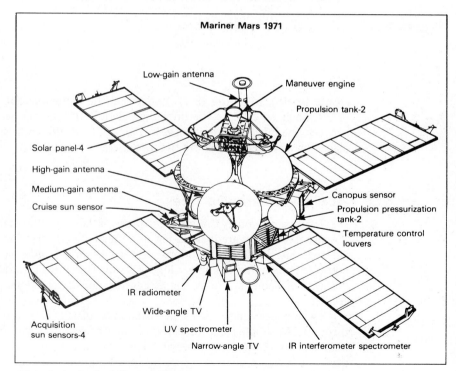

**Figure 19-16.**   Clearly Labeled Drawing.

From Edward Clinton Ezell and Linda Neuman Ezell, *On Mars: Exploration of the Red Planet* (Washington, D.C.: National Aeronautics and Space Administration, 1984), 166.

**U.S. Energy Sources, 1850-1980 (Percent)**

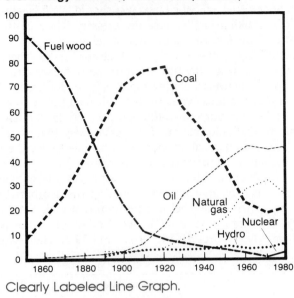

**Figure 19-17.**    Clearly Labeled Line Graph.

Courtesy of Oak Ridge Associated Universities.

clearly labeled. In Figure 19-17, notice that the writer was able to place most of the labels so close to the lines they identify that arrows were not needed between the labels and the lines. In Figure 19-18, notice that the labels are placed on a white background to make them stand out from the photograph.

<table>
<tr><td>Guideline</td><td rowspan="2">**Provide Informative Titles**</td></tr>
<tr><td>**7**</td></tr>
</table>

Guideline

**7**

## Provide Informative Titles

Titles serve much the same purposes as labels. Titles help readers find the visual aids they are looking for, and they help readers know what a visual aid contains. In communications written at work, a title typically includes both a number (for example, "Figure 3" or "Table 6-11") and a description ("Effects of Temperature on the Strength of M312" or "Control Panel of the DLP").

Titles appear above or below visual aids, usually in the same placement throughout a single communication. Long reports, proposals, and instruction manuals sometimes have special tables of contents that give the number, title, and location of every visual aid in the communication.

To help your readers understand and use your visual aids, you should make the descriptive part of your titles as brief, yet informative, as possible. Select a few key words that describe in specific terms what the visuals show:

Site Plan for the Creighton Landfill
Projected Increases in Peak Demand for Electricity over the Next Ten Years

Leave out redundant or vague words. Don't use a title like this:

Table 3
Table of Information about Three Amplifiers

That title would be more useful to readers if it were changed to this:

Table 3
Comparison of Three Amplifiers

There is a widespread convention that visual aids are numbered consecutively, either in one long sequence through the entire communication, or else with a new sequence beginning in each chapter (as is done in this book). Also, if some of the visual aids are tables, these are usually given a separate sequence from the sequence for all the other visual aids, which are called

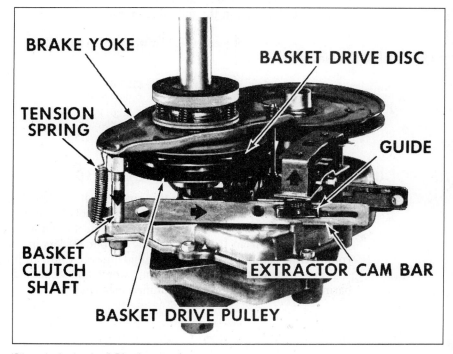

**Figure 19-18.**   Clearly Labeled Photograph.

Courtesy of Sears Roebuck and Company.

*figures.* Consequently, there will be a Table 1 and a Figure 1, a Table 2 and a Figure 2, and so on.

According to custom, the numbers assigned to figures are usually arabic (1, 2, 3, 4, and so on), while the numbers assigned to tables are either arabic or roman (1, 2, 3, 4, or I, II, III, IV). In communications where the numbering sequences begin anew in each chapter or section, the numbers contain two parts: the number of the chapter and the number of the figure or table within the sequence of that chapter. Thus, Figure 3-7 is the seventh figure in Chapter 3 and Figure 7-3 is the third figure in Chapter 7.

You may find that your employer or your client has rules about how titles are to be written and where they are to be placed. These rules might require, for instance, that you use arabic rather than roman numerals for your tables (or vice versa), and that you should place titles at the left margin above your visual aids rather than (for instance) centered below them.

Even if you are given no such regulations, your readers may have strong expectations concerning titles. To learn what these are, you can ask other people and scan communications similar to the one you are writing.

Finally, you should note that sometimes you *don't* need to provide a title for a visual aid. That might happen, for instance, when you are including a very short table in your text in a way that makes perfectly clear what it contains, as happens in the following example:

You will be very pleased to see how well our top four salespersons did in June:

| Chamberlain | $127,603 |
| Tonaka | 95,278 |
| Gonzales | 93,011 |
| Albers | 88,590 |

## INTRODUCTION TO GUIDELINES 8 THROUGH 10

The first seven guidelines in this chapter tell you how to plan and design visual aids. They focus on your visual aids in isolation from the rest of your communication. In contrast, the next three guidelines ask you to think about your visual aids from the point of view of your readers in the act of reading your *overall* communication. These guidelines tell how to integrate your visual aids with your prose so that your words and visual aids work together harmoniously to create a single, unified message.

### Guideline 8   Introduce Your Visual Aids in Your Prose

When readers read sequentially through a communication, they read one word then the next, one sentence then the next, and so on. When you want the next element they read to be a visual aid rather than a sentence or a paragraph, you need to take special care to direct the readers' attention to it. One way to do that is to introduce each of your visual aids just before you want your readers to look at it.

To make a proper introduction, you need to do two things: tell your readers that you want them to turn their attention from your prose to a visual aid, and indicate what they will find in the visual aid.

You can make your introduction in many different ways. You might, for example, give separate sentences to your direction and your description of the visual aid's contents:

> In the test, we found that the Radex decomposed several times more rapidly than Talon, especially at high humidities. See Figure 3.

You might combine the two elements in a single sentence:

> As Figure 3 shows, Radex decomposed several times more rapidly than Talon, especially at high humidities.

Or you might make your request to the readers in parentheses:

> Our test showed that Radex decomposed several times more rapidly than Talon, especially at high humidities (Figure 3).

Besides helping readers who are reading sequentially through a communication, such introductions also help readers who are flipping and skimming through the text. For example, after they have read the summary, introduction, and conclusion of a report, decision-makers sometimes flip and skim through its body to gain a general sense of what the writer presents there. Skimming readers often stop to read visual aids. If they find something interesting in one, they may want to read the writer's discussion of it. By skimming to locate the writer's introduction of the visual aid in the prose, they can quickly find that discussion.

Sometimes, your introduction to a visual aid will have to include information your readers need to understand or use the visual aid. For example, the writers of an instruction manual explained in this way how to use one of their tables:

> In order to determine the setpoint for the grinder relay, use Table 1. First, find the grade of steel you will be grinding. Then, read down column two, three, or four, depending upon the grinding surface you are using.

Whatever kind of introduction you make, place it at the exact point where you would like your readers to take their eyes off your prose to focus on your visual aid.

Guideline

9

## State the Conclusions You Want Your Readers To Draw from Your Visual Aids

A visual aid presents readers with a set of facts. If you want your readers to draw some particular conclusion from those facts, you should state that conclusion explicitly in your prose. Otherwise, the conclusion your readers derive might not be the same one you want them to draw.

Consider, for example, the way a writer discussed a graph of expected orders for his company's rubber hoses for the next six months. The graph showed that there would be a sharp decline in orders from automobile plants, and the writer feared that his readers might focus on that fact. However, the main point the writer wanted to make was quite different, so he wrote the following statement:

Conclusion the writer wanted the reader to draw.

As Figure 7 indicates, our outlook for the next six months is very good. Although we predict fewer orders from automobile plants, we expect the slack to be taken up by increased demand among auto parts outlets.

You might find it helpful to think of the sentences in which you explain a visual aid's significance as a special kind of topic sentence. Just as in the topic sentence at the head of a paragraph, you tell your readers the point to be derived from the various facts that follow.

Guideline

10

## Put Your Visual Aids Where Your Readers Can Find Them Easily

What happens when your readers come to a statement in which you ask them to look at a certain visual aid? Instead of going on to read the next sentence, as they would usually do, the readers lift their eyes from your prose and search for the visual aid. You want to make that search as short and simple as possible.

To do that, you must put your visual aids where your readers can locate them quickly. The ideal location is on the same page as the prose that accompanies the visual aid. Not only will the visual aid be easy to find, but your readers will be able to look back and forth between the prose and visual aid, if necessary.

You may not always have room for your visual aid on the same page as the prose that introduces it. If your communication has text on facing pages, try to put the visual aid on the page facing its accompanying prose. If your communication does not have facing pages of text, try to put your visual aid on the next page following its introduction. If you place the figure farther away than that (for instance in an appendix), you can help your readers by providing the number of the page on which the figure may be found:

A detailed sketch of this region of the brain is shown in Figure 17 in Appendix C (page 53).

## SUMMARY

Visual aids are an essential part of many communications written at work. To use them well, you need to follow the same strategy that you use when writing your prose: think about your readers' moment-by-moment reactions to your communication and create visual aids accordingly. This chapter presents ten guidelines that will help you affect your readers in the ways you desire.

1. *While planning your communication, look for places to use visual aids.* By including visual aids in your very first plans for your communication, you help yourself write efficiently and effectively.

2. *Choose visual aids appropriate to your purpose.* Most information can be conveyed in more than one type of visual aid. To decide which to use, think about the mental task you want your visual aid to enable your readers to perform. When you want to make a persuasive point, choose the type of visual aid that will convey the significance of your evidence most quickly and directly.

3. *Choose visual aids your readers know how to read.* Some types of visual aids are familiar to everyone. Others, however, require special training to read and you should employ them only when you are sure that your readers have that training.

4. *Make your visual aids easy to use.* Arrange the content of your visual aids to reflect the use you expect your readers to make of the aids.

5. *Make your visual aids simple.* Unnecessary detail makes extra work for your readers and obscures the important material.

6. *Label the important content clearly.* Determine what needs to be labeled by thinking about what your readers will want to learn from the visual aid. Then choose appropriate words for each label, place each label in a position that makes it easy to see, and (if necessary) draw a line from the label to the part it identifies.

7. *Provide informative titles.* Titles usually include a number for the visual aid and a short description of it. The description should be short, yet informative.

8. *Introduce your visual aids in your prose.* Your introduction should contain two elements: a brief request that the readers turn their attention from your prose to the visual aid, and an indication of what they will find in the visual aid. Place this introduction exactly at the point where you want your readers to refer to the visual aid.

9. *State the conclusions you want your readers to draw from your visual aids.* A visual aid is a collection of facts. If you want your readers to draw some particular conclusion from these facts, tell them what that conclusion is. Otherwise, they may miss the point you want to make.

10. *Put your visual aids where your readers can find them easily.* After reading your introduction to a visual aid, readers must find the visual aid. Help them do that by placing it on the same page as the introduction or on the facing or following page. If you are going to put it somewhere else (as in an appendix), include the page number for the visual aid in your introduction to it.

# CHAPTER 20

# Twelve Types
# of Visual Aids

## PREVIEW OF TYPES

**Tables**
**Bar Graphs**
**Pictographs**
**Line Graphs**
**Pie Charts**
**Photographs**
**Drawings**
**Diagrams**
**Flow Charts**
**Organizational Charts**
**Schedule Charts**
**Budget Statements**

n the preceding chapter, you read ten guidelines for using visual aids. Those guidelines can help you succeed with any of the thousands of types of visual aids used on the job. The guidelines apply to all the common types of visual aids, like tables and graphs, which anyone can create and interpret. They apply also to the more specialized types of visual aids, like those shown in Figure 19-13, which can be made and interpreted only by people with special training.

In this chapter, you will find additional, detailed advice about choosing and constructing twelve types of visual aids. These twelve types were selected because of their general helpfulness in the kinds of communications that most college graduates write on the job. You will learn when you can use each type to its best advantage and how you can design each effectively.

## THE ORGANIZATION OF THIS CHAPTER

So that this chapter will be as useful as possible to you, it is organized differently than are most of the other chapters in this book. Instead of being organized around guidelines, it is arranged in separate discussions of the twelve types of visual aids. Each discussion provides you with detailed advice about how to design the type of visual aid it describes. In addition, because you can present some kinds of information in two or more of these types, the chapter provides information about when to use each type of visual aid to its best advantage.

This organization allows you to use the chapter as a reference source in which you can quickly find complete information about the type of visual aid you want to study or use.

## A NOTE ABOUT COMPUTERS AND VISUAL AIDS

On the job—and perhaps even in school—you may be able to use computer programs to create your visual aids. Programs exist for making almost all the types described here. These programs can greatly simplify the work of making visual aids, and they can help you make visual aids that look much more polished than the ones most people can make on their own. Even when you are using those programs, however, you will still need to use the advice provided in this chapter.

That's partly because most of the computer programs for making visual aids leave the basic design decisions to you. For instance, if you are using a program to make a line graph, you will still have to decide which variable to place on the x-axis and which on the y-axis. You will still have to decide what intervals to use for your variables, and what your labels should say. This chapter provides you with advice about these essential decisions.

When you use computer programs that don't leave basic design decisions to you, you will have to determine whether or not the standard designs used by the programs are good ones. That's another task that this chapter can help you perform because its advice will enable you to tell a good design from a bad one. If you determine that a program's standard designs aren't good, you should obtain a different program or else make the visual aids some other way.

When using computer programs for visual aids, you may encounter one other difficulty. Many of these programs allow you to do many fancy things with your visual aids. In some situations, those fancy things can help you achieve your communication objectives. However, some writers become so enthralled with the programs that they forget about the purpose and readers of their communications. As a result, they make overly elaborate visual aids that diminish the effectiveness of their writing. By following the advice in this chapter and the guidelines in Chapter 19, you can avoid such difficulties.

## A NOTE ABOUT TITLES

An essential element of almost all visual aids is a title. Guideline 7 in the preceding chapter provides advice about how to write titles. The conventions about where to place titles vary somewhat. In typewritten communications, titles are almost always placed above tables and below all other visual aids. Generally, they are centered between the margins. In communications typeset by a printer, titles are always placed so that they are prominent and attractive, but their location varies considerably from communication to communication. Sometimes they are placed above the visual aids, sometimes below them, and sometimes within them. Within a single typeset communication, all tables share a consistent placement of titles and a consistent placement is also used for all other visual aids.

## TABLES

The table is one of the most versatile and widely used visual aids. When you sit down to breakfast, there is one on the side of your cereal box, telling you the nutritional value of the contents. When you read a technical or scientific report, you will probably find a table presenting the researcher's results. When you buy a stereo or piece of laboratory equipment, you will probably discover one in the owner's manual, telling the specifications and capabilities of your purchase. Tables are used so often because they can help writers achieve several common objectives:

- *To present groups of detailed facts in a concise and readable form.* To see how well suited a table is to presenting groups of detailed facts, compare Figure 20-1 with Figure 20-2. The prose in Figure 20-2 presents the same information that you would obtain by reading the *first three*

No. **146.** NATIONAL HEALTH EXPENDITURES: 1960 TO 1984

[Includes Puerto Rico and outlying areas]

| YEAR | TOTAL¹ | | | HEALTH SERVICES AND SUPPLIES | | | | | | | |
| | | | | Private | | | | Public | | | |
| | | | | | Direct patient payments | | Insurance premiums⁴ (bil. dol.) | Total⁵ | | Medical payments | |
| | Total (bil. dol.) | Per capita (dol.) | Per-cent of GNP² | Total³ (bil. dol.) | Total (bil. dol.) | Per-cent of total private | | Total (bil. dol.) | Per-cent of total health expend | Medicare (bil. dol.) | Public assistance (bil. dol.) |
|---|---|---|---|---|---|---|---|---|---|---|---|
| 1960 | 26.9 | 146 | 5.3 | 19.6 | 13.0 | 66.3 | 5.8 | 5.6 | 20.8 | (x) | .5 |
| 1965 | 41.9 | 207 | 6.1 | 29.5 | 18.5 | 62.8 | 10.0 | 8.9 | 21.3 | (x) | 2.1 |
| 1970 | 75.0 | 350 | 7.6 | 44.7 | 26.5 | 59.3 | 16.9 | 24.9 | 33.2 | 7.5 | 6.3 |
| 1971 | 83.5 | 386 | 7.7 | 48.9 | 28.1 | 57.4 | 19.3 | 28.4 | 34.0 | 8.3 | 8.1 |
| 1972 | 94.0 | 430 | 7.9 | 55.3 | 30.6 | 55.4 | 22.2 | 32.1 | 34.1 | 9.1 | 9.1 |
| 1973 | 103.4 | 469 | 7.8 | 60.7 | 33.3 | 54.8 | 24.7 | 35.8 | 34.6 | 10.1 | 10.3 |
| 1974 | 116.1 | 522 | 8.1 | 65.8 | 36.2 | 55.0 | 27.9 | 42.9 | 36.9 | 13.1 | 12.1 |
| 1975 | 132.7 | 591 | 8.6 | 73.0 | 38.1 | 52.2 | 33.2 | 51.3 | 38.6 | 16.3 | 15.1 |
| 1976 | 150.8 | 666 | 8.8 | 84.5 | 42.0 | 49.8 | 40.4 | 57.3 | 38.0 | 19.3 | 16.9 |
| 1977 | 169.9 | 743 | 8.9 | 96.7 | 46.4 | 48.1 | 48.0 | 64.0 | 37.7 | 22.5 | 18.9 |
| 1978 | 189.7 | 822 | 8.8 | 106.6 | 50.7 | 47.6 | 53.6 | 73.3 | 38.7 | 25.9 | 21.1 |
| 1979 | 214.7 | 920 | 8.9 | 120.4 | 55.8 | 46.4 | 62.0 | 83.8 | 39.0 | 30.3 | 24.3 |
| 1980 | 247.5 | 1,049 | 9.4 | 137.9 | 62.5 | 45.3 | 72.5 | 97.7 | 39.5 | 36.8 | 28.1 |
| 1981 | 285.2 | 1,197 | 9.6 | 159.0 | 70.8 | 44.5 | 84.8 | 113.1 | 39.6 | 44.7 | 32.3 |
| 1982 | 321.2 | 1,334 | 10.5 | 180.0 | 77.2 | 42.9 | 99.0 | 127.0 | 39.5 | 52.4 | 34.9 |
| 1983 | 355.1 | 1,461 | 10.7 | 200.2 | 86.4 | 43.1 | 109.7 | 139.6 | 39.3 | 58.8 | 37.7 |
| 1984 | 387.4 | 1,580 | 10.6 | 220.3 | 95.4 | 43.3 | 120.5 | 151.2 | 39.0 | 64.6 | 40.6 |

X Not applicable. ¹ Includes medical research and medical facilities construction. ² GNP = Gross national product; see table 718. ³ Includes other sources of funds not shown separately. ⁴ See footnote 3, table 147. ⁵ Includes other programs, not shown separately.

Source: U.S. Health Care Financing Administration, *Health Care Financing Review*, Fall 1985.

Figure 20-1. Table.

From U.S. Bureau of the Census, *Statistical Abstract of the United States,* 106th ed. (Washington, D.C.: U.S. Government Printing Office, 1986), 96.

*columns* of the table in Figure 20-1, together with its title and some of its notes. To present all the information contained in all eleven columns of the table, the prose passage would have to be about three times longer than it is. Even as it is, the prose is virtually unreadable because it is so dense with facts—facts that the table presents simply and elegantly. (You may find it helpful to review a similar example given in Figures 19-6 and 19-7.)

- *To enable readers to find particular facts rapidly.* You can demonstrate this point to yourself by using the prose in Figure 20-2 and then the table in Figure 20-1 to find out the total amount of money spent in the U.S. on health care in 1980. The difference between the table and the prose will be even more evident to you if you try to find and compare two or more facts. For instance, try to find out how the percentage of the gross national product spent on health care services changed from 1960 to 1980. (You may find it helpful to review a similar example given in Figures 20-6 and 20-7.)

- *To present extended comparisons economically and emphatically.* Tables can present information in a way that permits rapid and extended comparison. For example, Figure 20-3 shows a table that the

National Health Care Expenditures: 1960 to 1983

In the Fall 1985 issue of <u>Health Care Financing Review,</u> the U.S. Health Care Financing Administration published figures that show how much the nation spent on health care during selected years from 1960 through 1984. To calculate the total amount expended during these years, the Health Care Financing Administration included the following: direct private payments and insurance premiums for health care; public medical payments by medicare and public assistance programs; and expenditures for medical research and medical facilities construction.

In 1960, the U.S. spent $26.9 billion dollars, which was $146 per capita and 5.3% of the gross national product (GNP). (To learn what the gross national product was for 1960 and for each of the years mentioned below, see Table 718.) In 1965, it spent $41.9 billion, which was $207 per capita and 6.1% of the GNP. In 1970, it spent $75.0 billion, which was $350 per capita and 7.6% of the GNP. In 1971, it spent 83.5 billion, which was $386 per capita and 7.7% of the GNP. In 1972, it spent $94.0 billion, which was $430 per capita and 7.9% of the GNP. In 1973, it spent $103.4 billion, which was $469 per capita and 7.8% of the GNP. In 1974, it spent $116.1 billion, which was $522 per capita and 8.1% of the GNP. In 1975, it spent $132.7 billion, which was $591 per capita and 8.6% of the GNP.

In 1976, it spent $150.8 billion, which was $666 per capita and 8.8% of the GNP. In 1977, it spent $169.9 billion, which was $743 per capita and 8.9% of the GNP. In 1978, it spent $189.7 billion, which was $822 per capita and 8.8% of the GNP. In 1979, it spent $214.7 billion, which was $920 per capita and 8.9% of the GNP. In 1980, it spent $247.5 billion, which was $1,049 per capita and 9.4% of the GNP.

In 1981, it spent $285.2 billion, which was $1,197 per capita and 9.6% of the GNP. In 1982, it spent $321.2 billion, which was $1,334 per capita and 10.5% of the GNP. In 1983, it spent $355.1 billion, which was $1,461 per capita and 10.7% of the GNP. In 1984, it spent $387.4 billion, which was $1,580 per capita and 10.6% of the GNP.

**Figure 20-2.** Presentation in Prose of the Information in the First Three Columns of the Table Shown in Figure 20-1.

### Flux Cored Welding vs. Hi-Dep II
(1/4″-3/8″ Fillet and Downhand Butt Welds)

| | Cored Wire Gas-Shielded or Gasless | Solid Wire Hi-Dep II |
|---|---|---|
| Wire Deposition Rate | 18 lbs/hr | 20 lbs/hr |
| Deposition Efficiency | .85 | .98 |
| Weld Metal Deposition Rate | 15.3 lbs/hr | 19.6 lbs/hr |
| Weldor Duty Cycle | 35% | 40%* |
| Weld Deposited-8 hr day | 42.8 lbs | 62.7 lbs |

*No slag chipping required

**Figure 20-3.**    A Table Used To Enable the Reader To Make a Quick Comparison.
Courtesy of L-Tec Welding & Cutting Systems.

manufacturer of a special type of welding wire used to enable readers to compare several key features of its product with those of the most widely used type of welding wire.

The examples given so far point out some situations in which you can use tables to present *numerical* data. Tables are also an excellent means for conveying information in *words*. For example, tables are often used to display the information in the troubleshooting section of manuals (Figure 20-4) and in reports that compare the features of competing products or processes (Figure 20-5, on page 406).

## Rows and Columns

All tables display a collection of separate pieces of information arranged into rows and columns:

| | Column 1 | Column 2 |
|---|---|---|
| Row 1 | 32 | 318 |
| Row 2 | 66 | 467 |
| Row 3 | 118 | 599 |

In this tabular arrangement, every piece of information is defined in terms of its row and its column:

Number of orders (row 1) for last year (column 2)

Income (row 2) for this year (column 3)

|  | Two Years Ago | Last Year | This Year |
|---|---|---|---|
| Number of Orders | 32 | 66 | 118 |
| Income (in thousands of dollars) | 318 | 467 | 599 |

## Trouble Shooting Chart

| Problem | Cause | Remedy |
|---|---|---|
| **1** Engine fails to start | **A** Blade control handle disengaged | **A** Engage blade control handle. |
|  | **B** Check fuel tank for gas | **B** Fill tank if empty. |
|  | **C** Spark plug lead wire disconnected. | **C** Connect lead wire. |
|  | **D** Throttle control lever not in the starting position | **D** Move throttle lever to start position. |
|  | **E** Faulty spark plug | **E** Spark should jump gap between control electrode and side electrode. If spark does not jump, replace the spark plug. |
|  | **F** Carburetor improperly adjusted, engine flooded | **F** Remove spark plug, dry the plug, crank engine with plug removed, and throttle in off position. Replace spark plug and lead wire and resume starting procedures. |
|  | **G** Old stale gasoline | **G** Drain and refill with fresh gasoline. |
|  | **H** Engine brake engaged | **H** Follow starting procedure. |
| **2** Hard starting or loss of power | **A** Spark plug wire loose | **A** Connect and tighten spark plug wire. |
|  | **B** Carburetor improperly adjusted | **B** Adjust carburetor. See separate engine manual. |
|  | **C** Dirty air cleaner | **C** Clean air cleaner as described in separate engine manual. |
| **3** Operation erratic | **A** Dirt in gas tank | **A** Remove the dirt and fill tank with fresh gas. |
|  | **B** Dirty air cleaner | **B** Clean air cleaner as described in separate engine manual. |
|  | **C** Water in fuel supply | **C** Drain contaminated fuel and fill tank with fresh gas. |
|  | **D** Vent in gas cap plugged | **D** Clear vent or replace gas cap. |
|  | **E** Carburetor improperly adjusted | **E** Adjust carburetor. See separate engine manual. |
| **4** Occasional skip (hesitates) at high speed | **A** Carburetor idle speed too slow | **A** Adjust carburetor. See separate engine manual. |
|  | **B** Spark plug gap too close | **B** Adjust to .030". |
|  | **C** Carburetor idle mixture adjustment improperly set | **C** Adjust carburetor. See separate engine manual. |

**Figure 20-4.**   A Table Used To Provide Troubleshooting Information.

Courtesy of MTD Products Incorporated.

| Characteristic | Brake type | | | | |
| --- | --- | --- | --- | --- | --- |
| | Mechanical | Hydraulic | | Hydromechanical | |
| | | Down-acting | Up-acting | Conventional | Toggle |
| Operating speed | Good | Fair[a]-good | Good | Good | Fair |
| Ram levelness | Excellent | Good[a]-Excellent | Excellent | Excellent | Fair |
| Stroke accuracy | Excellent | Poor[a]-Excellent | Excellent | Excellent | Fair |
| Control, stroke reversability | Poor | Good | Excellent | Good | Good |
| Time for tool installation and setup | Fair | Good | Excellent | Good | Good |
| Setup skill required | High[b] | Medium | Low[c] | Medium | Medium |
| Overload resistance | Poor | Good | Good | Good | Good |
| Initial cost | High | Medium | Medium | High | Medium |
| Maintenance cost | High[d] | Medium | Low[e] | Medium-High | Medium |
| Share of market, percent | 10 | 45 | 8 | 1 | 35 |

**PRESS BRAKE TYPES COMPARED**

a. Conventional style; computerized controls upgrade capabilities.
b. Clutch slippage adjustment requires high skill level; remaining setup needs medium skill.
c. Operators quickly learn "feel" of the press.
d. Most wear, lubrication, and adjustment points.
e. Short hydraulic circuits ease fault or line tracing and maintenance.

**Figure 20-5.** A Table Used To Compare Various Types of Press Brakes.

From Larry Conley, "Progress in Press Brake Design," *Welding Design & Fabrication* 58(September 1985):54. Reprinted with permission from *Welding Design & Fabrication*, a Penton Publication, copyright 1985.

There are two kinds of tables, formal and informal. The difference between them lies in the way you provide your readers with the information they need to find, interpret, and use the facts presented in the rows and columns.

## Creating a Formal Table

In a formal table, to provide your readers with the information they need to understand and use the table, you use such devices as a title for the table and headings for the rows and columns.

On the job, strong conventions govern the placement of each element of a formal table. Figure 20-6 shows this arrangement in the abstract. You may want to compare that abstract presentation with the typical table shown in Figure 20-1.

To create a table that fits this conventional form, you should do the following:

Table Number

TITLE

|  | COLUMN HEADING | | |
|---|---|---|---|
| **STUB HEADING**[a] | Subheading | Subheading | Subheading[b] |
| Row Heading Subheading Subheading | | | |
| Row Heading Subheading Subheading | | | |
| Row Heading | | | |

Notes:
[a] Text of footnote *a*.
[b] Text of footnote *b*.

**Figure 20-6.** Structure of a Typical Table. (Note that the area containing the headings for the rows is called the *stub*.)

1. *Label the columns and rows.* If you are presenting information your readers will use to make comparisons, consider exactly what you want your readers to compare. Then label the *columns* with the names of the things being compared and label the *rows* with the specific points upon which your readers will make the comparison.

    For example, consider the table on the next page. It is designed to enable readers to compare a company's performance in 1985, 1986, and 1987. Therefore, the years are used to label the columns. The specific aspects of performance that readers can compare are output and profit, which are used to label the rows. Note how this arrangement allows readers to compare figures for the three years by using a normal left-to-right eye movement as they read across each row.

|                          | 1985 | 1986 | 1987 |
|--------------------------|------|------|------|
| Output (Thousand Tons)   | 137  | 148  | 159  |
| Profit ($ Million)       | 124  | 143  | 187  |

You can see other examples of the desirable arrangement of rows and columns in Figure 20-3, where the brands of welding wire are used to label the columns and the points upon which they are compared are used to label the rows. Similarly, in Figure 20-5, the types of brakes being compared are used to label the columns and the characteristics upon which they are being compared are used to label the rows.

2. *If the table presents numerical data, indicate the units.* Show the units of measurement (pounds, miles, per capita, and so on) in the labels for the appropriate columns and rows.

3. *Provide any special notes your readers need in order to understand and use your table.* If your notes are short enough, you can place them next to the appropriate title or headings. Otherwise, place them in footnotes at the bottom of your table. Use lowercase letters to label the footnotes if your readers might otherwise confuse a superscript footnote number with a mathematical exponent.

4. *Align the material you present in the columns.* If you are presenting numerical data, align your material on the units column or (if decimals are used) on the decimal point:

| | |
|---|---|
| 23,418 | 2.79 |
| 5,231 | 618.0 |
| 17 | 13.517 |

If you are presenting prose material, align it on the left-hand margin:

1. Turn the switch to "off."
2. Unfasten the lid.
3. Check the fan for damage.

Or center it within the column:

Acceptable
Marginally Acceptable
Unacceptable

5. *Arrange your columns and rows so your readers can easily find the particular label they want.* For example, you may put your labels in alphabetical order or arrange them chronologically. Or, as you think about the

ways your readers will use your table, you may devise some entirely different type of arrangement.

6. *Provide your readers with help in reading down your columns and across your rows.* This is an especially important concern in large tables. To provide that help you can use two techniques illustrated in Figure 20-1. Between columns and groups of columns, you can place vertical lines. Between set numbers of rows (for example, five or ten), you can place a blank line. Of course, if you are creating a table where the rows are grouped logically rather than arbitrarily, you could insert the blank lines between logical groups rather than at regular but arbitrary intervals.

## Creating an Informal Table

When you make an informal table, you place your table directly within a paragraph. You use the sentence or sentences that immediately precede the table to provide your readers with the information they must have to interpret it. As a result, you need no title, column headings, or notes.

Here is an example showing how an informal table is built into a paragraph.

Even more important, the sales figures demonstrate how our investment in technical research has helped us become more competitive in several of our divisions. The following figures show the increase in the market share enjoyed by each of the three major divisions since the same quarter last year:

| | |
|---|---|
| Strauland Microchips | 7% |
| Minsk Machine Tools | 5% |
| PTI Technical Services | 4% |

Use informal tables only when presenting a single column of facts in situations where you are sure that your readers will understand your table without a title, column headings, or notes. If you think that an informal table might confuse your readers, even momentarily, use a formal table instead.

When making an informal table, follow these conventions for arranging your information in rows and columns.

1. *Place the row labels in the left-hand column.*
2. *Align the information in your columns in the same way that you do in a formal table.* That is, align numbers on the zero point or decimal point, and align prose on the left or else center it in the column.

## Ineffective Use of Tables

Although tables are versatile, they can be misused.

Sometimes writers use tables when they could make their point much more effectively by using another type of visual display. This often happens when writers want to emphasize trends in numerical data. Later in this chapter you will find advice about choosing the best way to display such trends.

Also, writers sometimes crowd their tables by putting too much information in them. If your table contains several discrete types of information that your readers might want to examine at separate times, then break your table into smaller ones.

Finally, when making oral presentations, speakers sometimes use posters, overhead projectors, or 35-mm slides to present tables that have too much information to be readable from a distance, even a short distance. Large tables are not usually suited for oral presentations. If you must use a large table, consider presenting it in a handout.

## BAR GRAPHS

Like a table, a bar graph can present numerical quantities. Here, however, you represent the quantities with rectangles called *bars*. The greater the quantity, the longer the bar.

As Figure 20-7 demonstrates, bar graphs are very useful in situations where you want your readers to be able to compare quantities at a glance. Notice that because the bars are drawn to scale you can tell not only which quantities are bigger, but also how great the differences are.

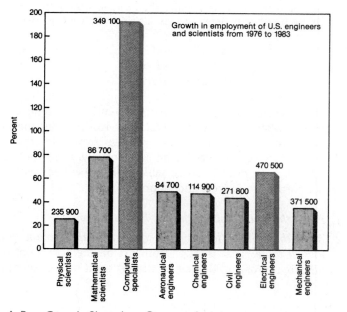

**Figure 20-7.**   A Bar Graph Showing Comparisons.

From "Engineering Employment Trends," *IEEE Spectrum* 22(October 1985):17. © 1985 IEEE. Reprinted with permission of the Institute of Electrical and Electronics Engineers.

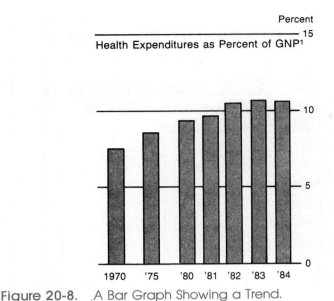

**Figure 20-8.**   A Bar Graph Showing a Trend.

From U.S. Bureau of the Census, *Statistical Abstract of the United States*, 106th ed. (Washington, D.C.: U.S. Government Printing Office, 1986), 94.

Also, when you are dealing with data from different time periods, you can use bar graphs to enable your readers to see trends at a glance. Figure 20-8 provides an example.

## Creating a Bar Graph

By following the steps and guidelines given below, you can create a bar graph that will be readily understandable by your readers.

1. *Draw the two axes of the graph.* Plan the length of the axis that shows quantities so that the longest bar will extend almost to the end of the axis.
2. *Label the axes.* There is no general convention about whether you should label the axes so that the bars will extend vertically or horizontally, although vertical bars are often used for height and depth, whereas horizontal bars are often used for distance, length, and time.
3. *On the axis that represents quantities, place labeled tick marks at regular intervals.* Usually, your graph will be less cluttered if you make these tick marks short (but still clearly visible), as shown in Figure 20-7. In some situations, however, your readers may find it easier to read your bar graph if you extend the tick marks as in Figures 20-8 and 20-9 (page 412).
4. *Arrange the bars in a way suited to your purpose.* For example, if you want your readers to discern the rank order of the quantities you are

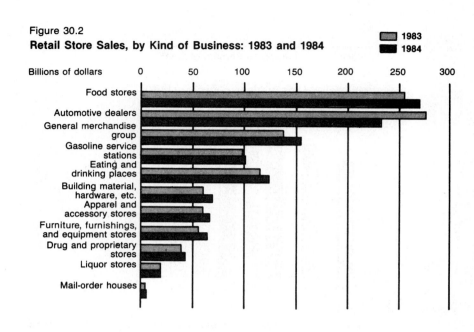

Figure 30.2
**Retail Store Sales, by Kind of Business: 1983 and 1984**

■ 1983
■ 1984

Source: Chart prepared by U.S. Bureau of the Census. For data, see table 1389.

**Figure 20-9.**    Multibar Graph.

From U.S. Bureau of the Census, *Statistical Abstract of the United States*, 106th ed. (Washington, D.C.: U.S. Government Printing Office, 1986), 773.

portraying, arrange the bars in order of length. If you want to enable your readers to make ready comparisons among subgroups of your bars, put those bars together.

5. *Make all bars the same width.* With bar graphs, readers are to compare the heights or lengths of the bars. If you vary the width, readers may compare the widths or areas of the bars and thereby draw an incorrect conclusion. Remember, for example, that if two bars have the same width but one is twice as long, it also has twice the area of the first. In contrast, if the second bar is twice as long *and* twice as wide, it will have four times as much area as the first.

6. *Label the bars.* If possible, place the labels next to the places where the bars extend from the axis (as in Figures 20-7 and 20-8). Generally, your readers will find it easier to use labels placed there than to use labels provided in a separate key.

An exception is the situation in which you are using the same groups of bars repeatedly, as in Figure 20-9. Then, use a distinctive pattern of lines or shading to distinguish each group of bars. Provide a key to these patterns *within* your graph.

7. *Provide numbers to indicate exact quantities only if your readers need such information.* Generally, numbers indicating the exact quantity rep-

resented by each bar (as in Figure 20-7) are not necessary and you can simplify your bar graph by omitting them.

## Avoid Misleading Your Readers

Bar graphs can be misleading if they do not include the zero point on the axis that shows quantity. To see how that happens, compare the two bar graphs shown in Figure 20-10. In the bar graph on top, the difference between the two bars seems to be very significant, because it amounts to one-

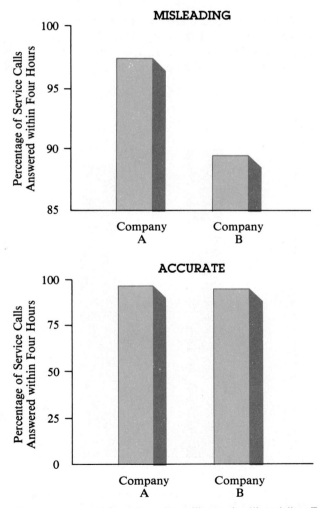

Figure 20-10. Comparison of Bar Graphs with and without the Zero Point.

Based on data provided in Edwin E. Mier and John Bush, "Rating the Long-Distance Carriers," *Data Communications* 13(August 1984):109. Copyright 1984 McGraw-Hill Inc. All rights reserved.

third of the length of the longest bar. In contrast, the difference in the graph on the bottom looks much less significant. Both graphs, however, represent the same data. The difference between the bars on the top is exaggerated because the quantity scale begins not at zero, as it should, but at 85%.

If you simply cannot use the entire quantity scale, then clearly indicate that fact to your readers. The conventional way to do that is to use two parallel slashes (called *hash marks*) to indicate a break in the quantity axis and to use the same technique to break the bars themselves (see Figure 20-11).

## PICTOGRAPHS

Pictographs are a special kind of bar graph in which the bars are replaced by drawings that represent the thing being described. Figure 20-12 provides an example in which the number of barrels of oil used per capita by the United States is represented by pictures of oil barrels. The chief advantage of

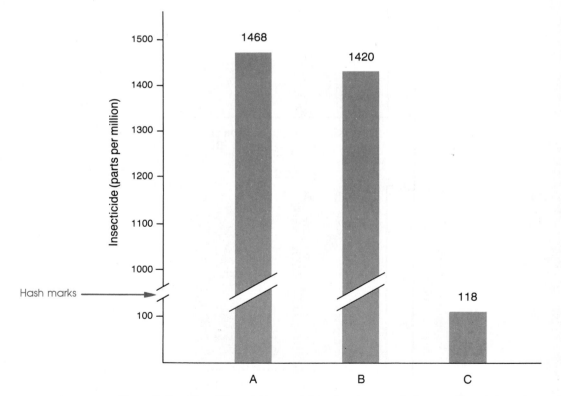

**Figure 7.** Quantity of insecticide remaining after the test solution was filtered through activated charcoal filters manufactured by A, B, and C.

Figure 20-11.    Bar Graph Using Hash Marks on an Axis and also on Bars.

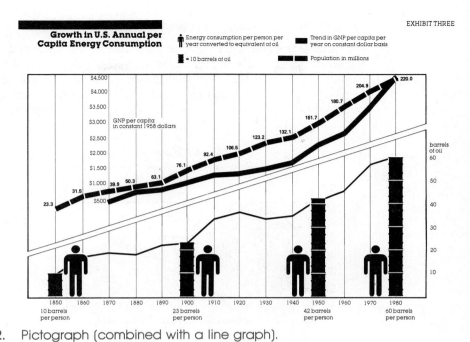

**Figure 20-12.**   Pictograph (combined with a line graph).

From Standard Oil Company, *Energy Adventures* (Cleveland, Ohio: Standard Oil Company, 1983), Exhibit 3. Courtesy of Oak Ridge Associated Universities.

the pictograph is that it lets you use the power of drawings to symbolize the concrete things you are talking about in your graph.

You will find pictographs to be especially useful to you in situations where you want to do one or both of the following things:

- *Emphasize the practical consequence of the data represented.* A pictograph that uses silhouettes of people to represent the workers who will be employed in a new plant helps to emphasize that the plant will bring a practical benefit to individual men and women in a community.
- *Make your data visually interesting and memorable.* Visual interest may be especially important when your communication is addressed to the general public. In some situations at work, however, your readers would expect a more abstract representation of information and consider pictographs to be inappropriate.

## Creating a Pictograph

The procedure for creating a pictograph is nearly identical to that for creating a bar graph. The difference is that you draw pictures instead of rectangles to represent quantities.

## Avoid Misleading Your Readers

Like bar graphs, pictographs can mislead the readers if they are not drawn properly. Consider, for example, the two pictographs shown in Figure 20-13, which both show the average percentage of an apple harvest that an apple grower should expect to be graded Extra Fancy. The pictograph on the top makes the percentage for golden delicious apples seem at least twice as

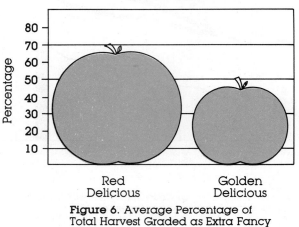

**Figure 6**. Average Percentage of
Total Harvest Graded as Extra Fancy

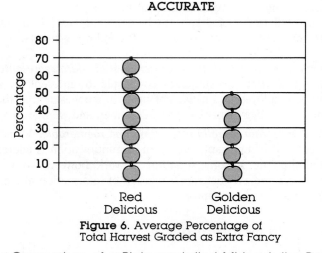

**Figure 6**. Average Percentage of
Total Harvest Graded as Extra Fancy

Figure 20-13.   Comparison of a Pictograph that Misleads the Reader with One that Does Not.

Based upon data from Paul J. Tvergyak, "Branching Out," *American Fruit Grower* 106(February 1986):56.

great as the percentage for red delicious, even though the actual difference is only 20%. That's because the picture for the golden delicious apples is larger in height *and* width, so that its *area* is much greater. In the pictograph on the bottom, the column for the golden delicious apples is greater in height alone. When you are using pictographs, you can avoid misleading your readers by keeping all of your pictures the same size, and using greater numbers of them to represent the greater quantity.

# LINE GRAPHS

A line graph shows how one quantity changes as a function of changes in another quantity. For example, the line graph in Figure 20-14 shows how changing the amount of an antifoam agent added to pulp changes the time it takes ink to penetrate paper made from the pulp.

You can use line graphs in many ways, including the following:

- *To help your readers rapidly interpret trends and cycles.* You can see how well suited line graphs are to this purpose if you look at Figure 20-15 (page 418), which compares a table and a line graph that present the same information. First, use the table to learn how the amount of intercity freight carried by railroads and by motor vehicles gradually

*FIGURE 1:* **Sizing response vs dosage of different antifoams.**

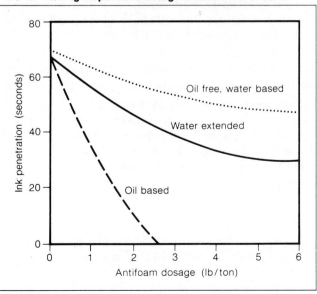

**Figure  20-14.**   Line Graph.

From Frank Poltenson, "Advanced Water-Based Antifoams Reduce Pitch and Deposit Formation," *Pulp & Paper* 60(February 1986):98. Used with permission.

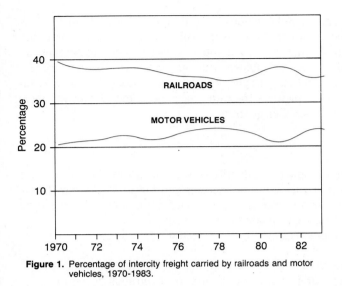

**Figure 1.** Percentage of intercity freight carried by railroads and motor vehicles, 1970-1983.

| Table 1 Percentage of Intercity Freight Carried by Railroads and Motor Vehicles, 1970–1983 | | |
|---|---|---|
| | **Railroads** | **Motor Vehicles** |
| 1970 | 39.82 | 21.28 |
| 1971 | 38.54 | 21.53 |
| 1972 | 38.21 | 21.68 |
| 1973 | 38.44 | 22.63 |
| 1974 | 38.52 | 22.38 |
| 1975 | 36.74 | 21.97 |
| 1976 | 36.33 | 23.16 |
| 1977 | 36.15 | 24.06 |
| 1978 | 35.18 | 24.28 |
| 1979 | 36.03 | 23.64 |
| 1980 | 37.47 | 22.32 |
| 1981 | 37.90 | 21.94 |
| 1982 | 35.81 | 23.21 |
| 1983 | 35.93 | 23.62 |

**Figure 20-15.**   Comparison of Line Graph and Table.

Based on data from U.S. Bureau of the Census, *Statistical Abstract of the United States,* 105th ed. (Washington, D.C.: U.S. Government Printing Office, 1985), 589.

changed between 1970 and 1983. To do that, you will need to search out a number of figures, and engage in a fair amount of addition and subtraction. Then, see how much faster it is for you to see those trends by using the line graph.

- *To emphasize trends.* For instance, you might use a line graph to portray the steady, dramatic increase in the productivity of your department.

- *To help your readers compare trends.* At work, you will sometimes want your readers to see how two or more trends compare with one another. For instance, Figure 20-16 provides a favorable impression of 304 stainless steel by showing that its price has risen much more slowly than either the price of finished steel mill products or the consumer price index.

## Creating a Line Graph

To create a readily understandable line graph, you should follow a procedure very similar to that used to create bar graphs.

1. *Draw the two axes of the graph.* Plan so that your longest quantity will just fit on its axis.

2. *Label the axes.* If you are trying to show how the variation in one thing (called the *independent variable*) affects another thing (called the *dependent variable*), place the independent variable on the horizontal axis. Time is usually treated as an independent variable, so it usually goes on the horizontal axis.

3. *On each axis place labeled tick marks at regular intervals.* Usually your graph will be less cluttered if you make these tick marks short (but still clearly visible). In some situations, however, your readers may find it easier to read your line graph if you extend the tick marks all the way across the graph. If you do that, make them with a *thinner* line than you use to represent the quantities you are describing. In that way, the heavier lines that represent your data will still stand out against the lighter grid lines.

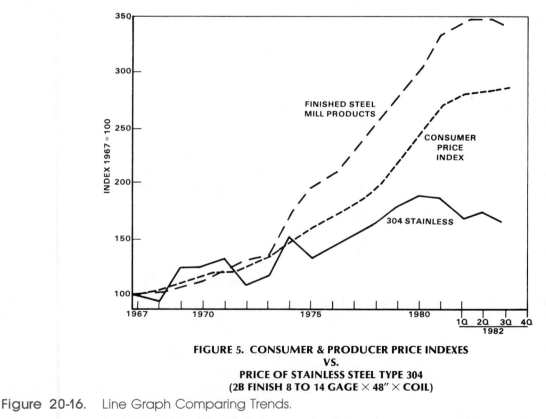

**FIGURE 5.  CONSUMER & PRODUCER PRICE INDEXES**
**VS.**
**PRICE OF STAINLESS STEEL TYPE 304**
**(2B FINISH 8 TO 14 GAGE × 48″ × COIL)**

Figure  20-16.   Line Graph Comparing Trends.

From Ronald S. Paul, "Materials Research and Technological Progress," *Fourth Technology Briefing: Focus on Materials Technology for the '80s* (Columbus, Ohio: Battelle Press, 1982), 11. Courtesy of R. K. Pitler, Allegheny Ludlum Steel Corporation.

4. *Plot the points on the graph and then join them with a line.* If you plot more than one line, distinguish between the lines in some way, for instance by using a solid line for one and various kinds of dashed lines for the others.

5. *Label your lines.* If possible, label the lines by placing the appropriate word or words right next to them. Readers find it easier to use labels located there than to use labels provided in a separate key.

6. *Provide numbers to indicate exact quantities only if your readers need such information.* Generally, numbers indicating the exact value of each point plotted on the graph are not necessary, and you can simplify your line graph by omitting them.

7. *Provide a zero point if one is needed to avoid misleading your readers.* In many fields writers do not use zero points because the specialists who read their graphs can interpret the data very well without those points. However, in other situations and for other readers, a zero point is as important

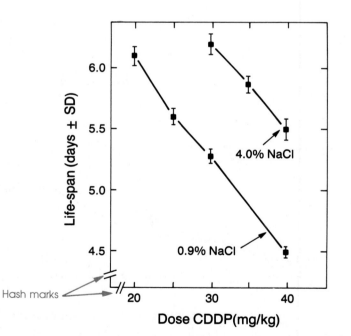

TEXT FIGURE 2. Effect of increasing CDDP (B.M. preparation) doses, dissolved in high- (4.0%) and low- (0.9%) NaCl solutions, on the life-span of mice. *Each point* represents the mean life-span for 6 mice.

**Figure 20-17.**   Hash Marks Used on Two Axes To Emphasize That the Data Do Not Begin at Zero.

From Steiner Aamdal, Oystein Fodstad, Olav Kaalhus, and Alexander Pihil, "Reduced Antineoplastic Activity in Mice of Cisplatin Administered with High Salt Concentration in the Vehicle," *Journal of the National Cancer Research Institute* 73(1984):745.

in a line graph as it is in a bar graph. If you want to emphasize to your readers that your data do not begin at zero, use hash marks on the appropriate axis or axes, as is done in Figure 20-17.

# PIE CHARTS

A pie chart is a circle divided into wedges, like the slices of a pie. The circle represents the total amount of something, and each wedge represents the proportion of that total amount that is associated with some subpart of the total. For instance, the circle might represent all the students in your writing class, and each wedge might represent the percentage of the students in a particular major. The larger the percentage, the larger the wedge. Figure 20-18 shows a pie chart.

A pie chart is ideal when you want your readers to compare relative proportions of a whole—for example, the portion of the department budget that goes for supplies in comparison with the portions that go for salaries and other uses, or the portions of the company's total sales that derive from each of its major product lines.

### Sources of Fat in Our Diet

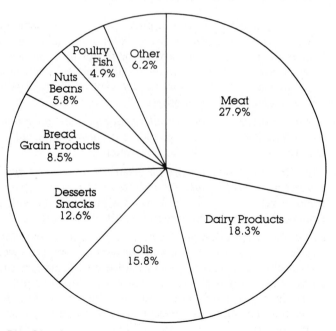

**Figure 20-18.**   Pie Chart.

Data from Bonnie Liebman, "What America Eats," *Nutrition Action Healthletter* 12(December, 1985):4. Reprinted from *Nutrition Action Healthletter*, which is available from the Center for Science in the Public Interest, 1501 16th Street, N.W., Washington, D.C., 20036, for $20/yr; copyright 1986.

### Creating a Pie Chart

Once you have drawn your circle, you can complete your pie chart by performing the following steps:

1. *Slice the circle into wedges.* Start the largest wedge at the twelve o'clock position. Measure clockwise so that each wedge occupies a portion of the circle's circumference proportional to the amount of the total it represents. Place the wedges in descending order of size. By arranging the wedges in this way, you help the readers determine the rank order of the wedges and compare the relative sizes of particular wedges. Of course, for some purposes, you may want to arrange your slices in some other order.

2. *Label each wedge with its title and percentage.* Depending upon the size of the wedge, you might place the labels inside the circle or outside.

## CHOOSING THE BEST VISUAL DISPLAY FOR NUMERICAL DATA

You can represent numerical data in any of the five types of visual displays described so far in this chapter: tables, bar graphs, pictographs, line graphs, and pie charts. To choose among them, you need to think about how you want your visual aids to affect your readers.

- *Use a table when you want to direct your readers' attention to individual pieces of numerical data.* However, do not use tables if you want the readers to emphasize comparisons or trends in numerical data. The other types of visual display accomplish that objective much more effectively. The only exception occurs when you want to compare only two or three pieces of data with one another.

- *Use a bar graph or a pictograph when you want your readers to compare quantities.* Bar graphs and pictographs show clearly not only which quantity is larger but also by how much.

- *Use a bar graph or a line graph when you want your readers to observe a trend in one set of data.* The plotted line of a line graph displays the trend or relationship you wish to draw attention to. With a bar graph, readers can mentally draw such a line to link the ends of the bars.

- *Use a line graph when you want your readers to compare two or more trends.* Often, you will want your readers to compare several trends with one another. For instance, you might want to compare the rise in sales for your company over the past five years with the dip in sales by each of your four major competitors over the same period. Line graphs are much better than bar graphs at presenting such potentially complicated and confusing arrays of information simply and clearly.

- *Use a pie chart to show generally how something is divided up.* Pie charts give a *general* idea of the relative sizes of the subcategories.

However, it is difficult for readers to compare accurately the relative sizes of the wedges of a pie. Therefore, where it is important to convey a *specific* comparison of the quantities in the subcategories, a bar graph is a better choice than a pie chart. See Figure 20-19.

## PHOTOGRAPHS

With a photograph, you can show your readers exactly what they would see if they personally were to look at an object. Because photographs can provide a realistic view of an object's appearance, they are valuable in helping you achieve a variety of communication purposes. Here are some examples:

- *To show the appearance of something the readers have never seen.* Perhaps it is the inside of a human heart, or the surface of one of the moons of Saturn, or the result of some special research technique, or a new product your company manufactures. See Figure 20-20 (page 425).
- *To show the condition of something.* Photographs can help portray the condition of an object when that condition is indicated by the object's appearance. Maybe you want to show the result of a treatment for a skin ailment, the damage to sheet metal caused by improper handling, or the pollution caused by a failure at a sewage treatment plant. See Figure 20-21 (page 425).
- *To help the readers find something.* In an instruction manual, for instance, you might show the insides of a control box, labeling each electrical relay the readers might need to repair. See Figure 20-22.
- *To help the readers recognize something.* For instance, in a lab manual for operators of a steel mill, you might want to include a series of photographs that would enable them to identify the imperfections that they might encounter in sheet steel. See Figure 20-23 (page 427).
- *To highlight an object's attractive or impressive appearance.* You might want to do that, for instance, in a sales brochure.
- *To communicate with special impact.* For instance, there is no better way to emphasize the devastation caused by a tornado than to show a photograph of the stricken area.

As useful as photographs are, they do have the following important disadvantages:

- Photographs show irrelevant as well as relevant detail.
- Photographs often fail to emphasize the most important material shown in them.
- It often is difficult for the writer to show the appropriate angle of view in a photograph.

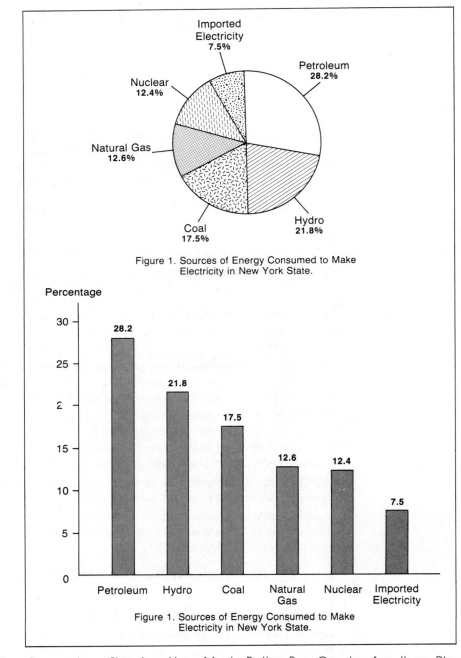

Figure 1. Sources of Energy Consumed to Make
Electricity in New York State.

Figure 1. Sources of Energy Consumed to Make
Electricity in New York State.

**Figure 20-19.**   Comparison Showing How Much Better Bar Graphs Are than Pie
Charts at Showing the Relative Sizes of Quantities.
Data from The Nelson A. Rockefeller Institute of Government, *1983–84 New York State Statistical
Yearbook* (New York: Nelson A. Rockefeller Institute, 1984), 271.

**Figure 20-20.**   Photograph of the Stress Pattern in a Gear Tooth.

Courtesy of Xtek, Inc., Cincinnati, Ohio.

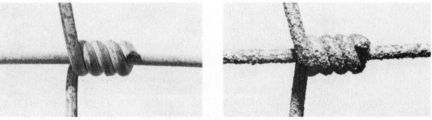

*Actual unretouched photographs of fences installed side by side in 1964. A to Z Fence, on the left, still looks good after standing 11 years in a mild industrial atmosphere. Right beside it, under identical conditions, ordinary galvanized fence has become heavily corroded.*

**Figure  20-21.**   Photograph Used To Show the Condition of Something.

Courtesy of Midwestern Steel Division, Armco Inc.

## Creating a Photograph

When making a photograph that will help you communicate your message, you will need to consider such matters as the following:

1. *Choose an appropriate angle of view.* For instance, if you are intending to help your readers recognize or find something, use the same angle of view that the readers would take when looking for it.

2. *Plan and trim your photograph to eliminate unnecessary or distracting detail.* The camera will record whatever is in its field of view, including even the most irrelevant objects. If objects near the one you want to show are irrelevant, remove them or screen them from view before taking the photograph. Often the photograph will include unnecessary material above, below, or to the sides of your subject. You can crop (or trim away) that extra material to focus your readers' attention more sharply on the thing you want to show them.

3. *Ensure that all the relevant parts show clearly.* Don't let important parts be hidden (or half-hidden) behind other parts. Be sure that the important parts are in focus.

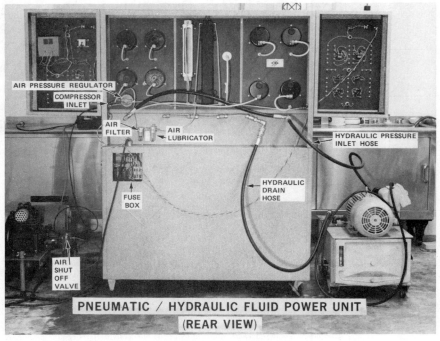

**Figure 20-22**    Photograph To Help the Readers Find Something.
Courtesy of Manufacturing Engineering Department, Miami University (Ohio).

**STRETCHER STRAINS (LUDER'S LINES)**—Stretcher strains are irregular surface patterns of ridges and valleys which develop during drawing.

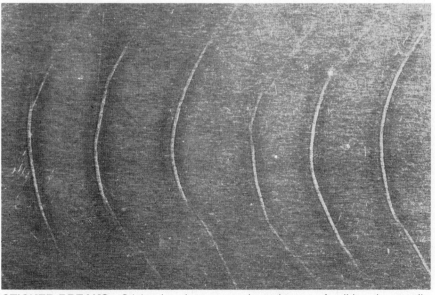

**STICKER BREAKS**—Sticker breaks are arc-shaped types of coil breaks usually located near the middle of the sheet.

Figure 20-23.  Photographs Used To Help the Readers Recognize Something.

From Society of Automotive Engineers, "Classification of Major Visible Imperfections in Sheet Steel," *SAE Handbook* (Warrendale, Pennsylvania: Society of Automotive Engineers, 1980), Part 1, 4.24–4.25.

4. *Provide whatever labels your readers will need.* Be sure that all labels stand out sufficiently from the background of the photograph. One way to do that is to place them on a white strip of paper, which you then paste down on the photograph. For an example, look at the labels in Figure 20-22. Imagine how difficult they would be to read—or even to detect—if they had not been placed on the white background.

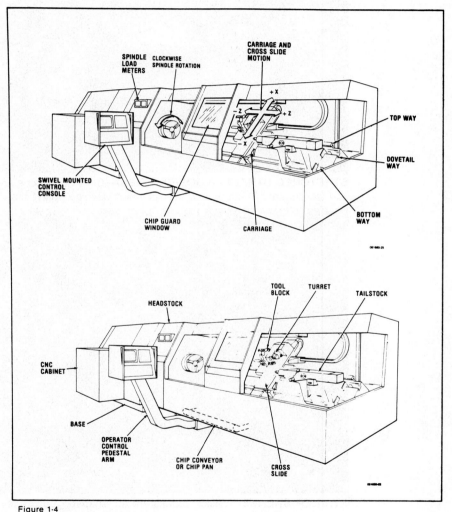

Figure 1-4
Machine Orientation; 3000 Turning Center

**Figure 20-24.**    Drawing that Shows the External View of Something.

From White Consolidated, *Operator's Manual, 3000 Turning Center, SWINC System CNC* (Cincinnati, Ohio: White Consolidated, 1982), 5. Courtesy of American Tool, Incorporated.

## DRAWINGS

There are many kinds of drawings. Here are a few common examples:

- *External View.* These drawings enable your readers to readily identify the major parts of an object. See Figure 20-24.
- *Internal View.* These drawings show the parts of something in a way that a camera cannot. For example, one common way to show the interior of something is to draw what the readers would see if all or part of the object's outside covering were removed (Figure 20-25). This type of view is called a *cutaway.* Another way to show an interior is to show what your readers would see if the object were sliced open (Figure 20-26, page 430). This type of drawing is called a *cross-section.*
- *Exploded View.* These drawings show each separate part of an object and also indicate how the parts are assembled. Exploded views are often included in instruction manuals and in catalogs people use to order parts. See Figure 20-27 (page 431).

C. Twin overhead camshafts are cogged-belt driven and act on four valves per cylinder via hydraulic lifters enclosed in inverted bucket-type tappets.

**Figure 20-25.**   Drawing that Shows an Object with Part of Its Covering Removed.

From Jack Yamaguchi, "Active Suspension Incorporates Rear Wheel Steering Effect," *Automotive Engineering* 93(October, 1985):107. Reprinted with permission. © *Automotive Engineering* magazine, October 1985, Society of Automotive Engineers, Inc. Artwork courtesy of Nissan Motor Company, Ltd.

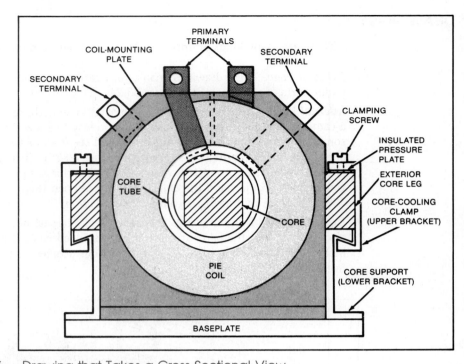

**Figure 20-26.**    Drawing that Takes a Cross-Sectional View.

From "High Efficiency, Low-Weight Power Transformer," *NASA Tech Briefs* 9(Fall 1985):156.

## Creating a Drawing

To make effective drawings, do the following:

1. *Choose the angle of view that your readers will find most helpful.* In many cases, this is a "three-cornered" view showing, for example, two sides and the top of an object. This view helps readers grasp at a glance what the object looks like and how its major parts fit together. Figure 20-28 (page 432) shows how much easier it is to understand the appearance and structure of an object when it is shown in a three-cornered view than when each of the three sides is shown separately. (Note, however, that diagrams showing each of the sides separately are useful for certain purposes.)

   If you are preparing your drawing for a set of instructions, show the object from the same point of view that the readers will have when working with the object.

2. *Leave out insignificant and distracting details.* Usually, your objective will not be to produce a realistic portrait of the object but to show its significant parts. In situations where you want to make a relatively realistic

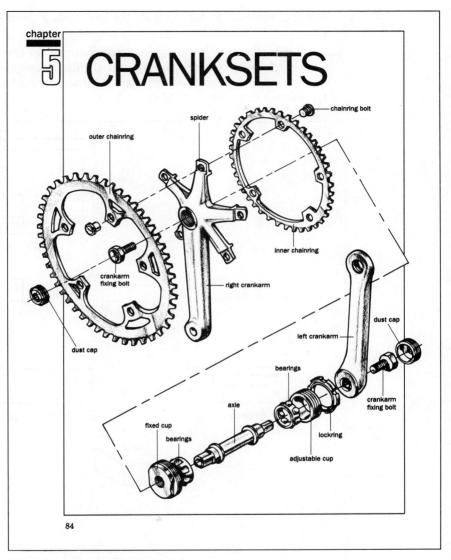

**Figure 20-27.**   Exploded View.

Reprinted from *Bicycling Magazine's Complete Guide to Bicycle Maintenance and Repair* (Emmaus, Pennsylvania: Rodale Press, 1986), 84. © 1986 by Rodale Press, Inc. Permission granted by Rodale Press, Inc., Emmaus, PA 18049.

drawing, you may be able to trace the important elements from a photograph.

3. *Emphasize the most important details.* You can do this in a variety of ways, such as by drawing key features slightly larger than they actually are, or by drawing them with a wider line than you use for other parts, or by pointing them out with labels.

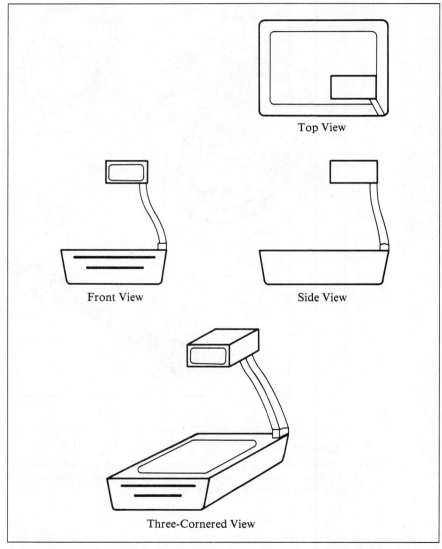

Front View                     Side View

Three-Cornered View

**Figure 20-28.**    Comparison of a Three-Cornered View with Three Separate Views of the Same Object.

4. *Provide the labels your readers need to understand the drawing.* Place the labels so that they are easy to see. Usually that means placing them outside the outline of the object itself, as is done in Figure 20-24. Be sure that the lines you draw from your labels to the parts they designate are clearly distinguished from the lines you use to draw the object itself. You can do that by having the labeling lines run at different angles from the lines of the object, or by drawing them with a wider (or thinner) line than you use for the object, or by placing arrowheads at the ends of the labeling lines.

## CHOOSING BETWEEN PHOTOGRAPHS AND DRAWINGS

Photographs and drawings both show the appearance and structure of something. Here are some guidelines for choosing which of these visual aids to use:

- *Use a photograph when you want to show in full and realistic detail what something really looks like.* A photograph shows what a reader's eye might see. It has an immediacy and impact that a drawing cannot capture.

- *Use a drawing when you want to focus your readers' attention on a few details.* A photograph will show everything that is in front of the lens when you take the picture. It makes no distinction between significant and insignificant detail. In contrast, a drawing shows only the details you choose to show. Further, you can draw in a way that emphasizes the important points.

- *Use a drawing when you cannot show some significant part with a photograph.* A camera cannot always see things in the way you want to show them. Maybe one part partially hides another, or the significant part is in a position that prevents you from taking a photograph that shows the most appropriate angle of view.

## DIAGRAMS

A diagram is much like a drawing except that it relies heavily on symbols to communicate. Thus, whereas drawings are tied to the appearance of the things they portray, diagrams depict subjects more abstractly.

Many specialized fields have developed their own symbol systems for diagrams. In civil engineering there are elevations and blueprints, in electronics there are schematics, and in computer science there are data flow diagrams. Figure 19-8 (in the preceding chapter) shows four types of specialized diagrams. Each requires special training to construct and to read. Those types of diagrams are not treated in this book. You will learn about the ones that are useful in your field from your coursework in your major.

For many diagrams, however, you do not need a specialized system of symbols. You can design an effective diagram by relying solely on your own creativity. Figures 20-29 and 20-30 show diagrams that communicate effectively without using special symbols. The following paragraphs provide general advice about making diagrams. In addition, three kinds of diagrams are used so widely in the workplace that they are treated separately in later sections of this chapter. They are flow charts, organizational charts, and schedule charts.

## Creating a Diagram

Because diagrams can be used for so many different purposes and because you can draw so heavily upon your creativity when designing them, there are not many specific guidelines for making diagrams. The general suggestions—to think about your readers and your purpose and to use your common sense—are already quite familiar to you. Nevertheless, the following suggestions will help:

1. *Decide exactly what you want to show.* What are the objects or events involved? What are the important relationships among them? Making these preliminary decisions can take a substantial amount of thought.

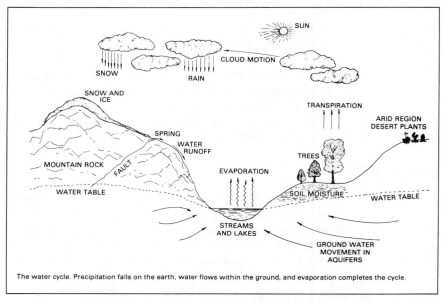

The water cycle. Precipitation falls on the earth, water flows within the ground, and evaporation completes the cycle.

**Figure 20-29.** Diagram of the Water Cycle that Does Not Use Specialized Symbols.

From Raymond L. Murray, *Understanding Radioactive Wastes* (Columbus, Ohio: Battelle Press, 1982), 81. Courtesy of Battelle Memorial Institute.

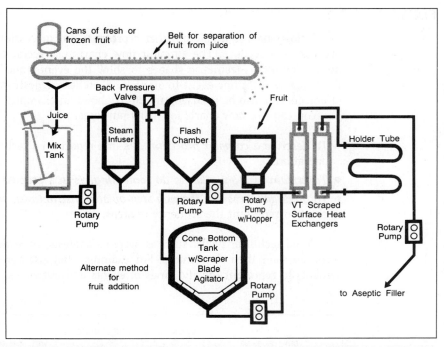

**Design for continuous UHT processing of fruits and fruit fillings (showing alternative method of adding fruit to juice) reduces product exposure to UHT for improved flavor and texture in end product such as yogurt and ice cream.**

**Designed by APV Crepaco, Inc.**

**Figure 20-30.**   Diagram that Does Not Rely on Specialized Symbols To Show a Design for Processing Fruits and Fruit Fillings at Ultra-High Temperatures (UHT).

From "Upgrades for Continuous UHT Processing of Viscous Products," *Food Engineering* 58(August 1986):119.

2. *Create an appropriate means of representing your subject.* You can represent objects and events with geometric shapes. Or you can represent objects with sketches that suggest their appearance. You can show relationships among objects and events by the way you place the representative shapes and by drawing arrows between them. Figures 20-29 and 20-30 illustrate these techniques. For ideas about how to represent your material, you may find it helpful to look at communications similar to yours that other people have written.

3. *Provide the explanations your readers need in order to understand your diagram.* You may provide this in the diagram itself, in a separate key, in the title, or in the accompanying text.

## FLOW CHARTS

A flow chart is a diagram that represents the succession of events in a process or procedure. The simplest flow charts use rectangles, circles, diamonds, or other geometric shapes to show the events, and arrows to show the progress from one event to another. Sketches suggesting the appearance of objects can also be used. Figure 20-31 shows an example.

You can use flow charts for many purposes.

- *To describe a complex procedure,* as in a report on a technical process or operation.
- *To explain how you plan to do something,* as in a proposal.
- *To guide your readers through step-by-step instructions,* by using a flow chart to represent the sequence of steps.

Some technical fields, such as systems analysis, have developed special techniques for their flow charts. For example, they use agreed-upon sets of symbols to represent specific kinds of events and outcomes, and they follow

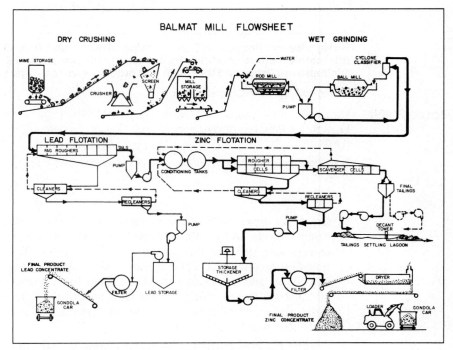

**Figure 20-31.**    General Flow Chart.

Courtesy of St. Joe Resources Company.

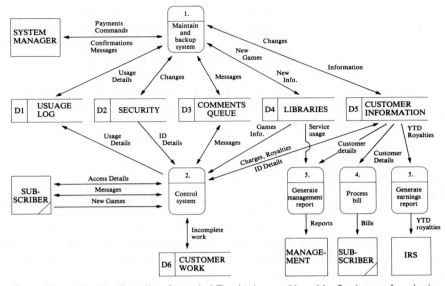

**Figure 20-32.**   Flow Chart Illustrating the Special Techniques Used in Systems Analysis.

From William S. Davis, *Systems Analysis and Design: A Structured Approach* (Reading, Massachusetts: Addison-Wesley, 1983), 187. © 1983, Addison-Wesley Publishing Company, Inc., Reading, Massachusetts. Reprinted with permission.

agreed-upon rules for arranging these symbols on the page. Figure 20-32 shows a flow chart that follows the conventions developed by systems analysts. If you are in a field that uses such specialized flow charts, include them in communications written to your fellow specialists. Remember, however, that you should not use them when writing to people outside your specialty, who will probably not know how to read them.

## Creating a Flow Chart

The general procedure for creating a flow chart is as follows:

1. *Decide which specific activities and outcomes to represent.* This preliminary planning is essential to making an effective flow chart.
2. *Draw a box or other appropriate shape for each.* If you have several similar activities or outcomes, use the same shape for all of them. Make all the boxes and other various shapes approximately the same size.
3. *Label each box.* Place the label inside the box.
4. *Arrange the boxes so that activity flows from left to right, or from top to bottom, or both.* This corresponds with your readers' normal eye move-

ments when reading. If your flow chart requires more than one line of shapes and arrows, begin the second and subsequent lines, like the first line, at the left-hand margin or top of the figure (as in Figure 20-31).

5. *Draw arrows between the boxes to indicate the flow of activity.*

As you create a flow chart, be patient. It may take you several drafts to get the boxes the right size, to place the labels neatly in them, and to arrange the shapes and arrows in an attractive and readily understandable way.

## ORGANIZATIONAL CHARTS

An organizational chart uses rectangles and lines to represent the arrangement of people and departments in an organization. As Figure 20-33 shows, an organizational chart reveals the organization's hierarchy, indicating how the smaller units (such as departments) are combined to create larger units (such as divisions). It also indicates who reports to whom and who gives direction to whom.

You can use an organizational chart in the following ways:

- *To show the scope and arrangement of an organization.*
- *To show the formal lines of authority and responsibility in an organization.*
- *To show how work on a particular project will be managed,* for instance when you are writing a proposal.
- *To provide a map of an organization,* so that readers can readily locate the particular people they want to contact.

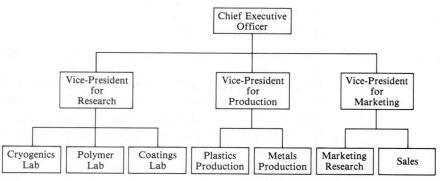

**Figure 20-33.**   Organizational Chart.

## Creating an Organizational Chart

To create an organizational chart, do the following:

1. *Draw and label boxes for each relevant part of the organization.* In some charts, you do not need to show every part of the organization, but only those parts and subparts relevant to your particular readers.

2. *Arrange the boxes into a pyramid.* Place the box for the highest level (for example, the company president) at the top and the boxes for the lowest level at the bottom. This will give your chart the shape of a pyramid.

3. *Draw lines to represent the lines of authority and responsibility.*

4. *Use a different kind of line to indicate special relationships.* For instance, you might use a dashed line to indicate close consultation or cooperation between groups that are not connected by direct lines of authority and responsibility.

# SCHEDULE CHARTS

A schedule chart identifies the major steps in a project and tells when they will be performed. As Figure 20-34 illustrates, a schedule chart enables

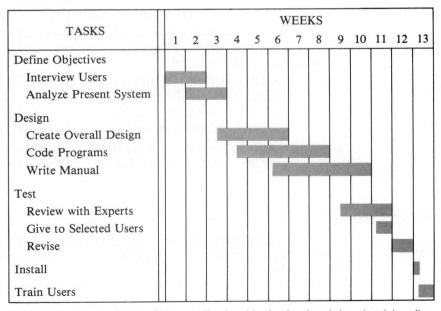

Figure 20-34.   Schedule Chart. (Note: This particular kind of schedule chart is often called a *Gantt chart*.)

readers to see what will be done, when each activity will be started and completed, and when more than one activity will be in progress simultaneously.

Usually, schedule charts are used in project proposals to show the proposer's plan of work. You can also use schedule charts in progress reports to show what you have accomplished and what you have left to do, and you can use them in final reports to describe the process you followed while working on a project that you have now completed.

### Creating a Schedule Chart

To create a schedule chart, do the following:

1. *Draw the two axes of a graph.*
2. *Along the vertical axis, list the major steps in your project.* Usually, you will help your readers follow your project most easily if you divide your overall project into four, five, or six major steps. Arrange the steps in chronological order, with the first at the top.

   If your readers will want the additional detail, list the principal subparts of each major step. As you do that, be sure to arrange your overall list so that the major tasks are most prominent. For instance, you can put the major tasks in capital letters, while capitalizing only the first letters of the substeps. You can also indent the substeps.
3. *Along the horizontal axis, draw a scale that covers the full duration of the project.* Use tick marks to indicate the major periods of the project schedule (usually weeks or months).
4. *Label the horizontal axis.* If each tick mark represents a specific date, place that date by the mark. If the schedule is not tied to specific dates, place the labels between the intervals to indicate, for instance, the first week of the project, the second week of it, and so on (see Figure 20-34).
5. *Draw a horizontal line or rectangle to represent the duration (including the starting and ending dates) of each step and substep.*

## BUDGET STATEMENTS

A budget statement is a table that shows how money will be gained or spent. It may be very simple or very elaborate. Figures 20-35 and 20-36 show two examples.

On the job, you can use budgets in the following situations:

● *To explain the expenses involved with a project or purchase.* You might do this when requesting funds (Figure 20-35) or when reporting on the feasibility of a particular course of action.

The most efficient and accurate way to obtain the readings in this experiment is to use electronic instrumentation that will enter the results directly into a computer. To do that, we will need to purchase the following equipment:

| | |
|---|---|
| HPIP Converter | $3,200 |
| Digital Plotter | 1,950 |
| Current Source | 3,000 |
| Digital Thermometer | 1,100 |
| Miscellaneous Components to Create Interface Board | · 375 |
| | $9,625 |

**Figure 20-35.**   Simple Budget Statement.

PROJECT COSTS

| | |
|---|---|
| Equipment | |
| 4 KRN 3781 Robots ($37,000 apiece) | $148,000 |
| 1 Microcomputer with hard disk | 6,500 |
| 4 Power supplies ($5,100 apiece) | 20,400 |
| | |
| Construction (includes labor and supplies) | |
| Rewiring | 7,100 |
| Constructing pads for mounting robots | 3,000 |
| | |
| Initial Programming (ten days at $125 per day) | 1,250 |
| | |
| Travel for Installation Personnel | |
| Air fare (two round trips) | 400 |
| Car rental | 200 |
| Living expenses (two people for five days at $150 per day) | 1,500 |
| | |
| TOTAL | $188,350 |

**Figure 20-36.**   Detailed Budget Statement.

- *To summarize the savings to be realized by following a recommendation you are making.* Because profits are a major goal of many organizations, budget statements of this type are common in the workplace.
- *To report the costs that have been incurred by a project for which you have responsibility,* as in Figure 20-36.
- *To explain the sources of revenue associated with some project or activity,* as when reporting your department's sales of products and services.

### Creating a Budget Statement

To create a budget statement, do the following:

1. *Divide your page into two vertical columns.*
2. *In the left-hand column, list the major categories of expense.* If your readers will want the additional detail, list the principal expenses found under each major category. Be sure to arrange your overall list so that the major categories are most prominent.
3. *In the right-hand column, write the amount of each expense.* Align these amounts on the decimal points.
4. *Clearly indicate the total.*
5. *If you want to show income also, repeat Steps 2 through 5 for it.*

## EXERCISES

1. Figure 20-37 shows a table containing information about enrollments in institutions of higher education in the United States. A writer might use that information in a variety of ways, including the four listed below. For each of those four uses, decide whether the writer should leave the information in a table or present the relevant facts in a bar graph, pictograph, line graph, or pie chart.

   A. To show how steadily and dramatically the number of women over age 35 enrolled in college has risen since 1970 and is expected to continue to rise through 1993.

   B. To provide a reference source in which researchers can find the numbers of men (or women) in a particular age group who were enrolled in college in a particular year.

   C. To enable readers to make a quick, rough comparison of the 1983 enrollments of men in each age group with the 1983 enrollments of women in each age group.

   D. To allow educational planners to see what proportion of the males (or females) who will enroll in 1993 are expected to be from each of the age groups. (To present this information, the writer would have to perform some mathematical calculations based upon the data supplied in Figure 20-37.)

2. In response to a survey, 13,586 employees of the General Electric Company who had earned college degrees each listed the college courses they found most valuable in their careers. They also listed courses they found least valuable. You are to present the results (which are given below) first in a table and then in a bar graph that would be useful to college juniors and seniors to help them decide what courses to take next term.

Begin by imagining a typical student preparing to register. The student has a course catalog, a schedule showing when classes are available, and the table or bar graph you are preparing. Your job is to decide how to make a table (and then a bar graph) that will present the survey results as helpfully as possible.

In the paragraphs below, the survey results for engineering graduates are presented separately from those for nonengineering graduates. You will have to decide whether you will help your readers more by making one table or two (and one bar graph or two). Be prepared to show your table(s) and bar graph(s) to your classmates and explain why you designed each the way you did.

When asked which courses they found most valuable in their careers, 55.21% of the engineering graduates included physics in their list of valuable courses; 53.84% engineering; 21.60% economics; 58.40% English communication; and 72.21% mathematics. Of the nonengineering graduates, 43.67% listed business; 73.68% English communication, 55.59% economics;

NO. **255.** ENROLLMENT IN INSTITUTIONS OF HIGHER EDUCATION BY SEX, AGE, AND ATTENDANCE STATUS, 1970 TO 1983, AND PROJECTIONS, 1988 AND 1993

[As of **fall**]

| SEX AND AGE | NUMBER (1,000) | | | | | PERCENT PART-TIME | | | | |
|---|---|---|---|---|---|---|---|---|---|---|
| | 1970 | 1975 | 1983 | 1988 | 1993 | 1970 | 1975 | 1983 | 1988 | 1993 |
| Total | 8,581 | 11,185 | 12,465 | 12,141 | 11,676 | 32.2 | 38.8 | 42.9 | 46.1 | 48.3 |
| Male | 5,044 | 6,149 | 6,024 | 5,909 | 5,641 | 30.5 | 36.1 | 38.8 | 41.8 | 43.7 |
| 14 to 17 years | 129 | 126 | 91 | 92 | 79 | 3.9 | 13.5 | 13.0 | 13.0 | 13.9 |
| 18 and 19 years | 1,349 | 1,397 | 1,283 | 1,171 | 1,019 | 6.2 | 9.2 | 10.8 | 10.8 | 10.7 |
| 20 and 21 years | 1,095 | 1,245 | 1,205 | 1,033 | 966 | 9.6 | 15.4 | 15.2 | 15.4 | 15.3 |
| 22 to 24 years | 964 | 1,048 | 1,148 | 1,041 | 988 | 32.6 | 34.5 | 34.7 | 34.9 | 34.8 |
| 25 to 29 years | 783 | 1,123 | 1,087 | 1,150 | 1,022 | 58.2 | 57.8 | 59.7 | 59.7 | 59.7 |
| 30 to 34 years | 308 | 557 | 597 | 676 | 710 | 76.6 | 67.0 | 72.6 | 72.5 | 72.5 |
| 35 years and over | 415 | 654 | 613 | 744 | 855 | 81.9 | 76.8 | 85.3 | 85.2 | 85.3 |
| Female | 3,537 | 5,036 | 6,441 | 6,232 | 6,035 | 34.6 | 42.1 | 46.9 | 50.2 | 52.6 |
| 14 to 17 years | 129 | 152 | 123 | 121 | 104 | 9.3 | 12.5 | 8.3 | 9.1 | 8.7 |
| 18 and 19 years | 1,250 | 1,388 | 1,427 | 1,266 | 1,101 | 8.8 | 10.6 | 12.6 | 12.5 | 12.5 |
| 20 and 21 years | 785 | 998 | 1,187 | 989 | 922 | 16.3 | 19.8 | 19.1 | 19.1 | 19.2 |
| 22 to 24 years | 493 | 706 | 938 | 818 | 773 | 53.1 | 54.4 | 46.7 | 46.7 | 46.7 |
| 25 to 29 years | 292 | 651 | 945 | 960 | 849 | 72.6 | 66.5 | 68.3 | 68.3 | 68.3 |
| 30 to 34 years | 179 | 410 | 717 | 782 | 812 | 84.4 | 76.8 | 80.8 | 80.8 | 80.8 |
| 35 years and over | 409 | 730 | 1,103 | 1,294 | 1,474 | 85.6 | 85.6 | 85.1 | 85.1 | 85.1 |

Source: U.S. National Center for Education Statistics, *The Condition of Education,* annual.

**Figure 20-37.**   Table for Use in Exercise 1.

From U.S. Bureau of the Census, *Statistical Abstract of the United States,* 106th ed. (Washington, D.C.: U.S. Government Printing Office, 1986), 151.

25.55% psychology; 25.00% physics; 53.24% mathematics; and 33.08% accounting.

When asked which courses they found least valuable in their careers, 1.99% of the engineering graduates listed miscellaneous humanities; 59.84% foreign language; 12.11% miscellaneous science; 14.74% miscellaneous business; 46.81% history; 20.42% government; 25.10% chemistry; 45.16% engineering; and 23.32% economics. Of the nonengineering graduates, 22.09% indicated miscellaneous humanities; 55.30% miscellaneous business; 52.24% foreign language; 52.45% history; 48.15% miscellaneous science; 18.78% government; 18.04% chemistry; and 17.70% physics. [From Educational Relations Service, General Electric Company. *What They Think of Their Higher Education* (New York: General Electric Company, 1957).]

3. Figure 20-38 presents a table showing how much crude oil was produced by the world's leading producers in the years 1980 through 1984. Using that information, make the following visual aids. Be sure to provide all appropriate labels and a title for each.

A. A bar graph showing how much crude oil was produced in 1984 by each of the five countries that produced the most in that year.

B. A pictograph showing how much crude oil was produced by Nigeria in each of the five years. Because the sample pictograph in Figure 20-12 uses barrels to represent quantities of oil, use some other symbol for your pictograph.

C. A line graph that shows trends in the crude oil production in the following countries: United States, United Kingdom, China, Mexico, and Iraq.

**Table 37.—Leading world producers of crude oil[1]**

(Million 42-gallon barrels)

| Country | 1980 | 1981 | 1982 | 1983[P] | 1984[e] |
|---|---|---|---|---|---|
| U.S.S.R | 4,434 | 4,475 | 4,503 | 4,528 | [2]4,506 |
| United States | 3,146 | 3,129 | 3,157 | 3,171 | [2]3,250 |
| Saudi Arabia[3] | 3,614 | 3,580 | 2,366 | 1,834 | [2]1,663 |
| Mexico | 708 | 844 | 1,002 | 973 | 983 |
| United Kingdom | 582 | [r]649 | 741 | 817 | [2]893 |
| China | 773 | 739 | 745 | 774 | [2]836 |
| Iran | 550 | 692 | 873 | 892 | [2]791 |
| Venezuela | 793 | 768 | 692 | 657 | [2]658 |
| Canada | 523 | 468 | 464 | 495 | [2]560 |
| Indonesia | 577 | 585 | 488 | 490 | [2]517 |
| Nigeria | 753 | 525 | 472 | 452 | 502 |
| United Arab Emirates (Abu Dhabi, Dubai, Sharjah) | 624 | 548 | 456 | 409 | 424 |
| Iraq | 969 | 326 | 310 | [e]400 | 410 |
| Libya | 670 | 408 | [e]418 | 402 | [2]391 |
| Total | 18,716 | [r]17,736 | 16,687 | 16,294 | 16,384 |
| Other | [r]3,160 | [r]2,892 | 2,833 | 3,120 | 3,402 |
| Grand total | [r]21,876 | [r]20,628 | 19,520 | 19,414 | 19,786 |

[e]Estimated.   [P]Preliminary.   [r]Revised.
[1]Table includes data available through Oct. 9, 1985.
[2]Reported figure.
[3]Includes the country's share of production from the Kuwait-Saudi Arabia Partitioned Zone.

**Figure 20-38.**   Table for Use in Exercise 3.

From U.S. Department of the Interior, *Minerals Yearbook, 1984* (Washington, D.C.: U.S. Government Printing Office: 1984), vol. III, 34. Courtesy of the Bureau of Mines.

D. A pie chart that shows the proportions of the world's crude oil produced in 1982 by the following countries. For your convenience, the correct percentages are provided below:

| | |
|---|---|
| Canada | 2% |
| China | 4% |
| Mexico | 5% |
| Saudi Arabia | 12% |
| United Kingdom | 4% |
| United States | 16% |
| U.S.S.R | 23% |
| All other countries | 34% |

4. In textbooks, journals, or other publications related to your major, find and photocopy three photographs. For each photograph, answer the following questions:

A. What does the photograph show, who is the intended audience, and how is the reader supposed to use or be affected by the photograph?

B. What angle of view has the photographer chosen—and why?

C. What, if anything, has the writer done to eliminate unnecessary detail?

D. Do all the relevant parts of the subject show? If not, what is missing?

E. Has the writer supplied helpful labels? Would any additional labels help? Are any of the labels supplied by the writer unnecessary?

F. Could the writer have achieved his or her purpose more effectively by using a drawing or diagram instead of a photograph? Why or why not?

5. Create a drawing that you could include in a set of instructions for operating one of the following pieces of equipment. Tell who your readers are and, if it isn't obvious, how they will use the product. Be sure to label the significant parts and include a figure title.

- Typewriter.
- Power lawnmower.
- Clock radio.
- An instrument or piece of equipment used in your field.
- The subject of a set of instructions you are preparing for your writing class.
- Some other object that has at least a half-dozen parts that are important to show in a set of instructions.

6. Draw a diagram that you might use in a report, proposal, instruction manual, or other communication that you would write on the job. The

diagram might be an abstract representation of some object, design, process, or other subject. You may create a diagram that requires special symbols, but you do not have to. Tell when you might include the diagram in something you would write on the job. Also, tell what your readers would use the diagram for. Be sure to provide all appropriate labels and a figure title.

7. Create a flow chart on one of the following processes and procedures. Tell when you might use the flow chart and why. Be sure to provide appropriate labels and a figure title. Do not show more than sixteen steps. If your process has more, show the major steps, not all the substeps.

   • Applying for admission to your college.

   • Changing the spark plugs in a car.

   • Making paper in a paper mill. Begin with the trees in the forest.

   • Preparing to make an oral presentation in class or on the job. Your presentation should include visual aids. Start at the point where you decide that you want to speak or are given the assignment to do so.

   • A process or procedure that you will describe in a communication you are preparing for your writing class.

   • Some process or procedure that is commonly used or important in your field.

8. Create an organizational chart for some organization that has at least three levels. You may choose any type of organization, including a club you belong to or a company that employs you, your friends, or your parents. You can also visit some office, store, or business, asking them to provide you with information about their organizational hierarchy.

9. Make a schedule chart that describes your schedule for working on an assignment you are preparing in your writing class. The chart should cover the period from the date you received the assignment to the date you turn the assignment in. Be sure to include all your major activities, such as planning your work, gathering your information, bringing a draft to class for review, and so on.

10. Create a budget statement from the following data concerning the monthly costs that an electronics company would incur if it opened an additional service center in a new city. Remember to group related expenses and to provide a total.

Salary for center manager $2450. Rent $600. Business tax (pro rated) $50. Electricity (year-round average) $180. Water $45. Receptionist's salary $700. Office supplies $200. Car lease $220. Salary for technician $1800. Salary for technician's assistant $1300. Supplies for technician $300. Travel for monthly trip to main office by center manager $300. Car driving expenses $250. Telephone $130. Depreciation on equipment $250.

# Designing Pages

## PREVIEW OF GUIDELINES

1. Create a grid to serve as the visual framework for your pages.
2. Use the grid to coordinate related visual elements.
3. Use the same design for all pages that contain the same types of information.

Y ou build your written messages out of *visual* elements. These visual elements are dark marks printed on a lighter background: words and sentences and paragraphs; drawings and graphs and tables. They are *seen* by readers before they are read and understood.

## THE IMPORTANCE OF GOOD PAGE DESIGN

When you type or print your communications, you can arrange the visual elements in many different ways—using wide margins or narrow ones, aligning your text in one column or in two or more columns, placing your prose and illustrations here or there on the page. Your page design—the way you arrange your visual elements—can greatly affect the success of your communications. Good page design helps you achieve your purposes in the following ways:

- *Good page design helps readers read efficiently.* To read a page, you (and your readers) move your eyes from one place to another. Sometimes you move your eyes in a simple, rhythmic pattern, as when reading sequentially through a text—reading one phrase, then the next, and so on. However, you often move your eyes in more complicated patterns, as when scanning for certain facts, when comparing facts presented in different paragraphs, or when looking back and forth between a paragraph and a drawing that illustrates it. Poorly designed pages confuse and frustrate readers by failing to tell them where to focus their eyes next. Well-designed pages guide the readers' eyes around the page so that the readers can read easily and efficiently.

- *Good design emphasizes the most important contents.* By varying such things as the size and placement of the visual elements on the page, a good design establishes a visual hierarchy. With this hierarchy, you can guide your readers' attention first to one element, then another, and you can emphasize the most important material by drawing the readers' eyes there first.

- *Good design encourages readers to feel good about a communication.* Page design affects readers' feelings. You've surely seen pages—perhaps in a textbook or a set of instructions—that struck you as uninviting, even ugly. As a result, you may have been reluctant to read them. And undoubtedly you've seen other pages that you have found to be attractive. These pages you've probably approached more eagerly, more receptively. A good design creates a favorable impression on its readers, making them feel good toward both the communication and its writer.

This chapter describes an easy-to-follow method you can use at work to design pages that will affect your readers in the ways you want them to.

# SINGLE-COLUMN AND MULTICOLUMN DESIGNS

Page designs are classified according to the number of columns they have. You are already quite accustomed to using a single-column design because it is the one used in college term papers. You type from one margin to the other, and you center most other material, such as tables and indented quotations, between the margins. In multicolumn formats, the area between the margins is divided into two or more columns. Multicolumn formats are used in many popular magazines, such as *Time, Sports Illustrated,* and *Mademoiselle,* as well as in many professional periodicals, such as *Textile World, International Journal of Biochemistry,* and the various *Proceedings of the Institute of Electrical and Electronics Engineers.* Figure 21-1 (pages 450–453) shows multicolumn page designs from a variety of publications.

For much of the writing you do at work, a single-column page design will work quite well. In fact, in most situations it's what your readers and employers will expect. There are, however, some situations in which you will be able to write much more effectively by using multicolumn designs. Why? First, people can read shorter lines of type more easily than they can read longer lines. When two or more columns are used on a page, the lines are shortened and reading becomes easier. Second, in multicolumn formats, writers have more flexibility to create varied and interesting designs and thereby to direct their readers' attention to the material they want to emphasize. Third, by using two or more columns, writers can often help their readers see readily the relationships between different pieces of information, as in instruction manuals where the text describing each step and the illustration showing that step can be placed directly opposite each other in adjacent columns.

For these reasons, multicolumn formats are especially useful in instructions, technical bulletins, product brochures, company newsletters, and advertisements. They are also useful in sections of reports and proposals where you can present your information in tabular form.

Because you are already familiar with one-column page design, this chapter concentrates on multicolumn designs.

# INTRODUCTION TO GUIDELINES 1 AND 2

The first two guidelines describe a two-step strategy that graphic designers use to design pages. (Graphic designers are trained professionally in the art of page design.) To follow these steps, you need to adopt the graphic designers' way of looking at a page. According to this view, every page has four kinds of material:

- *Text*. Your sentences and paragraphs.
- *Headings and Titles.*

# Why Whales Leap

*The action, which is called breaching, seems to be purposeful. It is associated with the social aspects of whale life and probably serves in communication*

by Hal Whitehead

A whale's leap from the water is almost certainly the most powerful single action performed by any animal. It is called breaching, a term that whalers of the 18th and 19th centuries gave this dramatic activity and that present-day investigators of the phenomenon have retained. Considering the great bulk and weight the whale must lift in breaching, one wonders why the animal does it.

A breach provides the only opportunity most human observers have to see an entire whale, and it has inspired a wide variety of impressions. Thus J. N. Reynolds, recounting for readers of *The Knickerbocker* in 1839 the adventures of whalers in the Pacific, wrote: "Occasionally, a huge, shapeless body would flounce out of its proper element, and fall back with a heavy splash; the effort forming about as ludicrous a caricature of agility, as would the attempt of some over-fed alderman to execute the Highland fling." To Herman Melville the breach was sublime. "Rising with his utmost velocity from the furthest depths," Melville wrote in *Moby Dick*, "the Sperm Whale thus booms his entire bulk into the pure element of air, and piling up a mountain of dazzling foam, shows his place to the distance of seven miles and more. In those moments, the torn, enraged waves he shakes off, seem his mane."

The whalers of earlier centuries, searching for their quarry in slow sailing vessels, had many opportunities to observe the whales they were trying to catch. For years the anecdotes told by such men formed the basis of what was known of breaching and other kinds of whale behavior. Among the explanations of breaching they proposed, somewhat anthropomorphically, were feeding, stretching, amusement, being chased by swordfish and an "act of defiance," which was presumably directed at the whalers.

In the past few years scientific obser-

vations of whales in the open ocean have begun to yield useful quantitative data on many aspects of their behavior, including the breach. Roger Payne of the U.S. World Wildlife Fund and his associates have contributed many insights through their long study of southern right whales (*Eubalaena australis*) off the Valdés Peninsula in Argentina. Other important studies include work with gray whales (*Eschrichtius robustus*) off Baja California by Kenneth S. Norris of the University of California at Santa Cruz and a number of other investigators and observations of humpback whales (*Megaptera novaeangliae*) off Hawaii by James D. Darling of the University of California at Santa Cruz, Peter Tyack of the Woods Hole Oceanographic Institution and others. My own work has also been mainly with humpbacks that are in the western North Atlantic off Newfoundland during the summer and on Silver Bank in the West Indies during the winter.

Such long-term observations are crucial to an understanding of breaching because the phenomenon is generally rare. Most whales are seldom seen to breach. Hence it usually takes many years to witness even a moderate number of breaches. In this respect the research on Silver Bank was particularly important. Humpback whales from the western North Atlantic congregate there during the winter months for mating and calving. They reach a density approximating one whale per square kilometer. Many of them breach: during our transects of some 200 kilometers across the bank for the purpose of estimating the size of the population we saw breaching in about 20 percent of the pods (usually containing from one whale to four whales) we sighted.

A leap by a humpback entails the lifting of as much biomass as would be accounted for by 485 people weigh-

ing an average of 68 kilograms (150 pounds) each. The largest humpbacks reach lengths approximating 15 meters (49 feet) and weigh 33 metric tons (72,765 pounds).

The breaches of the humpback and of other whales known to breach range from a full leap clear of the water to a leisurely surge in which only half of the body emerges. In more than a fourth of the breaches by humpbacks at least 70 percent of the animal comes out of the water, but it is rare for the entire whale to be seen above the surface. Humpbacks breach at all angles up to 70 degrees with respect to the surface of the sea.

Payne has observed the breaching process while watching southern right whales from cliffs or small airplanes. The whale swims horizontally until it has developed enough speed. Then it tilts its head upward and raises its flukes, or tail. These actions convert the horizontal momentum into vertical momentum and the whale emerges from the water. Because of the horizontal approach, a whale can breach in water that is only a few meters deep.

Whales perform other actions that superficially resemble breaching. One of them is lunging. In this maneuver the whale thrusts no more than 40 percent of its body through the surface. A lunge can be executed horizontally, vertically or at any angle between those extremes. The whale can be oriented so that its dorsal surface or ventral surface is uppermost or so that it is lying on its side. Whales often are seen closing their jaws while lunging, sometimes ingesting a mouthful of plankton or small fish. Lunging is therefore usually considered to be associated with feeding. Humpbacks, however, can be seen to lunge as they try to outmaneuver one another in large groups, for example when from two to 10 males compete for access to a female among them. Lunging, then, happens when a whale breaks the surface as an unin-

84

---

**A**

**Figure 21-1.**   Page Designs Used by Typical Popular and Professional Periodicals.

JUNE

# GARDENER'S LOG

## LOW-MAINTENANCE GARDENING

In June the spring showers of April and May often give way to dry, sultry days. Here's how to give your garden the water it needs without leaving you high and dry for free time.

### Timesaving tricks

- Cut down on watering chores by locating container-grown plants where they'll receive morning sunlight and afternoon shade.
- Use plastic instead of clay pots for better moisture retention. Cut holes in bottoms of plastic pots for drainage.
- Place smaller pots inside larger pots, then fill the space between the two containers with vermiculite. This will keep the soil in the inner pot cooler to reduce moisture loss.
- Minimize a plant's total water needs by harvesting crops as soon as they ripen.
- Plant seeds of second crops in wide bands instead of in single rows. Plants will grow closely together to shade the soil and keep roots cool. They also will choke out any weeds that compete for soil moisture.
- Mulch plants to keep soil moisture and temperature constant.

### Make your own rain

Chances are you'll have to supplement the rainfall to keep your garden thriving. These watering methods will do the work for you.

- *Overhead sprinklers* can be moved around to water

**GLADIOLUS**

*Illustration: S. D. Schindler*

The thirsty earth soaks up the rain,
And drinks, and gapes for drink again.
The plants suck in the earth, and are
With constant drinking fresh and fair.
—*Abraham Cowley*

areas of your garden that need it most. Because some moisture will evaporate before it hits the ground, however, sprinklers are less efficient than other watering systems. If you use a sprinkler, water during the cooler morning hours, when less moisture will dissipate. Early watering also lets foliage dry before evening to prevent fungus diseases.

- *Furrow irrigation* works best on a flat garden. To set up a furrow system, dig an 8-inch-deep by 12-inch-wide trench between rows of plants and regularly fill it with water.
- A *drip irrigation* system consists of a network of PVC pipes that have small holes punched in them. When they're placed alongside rows of vegetables, the pipes slowly release water at ground level. Because very little water goes to waste, this method is one of the most efficient ways to irrigate. You can buy drip irrigation kits at many garden supply stores.

To get the same results with less initial expense, fill half-gallon wine bottles with water and upend them in the soil near plants. Water will drain out slowly to provide a steady supply of moisture. Refill the bottles every two or three days.

### GOBS OF GLADS

You can have fresh bouquets of gladiolus all summer by planting corms every two weeks until mid-July. Blooming for at least 10 days, plants produce 1- to 5-foot-tall stalks that are smothered with solid or two-tone blossoms in almost every color imaginable.

Gladiolus thrive in a well-drained soil that gets full sun. For best color impact, grow them in rows or in clumps of at least four. Most varieties should be planted 4 inches deep and 6 inches apart; plant taller types 6 inches deep.

Varieties that grow over 2 feet tall may require staking. In the North, dig gladiolus corms in the fall and store them in a cool, well-ventilated place.

24

**B**

**Figure 21-1.**  *Continued*

# Status and Prospects of Surface Mount Technology

John W. Balde
Interconnection Decision Consulting
Flemington, New Jersey

*It is clear that surface mount technology can deliver higher packing densi-
ty and lower cost, but it is also clear that there are problems to be solved.
The progress in solving these problems, and the new developments that
will open up the applications and capability of surface-attach, are discussed. Particular reference is made to control of the soldering process,
to new technology to remove heat from leaded packages, and to the particular suitability of chip carriers for packages with high I/O count and
10 mil pad spacing.*

THE USE OF SURFACE-ATTACH TECHNOLOGY is no longer a
choice, but an inevitable design need for competitive future cost-
performance systems. It is not an option, because there is no other
technology that can deliver so high a density at so low a cost. Pin
grid array technology cannot do it because the cost and technological difficulty of the required multilayer boards are beyond the reach
of all but the largest main frame manufacturers. Multilayer co-
fired ceramic technology cannot do it, not only because of the high
level technology needed for the complex ceramics, but also because
the speed penalties are considerable. Surface-attach promises the
best cost-performance compromise. It does, however, require
shifting to a new technology that may be painful and may present
opportunities for error. Let us examine some of the requirements to
participate in this new technology.

First, surface-attach requires abandonment of manual assembly
processes. It is not a technology that involves assembly by low cost
labor in Indonesia or in Mexico. The handling and placement of
small resistors and capacitors, and the precise location and down-
ward positioning of IC packages both require machine performance. This is a technology of machines, of automatic screen printing,
and of mass assembly. Assembly and soldering of single ICs using
hot air or lasers may still be possible, but given the fairly low cost of
whole-board soldering, the alternatives are now less than attractive.

That's the bad news. The good news is that through the use of the
placement machines and mass soldering, labor costs will decrease
considerably. For the U.S., this provides the opportunity to pull
back assembly operations that had gone overseas. Even for Japan,
it reduces the threat from Korea and Taiwan. Typically, moving to
a new technology is not done for cost reduction purposes but for
superior system performance. Once that move is accomplished, the
lower cost is a much appreciated benefit.

For cost-performance electronics, surface-attach is not always a
viable alternative to pin grid array. If most of the board remains
with DIP components, the use of a pin grid array for the new VLSI
chips is an acceptable strategy, as is the use of leadless packages in

Fig. 1—Leadless chip carrier military board design, with matching
TCE materials. (RCA)

sockets. If over 90% of the boards are assembled by doing business
as usual, it makes little difference how the VLSI chips are mounted.
It is only when the whole mounting technology is converted to
surface-attach that the reduction in area, the simplification of the
printed circuit interconnection board, and the shift to automatic
assembly occurs. It is this application of surface mounting
technology that will be examined in this report. An understanding
of some of the difficulties encountered in making the shift to new
operations will be helpful. Make no mistake, there are problems in
converting from familiar and long understood operating techniques to a new set of operating procedures. These problems are
severe, and are discussed below.

**Leadless IC Packages and Their Limitations**

The first surface-attach IC packages to come into widespread use
were the leadless ceramic chip carriers. They fulfilled an important
and continuing need in military applications (Fig. 1), but are now

**c**
Figure 21-1.  *Continued*

# INDUSTRY NEWSMAKER

ELECTRONIC DESIGN EXCLUSIVE

# VME card gears VAX for data acquisition

*The MicroVAX II and other computers can breeze through real-time applications with a new translator that hastens the I/O process.*

### BY STEPHAN OHR

Equipping a MicroVAX II for real-time data acquisition or industrial process control has never been easy. The I/O bus of Digital Equipment's minicomputers is generally too slow to support real-time data transfers. Moreover, it has trouble recognizing any remote intelligent devices that can preprocess data.

But a new VME-to-VAX translator from Creative Electronic Systems SA has cleared the path. The SBC8221 is a single-board computer that resides on a VMEbus backplane, but furnishes a ribbon-cable connection to DEC's Q22 bus (Fig. 1). The bus, which multiplexes a 16-bit data bus with a 22-bit address bus, serves as the primary I/O channel for DEC's MicroVAX II and LSI-11 computers. The new translator thus forms the foundation for a remote data-acquisition system using the MicroVAX II or other VAX minicomputers as the host.

A separate card, which rides piggyback on one side of the translator card, sets up the interface with the Q22 bus (Fig. 2). If that DRV1B card is exchanged for a DR11-WA card, the translator can communicate over the Unibus—a non-multiplexed address and data bus, which is the primary I/O channel for the VAX-11/730, 750, and 780 machines, as well as for the PDP-11.

With either interface card, the translator can be an unintelligent node in the bus address space or a master for a VMEbus data-

acquisition system. In a third case, it can serve as the slave to a VMEbus master—a 68020 CPU, for example—to send data to disks, tapes, printers, terminals, and other peripherals on the Q22 bus, or messages to its VAX hosts.

In all cases, the translator's operation is meant to be transparent and painless to the DEC user. The board includes all software drivers, so that data-acquisition software can be developed on the VAX and down-

loaded into VMEbus satellites.

The translator is intended to capitalize on the large, worldwide installed base of both VAX and VME hardware. Indeed, DEC's recently introduced VAX-BI bus aims at data-acquisition applications with high data-transfer rates. But the founder and technical director of Creative Electronic Systems (CES), François Worm, predicts that it may be some time before dedicated hardware for the VAX-BI bus appears.

Mapping VME hardware directly onto the DEC Q22 bus is hindered by the technical problems. First, the bus does not accommodate multiple masters; it is strictly under the control of the VAX host. The VAX usually responds to interrupts, but it does not recognize intelligent devices in its I/O channels. Thus, any sort of front-end processor for data acquisition must disguise its intelligence. The processor-cannot battle the host for control of the bus.

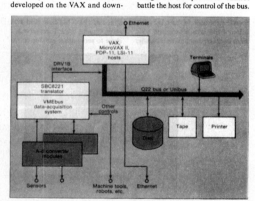

**1. The SBC8221 translator puts multiple CPUs on the DEC Q22 bus. With the translator, a data acquisition system or process control system can function simply as a bus address for a MicroVAX II or as a Unibus address for a VAX11/730, 750, or 780. Alternatively, the VME system can control peripheral devices on the Q22 bus.**

**D**

**Figure 21-1.** *Continued*

- *Visual Aids.* Tables, graphs, drawings, and so on.
- *White Space.* The blank areas of the page that contain no printed information.

When looking at a page, graphic designers see each of these elements in the abstract—as one of the building blocks for a page—not as particular words or illustrations. You need to do the same thing to follow the two-step strategy for designing pages that is explained in the first two guidelines.

<table>
<tr><td>Guideline<br><br>1</td><td>

## Create a Grid To Serve as the Visual Framework for Your Pages

As you design your pages, your aim is to create simple, clear, consistent, and meaningful relationships among the prose, visual aids, and headings and titles. The first step in doing that is to decide which areas on your page will be occupied by your visual materials and which will be devoted to the white space that defines the borders of those materials. Graphic designers do that by drawing a *grid* of vertical and horizontal lines. The designers place their visual materials within the rectangles formed by these lines. The areas around the rectangles are left to white space.

</td></tr>
</table>

### A Grid You Know Well

Actually, the process of creating a grid is already familiar to you. You use a grid every time you type a term paper. When you establish your four margins (top, bottom, left, right), you create a *communication area* to be occupied by your prose and other visual materials, as shown in Figure 21-2. The margins are devoted to white space.

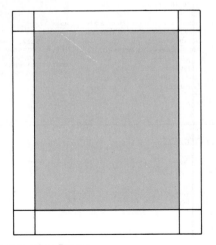

Figure 21-2.  Communication Area of a Page.

You can create more complex grids for your term papers by adding other grid lines. For example, if you wish to include indented lists or indented blocks of quotations, you establish additional grid lines for the left-hand and right-hand borders of that material. Similarly, if you indent the first line of each paragraph, you are in effect creating another grid line that defines the left-hand border of the first lines of all your paragraphs. Figure 21-3 shows the grid lines of a typical term paper.

## Creating a Multicolumn Grid

To create a multicolumn grid, you follow a process similar to the one you use to create the single-column grid for a term paper. First, establish your margins to define the communication area of your page. Next, divide this communication area by means of narrow, vertical *gutters* of white space. To be effective, gutters must be at least as wide as three letters of your text. Figures 21-4 and 21-5 show pages with gutters for two-column and three-column grids, respectively.

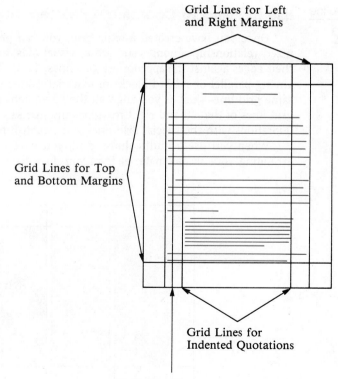

Grid Lines for Left
and Right Margins

Grid Lines for Top
and Bottom Margins

Grid Lines for
Indented Quotations

Grid Line for
Paragraph Indention

**Figure 21-3.**   Grid Lines of a Typical Term Paper.

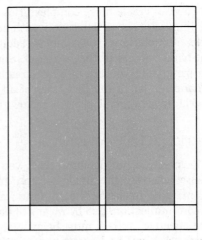

**Figure 21-4.**   A Single Gutter Dividing the Communication Area into Two Columns.

## Use the Grid To Coordinate Related Visual Elements

Once you have created a basic grid, you can provide a clear, well-organized relationship among your prose, visual aids, and headings by aligning their edges against the appropriate grid lines. To establish a visual connection among parallel or related blocks of material, you simply align them with the same grid line—just as you align all the paragraphs in a term paper with the grid lines of the left and right margins, and just as you align all the indented quotations with the special grid lines you establish for them.

When you use a multicolumn grid, you have much more flexibility for arranging your visual material than you do with a single-column grid. The

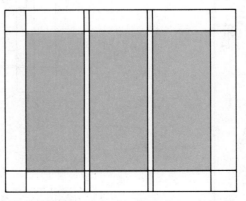

**Figure 21-5.**   Two Gutters Dividing the Communication Area into Three Columns.

following paragraphs provide some suggestions about ways you can use that flexibility while still maintaining a clear, consistent page design.

## Aligning Text and Visual Aids

In multicolumn designs, you can keep your text and visual aids within a single column or you can extend some elements across one or more additional columns. Notice that in the two examples on the right-hand side of Figure 21-6, the visual elements that cross a gutter go all the way across the next column. In general, when you extend your prose or a visual aid across grid columns, it should occupy the full width of each column.

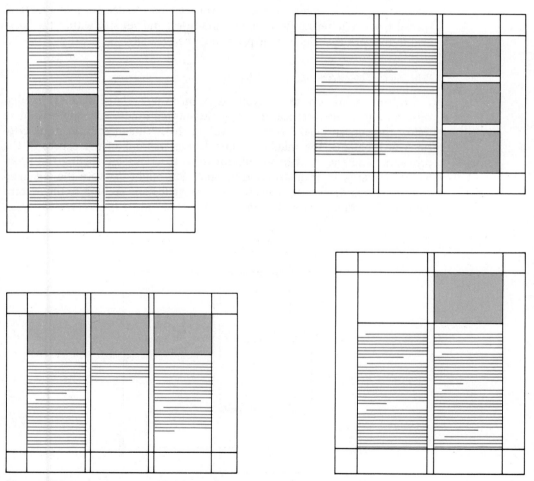

**Figure 21-6.**  A Variety of Placements of Text and Visual Aids within Multicolumn Formats.

## Aligning Titles and Headings

You can use your grid pattern to establish a visual relationship between your headings and titles and the other elements on your page. To do that, place the headings and titles centered, flush left, or flush right in the appropriate grid columns. A title or heading is *flush left* if it abuts the left-most grid line of the column or columns it labels; it is *flush right* if it abuts the right-most grid line of that column or columns. Figure 21-7 shows headings centered, flush left, and flush right in a single column.

Figure 21-8 shows two page designs in which major titles or headings span more than one column. To see other examples of the placement of headings and titles, look at the other sample page designs in the figures and exercises of this chapter. Notice that you can use larger sizes of type and special types of lettering for headings and titles, and yet stay within the basic grid framework that holds your pages together visually.

## Aligning Related Material in Adjacent Columns

In some communications, you may want to put related material in adjacent columns. For instance, in a troubleshooting guide, you might want to follow the conventional format of devoting the left-hand column to descriptions of problems, the middle column to explanations of the causes of the problems, and the right-hand column to instructions for solving the problems. Similarly, in an instruction manual, you might dedicate the left-hand column to text describing each step and the right-hand column to the accompanying illustrations of each step.

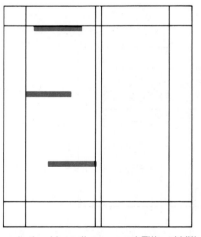

**Figure 21-7.**   Alternative Placements for Headings and Titles Within a Column.

In those situations, you want to help your readers quickly match the related material in the adjacent columns—Problem A with its cause and solution, Instruction 17 with its illustration. To do that, you can establish *horizontal gutters.*

Figure 21-9 (page 460) shows two strategies for using horizontal gutters. In each case, the corresponding materials in adjacent columns use the same top grid line. You can determine the placement of the grid lines in either of two ways. You may place them at regular intervals, as in the left-hand example in Figure 21-9. Or, you may place them at the end of the longest entry in each cross-column set of associated entries; such placement is shown in the right-hand example in Figure 21-9.

As the examples show, using horizontal grid lines to align associated materials in adjacent columns can leave considerable white space on the page. However, this white space is not wasted space. It helps readers see which text goes with which illustrations. Compare the two page designs shown in Figure 21-10 (page 460). In their left-hand columns, both pages contain instructions for three steps in a procedure; in their right-hand columns, both contain the three illustrations that accompany those steps. The difference between the two pages is that one uses horizontal gutters to coordinate each step with its accompanying illustration, whereas the other does not. Although the uncoordinated design has the apparent advantage of providing room at the bottom of the left-hand column to print additional steps, it is more difficult for readers to use. That page would be even *more* difficult if the steps added to the bottom of the left-hand column had accompanying illustrations that readers would have to turn the page to see.

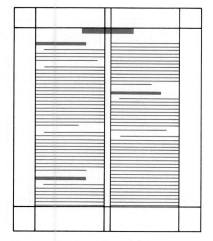
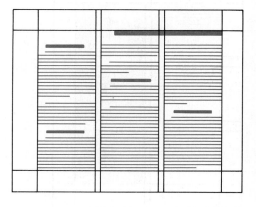

**Figure 21-8.**   Some Alternative Placements for Headings and Titles that Span Columns.

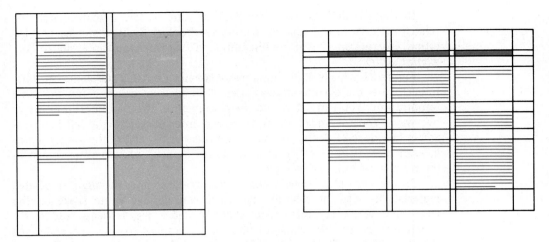

**Figure 21-9.** Two Ways of Coordinating Information by Means of Horizontal Gutters.

It is important for you to note that the use of horizontal gutters applies mainly to communications where each block of text has an associated illustration. In communications where you will include illustrations for some pieces of text but not for others, you may need to consider page designs where horizontal gutters are *not* used to align corresponding text and illus-

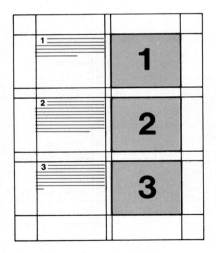

Coordinated

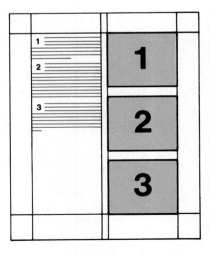

Uncoordinated

**Figure 21-10.** Comparison of a Page that Coordinates Text and Illustrations by Use of Horizontal Gutters with a Page that Does Not.

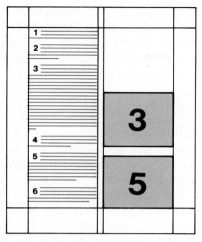

**Figure 21-11.**  Sample Design for a Page where Some Blocks of Prose Do Not Have Associated Illustrations.

trations. One example of such a design is shown in Figure 21-11. This design uses a two-column grid like that in Figure 21-10. Notice that although the illustrations are placed across from their associated steps, the associated steps and illustrations do not necessarily use the same top grid line.

Another alternative design (Figure 21-12) mixes text and illustrations in both columns, placing each illustration immediately after the text that concerns it.

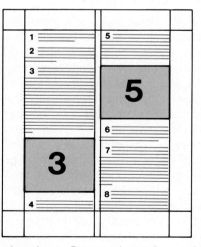

**Figure 21-12.**  Another Sample Design for a Page where Some Blocks of Prose Do Not Have Associated Illustrations.

In sum, where your communication includes both text and illustrations (or other kinds of associated materials), you should consider carefully your strategy for coordinating the two. In many situations you can coordinate them by means of horizontal gutters, although other alternatives may be more effective in other situations.

## A Note on Creativity

As you can see from looking at the various figures in this chapter, using grids allows you considerable room to creatively design pages that will help you achieve your communication objectives. Experienced graphic designers exercise even greater freedom by occasionally ignoring grid lines with elements they desire to emphasize, while keeping the rest of their design within their grid framework. They usually do that when making elaborate designs, such as advertisements. Of course, you can experiment in that way too, although for any of the things you are likely to write at work, you will have a perfectly effective design if you rely entirely upon the grid framework.

# INTRODUCTION TO GUIDELINE 3

The two guidelines you have just read explain a two-step procedure you can use to design a single page effectively. The next guideline concerns multi-page communications and provides advice about the relationships you establish among the designs of the various pages in them.

Guideline

3

## Use the Same Design for All Pages That Contain the Same Types of Information

In the visual design of a communication, consistency is both aesthetically pleasing and functional. Pages built upon the same grid pattern make an attractive group because of the visual harmony among them. Further, pages using the same grids help readers read efficiently, a point illustrated in Figure 21-13, which shows two pages from an instruction manual for servicing motorcycles. The student who wrote the manual used this same layout for a sequence of twenty pages. In that layout, readers can find the name of the test in the same places on every page and its purpose in the same places on every page; similarly, readers can find the expected result of the diagnostic activity and also the way to correct problems. Therefore, when turning to a new page readers can readily find the information they want without searching for it: they can simply direct their eyes to the appropriate place on the page.

Some documents contain many different kinds of information. For them, it is not reasonable to use the same page design for *every* page. For example, the motorcycle manual just mentioned has an introduction and also a page of specifications, each needing its own page design. However, even when you

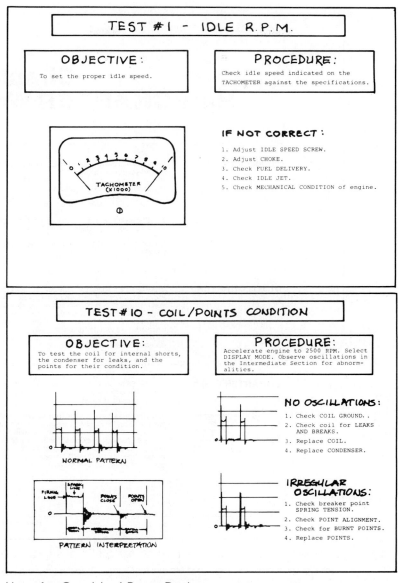

**Figure 21-13.** Use of a Consistent Page Design.

use more than one page design in a single communication, you can provide visual unity to your writing by using *related* designs. For example, if you design a major section of your communication so that the long side of the paper (the 11-inch side) is at the top, use the same page orientation through-

# Chapter 3
# How CHARTER Works

**Introduction**

The purpose of this chapter is to describe the three activities you complete to produce a graph:

1.  Entering the **data** to be graphed

2.  Specifying the **format** of the graph

3.  **Displaying** the graph on your workstation screen or **printing** it on the 5577 High-Density Matrix Printer

In this chapter you will also learn several ways to use CHARTER in your office.

**What is CHARTER?**

CHARTER is a powerful software tool that allows you to display and print black and white business graphs **electronically** rather than manually. With electronic drawing, you can **quickly** see what type of graph (line, bar, pie, word) is most effective for a presentation or report because you can **easily** produce the same information in a variety of graph forms without time-consuming hand drawing (much as you can reformat a word processing document without extensive retyping).

**A CHARTER Graph**

A CHARTER graph consists of two components: a **data list** and a **format.** This is similar to the way a word processing document consists of text and format lines.

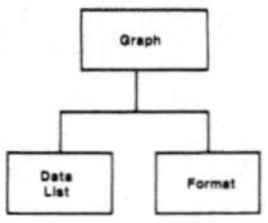

The data list is a group of tables that you reference to plot (draw) the data points for any chart in the graph. The format is defined by values that specify the appearance of elements in the chart, for example, chart type, title, border, size, position on the page, type of line, and labels. You specify the format values in fields in the format menus. When you tell CHARTER to draw a graph, the system applies the format values to the data.

How CHARTER Works    3-1

**Figure 21-14.** Consistent Use of a Single Design for All Pages that Begin Chapters. From Wang Laboratories, *Instruction Manual for Charter Graphics Program* (Lowell, Massachusetts: Wang Laboratories). Used with permission.

out. Otherwise, your readers will need to turn the book to this side and that as they proceed from section to section. Similarly, if you use a two-column page design in one major section, use a two-column design in all sections.

A special instance of related pages arises in long documents (such as reports, proposals, and instruction manuals) that contain more than one ma-

# Chapter 6
# Advanced Operations

**Introduction**

The purpose of this chapter is to teach you advanced Wang CHARTER operations. These operations include:

- Producing a line chart, a bar chart, and a pie chart using the **same** data set

- Adding labels, legends, and annotations to charts

- Positioning multiple charts on one page

- Adding sophisticated attributes to charts

- Converting a table in a word processing document into a data set

- Tailoring the format menus so that some fields do not appear on the screen

- Storing copies of your graphs

**Remember**

The HELP key displays information about a specific menu or field.

The EXIT key lets you exit quickly from any format menu. When you return to the graph, the system displays the menu from which you exited.

Press CANCEL to move back through the menus.

Use GO TO to move from chart to chart, axis to axis, plot to plot, line to line, or bar set to bar set.

To go directly to the X- or Y-axis of a chart, press GO TO (from the X-Y menu, X-Axis menu, or Y-Axis menu), enter the chart number, and type "x" or "y".

Press ERASE or COMMAND and DELETE to remove all characters from the current cursor position.

You do not have to press EXEC after entering data in a menu.

You can display an index quickly by pressing the INDEX key from any Graph, Library, or Volume field.

Advanced Operations    6-1

**Figure 21-14.** *Continued*

jor section. In these documents, writers often treat each section as a new chapter; that is, they begin each section at the top of a new page and they put chapter titles on these pages. Consequently, all first pages of chapters look different from the pages that don't begin chapters—but all the first pages look like one another. See Figure 21-14 for some examples.

# PRACTICAL PROCEDURES FOR DESIGNING PAGES

You do not need any special technical skills to follow the three guidelines for designing your pages. Here are some practical procedures that will help you design pages by using resources available to almost everyone: a pen, a typewriter, press-on letters, scissors, clear tape, and a photocopy machine.

The first four steps described below will help you create an effective grid and basic design for your pages:

1. *Determine the amounts of text and illustration you will include.* If you are coordinating specific blocks of text or visual aids across adjacent columns by using horizontal gutters, find the longest passage of text that will accompany one illustration. Use that passage to determine the amount of text space you will provide for the prose that accompanies the other illustrations.

2. *Draw thumbnail sketches of a variety of grid patterns.* Figure 21-15 shows nine thumbnail sketches exploring ways of designing a single page of a particular communication. Sketches like these can help you quickly assess the various grid patterns that you might use. Such assessment is important because there is no "correct" or "best" design for all pages.

   Even when making your thumbnail sketches, keep in mind the purpose and readers of your communication, so that you will know what material to emphasize and what relationships to make clear. Think, too, about how much detail your illustrations will include so that you can plan to make your illustrations large enough to be legible. If you are planning to use coordinated columns, keep in mind the size of your illustrations and the lengths of your text entries so that you can determine whether or not to place your horizontal grid lines at fixed intervals.

3. *Pick the best thumbnail sketch for your purpose.* Make this judgment by thinking how your readers will react, moment-by-moment, as they read a communication presented with each design.

4. *Make a full-size mock-up page.* Fill in the illustration areas with colored paper. Indicate lines of text with straight lines of the appropriate length, and indicate headings with heavier lines. If the design you have selected still seems to work, proceed to the next step; if not, try a grid represented by one of your other thumbnail sketches.

Once you have created a mock-up for an effective design, you are ready to begin making your finished pages. The following steps tell you how to do that:

5. *On a blank sheet of paper, carefully draw the grid in dark pen.*
6. *Tape the grid down on a desk or table.*

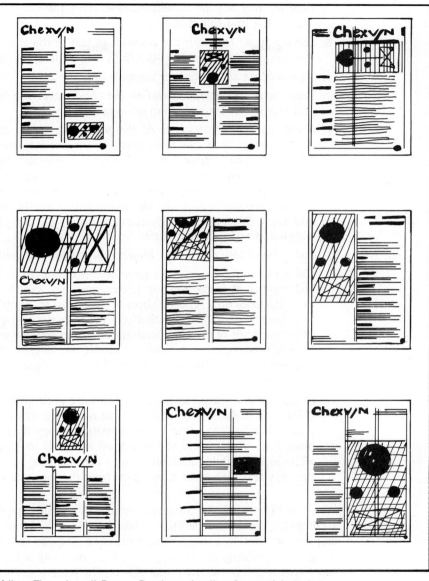

**Figure 21-15.**   Nine Thumbnail Page Designs for the Same Material.

 Courtesy of Professor Joseph L. Cox III.

7. *Type the text of your communication.* Type the text columns exactly as wide as the grid columns.

8. *Cut the typed text.* Cut the columns of text at the appropriate places so that the various text segments will fit into their grid areas.

9. *Tape a blank sheet of paper squarely over the grid.*

10. *Tape the text segments neatly on the blank page.* Use as your guide the grid that shows through from underneath.

11. *Create your illustrations to fit the grid areas.* If you are creating your own drawings, you can draw them to the needed size. If it would be difficult for you to draw that small, draw the illustrations larger and then use a photocopy machine to reduce them to the appropriate size. Often, you can make your page appear crisper if you draw a box around your illustrations; the lines of the box would align directly on the appropriate grid lines. See Figure 21-16 (pages 470-471) for comparisons of pages with boxed and unboxed figures.

12. *Cut out and tape down your illustrations as you did your text segments.*

13. *Use press-on letters to make your headings and titles.* Unless you have had practice at using press-on letters, you may want to make the headings and titles on a separate sheet of paper, cut them out, and tape them onto your page in the way that you did with the text and illustrations. Other methods of creating headings and titles include using Croy lettering machines or other devices that enable you to print strips of letters in large-sized type. (For more information on the use of press-on letters, see Chapter 14.)

14. *Photocopy your taped-up pages.* If you use a good-quality photocopy machine, the finished product will have much the same appearance as a page printed on an offset press. Sometimes copy machines will record black lines at the edges of taped-down materials. If that happens, use a white masking fluid (available at most copy centers) to cover the lines on the copy and then make a new copy from the touched-up version.

At work, you may be fortunate enough to receive page-design help from a computer system for *desktop publishing.* Such a system permits you to prepare the prose by using a word-processing program and the visual aids by using a computer-graphics program. Then, by using a third program, you can combine the prose and visual aids on the same page, arranging them as you think best. Even if you are able to use such a computer system, you will still have to create the actual design of your pages. Therefore, the guidelines in this chapter will remain useful to you.

You may also be fortunate enough to have the assistance of a graphic artist. This is especially likely if you write reports to stockholders, instructions to be packed with your company's products, or other communications in which your employer wants to make the best impression it can. A graphic artist will have much more sophisticated equipment than the scissors and tape described above, as well as considerable training, experience, and talent. Even so, you will benefit from knowing the principles of page design, because that knowledge will help you work smoothly and productively with the graphic artist.

## SUMMARY

This chapter presents three guidelines for designing pages that will help you to emphasize your most important material, enable your readers to read efficiently, and encourage your readers to be receptive to your message.

1. *Create a grid to serve as the visual framework for your pages.* Begin by establishing your margins, which define the communication area of your page. Draw additional grid lines to define other borders and gutters.

2. *Use the grid to coordinate related visual elements.* Coordinate associated elements in adjacent columns by aligning their tops on the same horizontal grid line. Generally, place titles and headings flush left, flush right, or centered between the appropriate grid lines.

3. *Use the same design for all pages that contain the same types of information.* Consistency in page design is esthetically pleasing and it helps readers find quickly the information they want.

## EXERCISES

1. Describe the page design used in each of the sample pages shown in Figure 21-1. Identify the number of columns; the placement of prose, headings, and visual aids; and other important features of the design.

2. Figure 21-17 (pages 472-477) shows five package inserts for a prescription medication. All contain the same information, but each was prepared by a different graphic designer. Describe the page designs used in each. Guess at the purpose each designer had in mind. Which design do you think works best? Worst? Why?

3. Look again at Figure 21-15. It contains nine thumbnail sketches, all for the same material. Describe each design in terms of the number of columns, the placement of the title and headings, and the location of the illustration. For each, state what element of the page (title, illustration, headings, and so on) receives primary emphasis.

4. Find three different page designs. (Use no more than one from a popular magazine; look at instructions, insurance policies, leases, company brochures, technical reports, and the like.) If your instructor asks, find all

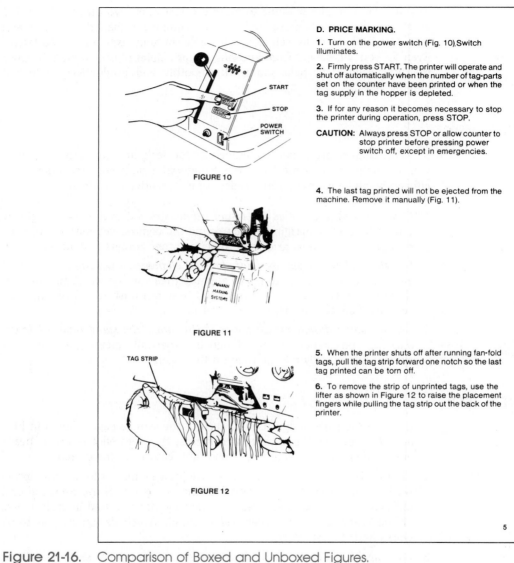

**D. PRICE MARKING.**

**1.** Turn on the power switch (Fig. 10).Switch illuminates.

**2.** Firmly press START. The printer will operate and shut off automatically when the number of tag-parts set on the counter have been printed or when the tag supply in the hopper is depleted.

**3.** If for any reason it becomes necessary to stop the printer during operation, press STOP.

**CAUTION:** Always press STOP or allow counter to stop printer before pressing power switch off, except in emergencies.

**4.** The last tag printed will not be ejected from the machine. Remove it manually (Fig. 11).

**5.** When the printer shuts off after running fan-fold tags, pull the tag strip forward one notch so the last tag printed can be torn off.

**6.** To remove the strip of unprinted tags, use the lifter as shown in Figure 12 to raise the placement fingers while pulling the tag strip out the back of the printer.

START
STOP
POWER SWITCH

FIGURE 10

FIGURE 11

TAG STRIP

FIGURE 12

5

**Figure 21-16.** Comparison of Boxed and Unboxed Figures.

From Monarch Marking, *Operating Instructions for Model 100 Dial-A-Pricer Printer* (Dayton, Ohio: Monarch Marking Systems, Inc.).

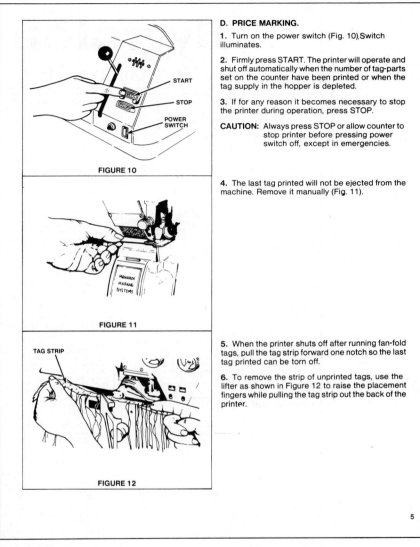

**FIGURE 10**

**FIGURE 11**

TAG STRIP

**FIGURE 12**

**D. PRICE MARKING.**

**1.** Turn on the power switch (Fig. 10),Switch illuminates.

**2.** Firmly press START. The printer will operate and shut off automatically when the number of tag-parts set on the counter have been printed or when the tag supply in the hopper is depleted.

**3.** If for any reason it becomes necessary to stop the printer during operation, press STOP.

**CAUTION:** Always press STOP or allow counter to stop printer before pressing power switch off, except in emergencies.

**4.** The last tag printed will not be ejected from the machine. Remove it manually (Fig. 11).

**5.** When the printer shuts off after running fan-fold tags, pull the tag strip forward one notch so the last tag printed can be torn off.

**6.** To remove the strip of unprinted tags, use the lifter as shown in Figure 12 to raise the placement fingers while pulling the tag strip out the back of the printer.

5

**Figure 21-16.**   *Continued*

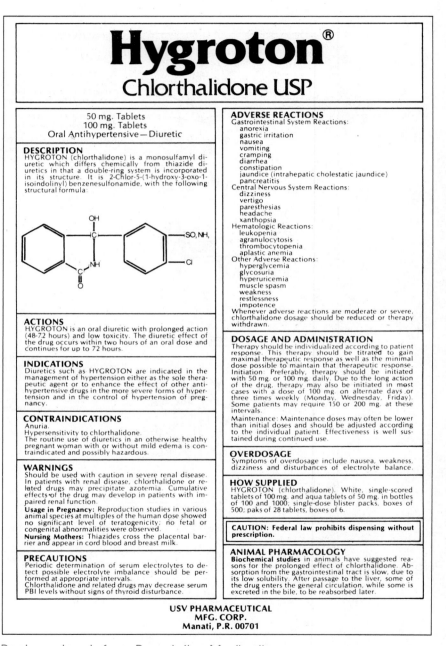

# Hygroton®

## Chlorthalidone USP

50 mg. Tablets
100 mg. Tablets
Oral Antihypertensive — Diuretic

### DESCRIPTION
HYGROTON (chlorthalidone) is a monosulfamyl diuretic which differs chemically from thiazide diuretics in that a double-ring system is incorporated in its structure. It is 2-Chlor-5-(1-hydroxy-3-oxo-1-isoindolinyl) benzenesulfonamide, with the following structural formula:

### ACTIONS
HYGROTON is an oral diuretic with prolonged action (48-72 hours) and low toxicity. The diuretic effect of the drug occurs within two hours of an oral dose and continues for up to 72 hours.

### INDICATIONS
Diuretics such as HYGROTON are indicated in the management of hypertension either as the sole therapeutic agent or to enhance the effect of other antihypertensive drugs in the more severe forms of hypertension and in the control of hypertension of pregnancy.

### CONTRAINDICATIONS
Anuria.
Hypersensitivity to chlorthalidone.
The routine use of diuretics in an otherwise healthy pregnant woman with or without mild edema is contraindicated and possibly hazardous.

### WARNINGS
Should be used with caution in severe renal disease. In patients with renal disease, chlorthalidone or related drugs may precipitate azotemia. Cumulative effects of the drug may develop in patients with impaired renal function.
**Usage in Pregnancy:** Reproduction studies in various animal species at multiples of the human dose showed no significant level of teratogenicity; no fetal or congenital abnormalities were observed.
**Nursing Mothers:** Thiazides cross the placental barrier and appear in cord blood and breast milk.

### PRECAUTIONS
Periodic determination of serum electrolytes to detect possible electrolyte imbalance should be performed at appropriate intervals.
Chlorthalidone and related drugs may decrease serum PBI levels without signs of thyroid disturbance.

### ADVERSE REACTIONS
Gastrointestinal System Reactions:
  anorexia
  gastric irritation
  nausea
  vomiting
  cramping
  diarrhea
  constipation
  jaundice (intrahepatic cholestatic jaundice)
  pancreatitis
Central Nervous System Reactions:
  dizziness
  vertigo
  paresthesias
  headache
  xanthopsia
Hematologic Reactions:
  leukopenia
  agranulocytosis
  thrombocytopenia
  aplastic anemia
Other Adverse Reactions:
  hyperglycemia
  glycosuria
  hyperuricemia
  muscle spasm
  weakness
  restlessness
  impotence
Whenever adverse reactions are moderate or severe, chlorthalidone dosage should be reduced or therapy withdrawn.

### DOSAGE AND ADMINISTRATION
Therapy should be individualized according to patient response. This therapy should be titrated to gain maximal therapeutic response as well as the minimal dose possible to maintain that therapeutic response.
Initiation: Preferably, therapy should be initiated with 50 mg. or 100 mg. daily. Due to the long action of the drug, therapy may also be initiated in most cases with a dose of 100 mg. on alternate days or three times weekly (Monday, Wednesday, Friday). Some patients may require 150 or 200 mg. at these intervals.
Maintenance: Maintenance doses may often be lower than initial doses and should be adjusted according to the individual patient. Effectiveness is well sustained during continued use.

### OVERDOSAGE
Symptoms of overdosage include nausea, weakness, dizziness and disturbances of electrolyte balance.

### HOW SUPPLIED
HYGROTON (chlorthalidone). White, single-scored tablets of 100 mg. and aqua tablets of 50 mg. in bottles of 100 and 1000; single-dose blister packs, boxes of 500; paks of 28 tablets, boxes of 6.

CAUTION: Federal law prohibits dispensing without prescription.

### ANIMAL PHARMACOLOGY
**Biochemical studies** in animals have suggested reasons for the prolonged effect of chlorthalidone. Absorption from the gastrointestinal tract is slow, due to its low solubility. After passage to the liver, some of the drug enters the general circulation, while some is excreted in the bile, to be reabsorbed later.

USV PHARMACEUTICAL
MFG. CORP.
Manati, P.R. 00701

**A**

**Figure 21-17.** Package Inserts for a Prescription Medication.

Courtesy of USV Pharmaceuticals.

# HYGROTON®   Chlorthalidone USP

50 mg. Tablets    Oral Antihypertensive-
100 mg. Tablets   Diuretic

**Description.** HYGROTON (chlorthalidone) is a monosulfamyl diuretic which differs chemically from thiazide diuretics in that a double-ring system is incorporated in its structure. It is 2-Chlor-5-(1-hydroxy-3-oxo-1-isoindolinyl) benzenesulfonamide, with the following structural formula:

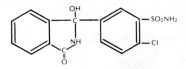

**Action.** HYGROTON is an oral diuretic with prolonged action (48-72 hours) and low toxicity. The diuretic effect of the drug occurs within two hours of an oral dose and continues for up to 72 hours.

**Indications.** Diuretics such as HYGROTON are indicated in the management of hypertension either as the sole therapeutic agent or to enhance the effect of other antihypertensive drugs in the more severe forms of hypertension and in the control of hypertension of pregnancy.

**Contraindications.** Anuria. Hypersensitivity to chlorthalidone. The routine use of diuretics in an otherwise healthy pregnant woman with or without mild edema is contraindicated and possibly hazardous.

**Warnings.** Should be used with caution in severe renal disease. In patients with renal disease, chlorthalidone or related drugs may precipitate azotemia. Cumulative effects of the drug may develop in patients with impaired renal function.

*Usage in Pregnancy:* Reproduction studies in various animal species at multiples of the human dose showed no significant level of teratogenicity; no fetal or congenital abnormalities were observed.

*Nursing Mothers:* Thiazides cross the placental barrier and appear in cord blood and breast milk.

**Precautions.** Periodic determination of serum electrolytes to detect possible electrolyte imbalance should be performed at appropriate intervals. Chlorthalidone and related drugs may decrease serum PBI levels without signs of thyroid disturbance.

**Adverse Reactions.**
*Gastrointestinal System Reactions:*
- anorexia
- gastric irritation
- nausea
- vomiting
- cramping
- diarrhea
- constipation
- jaundice (intrahepatic cholestatic jaundice)
- pancreatitis

*Central Nervous System Reactions:*
- dizziness
- vertigo
- paresthesias
- headache
- xanthopsia

*Hematologic Reactions:*
- leukopenia
- agranulocytosis
- thrombocytopenia
- aplastic anemia

*Other Adverse Reactions:*
- hyperglycemia
- glycosuria
- hyperuricemia
- muscle spasm
- weakness
- restlessness
- impotence

Whenever adverse reactions are moderate or severe, chlorthalidone dosage should be reduced or therapy withdrawn.

**Dosage and Administration.** Therapy should be individualized according to patient response. This therapy should be titrated to gain maximal therapeutic response as well as the minimal dose possible to maintain that therapeutic response.

*Initiation:* Preferably, therapy should be initiated with 50 mg. or 100 mg. daily. Due to the long action of the drug, therapy may also be initiated in most cases with a dose of 100 mg. on alternate days or three times weekly (Monday, Wednesday, Friday). Some patients may require 150 or 200 mg. at these intervals.

*Maintenance:* Maintenance doses may often be lower than initial doses and should be adjusted according to the individual patient. Effectiveness is well sustained during continued use.

**Overdosage.** Symptoms of overdosage include nausea, weakness, dizziness and disturbances of electrolyte balance.

**How Supplied.** HYGROTON (chlorthalidone). White, single-scored tablets of 100 mg. and aqua tablets of 50 mg. in bottles of 100 and 1000; single-dose blister packs, boxes of 500; Paks of 28 tablets, boxes of 6.

*Caution:* Federal law prohibits dispensing without prescription.

**Animal Pharmacology.** *Biochemical studies in animals have suggested reasons for the prolonged effect of chlorthalidone. Absorption from the gastrointestinal tract is slow, due to its low solubility. After passage to the lever, some of the drug enters the general circulation, while some is excreted in the bile, to be reabsorbed later.*

USV Pharmaceutical Mfg. Corp.
Manati, PR  00701 • (914) 779-6300

**B**

**Figure 21-17.**   *Continued*

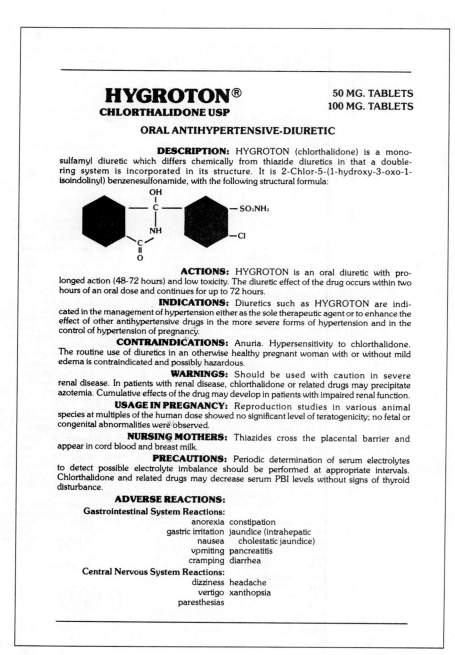

# HYGROTON®
## CHLORTHALIDONE USP

**50 MG. TABLETS**
**100 MG. TABLETS**

### ORAL ANTIHYPERTENSIVE-DIURETIC

**DESCRIPTION:** HYGROTON (chlorthalidone) is a mono-sulfamyl diuretic which differs chemically from thiazide diuretics in that a double-ring system is incorporated in its structure. It is 2-Chlor-5-(1-hydroxy-3-oxo-1-isoindolinyl) benzenesulfonamide, with the following structural formula:

**ACTIONS:** HYGROTON is an oral diuretic with pro-longed action (48-72 hours) and low toxicity. The diuretic effect of the drug occurs within two hours of an oral dose and continues for up to 72 hours.

**INDICATIONS:** Diuretics such as HYGROTON are indi-cated in the management of hypertension either as the sole therapeutic agent or to enhance the effect of other antihypertensive drugs in the more severe forms of hypertension and in the control of hypertension of pregnancy.

**CONTRAINDICATIONS:** Anuria. Hypersensitivity to chlorthalidone. The routine use of diuretics in an otherwise healthy pregnant woman with or without mild edema is contraindicated and possibly hazardous.

**WARNINGS:** Should be used with caution in severe renal disease. In patients with renal disease, chlorthalidone or related drugs may precipitate azotemia. Cumulative effects of the drug may develop in patients with impaired renal function.

**USAGE IN PREGNANCY:** Reproduction studies in various animal species at multiples of the human dose showed no significant level of teratogenicity; no fetal or congenital abnormalities were observed.

**NURSING MOTHERS:** Thiazides cross the placental barrier and appear in cord blood and breast milk.

**PRECAUTIONS:** Periodic determination of serum electrolytes to detect possible electrolyte imbalance should be performed at appropriate intervals. Chlorthalidone and related drugs may decrease serum PBI levels without signs of thyroid disturbance.

**ADVERSE REACTIONS:**

**Gastrointestinal System Reactions:**

anorexia constipation
gastric irritation jaundice (intrahepatic
nausea cholestatic jaundice)
vomiting pancreatitis
cramping diarrhea

**Central Nervous System Reactions:**

dizziness headache
vertigo xanthopsia
paresthesias

**c**

**Figure 21-17.** *Continued*

# Hygroton®
chlorthalidone USP

| 50 mg. Tablets<br>100 mg. Tablets | **Oral**<br>**Antihypertensive-Diuretic** | |
|---|---|---|
| **Description** | **Hygroton** (chlorthalidone) is a monosulfamyl diuretic which differs chemically from thiazide diuretics in that a double-ring system is incorporated in its structure. It is 2-Chlor-5-(1-hydroxy-3-oxo-1-isoindolinyl) benzenesulfonamide, with the following structural formula: | |
| **Actions** | **Hygroton** is an oral diuretic with prolonged action (48-72 hours) and low toxicity. The diuretic effect of the drug occurs within two hours of an oral dose and continues for up to 72 hours | |
| **Indications** | Diuretics such as **Hygroton** are indicated in the management of hypertension either as the sole therapeutic agent or to enhance the effect of other | antihypertensive drugs in the more severe forms of hypertension and in the control of hypertension of pregnancy |
| **Contraindications** | Anuria<br>Hypersensitivity to chlorthalidone | The routine use of diuretics in an otherwise healthy pregnant woman with or without mild edema is contraindicated and possibly hazardous |
| **Warnings** | Should be used with caution in severe renal disease. In patients with renal disease, chlorthalidone or related drugs may precipitate azotemia | Cumulative effects of the drug may develop in patients with impaired renal function |
| Usage in Pregnancy: | Reproduction studies in various animal species at multiples of the human dose showed no significant level of teratogenicity. no fetal or congenital abnormalities were observed | |
| Nursing Mothers: | Thiazides cross the placental barrier and appear in cord blood and breast milk | |
| **Precautions** | Periodic determination of serum electrolytes to detect possible electrolyte imbalance should be performed at appropriate intervals | Chlorthalidone and related drugs may decrease serum PBI levels without signs of thyroid disturbance |

| **Adverse Reactions**<br>Gastrointestinal<br>System Reactions: | anorexia<br>gastric irritation<br>nausea | vomiting<br>cramping<br>diarrhea | constipation<br>jaundice (intrahepatic<br>cholestatic jaundice) | pancreatitis |
|---|---|---|---|---|
| Central Nervous<br>System Reactions: | dizziness<br>vertigo<br>paresthesias | headache<br>xanthopsia | | |
| Hematologic Reactions: | leukopenia<br>agranulocytosis | thrombocytopenia<br>aplastic anemia | | |
| Other Adverse Reactions: | hyperglycemia<br>glycosuria<br>hyperuricemia<br>muscle spasm | weakness<br>restlessness<br>impotence | Whenever adverse reactions are moderate or severe. chlorthalidone dosage should be reduced or therapy withdrawn. | |

| **Dosage and Administration** | Therapy should be individualized according to patient response. This therapy should be titrated to gain maximal therapeutic response as well as | the minimal dose possible to maintain that therapeutic response. |
|---|---|---|
| Initiation: | Preferably, therapy should be initiated with 50 mg or 100 mg. daily. Due to the long action of the drug, therapy may also be initiated in most cases with a dose of 100 mg. on alternate days or | three times weekly (Monday, Wednesday, Friday) Some patients may require 150 or 200 mg. at these intervals. |
| Maintenance: | Maintenance doses may often be lower than initial doses and should be adjusted according to the | individual patient. Effectiveness is well sustained during continued use. |
| **Overdosage** | Symptoms of overdosage include nausea, weakness, dizziness and disturbances of electrolyte balance | |
| **How Supplied** | **Hygroton** (chlorthalidone). White, single-scored tablets of 100 mg. and aqua tablets of 50 mg. in | bottles of 100 and 1000; single dose blister packs. boxes of 500. Packs of 28 tablets. boxes of 6 |
| Caution: | Federal law prohibits dispensing without prescription | |

**D**

**Figure 21-17.** *Continued*

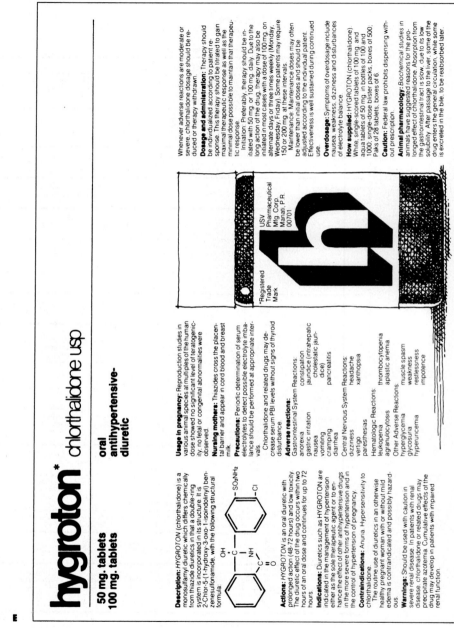

**Figure 21-17.** *Continued*

# HYGROTON®
## chlorthalidone usp | 50 mg Tablets, 100 mg Tablets | Oral Antihypertensive-Diuretic

**DESCRIPTION**
HYGROTON (chlorthalidone) is a monosulfamyl diuretic which differs chemically from thiazide diuretics in that a double-ring system is incorporated in its structure. It is 2-Chlor-5-(1-hydroxy-3-oxo-1-isoindolinyl) benzenesulfonamide, with the following structural formula:

**ACTIONS**
HYGROTON is an oral diuretic with prolonged action (48-72 hours) and low toxicity. The diuretic effect of the drug occurs within two hours of an oral dose and continues for up to 72 hours.

**INDICATIONS**
Diuretics such as HYGROTON are indicated in the management of hypertension either as the sole therapeutic agent or to enhance the effect of other antihypertensive drugs in the more severe forms of hypertension and in the control of hypertension of pregnancy.

**CONTRAINDICATIONS**
Anuria. Hypersensitivity to chlorthalidone. The routine use of diuretics in an otherwise healthy pregnant woman with or without mild edema is contraindicated and possibly hazardous.

**WARNINGS**
Should be used with caution in severe renal disease. In patients with renal disease, chlorthalidone or related drugs may precipitate azotemia. Cumulative effects of the drug may develop in patients with impaired renal function.

USAGE IN PREGNANCY: Reproduction studies in various animal species at multiples of the human dose showed no significant level of teratogenicity; no fetal or congenital abnormalities were observed.

NURSING MOTHERS: Thiazides cross the placental barrier and appear in cord blood and breast milk.

**PRECAUTIONS**
Periodic determination of serum electrolytes to detect possible electrolyte imbalance should be performed at appropriate intervals.

Chlorthalidone and related drugs may decrease serum PBI levels without signs of thyroid disturbance.

**ADVERSE REACTIONS**
Gastrointestinal System Reactions:
anorexia / constipaton
gastric irritation / jaundice
nausea / (intrahepatic
vomiting / cholestatic
cramping / jaundice)
diarrhea / pancreatitis

Central Nervous System Reactions:
dizziness / headache
vertigo / xanthopsia
paresthesias

Hematologic Reactions:
leukopenia / thrombocytopenia
agranulocytosis / aplastic anemia

Other Adverse Reactions:
hyperglycemia / muscle spasm
glycosuria / weakness
hyperuricemia / restlessness
impotence

Whenever adverse reactions are moderate or severe, chlorthalidone dosage should be reduced or therapy withdrawn.

**DOSAGE AND ADMINISTRATION**
Therapy should be individualized according to patient response. This therapy should be titrated to gain maximal therapeutic response as well as the minimal dose possible to maintain that therapeutic response.

Initiation: Preferably, therapy should be initiated with 50 mg or 100 mg daily. Due to the long action of the drug, therapy may also be initiated in most cases with a dose of 100 mg on alternate days or three times weekly (Monday, Wednesday, Friday). Some patients may require 150 or 200 mg at these intervals.

Maintenance: Maintenance doses may often be lower than initial doses and should be adjusted according to the individual patient. Effectiveness is well sustained during continued use.

**OVERDOSAGE**
Symptoms of overdosage include nausea, weakness, dizziness and disturbances of electrolyte balance.

**HOW SUPPLIED**
HYGROTON (chlorthalidone). White, single-scored tablets of 100 mg and aqua tablets of 50 mg in bottles of 100 and 1000; single-dose blister packs, boxes of 500. Paks of 28 tablets, boxes of 6. CAUTION: Federal law prohibits dispensing without prescription.

**ANIMAL PHARMACOLOGY**
Biochemical studies in animals have suggested reasons for the prolonged effect of chlorthalidone. Absorption from the gastrointestinal tract is slow, due to its low solubility. After passage to the liver, some of the drug enters the general circulation, while some is excreted in the bile, to be reabsorbed later.

USV PHARMACEUTICAL MFG. CORP.
Manati, P.R. 00701

F

**Figure 21-17.** *Continued*

the samples in documents related to your major. Photocopy one page illustrating each design. Then, for each page, describe what you think is the purpose of the document and discuss the specific features of the page design that help or hinder the document from achieving that purpose.

5. This exercise will provide you with practice at evaluating and improving an existing page design (Figure 21-18).

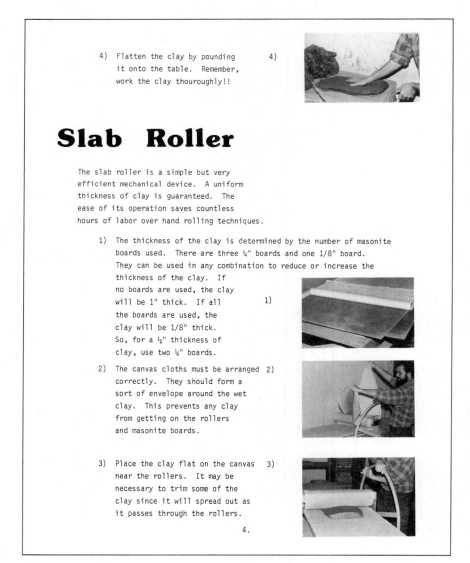

**Figure 21-18.** Sample Page for Use with Exercise 5.

A. List the ways in which the design of the page shown in Figure 21-18 could be improved.

B. Redesign the page by following Steps 1 through 4 of the "Practical Procedures for Designing Pages." Specifically, do the following:

- Create three alternative thumbnail sketches for the page.
- Create a full-size mock-up of the best design you can create.

C. Display your mock-up on the wall of your classroom along with the mock-ups prepared by the other students in your class. Decide which mock-ups work most effectively. Discuss the reasons.

6. Find an example of an ineffective page design. If your instructor requests, this might be from a set of instructions or from a document related to your major. Following Steps 1 through 4 of the "Practical Procedures for Designing Pages," draw thumbnail sketches and then make a mock-up of an improved page design for communicating the same material. Write a brief explanation of how you changed the original design—and why.

7. Create a multicolumn page design for a course project you are now preparing, or redesign an earlier project using a multicolumn design.

# Evaluating

When you finish drafting a communication, you stand at a crossroads. One course you might take is to *assume* that you have written effectively, so that your employer will applaud your writing and your intended readers will react to it in exactly the way you want. If you make that assumption, you are ready to put your draft in the company mail, deliver it to the print shop, or entrust it to the postal service.

Alternatively, you can undertake a conscious, systematic effort to *find out* how good your draft is and how you might improve it. If you do that, you are *evaluating* your draft. The following paragraphs provide a general introduction to evaluation, and the next three chapters discuss the major types of evaluation used on the job:

- *Checking.* Using various techniques and aids, you carefully examine your draft by yourself (discussed in Chapter 22).

- *Reviewing.* You submit your draft to someone else for his or her advice (discussed in Chapter 23). At work, your reviewers will usually be people who are relatively close to you in your own organization (for instance, other people in your department, your boss, and your boss's boss).

- *Testing.* You ask one or more people to read and use your draft in the same way your target audience will read and use your finished communication (discussed in Chapter 24).

## DRAFTS

All the guidelines presented in the three chapters that follow assume that you are preparing more than one draft of your communication. Often, in fact, you will prepare many drafts, beginning with a very bumpy "rough draft" and gradually refining and polishing your communication until you have produced your "final draft," the one you will send to your intended readers.

What separates one draft from another, of course, is the work you do at evaluating and revising the earlier draft to improve it. Thus, in your writing process, evaluation and revision alternate with one another for as many drafts as you prepare. Much of this evaluation will be the checking you do yourself, but you can also show your drafts to other people for reviewing and testing.

Sometimes at work your first draft will be your final one, except perhaps for a very quick effort to correct the most obvious and serious errors and weaknesses. That will usually happen when you are writing a short communication that needs to be sent right away to people in your own organization. Generally it is best to avoid such situations if you can. As soon as you receive an assignment, plan a writing schedule that includes, if possible, time for more than one draft.

## THE TWIN GOALS OF EVALUATION

You will be able to understand the important role that evaluating will play in the writing you do at work if you consider the twin goals of evaluation in the workplace: quality control and writing improvement.

## Quality Control

To employers, evaluation is a form of quality control. It is similar to the elaborate procedures that manufacturing plants devise for ensuring that their products meet company standards. Whether evaluation takes the form of checking, reviewing, or testing, it is a procedure for determining whether or not a draft achieves the minimum acceptable level of quality.

Usually, you will find in your evaluation that your draft is good enough to send. But that does not mean that you should abandon evaluating altogether any more than a manufacturing plant should abandon its quality control program because 98% of its coffeepots are meeting its standards. The factory wants to prevent any substandard coffeepots from reaching consumers, so it must check each one to be sure it is satisfactory. Similarly, neither you nor your employer wants you to send any unsatisfactory communications, so you should evaluate all your writing.

## Writing Improvement

The second goal of evaluation in the workplace is the improvement of writing. Most employers want their employees to write communications that are better than "good enough." They feel that even an instruction manual that is usable should be improved, even a proposal that will succeed should be polished. Through evaluation, writers identify ways they can improve their writing even if their communications already meet minimum standards.

Presumably, your employer's goals for evaluation will be yours also. You will want to ensure that all your communications achieve the minimum level of quality necessary to be sent and you will want them to exceed that minimum by enough to bring you favorable recognition.

## THE THREE QUESTIONS OF EVALUATION

To evaluate a draft, you need some standard against which to assess it. Whether you are evaluating to see if your communication is good enough for immediate delivery or to find ways of improving an already satisfactory draft, you should derive your standards of evaluation from the same source: the objectives set when you first started work on your communication. If you filled out the Worksheet for Defining Objectives (Figure 5-4), you should pull it out as you prepare to evaluate. If you took additional notes concerning your readers' and employer's expectations about your communication (see Chapter 6), you should take them out also. Even if you noted your objectives mentally, rather than in writing, you should review them now. Those objectives describe what you tried to achieve when drafting. They should also form the standard against which you evaluate the result.

To apply your objectives to the task of evaluating, you can ask the following three questions:

- *Will my draft do what it is supposed to do?* Review your descriptions of the tasks you want your communication to help your readers perform and the way you want your communication to alter their attitudes. You should evaluate the extent to which your draft is likely to succeed in doing those things.

- *Could my draft harm the organization?* One of your overall objectives in writing is to serve your employer's best interests. That's why you will take the trouble, when defining your objectives, to learn about any communication policies your employer has established to guard against the harm an employee can cause by saying the wrong thing. Use your knowledge of those policies, together with advice from other people, to be sure your communication won't accidentally open your employer to a lawsuit, needlessly offend another department, tip off a competitor about a new product that your employer is developing, or bring harm to your employer in some other way.

- *Does my draft conform to the writing style used in my organization?* When writing at work, one of your objectives will be to follow your employer's preferences and regulations concerning the style and appearance of written communications. As part of your work at defining objectives, you will find what those preferences and regulations are. When evaluating, you should determine whether or not you have followed them successfully.

As you answer these three questions, remember to evaluate your entire communication—the visual aids as well as the prose, the page design as well as the spelling and grammar. All aspects of your communication contribute to its effect on your readers and its consequences for your employer, so all need to be examined carefully.

## THE NEED FOR EVALUATION

While reading the preceding paragraphs, you may have been asking a few questions: "Why should I undertake a special effort to evaluate my drafts? If I have carefully followed the guidelines given in this book for defining objectives, planning, and drafting, won't my communication be just fine already? Won't special efforts at evaluating simply waste my time?" Basically, the answer is "No." In fact, your efforts at evaluating will benefit you and your employer almost every time you write because

drafts are prone to several kinds of difficulties.

## Drafting Is Guesswork

First, no matter how carefully you have crafted your draft, it really constitutes a guess about how you can achieve the results you desire. You *guess* that by presenting *this* material in *this* manner you will affect your readers in a particular way. It is possible, however, that you have guessed incorrectly, at least in some small (but perhaps crucial) respect. When drafting a proposal, you may have accidentally left out a piece of information essential to your readers. Or, you may have assumed that the readers are familiar with some concept that they really need to have explained.

Second, although you may have guessed well about your overall strategy, you may have slipped in carrying out that strategy. Every writer slips sometimes. The explanation that seemed so plain to us turns out to be murky to our readers. The phrasing we thought would sound conciliatory seems to the readers to be sarcastic.

Evaluation is a means of locating such problems so you can correct them *before* you send your communication to your intended readers.

## Your Employer's Concerns

When drafting, you are guessing not only about how your readers will react to your communication but also about how your communication will affect your employer's organization. If you work for a large organization, for example, you probably won't be able to keep track of all its activities. Therefore, when drafting, you may overlook problems that your communication could create for some other department, project, or plan. Likewise, unless you are an attorney, you may not be able to see all the legal implications of what you write.

Evaluation, particularly reviewing, can help you find and fix aspects of your drafts that might harm your employer's organization. In fact, because they are aware that any employee has only a limited ability to see the organizational consequences of the things he or she writes, most employers *require* that certain kinds of communication be reviewed before being sent. Employers are especially careful with communications sent outside the organization. People designated as reviewers typically include the writer's manager, other specialists in the writer's field, and members of other departments affected by the communication's contents. Sometimes, reviewers also include the company's business officers and attorneys.

When you are new on the job, evaluation can help you in another way. Until you have been at work for a while, you may have difficulty remembering all the details of your employer's preferences about the layout of title pages, the placement of page numbers, and so on. Through careful checking or by enlisting someone else's aid as a reviewer, you can identify places where you have deviated from the style your employer prefers.

## LEARNING FROM EVALUATION

In addition to helping you produce effective communications that serve your employer's best interests and conform to your employer's stylistic preferences, evaluating can help you learn to write better. It can help by providing two things essential to learning: detailed information about where your writing does and does not work, and insights into ways you can overcome specific problems that you are having as a writer.

Because evaluating is such a powerful learning activity, many organizations assign someone (usually a supervisor) to work with new employees on their writing. Likewise, many writing instructors require students to engage in evaluation activities, such as discussing one another's drafts in class or submitting drafts to the instructor for review.

## SUMMARY

Evaluation is a form of quality control. In it, you assess the quality of your drafts, trying to answer three questions:

- Will my draft do what it is supposed to do?
- Could my draft harm the organization?
- Does my draft conform to the writing style used in my organization?

By answering those three questions you can learn whether or not your communication is good enough to send to its intended readers. Just as important, you can learn how to improve your communication, even if it is already quite good. And you can learn to become a better writer.

The next three chapters discuss three forms of evaluation that you can use in class and on the job: checking your draft yourself, giving your draft to reviewers in your organization, and testing your draft by asking one or more people to read and use it in the same way your target audience will read and use it.

## CONTENTS OF PART VI

**Chapter 22**
**Checking**

**Chapter 23**
**Reviewing**

**Chapter 24**
**Testing**

# Checking

## PREVIEW OF GUIDELINES

1. Play the role of your readers.
2. Consider the consequences of your communication for your employer.
3. Let time pass before you check.
4. Read your draft more than once, concentrating on different aspects of your writing each time.
5. Use checklists.
6. Read your draft aloud.
7. When checking for spelling, focus on individual words.
8. Double-check everything you are unsure about.
9. Use computer aids to find, but not to cure, possible problems.

Diane is experiencing one of the most excruciating of all feelings. Ten minutes ago, she mailed an important report to a client. Now, as she puts a photocopy of it in her file, her eye catches an obvious error. How could she have missed it? She read and reread that report before mailing it, looking diligently for errors. And yet she let this one get past her.

Diane's experience is one we have all shared. That's because it is very difficult for any of us to locate problems in our own writing. Of course, we get plenty of practice at checking our writing. We are expected to do so at both school and work. Still, we let blunders slip by all too often.

This chapter presents nine guidelines that will help you overcome the chief obstacles to successful checking, which is the evaluation of your drafts that you do by yourself, without anybody else's help.

## INTRODUCTION TO GUIDELINES 1 AND 2

The first two guidelines concern your general strategy for checking. They remind you that in addition to examining your draft from the point of view of its correctness in spelling, punctuation, and grammar, you should also consider it from the points of view of your readers and your employer.

Guideline
1

### Play the Role of Your Readers

Like any other writing activity, checking should be reader-centered. You should determine how understandable, useful, and persuasive your draft will be to your readers.

To do that you must see your communication as your readers will see it, something you will have to concentrate very hard to accomplish. The best way to begin is to review the descriptions of your purpose and readers that you developed when defining your communication's objectives (see Chapters 4 and 5). Then, as you check your draft, try to simulate your readers' attitudes and moment-by-moment reading activities. If you have written a recommendation, think of the questions and objections that your readers will raise. If you have written an instruction manual, try to follow the steps just as your readers would. To the extent possible, read in the way your readers will read.

Guideline
2

### Consider the Consequences of Your Communication for Your Employer

When checking, you must take not only your readers' point of view, but also your employer's. The things you say in writing can affect your employer's organization greatly. If you write your proposal well, you can bring in a big contract. If you accidentally slip in your writing, you can bring about a bitter internal feud or an expensive lawsuit. When you check any draft that you write at work, you must consider its consequences for your employer by thinking about the effects it will have on other people and departments in your organization, the commitments it makes on behalf of your employer, and the extent to which it complies with your employer's communication policies.

## Effect on the Organization

Many things you write at work will affect other people and departments in your employer's organization, especially if you are making a recommendation or providing information that will be used to make a decision. Ask yourself what effect your communication will have on those people and departments and what responses they will have to what you say. By anticipating the conflicts and objections that your communication might stir up, you can reduce their severity or, perhaps, avoid them altogether.

## Commitments

When you submit a proposal outside your organization, you will be promising that your employer will provide the materials or perform the work you describe. When you move to a managerial position and write policies, you will often be promising that in certain situations your employer will do certain things. As you check drafts in which you make such commitments, you should ask whether you have the authority to do so, whether the commitments are in the best interest of your employer, and whether they are the kind of commitments that your employer wants to make.

## Compliance with Communication Policies

To view your draft from your employer's point of view, you must also see whether the draft complies with organizational policies governing communications. Employers often have special policies about communications on sensitive topics. For example, your employer may not want outside people (particularly competitors) to learn the details of a project you are working on. This could happen if you are developing a new product or service that will give your employer a competitive advantage in the marketplace. Similarly, for legal reasons, your employer may want to restrict when and how you convey information about some aspect of your work. This might happen, for instance, if you are working on a patentable project or if you are working on a project involving materials regulated by a government body, such as a state or federal environmental protection agency. Whatever the reason for your employer's policy, be sure that your communication complies with it.

# INTRODUCTION TO GUIDELINES 3 THROUGH 9

The two guidelines you have just read concern your general strategy for checking drafts. The remaining seven guidelines suggest specific techniques that will help you check thoroughly and effectively.

Guideline

## 3

## Let Time Pass before You Check

As you read in the introduction to Part VI, one of the chief reasons we find it difficult to check our own writing is that we are so "close" to it. We have the disadvantage of knowing what we *meant* to say. As a result, when we check for errors and mistakes, we often see what we intended to write down, not what is actually on the page. A word is misspelled, but we see the correct spelling that we intended to use. A paragraph is cloudy, but we clearly see the meaning we intended to convey.

To some extent, you can overcome the difficulties arising from being so close to your own writing by letting time pass between drafting and checking. In this way, you "distance" yourself from what you have written. You forget some of your intentions and thereby increase your ability to spot weaknesses in your draft.

How much time should you let pass? When you are working on a brief document, a few minutes may provide you with all the time you need. Hours and days are better for longer, more complex documents. Many people find it helpful to wait through a night's rest, so they can check in the morning, when they are sharpest mentally. Of course, you must carefully schedule your work to leave time to set your draft aside for a while.

Guideline

## 4

## Read Your Draft More Than Once, Concentrating on Different Aspects of Your Writing Each Time

A second cause for the difficulties we all have checking is summed up in the old adage, "You can't do two things at once." According to researchers who study the ways humans think, to "do" something requires "attention." Some activities (like walking) require much less attention than others (like solving algebra problems). We can attend to several more-or-less automatic activities at once. However, when we are engaged in an activity that requires us to *concentrate,* we have difficulty doing anything else well at the same time.[1]

This limitation on attention has important consequences. When you check your draft, you must attend to many different things, each requiring concentration. To check spelling, you must concentrate on the letters in your words. To check the consistency of your headings, you must concentrate on their appearance and phrasing. To check your clarity, you must concentrate on the unfolding meaning of your prose. When you concentrate on one kind of checking, you diminish your ability to concentrate on the others.

You can overcome the limitations on attention by reading through your draft more than once. You should read it through at least twice, once for *substantive* matters (like clarity of meaning) and once for *mechanical* matters (like correctness of punctuation and consistency in the format of headings). If you have the necessary time, you might read through more than twice, sharpening the focus of each reading even more.

## How To Check for Substantive Matters

When reading through your draft to check for substantive matters, like clarity, follow the first two guidelines in this chapter. Read in the way your readers will read. As you read, consider also the consequences for your employer of the things you are saying.

## How To Check for Mechanical Matters

When checking for mechanical matters, use a reading pattern that your readers will *not* use. Go slowly to examine each detail of your draft. Flip back and forth among your pages as necessary to check for consistency among headings, figure titles, and so on. Pay careful attention to each of the following:

- *Correctness.* Be sure that you are correct in matters where there is a clear right and wrong (as with spelling and punctuation).
- *Consistency.* Where two or more items should have the same form, be sure that they do. For instance, be sure that parallel tables look the same. Where items are in sequence, be sure that the sequence is followed. For example, be sure that the figure after Figure 3 is correctly numbered as Figure 4. Where you ask your readers to look elsewhere, be sure the reference is accurate. For example, if you tell your readers to look on page 10 for a particular drawing, be sure that the drawing is there, not on page 11.
- *Conformity.* Be sure that you have followed all of your employer's policies concerning the mechanical aspects of writing, such as the width of margins, the use of abbreviations, and the contents of title pages. This will be relatively easy for you to do if your employer has put these policies in writing. If your employer hasn't, remember that your communication must still conform to your employer's informal policies on writing.
- *Attractiveness.* Be sure your communication looks appealing and neat.

## Guideline 5   Use Checklists

To be sure that you devote sufficient attention to each important aspect of your writing, use checklists. Read through your communication once for each item or small group of items on your list.

When creating a checklist, you might use the following sources:

- Any style guide or other directive issued by your employer about how communications should be written.
- The checklist shown in Figure 22-1.
- Notes or recollections of what you were asked to do (if you are working on a communication that someone else asked you to write).

You may also create your checklist based upon the guidelines given in this book. When you are deciding which guidelines to include on your personal checklist, review the comments that your instructor, your classmates, or your supervisor has made on communications you have written recently. Include guidelines for writing activities that give you difficulty, but omit those for activities you have mastered.

## Guideline 6 Read Your Draft Aloud

During one reading of your draft, read aloud. Reading aloud helps you distance yourself from your draft. To speak your words aloud, you must process them mentally in a somewhat different way than when you read silently. Also, when you read your draft aloud, you *hear* it as well as see it. These small shifts may seem trivial, but if you are like most writers, you will find that they give you a new view of your draft, so that you are able to spot problems you might otherwise have overlooked.

Reading aloud can be particularly useful in helping you find sentences and passages that are confusing or graceless. Where you stumble over your own words, you know that your readers are likely to have difficulties, too. Reading aloud will also help you determine whether you have succeeded in writing in a respectful, professional voice (see General Guideline 2 in Chapter 2). There is no better way to find out how your communication will "sound" to your readers than to hear it read aloud.

## Guideline 7 When Checking for Spelling, Focus on Individual Words

When we read, we usually do so without looking closely at each individual word. This may surprise you. Nevertheless, research shows that we usually read by the phrase and sentence. We identify every word, but we do not look closely enough at each one to inspect all its letters.[2] This presents special difficulties for you when you try to check your communications for misspellings and typographical errors.

To overcome this difficulty, you must slow down your reading so that you see each letter in each word. Some people can do that by applying their will power. Many cannot. After reading one or two dozen words, they begin reading the *meaning* of their communication, not the letters of their words.

Worksheet
# CHECKING YOUR DRAFT

**Project** _____ **Date** _____

### Check from Your Readers' Point of View

_____ Will your communication be easy for your readers to use?

_____ Will it persuade your readers?

### Check from Your Employer's Point of View

_____ Consider the effects of your draft on your employer's organization.

   _____ Do you know which people and departments will be affected?

   _____ Do you know *how* they will be affected?

   _____ Do you know what their reactions will be?

_____ Review the commitments you make on behalf of your employer.

   _____ Do you have the authority to make these commitments?

   _____ Are they in your employer's best interest?

   _____ Are they commitments your employer wants to make?

_____ Be sure that your communication complies with any special policies your employer has concerning statements on the subjects you discuss.

### Check for Mechanical Matters

_____ Are the spelling, grammar, and punctuation correct?

_____ Is your draft consistent in the numbering of figures, appearance of headings, and similar matters?

_____ Does your draft conform to the style your employer prefers? (If your employer has a style guide, did you follow it?)

_____ Is your draft neat and attractive?

**Figure 22-1.**   Worksheet for Checking Drafts

If that happens to you, you might try one of the following strategies:

- *Read backwards.* By reading your draft backwards, you take your words out of the sequence in which they form larger meanings. Consequently, you have an easier time concentrating on each word and the letters that compose it.
- *Use a peephole card.* In an index card, cut a hole just large enough to see one long word through. You can check for spelling errors by moving this card across your page, looking at each word or two in isolation from the words that surround them.

Computers can also help you concentrate on individual words when proofreading. If you prepare your drafts with a word-processing program, you may have another program that serves as a spelling checker. Such a program checks each word in a communication against a list of words in a dictionary stored in the program. If the program finds your word in its dictionary, it passes along to the next word. However, if it does not find the word, it places the word in a list of *possible* misspellings that it will show to you.

As you can see, these programs don't tell you which words you misspelled but only lists some words that *might* be misspelled. The lists they show you include many words, such as people's names and technical words, that are spelled correctly but that don't appear in their dictionaries. Further, the lists omit words that you mistyped in a way that makes them look like other words. For instance, if you mistype *take* so that it is *rake,* a spelling program will consider that the word is spelled correctly.

Despite these shortcomings, computerized spelling checkers are helpful. They enable you to check your entire draft very quickly for possible misspellings. You must remember, however, that even after using a spelling checker, you still need to proofread your writing.

Guideline

8
## Double-Check Everything You Are Unsure About

When checking a draft, all of us sometimes pause to ask ourselves such questions as, "Is that word spelled correctly?" "Is that number accurate?" "Is that precisely what the lab technician told me?" Because checking drafts is a demanding and difficult task, we may cringe at the thought of getting out the dictionary, looking back through our calculations, or calling up the technician. We may be tempted to answer our questions with, "Yeah, sure. No need to double-check." Unfortunately, whenever we ignore our own doubts, we are ignoring one of the primary signals that there may be a problem. When checking, trust the instinct that made you hesitate in the first place.

Such double-checking can be especially irksome if you are shaky about mechanical matters, such as spelling and punctuation. You may need to look up the same word again and again over the years: *seperate* or *separate?* con-

*ceiveable* or *conceivable?* (The latter in both cases.) You may not learn how to spell a word by looking it up repeatedly in the dictionary, but you might learn that you do not know how to spell it. Such knowledge is valuable to you—if you use it to double-check.

Guideline

**9**

## Use Computer Aids To Find, but Not To Cure, Possible Problems

People have proposed many aids to make checking easier. Some are now available in computer programs you can use to check drafts prepared with word-processing software. These aids can be helpful *if* you use them carefully.

### Readability Formulas

Readability formulas are one kind of checking aid. They calculate a single number that, according to their creators, represents the difficulty of a piece of writing. As you can see by looking at the popular readability formulas shown in Figure 22-2, they typically base their ratings on two factors: sentence length and word length. They assume that bigger words and longer sentences necessarily make reading more difficult.

But that's not necessarily the case. Although using needlessly long words does make for difficult reading, many long words (like *excitement*) are much easier for most readers to understand than are many short words (like *erg,* a term from physics). Similarly, long sentences can be written well and short ones written poorly.

Furthermore, readability formulas won't help you identify problems that arise from a myriad of other causes that have nothing to do with sentence length and word length—causes such as poor organization, poor use of topic sentences, or poor use of headings. And the formulas don't even give you a procedure for curing the problems they do find. George R. Klare, one of the leading authorities on readability formulas, says that trying to cure writing problems simply by using shorter sentences and shorter words is like trying to warm up your house by holding a match under a thermometer.[3] It simply doesn't work. The research in support of his position is overwhelming.[4]

### Other Computerized Checking Aids

There are also computer programs that analyze, to a limited extent, the grammar of a draft. Commonly, these aids look for grammatical forms that are often signs of weak prose. Examples include the use of the passive voice instead of the active voice, and the use of the verb *to be* rather than action verbs.

The limitations of these aids are much like the limitations of readability formulas. Although the aids can identify a passive verb, for example, they

## FLESCH READING EASE SCALE

> Reading Ease $= 206.835 - .846\text{wl} - 1.015\text{sl}$
>
> where wl = number of syllables per 100 words
> sl = average number of words per sentence

The resulting figure will fall on a scale between 0 and 100, with 100 representing the easiest reading.

## GUNNING FOG INDEX

> Reading Grade Level = .4 (ASL + %PW)
>
> where ASL = average sentence length
> (number of words)
> %PW = percentage of words with more
> than two syllables

The resulting figure will give a number that represents the grade level for which the writing is appropriate (for example, 1 = first grade, 13 = college freshman).

**Figure 22-2.**    Two Popular Readability Formulas.

The Flesch Reading Ease Scale is from Rudolf Flesch, "A New Readability Yardstick," *Journal of Applied Psychology,* 32 (1948):221–33. The Gunning Fog Index is from Robert Gunning, *The Technique of Clear Writing,* (New York: McGraw Hill, 1952 [revised 1968]), 38.

cannot distinguish a good use of the passive voice from a poor use of it (see Guideline 3 in Chapter 16 and Guideline 1 in Chapter 17).

In sum, readability formulas and similar aids can help you check your drafts by pointing out places that might have problems. However, the formulas may indicate problems where there are none, they may overlook problems, and they do not tell you how to correct the problems they find.

## SUMMARY

The guidelines in this chapter are intended to help you overcome the difficulties we all have in checking our own work.

1. *Play the role of your readers.* Like writing, checking should be a reader-centered activity. Before checking, review your descriptions of your purpose and readers. As you check, try to simulate your readers' attitudes and reading activities.

2. *Consider the consequences of your communication for your employer.* Some things to keep in mind include: (1) your communication's effect on other people and departments in your organization, (2) the commitments your communication makes on behalf of your employer, and (3) your communication's compliance with the policies of your organization that govern your subject.

3. *Let time pass before you check.* One of the reasons we have so much difficulty checking our own work is that we are so "close" to it. By waiting awhile after drafting and before checking, you can "distance" yourself from what you have written.

4. *Read your draft more than once, concentrating on different aspects of your writing each time.* When you check your drafts, you must attend to many different things, such as correctness, consistency, conformity to your employer's preferences, and attractiveness. Focusing on one aspect of writing distracts you from the others. Therefore you should read through your drafts *at least* twice, once for substantive matters and once for mechanical ones.

5. *Use checklists.* Using a checklist is one way to be sure you devote sufficient attention to each of the many aspects of your writing. To develop your checklist, you might use style guides or other materials from your employer, notes or recollections of your assignment, or the guidelines in this book.

6. *Read your draft aloud.* Reading aloud helps you distance yourself from your writing, and helps you find out how your communication will "sound" to its readers.

7. *When checking for spelling, focus on individual words.* To check for spelling, you need to slow down your reading so that you see each letter in each word. You can do this (1) by reading backwards or (2) by reading individual words through an index card with a small hole cut in it.

8. *Double-check everything you are unsure about.* Resist the temptation to ignore your doubts. Trust the instinct that made you hesitate in the first place.

9. *Use computer aids to find, but not to cure, possible problems.* Such aids may help you by pointing out places in your draft that might have problems. However, they may indicate problems where there are none, they may overlook problems, and they do not tell you how to correct the problems they find. Use them carefully.

## NOTES

1. John R. Anderson, "Attention and Sensory Information Processing," *Cognitive Psychology and Its Implications,* 2nd ed. (New York: W.H. Freeman, 1985), 40–8.
2. Frank Smith, "Word Identification," *Understanding Reading*, 3rd ed. (New York: Holt Rinehart and Winston, 1982), 120–34.
3. George R. Klare, "Readable Technical Writing: Some Observations," *Technical Communication,* 24 (Second Quarter 1977): 2.
4. Jack Selzer, "What Constitutes a 'Readable' Technical Style?" in *New Essays in Technical and Scientific Communication,* ed. Paul V. Anderson, R. John Brockmann, and Carolyn R. Miller (Farmingdale, New York: Baywood, 1983), 71–89.

## EXERCISES

1. Following the guidelines in this chapter, carefully check a draft you are preparing for your class. Then give your draft to one or more of your classmates to review. Make a list of the problems they find but you overlooked. For three of those problems, try to explain why you missed it. If possible, pick problems that have different explanations.

2. Make a personal checklist of the aspects of writing that you feel you need to examine most carefully when you check your own drafts. Then, do one of the following things:

   A. Compare your list with the lists prepared by other students in your class. This activity will give you a sense of the kinds of writing tasks that generally give problems to writers, and it will help you see which tasks are more difficult—or simpler—for you than for most others. There are no "correct" answers in this exercise. Its purpose is to make you more aware of your strengths as a writer and to help you identify areas of writing in which you should concentrate your efforts at improvement.

   B. Give one or more of your classmates a draft of an assignment you are working on. After they have reviewed your draft, compare their suggestions with the items on your checklist.

   Whichever task you perform, use the results to modify your list as seems appropriate.

3. The objective of this exercise is to help you become more aware of what you do when you check a draft. The results may help you improve your approach to this important activity.

   Your task is to record your actions and results as you check a draft of something you are writing. Check once for substantive matters and then once for each of the four mechanical matters listed in Figure 22-1. As you perform each of these checking activities, list the problems you find and ideas for improvement that occur to you. (*Note:* It is normal to observe some problems that you are not looking for specifically. For example, you might discover a lack of consistency in the headings when you are checking the spelling.)

When you are done checking, examine your list. What kinds of generalizations can you make about your approach to checking?

4. The memo shown in Figure 22-3 (page 498) has 23 misspelled or mistyped words. Find them by using this procedure:

A. Read the communication from front to back.

B. If you don't find them all *on your first check,* look for the rest using one of the procedures mentioned in the discussion of Guideline 7 (that is, read backwards or use a card with a peephole).

C. Unless you found all the words on your first reading (front to back), explain why you missed each of the words you overlooked on that reading.

D. Based on your performance in this exercise, state in a sentence or two your advice to yourself about how to improve your reading for misspellings and typographical errors.

5. In addition to its many misspellings, the memo in Exercise 4 has several other problems such as inconsistencies and missing punctuation. Find as many of these problems as you can.

Martimus Corporation
Interoffice Memorandum

February 19, 19—

From: T. J. Mueller, Vice President for Developmnet

To:    All Staff

RE:    PROOFREADING

Its absolutely critical that all members of the staff carefully
proofread all communications they write. Last month we learned
that a proposal we had submited to the U.S. Department of
Transportation was turned down largely becasue it was full of
careless errors. One of the referees at the Department commented
that, "We could scarcely trust an important contract to a company
that cann't proofread it's proposals any better than this. Errors
abound.

We received similar comments on final report of the telephone
technology project we preformed last year for Boise General.

In response to this widespread problem in Martinus, we are
taking the three important steps decsribed below.

I. TRAINING COURSE

We have hired a private consulting firm to conduct a 3-hour
training course in editing and proofreading for all staff members.
The course will be given 15 times so that class size will be held to
twelve participants. Nest week you will be asked to indicates times
you can attend. Every effort will be made to accomodate your
schehule.

II. ADDITIONAL REVIEWING.

To assure that we never again send a letter, report, memorardum,
or or other communication outside the company that will
embarrass us with it's carelessness, we are estabishing an
additional step in our review proceedures. For each communication
that must pass throug the regular review process, an additionnal
step will be required. In this step, an appropriate person in each
deparmtent will scour the communication for errors of expression,
consistancy, and correctness.

III. WRITING PERFORMANCE TO BE EVALUATED

In addition, we are creating new .personnel evaluation forms to be
used at annual salary reviews. They include a place for evaluating
each employees' writing.

III. Conclusion

We at Martimus can overcome this problem only with the full
cooperation of every employe. Please help.

**Figure 22-3.** Memo for Use in Exercise 4.

# Reviewing

## PREVIEW OF GUIDELINES

### Having Your Writing Reviewed

1. Think of your reviewers as your partners.
2. Tell your reviewers about the purpose and readers of your communication.
3. Tell your reviewers what you want them to do.
4. Stifle your tendency to be defensive.
5. Ask your reviewers to explain the reasons for their suggestions.
6. Take notes on your reviewers' suggestions.

### Reviewing the Writing of Other People

1. Think of the writers as people you are helping, not people you are judging.
2. Ask the writers about the purpose and readers of their communication.
3. Ask the writers what they want you to do.
4. Review from the point of view of the readers.
5. Begin your suggestions by praising strong points in the communication.
6. Distinguish matters of substance from matters of personal taste.
7. Explain your suggestions fully.
8. Determine which revisions will most improve the communication.

One way to evaluate your drafts is to give them to other people for their advice. This method of evaluation, called *reviewing*, is a practice widely used in the workplace. It is also employed by many writing instructors, including (perhaps) your own. This chapter presents six guidelines that will help you obtain the best possible advice from the people who review your drafts. The better their advice, the more you can improve your communication and your writing skills.

In addition, this chapter presents a second set of guidelines that will help you when you are asked to review someone else's drafts. Although advising other people about *their* drafts is not one of the steps in writing your own communications, it will be one of your important responsibilities on the job, so you will benefit from studying it now.

Both sets of guidelines—the one for having your drafts reviewed and the one for reviewing other people's drafts—are as useful in the classroom as in the workplace.

## REVIEWING IN THE WORKPLACE

Eight months ago, Jim graduated from college with a degree in chemistry. This morning, after days of hard work, he finished drafting a five-page memo reporting on his chemical analysis of a new metal alloy his company is developing. However, Jim has not mailed that memo yet. Instead, he is now meeting with his boss to discuss it. Overall, Jim is pleased to learn, his boss likes what he has written, but she asks Jim to make a few changes. She wants him to emphasize some of the data and to delete an explanation she thinks is unnecessary.

Jim has had similar meetings with his boss concerning every report he has written since he began work. Usually, Jim gives his draft to his boss a day or two before the meeting. By the time he arrives for the meeting, she has made many marginal notes on the draft and sometimes has written additional notes on another sheet of paper. After Jim and his boss go over the notes, Jim revises his draft and resubmits it to her. Only after she approves the report will it be typed for the last time and sent to the intended readers.

Jim's situation is typical. At work, you will almost certainly submit much of your writing for review, especially early in your career. Almost everybody does. Employers require it. Even when you have been promoted to management, you will still be submitting your writing for review. And then, like Jim's boss, you will also have the additional responsibility of reviewing things written by other people.

### Why Employers Require Reviews

Through reviewing, your employer will be pursuing three goals: to ensure that the communications you write will succeed in doing the jobs they are intended to do, to protect against communications in which you might accidentally say something that could harm the organization, and to help you learn to write better. Because these are the same goals that you pursue when you check your drafts over yourself, you may be wondering why your employer will think that it is necessary to have someone else read over your drafts. Why can't you determine by yourself what your communications will say, just as you do with the papers you write in college?

At work, your writing will be viewed much differently than it is in college. When you write a report in college, you and your instructor think of it as *yours*. On the job, however, most of what you write will be only *partly* yours. It will also "belong" to your employer. Your employer will be paying you to write, and your employer stands to gain or lose according to the skill with which you write. If you write a proposal well, your employer may get the contract. If not, your employer may lose it. If you write an instruction manual well, your employer will obtain efficient and safe operations from the people who will use it. If not, these people may accidentally damage your employer's goods and equipment, or even injure themselves. In these and many other ways, your writing can affect the well-being of your employer's organization. That is why your employer will want someone else to review many of your communications before you send them to their intended readers.

## Formal Procedures for Reviewing

Because reviewing is so important to employers, many of them establish formal procedures for this activity. These procedures often require that a communication be reviewed by several people—including managers, other specialists in the writer's field, lawyers, and, sometimes, a professional editor.

Employers who have such procedures often use special forms (sometimes called *cover sheets* or *transmittal sheets*) to ensure that all reviews are carried out. Each of the designated reviewers must sign the form before the communication can be sent to the intended readers. Reviewers often send drafts back to the writer for revision. The writer must then make the revisions before the reviewer will sign the approval form. Figure 23-1 shows such a form.

It can be worth your while to seek a review even when none is required. To determine when such occasions occur, weigh the time and effort that you and your reviewers will spend on the review against the likelihood that the review will produce some significant improvement in your draft or in your writing ability. The discussion in the chapter on "Revising" (Chapter 25) provides a framework for balancing those considerations.

# REVIEWING IN THE CLASSROOM

Many writing instructors incorporate reviewing into their classes. For example, your instructor may ask to review a draft of an assignment you are

---

**MIAMI UNIVERSITY PROPOSAL TRANSMITTAL FORM**

**PRELIMINARY BUDGET REVIEW**—It is required that your budget be reviewed in advance of final submission of your proposal. **CONTACT:** Budget Office

---

**THE PROPOSAL**

Principal
Investigator _____     Department _____

Starting Date _____
Title of Project _____     Ending Date _____

New Project ☐
Funding Agency _____     Continuation ☐

Deadline Date for Submission _____     Postmark ☐     Receipt ☐

Human Subjects Yes ____ No ____ Live, Vertebrate Animals Yes ____ No ____ Recombinant DNA Yes ____ No ____

**BUDGET DATA**

| | Funding Agency | | | Miami U. Cost-Sharing | |
|---|---|---|---|---|---|
| | 1st year | Total | | 1st year | Total |
| Direct Costs ............... $ _____ | _____ | | $ _____ | _____ |
| Indirect Costs ............. $ _____ | _____ | | $ _____ | _____ |
| Total Costs ................ $ _____ | _____ | | $ _____ | _____ |

**APPROVALS**

1. Principal Investigator _____ Date     4. Dean, The Graduate School & Research _____ Date

2. Department Chair _____ Date     5. Executive Vice President for Academic Affairs _____ Date

3. Divisional Dean/Executive Director _____ Date     6. Budget Director _____ Date

**COMMENTS BY SIGNERS:**

**Figure 23-1.** Transmittal Form.

writing. Or, your instructor may ask you and your classmates to review each other's drafts.

You can benefit in several ways from these reviewing activities. Obviously, you will have a chance to improve your assignments before you turn them in for a grade. More important, you will gain insights into the strengths and weaknesses of your writing, insights you can use in your efforts to write better.

**RESEARCH OFFICE REVIEW**

CHECK RELEVANT ITEM AND COMMENT IF PROPOSAL HAS NOTEWORTHY FEATURE

**GENERAL**                                                            **RESEARCH OFFICE COMMENTS**

_____ Scope of the Proposal

_____ Long-Term Commitments

_____ Community Impacts

**PERSONNEL**

_____ The Principal Investigator

_____ Faculty or Postdoctoral Associates

_____ Graduate or Undergraduate Students

_____ Staff Time/Effort

**ACADEMIC RIGHTS**

_____ Freedom to Publish

_____ Patent Agreements, Copyrights
               or Rights in Data

_____ National Security Restriction

**FISCAL**

_____ Student Fees

_____ Staff Benefits and Indirect Cost Rates

_____ Cost Sharing

_____ Space

_____ Equipment

**COMPLIANCES**

_____ Human Subjects

_____ Care of Laboratory Animals

_____ Radiation Hazards

_____ Safety and Health

_____ Biosafety

_____ Computers/Word Processors

Reviewer _____

Date _____

Figure 23-1.   *Continued*

You also benefit in another way from reviewing the drafts of your class-mates. Reviewing well requires thought and practice, just as writing well requires thought and practice. By reviewing your classmates' drafts, you are preparing yourself for the many times on the job when you will be asked to look at someone else's drafts. Because reviewing the writing of others is an essential part of a manager's job, your accomplishments in this area can increase your chances of promotion to a management position.

## IMPORTANCE OF GOOD RELATIONS BETWEEN WRITERS AND REVIEWERS

When one person reviews another's draft, the human relationship between the writer and reviewer is very important. In this relationship, both people are influenced by their professional responsibilities and their personal feelings.

As employees, the writer and the reviewer both have professional obligations to fulfill. The writer is expected to seek advice and to *use* it. The reviewer is expected to provide *substantive* advice that will improve the communication and increase the writer's writing ability.

The personal feelings of the writer and reviewer are deeply intertwined with their sense of professional responsibility. In many cases, the dominant feeling of each person is insecurity. As writers, we all view our writing as something particularly close to us. Maybe that's because we invest so much of our personal mental effort and so much of our individual creativity in it. Maybe it's because we realize that in many settings we will be judged largely on our writing. In any event, when we give our writing to someone else for review we may fear that we will be embarrassed or even crushed by what we hear from our reviewer.

On the other side, reviewers are fearful also. When we must review someone else's writing, most of us are afraid of making bad suggestions. Through our suggestions, we might show that we have not understood something that would be plain to any other reader. Or, our suggestions might reveal our own weaknesses as writers. Also, as reviewers, most of us are afraid of hurting the writer's feelings. We fear that we will say something that will cause the writer to become angry or to dislike us. Unfortunately, if we give in to these fears and withhold our advice, we are letting down the writer who is depending upon us for assistance.

There is no surefire way to avoid the insecurities that beset writers and reviewers. However, your anxieties will become less troublesome as you gain experience in reviewing situations. Therefore, if you have the opportunity to participate in reviewing sessions in your writing course, you should take advantage of them by participating as fully as you can. This experience will help you significantly as you enter similar situations on the job.

## INTRODUCTION TO GUIDELINES FOR HAVING YOUR WRITING REVIEWED

This chapter presents two sets of guidelines. The first set concerns situations in which your writing is being reviewed. The six guidelines in this set all share two objectives:

- To encourage your reviewers to think hard about your writing and to share freely their insights.

• To provide your reviewers with information they need about your draft in order to review it as helpfully as possible.

Guideline

1

## Think of Your Reviewers as Your Partners

The most effective way to encourage your reviewers to work hard on your behalf is to bring a constructive attitude to your relationship with them. Unfortunately, your natural insecurity about the reviewing situation, and your natural desire to finish your project with as little additional work as possible, can tempt you to adopt unproductive attitudes instead. You might be tempted to see your reviewers as "against" you, as your judges, your critics, or as obstacles to the completion of your project. However, if you convey any of those attitudes, you are likely to communicate to your reviewers that you don't want their help. And, if they feel that their help isn't wanted, they may do as little as possible on your behalf. No doubt, they will still do something, because it is their professional responsibility to do so. But they may not give you the full value of their insights.

Thus, it is crucial that you communicate to your reviewers that you welcome their help and are grateful for their suggestions. One way to do that is to remind yourself that your reviewers are your partners. They have the same aims that you have: to make your communication better and to improve your writing ability. You might say to yourself, "By the time I completed the draft I gave my reviewer, I went as far as I could on my own in preparing a really effective communication. Now, my reviewer will give me some ideas for making it even better." That can be a helpful thing to say to yourself, even if your hesitancies make it difficult for you to say it in your heartiest voice.

Guideline

2

## Tell Your Reviewers about the Purpose and Readers of Your Communication

To write effectively, you must think constantly about your purpose and readers. For reviewers to do their best work on your behalf, they too must think constantly about your purpose and readers. Without knowing whom you are writing to, and why, your reviewers cannot easily judge whether your current draft is likely to achieve its purpose. Neither will they have the knowledge they need to suggest ways of making your draft more effective. Thus, when you ask your reviewers to "Please look this over," be sure to add, "It's written to such and such a *reader* and has such and such a *purpose.*"

Guideline

3

## Tell Your Reviewers What You Want Them To Do

In reviewing any draft, there are lots of things a reviewer might look at: organization, selection of material, accuracy, tone, spelling, page design, and so on. Unfortunately, it is very difficult to concentrate on all these aspects of a draft simultaneously. You can help your reviewers if you direct their attention to the particular features of the draft that you would most like their advice about.

To determine what to ask your reviewers to look at, you should consider such things as the areas of writing that you are most unsure about, the sections of the draft that you had the most trouble writing, and the sections of the draft that seem to you to be the most critical to its success.

## Stifle Your Tendency To Be Defensive

Your response to your reviewers' suggestions can determine how many ideas they will be willing to share with you. The more you seem to want, the more you are likely to get. Unfortunately, our natural reaction to suggestions about our writing is to search for arguments that explain why what we wrote is better than what the reviewers suggest. If you respond to your reviewers in this defensive way, however, they may decide to quit giving you advice, feeling that their suggestions are inevitably going to meet not with thanks, but with arguments. Thus, even if you *know* that a particular suggestion is not good, listen to it without argument. Perhaps the next suggestion will be excellent. Don't discourage your reviewers from making it.

On the other hand, in your efforts to stifle your defensiveness, don't suppress all dialogue with your reviewers. When your reviewers' suggestions seem to be based upon a misunderstanding of what you are trying to say or what you are trying to achieve, explain your meaning or aims. But do so in a way that shows that you are still open for advice. If your reviewers misunderstood what you were attempting to say or accomplish, it's likely that you need advice about how to revise anyway.

In sum, to get the full benefit of your reviewers' advice, stifle your defensiveness. Keep your reviewers talking.

## Ask Your Reviewers To Explain the Reasons for Their Suggestions

Sometimes, your reviewers' explanations of the reasons for their suggestions will be more valuable to you than the suggestions themselves. This can be especially true when your reviewers make suggestions that you *know* won't work. In such situations, you may be inclined to remain silent so that the reviewers will move on to their next point. However, reviewers make suggestions about things that don't seem right to them. Even when their suggestions seem useless to you, you can gain valuable insights by asking them what they thought was wrong. Then, you can devise your own solution to the difficulty.

You can also benefit from explanations of your reviewers' good suggestions. Perhaps your reviewers will have a way of explaining a certain piece of advice that will help you recognize and avoid the same kind of problem in the future. Or, perhaps, their explanation will help you think of an even better way of solving the problem they have helped you identify.

## Take Notes on Your Reviewers' Suggestions

Sometimes your reviewers will give you their suggestions in writing and sometimes they will give them in person. In the latter case, not only must you keep them talking but you must also remember what they say. Unfortunately, remembering what your reviewers are saying to you can be difficult. Your conversations with your reviewers may head off in many directions as you discuss various aspects of your draft. Further, most of us have difficulty remembering the details of conversations in which we feel a little bit on the spot, as most of us do when our writing is being reviewed. Therefore, unless you know from experience that you have very good recall for meetings with reviewers, you should take notes during these meetings.

Besides helping you remember what your reviewers have said, notetaking can also help you in another way. If you read your notes back to your reviewers, you will be able to see that you have correctly understood all of their points. This sort of check may also remind the reviewers of some additional, important points that they might make.

# SUMMARY OF GUIDELINES FOR HAVING YOUR WRITING REVIEWED

The six guidelines for having your work reviewed are intended to help you (1) establish and maintain a productive professional and personal relationship with your reviewers so that you can (2) gain from them the best advice they can give about how to improve your draft and your writing in general.

1. *Think of your reviewers as your partners.* Your reviewers have the same aims that you have: to make your communication better and to improve your writing ability.

2. *Tell your reviewers about the purpose and readers of your communication.* Your reviewers cannot provide good advice unless they know who your readers are and what you hope to accomplish by writing to them.

3. *Tell your reviewers what you want them to do.* To determine what to ask your reviewers to look at, consider such things as the areas of writing you are most unsure about, the sections of the draft that you had the most trouble writing, and the sections that seem to you to be most critical to its success.

4. *Stifle your tendency to be defensive.* If you respond to your reviewers' comments by arguing and being defensive, your reviewers may quit giving advice, feeling that their suggestions are not appreciated.

5. *Ask your reviewers to explain the reasons for their suggestions.* Even when your reviewers' suggestions do not seem useful to you, you can gain valuable insight by finding out what the reviewers thought was wrong. Perhaps then you can find a better solution to the problem.

6. *Take notes on your reviewers' suggestions.* Notetaking can help you remember what your reviewers suggest. Also, if you read your notes back to your reviewers, you can make certain that you have understood all of their points.

# INTRODUCTION TO GUIDELINES FOR REVIEWING THE WRITING OF OTHER PEOPLE

The following guidelines will help you work effectively when you are asked to review someone else's writing. They will be useful to you in the classroom as well as on the job.

Guideline

**1**

## Think of the Writers as People You Are Helping, Not People You Are Judging

Your success as a reviewer will depend largely upon your ability to make writers feel open and receptive to your suggestions. If writers think that you lack respect for their ideas or feelings, they will probably resist your suggestions. If they feel you are genuinely interested in helping them, they will be much more likely to accept your suggestions.

### Alternative Roles for a Reviewer

A writer's initial reaction to you as a reviewer may depend largely upon the role you have been asked to play by your employer. You may be asked to play the "coach," a person who merely *suggests* improvements to the writer. The writer is free to take your suggestions or ignore them. On the other hand, you may be asked to play the "gatekeeper." As a gatekeeper, you have been given by your employer the authority to *require* the writer to make the revisions you suggest. Only after he or she has done so can the writer send the communication to the intended reader.

Whether you are a coach or a gatekeeper, you should interact with the writer in the same way. That advice may surprise you. It may seem that when you are a coach you must be very considerate of the writer's feelings because you have to persuade the writer that your suggestions are worth taking, but that when you are a gatekeeper you can ignore the writer's feelings because the writer must follow your suggestions no matter how he or she feels about them.

Actually, to be an effective reviewer, you must always be considerate of the writer's feelings. When writers are forced to make changes they don't agree with, they become discouraged and bitter. Instead of trying to write better next time, they may give up. They may turn in sloppy, thoughtless work, saying to themselves that there is no point in working hard at their writing because whatever they write will be changed arbitrarily anyway. Also, when writers are required to make changes they don't agree with, they can build resentments towards their reviewers that carry over to other parts of their working relationships. Consequently, even when you are a gatekeeper you should work very hard at persuading writers that your suggestions are sound.

### How To Persuade Writers That You Want To Help

How can you persuade writers that you are trying to help them by pointing out weaknesses in their drafts? Most important, you can phrase your comments in a supportive manner. Instead of saying, "I think you have a problem in the third paragraph" or "Your third paragraph doesn't work," say, "I have a suggestion about how you can improve your third paragraph" or "Perhaps the third paragraph would be even more persuasive if you were to mention the following points."

You can also convey your desire to help by following several of the other guidelines discussed later in this chapter:

- Take the trouble to learn exactly what the writers are trying to accomplish (Guideline 2).
- Ask the writers what things they especially want you to look at in your review (Guideline 3).
- Praise the things the writers have done well (Guideline 5).
- Respect each writer's style and personal preferences where they do not conflict with your employer's policies or diminish the effectiveness of the communication (Guideline 6).
- Explain your suggestions fully, so that the writers don't think that you are making arbitrary suggestions that disregard their reasons for writing the way they did (Guideline 7).

### How You Benefit

Your efforts to act like a helper, not a judge, can assist you almost as much as they will help your writers. All of us instinctively draw back from doing things that will hurt or offend others. If, when you review, you see yourself as a judge, you may hesitate to offer advice because you may sense that your advice will not be welcome. You may find yourself thinking only superficially about the drafts you review, and you may even withhold some of the few suggestions you do come up with.

On the other hand, if you view yourself as a helper, you can anticipate that your efforts will be welcome and you can respond more eagerly to your responsibility to work hard on the writers' behalf, to think seriously about their drafts, and to convey to them all the advice that might help them.

## Ask the Writers about the Purpose and Readers of Their Communication

Guideline 2

There are two important reasons for you to ask the writers to tell you about the purpose and readers for the communications you review. The first is to indicate to the writers that you are interested in what they are trying to accomplish, that you want to see their communications from their point of view. The second is that you must have this information to tell where a communication works and where it doesn't, and to formulate good suggestions for improving it.

Your inquiries about purpose and readers will enable you to be especially helpful to those writers who usually take a writer-centered (not a reader-centered) approach to writing. If you gain information about purpose and readers from the writers and later use that information to explain your suggestions, you will have helped the writers learn an entirely new approach to the task of writing a communication, one that can improve their writing quickly and dramatically.

## Ask the Writers What They Want You To Do

Guideline 3

When you are reviewing you can look at many different aspects of a communication. Before you begin, you and the writer should agree about what you are to look for. Misunderstandings on this matter can waste your time. If a writer wants you to proofread only, or to look at one section only, you should find that out so that you don't devote time needlessly to other aspects or other sections of the communication.

Of course, your instructions about what to look for when reviewing will not always come solely from the writers. When you are asked by your employer to serve as a gatekeeper, you will also be told by your employer what kinds of things you are to look for. Even then, it is important for you to ask the writers if there is anything special you should consider. Writers often have a very good sense of what most needs attention in their drafts. Further, by asking the question, you are expressing your personal interest in helping the writers.

## Review from the Point of View of the Readers

Guideline 4

You need to review in a reader-centered way just as surely as you need to write in a reader-centered way. Therefore, when your employer or a writer asks you to consider the effectiveness of a communication, read it from the point of view of the readers. Try to imagine how the readers will use it, and then determine how effective it is from that point of view. Try to imagine the readers' attitudes, and then determine whether it is likely to persuade those particular individuals.

Guideline

5

## Begin Your Suggestions by Praising Strong Points in the Communication

When you convey your suggestions to writers, pay special attention to the way you open the conversation. Like the opening of any communication, the opening of a series of review comments is crucial in establishing the attitude that the people you are addressing will take toward what you have to say. By opening with praise, you let writers know that you appreciate the good things they have done, and you indicate your sensitivity to their feelings.

Unfortunately, it is easy to forget to open with praise. After all, the essential strategy of reviewing is to find weaknesses and make suggestions about how to improve them. So that you don't forget to begin with praise, you may find it helpful to develop the habit of planning your comments before you meet with the writers. In this way, you can be sure that you don't focus too quickly on what is wrong with the writers' writing.

The praise you give writers serves other important purposes as well. Some writers don't know what they do best any more than they know where their weaknesses lie. By praising the things they have done well, you are encouraging them to continue to do those things in their future writing, and you are reducing the chances that they will weaken one of the strong parts of their communication by revising it.

One very good way of explaining your suggestions is to point to strong places in a writer's draft. For example, if the writer has included a topic sentence in one paragraph but omitted one in another place, you could offer the former paragraph as an example of how to improve the latter one. In this way, you are considerate of the writer's feelings by indicating that you know that the writer understands the basic writing principle but slipped once in applying it. You are also helping the writer become more conscious of the need to use consistently the skills he or she already possesses.

Guideline

6

## Distinguish Matters of Substance from Matters of Personal Taste

We all have individual ways of expressing ourselves. These personal preferences guide the way we say things. When reviewing, you should be careful not to impose your personal preferences on writers. If you make suggestions that merely substitute your preferred way of saying things for the writers' preferred ways, you have not improved the communications. You have merely made them sound more like things written by you and less like things written by their real writers.

You can determine whether a change that you are thinking of suggesting involves a matter of personal taste or a matter of substance by examining the *reason* you would offer for the change. If all you can say is, "My way sounds better," you are probably dealing with a matter of taste. Your way of saying it sounds better to you because it is the way you like to say things. On the other hand, if you can offer a more objective reason, for instance an explanation based upon one of the guidelines in this book, you are dealing with a matter of substance.

One caution, however. Sometimes your sense that something doesn't sound right is your first clue that there is a problem in the writing. For example, when you are reading through a draft, you may stop at a sentence that doesn't sound right and find, after closer examination, that it has a grammatical error. Therefore, it is important to follow your instincts about how something sounds. The way to follow them is to see whether you can find an objective reason for your dissatisfaction. If so, tell the writer. If not, let the matter drop.

<table>
<tr><td>Guideline<br>7</td><td></td></tr>
</table>

## Explain Your Suggestions Fully

You should tell writers *why* you have made your suggestions. If you don't, they may assume that you are simply expressing a personal preference that makes no substantive difference to their communications. As a result, they may not take your suggestion, leaving the communications weaker than they might be.

Besides persuading writers to make needed changes, explanations help writers learn to write better. For example, imagine that you suggest rephrasing a sentence so that the old information is at the beginning and the new information is at the end (see Chapter 17). The writer may agree that your version is better but not know in general how to improve his or her skill at writing sentences unless you *explain* the principle you applied.

Some reviewers withhold explanations because they think that the reasons for the change are obvious. However, explanations that are obvious to a reviewer are not necessarily obvious to the writer. After all, if the problem had been obvious to the writer, the writer probably would have eliminated it before giving the draft to you for review.

<table>
<tr><td>Guideline<br>8</td><td></td></tr>
</table>

## Determine Which Revisions Will Most Improve the Communication

Sometimes you will review drafts that need only a few revisions, and sometimes you will review drafts that need many. In the latter cases, you should consider the possibility that the writers will not be able to make all the changes you could suggest. Perhaps the writers (like any of us) can face only a certain number of changes without feeling overwhelmed and defeated. Perhaps the writers have only a limited amount of time to make the revisions—less time than would be required to do everything that might be done to improve their drafts.

In every situation, writers should concentrate on making the revisions that will contribute most to improving their communications. As a reviewer, therefore, you should look over your own list of possible suggestions to see which ones will make the most difference. Convey those to the writers. Keep the rest of the suggestions to yourself. Or make it clear that they are less important than the others.

To help themselves in this effort, some reviewers begin by scanning a draft to plan what sorts of suggestions they might want to make. The alternative to such planning is simply to begin on page one and proceed through the draft, marking each problem as it occurs. Such a strategy can create difficulties if the problems on the first pages are small matters (like minor rephrasings of sentences) compared to large problems (in organization, let's say) that become apparent later on. In Chapter 25 you will find advice about how to judge the relative improvements that various kinds of revision can make in a communication.

## SUMMARY OF GUIDELINES FOR REVIEWING THE WRITING OF OTHER PEOPLE

The eight guidelines for serving as a reviewer will help you provide good advice that the writers are likely to use.

1. *Think of the writers as people you are helping, not people you are judging.* In practical terms, this means that you should make your suggestions in a way that indicates that you desire to help, not to criticize.
2. *Ask the writers about the purpose and readers of their communication.* You need this information to tell whether the communication works and to formulate good suggestions for improving it.
3. *Ask the writers what they want you to do.* You may waste time if you and the writer do not agree—before you begin reviewing—on which of the many aspects of the communication you are to review.
4. *Review from the point of view of the readers.* Just as it is important to *write* in a reader-centered way, it is important to *review* in a reader-centered way. When you review, try to put yourself in the readers' position.
5. *Begin your suggestions by praising strong points in the communication.* Like the opening of any communication, the opening of a series of review comments is crucial in influencing the attitude that the person you are addressing will take. By opening with praise, you let writers know that you appreciate the good things they have done.

6. *Distinguish matters of substance from matters of personal taste.* If you make suggestions that merely substitute your preferred way of saying things for a writer's preferred way, you have not improved the communication. To distinguish between matters of substance and matters of taste, examine the reasons for your suggestions.

7. *Explain your suggestions fully.* Explanations persuade writers to make needed changes and help them to learn to write better.

8. *Determine which revisions will most improve the communication.* A writer may not be able to make all of the changes you could suggest. Decide which suggestions will most improve the communication, and share those with the writer.

## EXERCISES

1. For one of the assignments you are preparing in your class, write a note telling the information a reviewer would need to have to give you good advice about how to revise your draft. Be sure to follow Guidelines 2 and 3 from the list of guidelines for having your writing reviewed. Then, participate in *one* of the following activities:

   A. In a group discussion, share your note with your classmates. Have them tell you whether or not you have given them all the information they would need in order to review your draft.

   B. Give your draft and your note to one or more of your classmates. Before they review your draft, ask them if they need additional information not provided in your note. Ask them again after they have read your draft.

2. This exercise will give you practice at taking notes on the suggestions your reviewer makes (see Guideline 6 for having your work reviewed).

   A. Exchange drafts with a classmate (perhaps while doing Exercise 1).

   B. Make notes on improvements you would like to suggest to your partner, while he or she writes down suggestions for you.

   C. Listen to your partner's suggestions, taking notes on them. (Your partner may refer to the suggestions he or she wrote down, but should not show them to you.)

   D. Compare the notes you took with the notes from which your partner spoke. Do the two sets of notes correspond closely? Did additional suggestions emerge from your discussion with your partner?

   E. Switch roles, so that you are the reviewer and your partner is the writer.

3. This exercise will help you develop your skills at delivering your suggestions to a writer in a constructive way.

A. Exchange drafts with a classmate. Also exchange information about the purpose and readers for the drafts.

B. Carefully read your partner's draft, playing the role of a reviewer. Your partner should be reading your draft in the same way.

C. Make your comments to your partner, following Guidelines 5 through 8 for reviewing the writing of other people.

D. Evaluate your success in delivering comments to the writer in a way that makes the writer feel comfortable while still providing him or her with substantive, understandable advice. Do this by writing down three specific things you think you did well when delivering your review comments and three ways in which you think you could improve. At the same time, your partner should make a similar list of observations about your delivery. Both of you should focus on such things as how you opened the discussion, how you phrased your various comments, and how you explained them.

E. Talk over with your partner the observations that each of you made.

F. Switch roles with your partner, so that he or she is now the reviewer of your draft. Repeat Steps 3 through 5.

4. In this exercise, you are to prepare review comments for a writer. Depending upon what your instructor asks you to do, you may present these comments orally or you may convey them in a memo to the writer. In either case, be sure to follow *all* the guidelines for reviewing other people's writing, including Guidelines 1, 4, 6, 7, and 8. Remember that *how* you present your comments to the writer can be as important as *what* you suggest. For the sake of this exercise, imagine that you are a coach (not a gatekeeper), so that the writer has the choice of following or not following your suggestions.

The memo you are to review is shown in Figure 9-2. It was written by Benjamin Bradstreet, who works for a firm that builds manufacturing plants in other countries. Yesterday, Ben received a call from Dick Saunders, who has been in the Philippines for the past 2 years. Dick has been overseeing the construction of a factory for manufacturing roof shingles. The shingles are made from bagasse, the fibers that remain after the sugar has been squeezed out of sugar cane.

Dick reached Ben while trying to call his own boss, Tom Wiley, who wasn't in. Consequently, Dick asked Ben to take notes on their conversation and to forward them to Tom.

The audience for Ben's note will include not only Tom, who will be interested in all the information it contains, but also several other people, who will want only certain pieces of that information. These other people include Tom's assistant, who will be responsible for sending the things Dick requests, and several other men and women in the sales department, who will need to talk with Dick on the phone or in person while he is in the States.

# Testing

## PREVIEW OF GUIDELINES

1. Ask your test readers to use your draft in the same ways your target readers will.
2. Learn how your draft affects your test readers' attitudes.
3. Learn how your test readers interact with your draft while reading.
4. Choose test readers from your target audience.
5. Use enough test readers to determine typical responses.
6. Use drafts that are as close as possible to final drafts.

An employee of an electronics company, Imogen has written a set of instructions for installing a car radio and tape deck made by her employer. She has checked her instructions carefully herself, and she has asked her boss and two engineers to review her draft. Through these efforts, she found many ways to improve her draft, which she now feels pretty good about.

However, Imogen still doesn't know one crucial thing: how well her instructions will *really* work for their intended readers, ordinary consumers who buy the radio and tape deck and want to install it themselves. When she checked her draft over, she *guessed* about how useful it would be to these readers. When her boss and the engineers reviewed her draft, they guessed, too.

One way Imogen can find out how good her guesses have been is to print up her instructions and begin packing them with the radio and tape decks. If it turns out that her guesses were good, that strategy will work well. However, if her guesses were not good she won't find out until after it is too late for her to change them. It would be desirable, therefore, for her to evaluate her guesses before the instructions are printed up in final form. Then, if any problems are discovered, she can revise her instructions before they reach the hands of her employer's customers.

Imogen can obtain that preliminary evaluation of her guesses by *testing* her instructions. This chapter presents six guidelines that Imogen—and you—can follow in order to create helpful, informative tests.

## BASIC STRATEGY OF TESTING

When you test a communication, you try it out in a way that will enable you to predict what will happen when your intended readers read it. To do that, you ask some people to serve as test readers who will read and use a draft of your communication in the way your real readers will read and use the final draft. Then, by means of various techniques, you gather information from your test readers that will serve as the basis for your prediction. If this information shows that your communication affects your test readers in the way you desire, then you predict that it will succeed with your real readers as well. On the other hand, if you find that your communication doesn't affect your test readers the way you want, then you predict that it will similarly fail with your real readers unless you revise it.

For your prediction to be good, two things must happen. First, your test must resemble the real situation as much as possible. Your test readers and test draft must closely resemble your intended readers and final draft. Similarly, you must have your test readers read in the same way your real readers will read. Differences in any of these areas will decrease the likelihood that your test results will accurately predict what will happen when your real readers receive your communication.

The second thing that must happen if your prediction is to be good is this: you must focus your test on the aspects of your communication that are most crucial to its success. If you don't, you may learn how well certain trivial features of your communication will work without learning anything at all about how well your communication will do in achieving its primary purpose. Thus, if you have written an informative report or booklet in which some points are much more important than others, your test should determine how well your communication teaches those key points to your test readers. If you have written a long instruction manual, be sure to find out whether or not your manual enables your readers to perform the most critical steps.

The guidelines in this chapter provide advice about how to make your tests resemble the real reading situations as closely as possible, and they contain advice that will help you focus your tests on the aspects of your communications that are most critical to their success.

## TWO MAJOR USES OF TEST RESULTS

As you read this chapter's guidelines, you will see that you will face a great many choices as you design and conduct a test. In making these choices, you will find it helpful not only to keep in mind the purpose, readers, and circumstances of the particular communication you are testing, but also to remember the overall use to which you will put the information you gather from your test. You will use the results to do the following:

- Determine if your communication works well enough to send to your intended readers.
- Gain ideas about how to improve your communication.

The next paragraphs discuss these uses more fully.

### Does Your Communication Work Well Enough?

One way to determine if your communication works well enough is to see if it meets some predetermined standard in the test of its effectiveness. For instance, before testing her instructions, Imogen might have decided that they would be good enough if a typical reader could use them to successfully install the radio and tape deck within two hours. For communications intended to teach, minimum test scores are often expressed in terms of percentage of correct answers. Thus, a person who wrote a description of the electromagnetic fields that surround the earth might say that the description achieved a satisfactory score if readers could answer correctly at least 75% of the questions on a test based on it. (Remember that the tests described in this chapter evaluate the communication, not its readers—a point you should emphasize to your test readers before they begin their reading.)

Sometimes, you may decide to test your communications in ways that do not produce numerical results. For instance, you might decide to give your communication to some test readers and then talk with those readers after they have finished reading. To interpret the results, you need to use good judgment and common sense. Did your readers seem to be able to use the communication easily? Were their problems, if any, minor ones?

In fact, you will often need to use good judgment and common sense to interpret the outcome of tests that produce numerical results. Suppose, for instance, that Imogen found that her readers needed three hours, not two, to install the radio and tape deck. If she observed that her readers worked steadily throughout without having any difficulties understanding or using her instructions, she might conclude that three hours were needed because of the difficulty of the task, not because the instructions were written poorly. On the other hand, if she observed that the readers stumbled through the procedure because they had trouble understanding and using the instructions, she might conclude that her instructions were not yet good enough.

## How Can You Improve the Communication?

For the most part, your primary aim in testing will not be to find out if your communication is good enough but to gain ideas about how to improve it. Generally, by the time you have worked on your communication long enough to prepare it for a test, the communication will work satisfactorily. If it doesn't, you will certainly want your test to suggest ways of improving the communication. But even if your test shows that your communication is good enough, you will still want to know how to make it better. For those reasons, this chapter emphasizes diagnostic testing—testing to determine where improvements can be made and how they might be carried out.

# WHEN SHOULD YOU TEST A COMMUNICATION?

Before starting to read this chapter, you may never have thought that you might want to test the communications you will write at work. Indeed, of the three methods of evaluation described in this book—checking, reviewing, and testing—testing is used least often. At work, writers almost always check their own work, and a great many communications are reviewed before they are delivered to their intended readers. In contrast, only a small proportion of the communications written at work are tested.

Testing is used less often than the other two methods of evaluation because testing a communication usually requires a substantial effort over and above the effort already spent on checking and reviewing it. For many kinds of communications, such as technical reports and routine memos, people rarely even consider the possibility of testing. They believe that the improvement that testing might bring would not be great enough to compensate for the time and effort testing requires. In contrast, step-by-step instructions are often tested—especially for consumer products. That's because writers and their employers believe that the tests are often easy to conduct and that it will be well worth the time and effort invested to learn of any problems the tests might uncover. Maybe testing will identify problems in the instructions that could affect the sales of the product itself, or maybe it will identify problems that could lead to product liability lawsuits.

At present, many communications that could benefit immensely from testing are not tested, a fact that a slowly growing number of organizations are beginning to realize. In your future job, you may be in a position to persuade your employer of the value of testing some of its communications. When should you urge that a communication be tested? The more important the communication is and the larger the number of people who will read it, the more reason there is to test it.

## THREE ELEMENTS OF A TEST

The guidelines in this chapter are organized into three groups, corresponding to the three basic elements of a test:

- The *test activities*—what you will ask your readers to do and the way you will gather information about their use of the draft. (Guidelines 1 through 3)
- The *people* who will read your communication in your test—what kinds of people you should ask to be the readers in your test, and how many you should have. (Guidelines 4 and 5)
- The *draft* you will test—what it should be like. (Guideline 6)

Guideline

**1**

### Ask Your Test Readers To Use Your Draft in the Same Ways Your Target Readers Will

When you draft, you create a communication designed to achieve a particular purpose. Use that purpose to design your test. How? If you followed the guidelines in Chapter 4, you defined your communication's purpose in terms of (1) the tasks you want it to enable your readers to perform while reading and (2) the way you want it to alter your readers' attitudes. The guideline you are now reading concerns testing related to your readers' tasks, and the next guideline provides advice about testing related to changes in your readers' attitudes.

At work, tests usually focus on the readers' efforts to carry out one or more of the following tasks:

- *Perform a procedure,* as when reading instructions.

- *Understand content,* as when trying to learn about something through reading.
- *Locate information,* as when looking for a certain fact in a reference manual.

For some communications, all three tasks are important. For example, the owner's manuals for some personal computers open with a section describing how computers work (the readers' chief task is to *understand*), proceed to step-by-step instructions (the readers' chief task is to *perform* a procedure), and conclude with a reference section (where the readers must *locate* the particular pieces of information they need).

You should test your communication's effectiveness in enabling your readers to perform each and all of the tasks crucial to its success. Although you may test a single communication for all three tasks, you should test each task in a different way. The following paragraphs describe appropriate tests for each task.

## Performance Test

To test a communication's effectiveness at enabling readers to perform a procedure, give the communication to a few test readers and watch them use it. As much as possible, assign your test readers the same tasks your target readers will perform and have them work in the same kind of setting. That way, you can learn how the communication will work for readers who have the same resources (for example, work space, lighting, tools) that your target readers will have—not any more nor any less. Also, as you watch your test readers, avoid intruding upon their work, either by getting in their way or by providing them with assistance that your target readers would not have.

### An Example Test

To see how you might design and conduct a performance test, consider what Imogen did to test the instructions (mentioned at the beginning of this chapter) for the car radio and tape deck. First, she asked two friends, Rob and Janice, to use her instructions to install the equipment in their own cars. (Her company gave them the equipment, considering that to be part of the expense of developing good instructions.) Imogen asked each of them to work alone because she imagined that most purchasers of the radio and tape deck would install it without help. She had Rob work at his home and Janice in the parking lot of her apartment building, where each would have the kind of work area and tools that would be available to most purchasers.

To gather information about how the instructions worked, Imogen did two things:

- She watched Rob and Janice throughout their work, taking notes about places where they had difficulties.

- She asked Rob and Janice about the instructions after they were done with the installation. She had prepared some questions in advance, and others emerged from her discussion with them.

While watching Rob and Janice, Imogen was careful not to intrude upon their work. When Rob began his work, he several times asked Imogen about various instructions before trying them. Instead of answering, Imogen urged him to do his best without her help. If she had instead begun to give oral instructions, she would no longer have been testing her writing. She did help both Rob and Janice at one point when it became clear that they could not understand the instructions at all. If she hadn't helped them at that point, they would have had to stop work, so that Imogen couldn't have found out how well the remaining instructions worked.

## Performance Test Laboratories

To provide test situations that are realistic and permit extensive observation that does not interrupt the reader, some companies have built special test facilities. Typically, these include two adjacent rooms. In the first room, the test readers work with the instructions. It is furnished to resemble the setting in which the target audience will use the instructions. For instance, if the instructions tell how to use a piece of office equipment, such as a typewriter or word-processing system, the room is outfitted like an office. If the instructions tell service personnel how to repair something, the room is furnished like the kind of shop that service personnel would work in.

The second room is an observation room. From it, the people conducting the test watch the readers and record information. They might watch through one-way mirrors or with the help of television cameras. When readers are stumped by the instructions they are trying to follow, they can use a telephone to call the observation room for help—as if they were calling the manufacturer's customer service number.

Although such test facilities are relatively rare, their existence helps to underscore the importance of designing tests in which you do the following:

- Have your test readers work in a situation that resembles as closely as possible the situation in which your readers will work.
- Gather information without interfering with the readers' activity.

## Adjusting Your Tests to Circumstances

Sometimes it may be impossible for you to ask the test readers to use your communication in exactly the way your target readers would. Imagine, for instance, that Imogen's instructions included a troubleshooting section designed to enable consumers to make some repairs to their radios and tape decks on their own. Target readers would use that section only when their radios and tape decks quit working properly. The instructions would tell them what symptoms to look for. Based upon the symptoms they found, readers would determine what the problem is and then fix it.

To test her troubleshooting section in the ideal way, Imogen would need several broken radios and tape decks, each with different symptoms. But it might be difficult for her to persuade her company to spend the time and money to make such broken equipment. In that case, she might prepare *written descriptions* of malfunctioning radios and tape decks. She would ask her test readers to diagnose the problem and describe the repair they would make, based upon the information contained in her descriptions.

At work, practical circumstances may require you to adjust your test procedures in similar, creative ways.

## Understandability Test

Sometimes, you may want to see if your readers understand a communication accurately. You can test your communication's understandability by having some people read your communication and then asking them questions about its subject matter.

For example, Norman has been assigned by his employer, an insurance company, to write a new, more comprehensible version of its automobile insurance policies. Before the company begins using Norman's version, however, the company wants to test it. Through a newspaper ad, Norman has arranged to have two dozen people read a draft and answer questions about it.

Understandability tests can be used to test understandability alone or to test memorability in addition. To test understandability alone, Norman would let his test readers refer to the insurance policy while answering the questions—as if they were taking an open-book test. To test memorability also, he would have them read the entire policy, set it aside, and then answer the questions without looking at it again. He might even introduce a long interval between their reading of the policy and their work on the questions.

To decide whether or not to test memorability as well as understandability, you must think about how your target readers will normally use your communication. Norman determined that most people read their insurance policies only when they have specific questions about their coverage and that they do not try to memorize the policy's contents. For that reason, he will not test for memorability. There are situations, however, for which a memorability test is very important. For example, Jason is writing a booklet to teach people cardiopulmonary resuscitation (CPR), which is a method for keeping people alive when they have stopped breathing and their hearts have stopped beating. Memorability would be very important because most people do not have their CPR instruction booklets with them when an emergency arises. Therefore, Jason will ask his test questions after his test readers finish reading the booklet and give it back to him.

The sections below describe three ways you can construct questions to test understandability alone or both understandability and memorability. The sample questions all refer to the following paragraph, which is from the liability portion of the automobile insurance policy that Norman drafted:

> In return for your insurance payments, we will pay damages for bodily injuries or property loss for which you or any other person covered by this policy becomes legally responsible because of an automobile accident. If we think appropriate, we may defend you in court at our expense against a claim for damages rather than pay the person making the claim. Our duty to pay or defend ends when we have given the person making a claim against you the full amount of money for which this policy insures you.

### Ask Your Test Readers To Recognize a Correct Paraphrase of Your Information

When you test your readers' ability to recognize a paraphrase of your communication, you see whether or not they have understood your communication well enough to recognize essentially the same meaning as expressed in different words. In your test questions, you should present paraphrases rather than direct quotations of your communication because readers may recognize the quotation through a simple act of memory, without having understood its meaning.

To test a reader's ability to recognize a paraphrase, you can use true/false questions. Here is such a question that Norman might use:

> *True/False:* We will defend you in court against a claim when we feel that is the best thing to do.

Or you can ask multiple choice questions:

> When will we defend you in court? (A) When you ask us to. (B) When the claim against you exceeds $20,000. (C) When we think that is the best thing to do.

### Ask Your Test Readers to Provide a Paraphrase of Your Communication

When you ask the your test readers to paraphrase your communication, you can find out whether or not the readers understand the material well enough to explain it in their own words to someone else. Here is a sample question:

> In your own words, tell when we will defend you in court.

Presumably, if the readers provide a correct paraphrase your communication is understandable, and if they don't, it isn't. However, you must use your judgment to determine whether an unclear paraphrase results from ineffective writing on your part or from the test readers' lack of skill at explaining things. For this reason, if you are going to use paraphrase questions, it is important for you to have more than one test reader. Then, you can identify readers who typically have difficulty expressing themselves.

### Ask Your Test Readers To Apply Your Information

A third way to test your communication's understandability is to ask your readers to apply the information you provide. By doing so, you can see if they can understand your communication well enough to figure out how they should use your information in a particular situation.

To test your readers' ability to apply information, you need to construct a fictional situation in which the information can be used. Here is a situation that Norman devised to test his readers' ability to understand one of the sections of the insurance policy.

> You own a sports car. Your best friend asks to borrow it so he can attend his sister's wedding in another city. You agree. Before leaving for the wedding, he takes his girlfriend on a ride through a park, where he loses control of the car and hits a hot dog stand. No one is injured, but the stand is damaged. Is that damage covered by your insurance? Explain why or why not.

## Location Test

At work, you will sometimes write things in which your readers will want to find specific pieces of information without reading (or rereading) the rest of your communication. To find that information, the readers may use many of your communication's features, including its headings, topic sentences, table of contents, and index. Even the way you have arranged your prose and visual aids on the page can affect the ease with which your readers can find what they are looking for. By conducting a location test, you can determine how effective all these features are at helping searching readers.

A location test is similar to a performance test: give your communication to the test readers and ask them to find specific pieces of information as rapidly as possible. As in a performance test, you should have the test readers work in a situation as close as possible to the one in which the target readers will read, and you should gather information about the test readers' performance without disrupting their natural reading activity.

When devising a location test, you must decide what kind of assignment you want to give your test readers. The most straightforward way is to ask them to find the pages on which specific pieces of information are presented. For instance, if Norman were to construct a location test for his insurance policy, one of his questions might be:

> On which page do you find information about the liability insurance that is included in this policy?

Alternatively, you can start your test readers on their search with a problem-solving question. To do that, describe a situation that they can solve only by finding a certain kind of information. Such questions are most helpful to you as a writer if they present problems similar to those your target readers will use your communication to solve. Here is a problem-solving question that Norman might ask:

> You are making preparations for a party. Because you are running late, you ask a friend to use your car to pick up some things at a bakery so you can finish putting up the decorations. On the way back, your friend hits a parked car while trying to avoid a child who has run into the street. Find the page that tells you whether or not we will pay for the damage to the other car.

Location tests can easily be combined with an understandability test. Simply ask the test readers to apply the information they find. For example, Norman could make an understandability question out of the sample just given by rewriting the final sentence to say, "Will we pay for damage to the other car?"

Similarly, you can easily combine a location question with a performance question. Imagine, for instance, that Imogen wanted to test the success with which readers could use her instructions to solve problems with their car radio and tape deck. She could do that by presenting her readers with her instructions and a broken car radio. Her assignment to the test readers would be to find the place in her instructions that tells how to fix the problem and then to do what it says.

In sum, to test your readers' ability to use the communications you write on the job, you can use one or more of the following: performance test, understandability test, and location test.

**Guideline 2**

## Learn How Your Draft Affects Your Test Readers' Attitudes

In the preceding guideline, you learned about ways to find out how well your communication enables your readers to perform their reading tasks. This guideline looks at the second element of purpose: the way your communication affects your readers' attitudes.

At work, tests usually focus on the readers' attitudes towards one or both of the following:

- The *subject matter* of the communication. For example, does this pamphlet by the National Cancer Institute change the readers' attitudes about prohibiting smoking in public places?
- The *quality* of the communication. For example, do readers think this instruction manual is easy to use, complete, and accurate?

## Compare Attitudes before and after Reading

To determine how your communication affects your test readers' attitudes, you must learn about their attitudes both before and after reading it. Consider, for example, the test that Janet might give to determine the effects of a brochure she wrote concerning the hazards of smoking. Suppose that one of the things she learned from her test was that after reading her brochure her test readers were moderately in favor of banning smoking in public places. Unless she knew what her readers' attitudes were *before* they read her brochure, she could not conclude that their attitude about a smoking ban was caused by what she wrote. Maybe her test readers felt exactly the same way before reading, and her brochure made no difference. Or maybe they were even more strongly in favor of such a ban, and her brochure made them more doubtful about the desirability of one.

One way to gather information about your readers' attitudes before and after reading is to ask them—at *both* times—the same set of questions about their attitudes. Another way is to ask the readers after reading how their present attitudes differ from those they held before reading. There is a danger, however, in asking readers to report their earlier attitudes: readers sometimes misremember their earlier attitudes.

In some situations, you may reasonably assume that you know what your readers' attitudes are before reading, without making a special effort to find out about them. That assumption can simplify your work at testing. However, it is usually safe to make assumptions about your readers' attitudes before reading only when you can be reasonably sure that your readers were neutral. That happens most often when the readers have not read or thought about your subject before reading your communication.

## How To Determine Attitudes

What can you do to learn about your readers' attitudes? Because attitudes are something you cannot see directly, you must ask your test readers to report them to you. Here are two ways to do that.

### Ask the Readers To Describe Their Attitudes

When you ask your test readers about their attitudes, focus on the specific aspects of your subject or communication most crucial to you. "Do you think that the rights of smokers would be unfairly ignored if smoking were banned in public places?" "Do you feel that these instructions are thorough? Clear? Helpful?"

At the end of a series of focused questions, you may want to ask a more open question, such as, "What else would you like to say about this subject (or this communication)?" An open question can uncover additional insights into your readers' attitudes and their reactions to your communication.

You may present your questions to your readers in writing or orally. In either case, prepare your questions in advance, so that you are sure you get the information you want.

*Ask the Readers a Series of Questions Constructed Around a Scale.*

When you ask your readers to answer questions on a scale, you are asking them to assign their feelings a number:

Circle the number that most closely corresponds to your feelings about this communication.

It is complete.         Disagree  1  2  3  4  5  6  7  Agree
It is easy to use.       Disagree  1  2  3  4  5  6  7  Agree

Questions of this sort are especially useful when you are going to gather information about your readers' attitudes both before and after they read your communication. To determine the effect of your communication, you can simply compare the number you receive before with the number you receive afterwards. Here, for example, are some questions Janet might use to test the effectiveness of her brochure about smoking:

Smoke in a room can         Disagree  1  2  3  4  5  6  7  Agree
harm the nonsmokers
who are there.

The rights of smokers       Disagree  1  2  3  4  5  6  7  Agree
would be unfairly ignored
if smoking were banned
in theaters, stores, offices,
and other public places.

A formal analysis of results obtained in this way requires the use of statistics, something outside the scope of this book. However, if you use ten or more readers you can gain at least a general idea about the effectiveness of your communication by comparing the average score before and the average score after reading. Don't be discouraged, by the way, if the change you see is small. Especially in situations where you are writing on topics that are important personally to your readers, you cannot expect that reading your communication once will shift their attitudes greatly. A small change is a reasonable goal and a commendable accomplishment.

## Learn How Your Test Readers Interact with Your Draft while Reading

Guideline

3

Through your tests, you should learn as much as possible about how to *improve* your drafts. For that reason, you should learn not only about the outcome of your test readers' reading but also about your test readers' moment-by-moment interaction with the text. The following paragraphs describe four ways you can do that.

### Sit beside Your Test Readers during the Test

This simple, direct method of gathering information about your readers' progress through your draft is particularly useful for performance and location tests. You need no special equipment, no special preparation. You simply pull up a chair next to your readers and watch.

While watching, you should try to remain as unobtrusive as possible. Simply by being there you have already changed your test readers' reading situation from the situation in which your target readers will read. You will magnify this difference if you begin to answer questions from your test readers or help them in any way. Postpone discussions until after the readers are done reading. (For advice about how to conduct after-reading discussions, see the suggestions below for interviewing test readers.)

### Videotape Your Test Readers

You can also observe test readers by using videotape equipment. Depending upon the placement of the camera and the nature of the readers' task, you can learn a great deal from such tapes.

Videotaping has several advantages over direct observation. Most important, it removes you from the setting. Second, videotaping provides a detailed record that you and each test reader can look at together. This is important because either one of you might have difficulty remembering the details of the reader's progress through a draft. Third, by using more than one videotape camera at a time, you can see the reader from several angles of view simultaneously. This can be especially helpful when the reader is following step-by-step instructions and therefore moves from place to place (from a desk to a machine, for instance).

### Ask Your Test Readers To Provide Reading-Aloud Protocols

In a reading-aloud protocol, the test readers don't read the text aloud but rather report continuously upon their thoughts and feelings as they read through your draft. The special advantage of such protocols is the insight they give into the readers' "invisible" mental processes. For example, without a protocol you might see (in person or on tape) that a reader is looking for something, but you would not know for sure what the person was trying to find—or why—unless the person told you.

Protocols can also help you discover good strategies for overcoming problems in your draft. For instance, in an experiment in which they collected reading-aloud protocols, Flower, Hayes, and Swarts found that when readers tried to understand difficult, impersonal prose from some government regulations, the readers reconstructed the statements in terms of stories in which particular people performed specific actions. Here, for example, is one part of the regulations the researchers used in their experiment. It concerns loans made by the Small Business Administration (SBA):

> Advance payments may be approved for a Section 8(a) business concern when the following conditions are found by SBA to exist:
>
> > (1.) A Section 8(a) business concern does not have adequate working capital to perform a specific Section 8(a) subcontract; and . . .

When trying to determine the meaning of those sentences, readers would transform them into something like the following:

> If the owner of a small business needs the money to fulfill a contract, the SBA will lend it. However, if the owner already has the needed money on hand, the SBA won't make her a loan so that she can use that money on hand to buy land or things like that.

By using read-aloud protocols to learn about the ways that readers rework such statements, Flower, Hayes, and Swarts learned a great deal about how to write the regulations more effectively.[1]

The simplest method of collecting reading-aloud protocols is to have the readers speak into a tape recorder. Or, if you are videotaping the test readers, you can create a sound track at the same time.

### Interview Your Test Readers after They Have Finished Reading

In interviews, you can ask your test readers for detailed information about their efforts to use your communication and their reactions to it. Usually, you will want to combine this technique for gathering information with one of the other three you have just read about. In interviews, you should focus on problems your readers have had, because those problems may indicate areas in which you can improve your draft.

There are several ways to be sure that you discuss each of the places that caused difficulties for the test reader. For instance, you can ask the readers to make marginal marks as they are reading at every point where they have difficulty. Alternatively, you can make a list of problem-causing spots by watching your readers read and marking a copy of your draft at every point where you see a reader make a mistake, flip through pages, or act in other ways that signal a problem. If you had your test readers make reading-aloud protocols, you can listen to each reader's tape with that person, stopping to talk at each mention of a problem.

During the interviews, you should also ask your test readers to tell you how they tried to overcome the problems they encountered. Their answers can be as helpful as reading-aloud protocols for suggesting ways you can improve troublesome parts of your draft. In addition, ask your test readers what they think you should do when you revise your communication. Test readers sometimes make excellent suggestions.

During interviews, you can also ask your test readers about their attitudes while reading. When inquiring about attitudes, you should ask not only what those attitudes are but about how they were shaped, moment by moment, as the readers read.

In sum, by learning about your test readers' interactions with your draft, you can gain a wealth of important information about where readers have problems with your text, about what causes those problems, and about how you might revise your draft to eliminate the problems.

# INTRODUCTION TO GUIDELINES 4 THROUGH 6

The guidelines you have just read concern the kind of things you ask your test readers to do and the way you will gather information about their interaction with your draft. The next two concern the people you choose as your test readers, and the third concerns the draft you give them.

Guideline
### 4
## Choose Test Readers from Your Target Audience

Different kinds of people will read the same communication differently. They bring different needs and expectations to it. They bring different attitudes and different levels of knowledge and interest in the subject. Consequently, the best way to test a draft is to present it to the kind of people it is written for. If you are writing to plumbers, have plumbers test the communication. If you are writing to adults suffering from asthma, have adults with asthma serve as your subjects. In short, obtain your test readers from your target audience.

If you can't do that, then seek individuals who are as similar as possible to the members of your target audience. In your class, for instance, you may be writing a report to engineers who design robots. If you can't obtain such engineers for your test readers, you might instead use engineering seniors who have had coursework in robotics.

Guideline
### 5
## Use Enough Test Readers To Determine Typical Responses

We all read differently. A reaction or problem that one reader has with a particular communication may not bother the next hundred readers. For that reason, you must use enough test readers to feel sure that the test results you obtain are typical, not just the reflections of personal idiosyncrasies.

How many readers are enough? That depends on several factors. Generally, there is little variation from reader to reader in performance tests involving step-by-step instructions. If you get two or three members of your target audience to serve as test readers, you may have enough. More variation among readers shows with written tests. Common practice, based upon practical experience, suggests that about a dozen readers is a good number for written tests. Another factor that determines the number of test readers is your own knowledge of your audience. If you know it well, you will be able to use your personal experience to distinguish responses that probably are typical from those that probably are not.

## Use Drafts That Are as Close as Possible to Final Drafts

Even when the value of testing is obvious to both you and your employer, you will want to test a communication only once. The draft to use for that test is a *pilot* draft—one that is as close as you can make it to the final draft. For that reason, testing should follow both checking and reviewing. And the draft you test should *look* as much as possible like a finished communication. It should be neat, it should have all the drawings and figures, and it should have the same page design (same size of margins, same number of columns of print per page, same arrangement of headings, and so on).

Unless you use a pilot draft for your testing, you may have difficulty telling whether the problems you find arose because of some general weakness in the planning or drafting of your communication or simply because you didn't include some material that you intend to add later.

## SUMMARY

Testing enables you to answer two important questions about something you are writing:

- Is your communication good enough to send to your intended readers?
- How can you improve your communication?

The six guidelines in this chapter will help you design and conduct tests that produce full and useful answers to those questions.

1. *Ask your test readers to use your draft in the same ways your target readers will.* Depending upon the purpose of your communication, you may give one or more of the following kinds of tests: performance tests, understandability tests, and location tests.

2. *Learn how your draft affects your test readers' attitudes.* Be sure to learn about your readers' attitudes before reading as well as after because you will need to compare the two to determine the effect of your communication. You can gather information about attitudes by asking your readers to describe them to you or by asking them a series of questions constructed around a scale.

3. *Learn how your test readers interact with your draft while reading.* Four ways to gather information are to (1) sit beside them during the test, (2) videotape them, (3) ask them for a reading-aloud protocol, in which they

talk aloud while reading the communication, and (4) interview them after they finish reading.

4. *Choose test readers from your target audience.* Different kinds of people will read the same communication differently. The best way to test a draft is to present it to the kind of people it is written for.

5. *Use enough test readers to determine typical responses.* Even two people who are members of the same target audience will read a communication differently from one another. The more test readers you use, the more confident you can be about the results you obtain.

6. *Use drafts that are as close as possible to final drafts.* You will get the most accurate idea of how your final draft will work if you test drafts that resemble it as closely as possible.

## NOTE

1. Linda Flower, John R. Hayes, and Heidi Swarts, "Revising Functional Documents," in *New Essays in Technical and Scientific Communication,* ed. Paul V. Anderson, R. John Brockmann, and Carolyn R. Miller (Farmingdale, New York: Baywood, 1983), 41–58.

## EXERCISES

1. Explain how you would test each of the following communications:
   A. A display in a national park that is intended to explain to the general public how the park's extensive limestone caves were formed.
   B. Instructions that tell home owners how to design and construct a home patio. Assume that you must test the instructions without having your test readers actually build a patio.

2. Following the guidelines in this chapter, test one of the communications you are preparing in your writing class.

# Revising

In the three chapters of Part VI, you learned three techniques for evaluating your written work: checking, reviewing, and testing. Each technique can provide you with many ideas about how to improve the drafts of a communication before you deliver the final version to your readers. The single chapter in Part VII will help you use those ideas effectively. It concerns the final activity in the writing process: revising.

## CONTENTS OF PART VII

**Chapter 25**
**Revising**

# Revising

## PREVIEW OF GUIDELINES

1. Determine how good your communication needs to be.
2. Determine how good your communication is now, before you revise it.
3. Determine which revisions will improve your communication most.
4. Plan to make the most important revisions first.
5. Plan when to stop revising.
6. Revise to become a better writer.

Mario has just returned to his office. On the top of his desk, which already is crowded with work, he drops three binders. Each contains a copy of the current draft of the first major report he has written at work. In its forty-five pages, Mario makes several far-reaching recommendations. Each of the three copies was edited by a different reviewer—his boss and two co-workers.

Mario sits down and begins to thumb quickly through the three binders. He is surprised at what he finds. Together, his reviewers have made well over one hundred comments. Mario looks at his calendar. In only three days he must finish the report. During those same days, he has one other project to complete, several others to work on, and a staff meeting to prepare for and attend. He cannot possibly make all the revisions his reviewers have suggested. Yet he clearly needs to improve his communication as much as possible in the time he has. What should he do?

Mario's situation resembles one that you are likely to face every time you write an important communication on the job. By evaluating a draft, you will discover many ways of improving it. You could make the communication much better by acting on all those insights. However, revising your communications will be only one of your responsibilities. When you revise, you take time away from your other projects and duties. Furthermore, at work you will often have deadlines for completing your communications that will prevent you from making as many revisions as you like.

In such situations, you need to make careful decisions about which of the revisions you might make are really worth making. The six guidelines in this chapter will help you make those decisions.

## WHEN REVISING, YOU ARE AN INVESTOR

Underlying the guidelines is a metaphor: at work you will be an investor. The resources you will have to invest will be your energy, talent, and time. After evaluating a communication, you must decide whether or not you should invest any of those resources in revising it. If you decide that you should invest, you determine how much of your resources you can afford to use in this way and you try to identify the particular revisions that will give you the greatest return—the biggest improvement—on your investment.

## TO REVISE WELL, FOLLOW THE GUIDELINES FOR WRITING WELL

As you read the preceding paragraphs, you may have been wondering whether, in addition to helping you decide which revisions to make, this chapter will tell you how to make those revisions—how to revise this or that sentence, how to improve a certain paragraph, how to reorganize a tangled section. This chapter does not provide that kind of advice, because that advice is presented in detail in the chapters on planning and drafting your communications (Parts III, IV, and V). Thus, for example, if you want to revise your sentences to make them more forceful, follow the guidelines on sentences (Chapters 16 and 17). If you want to revise a table or diagram to make it more useful, follow the guidelines for using visual aids (Chapters 19 and 20).

"Is that right?" you may be asking yourself. "Shouldn't a chapter on revising tell me how to rewrite a sentence, restructure a paragraph, and so on?" Not necessarily—at least not if you can find the most important advice about those very matters elsewhere in the same book, as you can in this one. Keep in mind that the ultimate purpose of revising is the same as the ultimate purpose of planning and drafting: to create a communication that will affect your particular readers in the way you desire. Furthermore, keep in mind that whether you are writing or whether you are revising, the principles and guidelines for good writing remain the same.

Why, then, is revising necessary at all? Haven't you already applied those same guidelines as best you can while drafting, so that revising would be a redundant activity? Revising is necessary because, when drafting, you make your best guess about how to follow the guidelines. Then you evaluate your draft to determine how well you guessed. Where your evaluation shows that you might improve your application of the guidelines, you need to apply them in a different way—not search for alternative guidelines.

In sum, to revise well, follow the guidelines for writing well.

## INTRODUCTION TO GUIDELINES 1 THROUGH 3

The guidelines in this chapter will help you decide what revisions to make in a communication by helping you make a plan for investing your energy, talent, and time in revising. The first three guidelines tell you what kinds of information you need to gather so that you can formulate a good plan.

Guideline
1

## Determine How Good Your Communication Needs To Be

The first step in developing a plan for revising is to determine your goal. Because your overall purpose in revising is to raise the quality of your draft, you can define your goal by asking, "Exactly how good do I want this communication to be?"

At first, that might seem like a peculiar question to ask. "After all," you may be thinking, "shouldn't I try to make *everything* I write as good as it can be?"

### Some Communications Need To Be Better than Others

Of course, all the communications you write at work will have to achieve at least a minimum level of quality. They will need to convey essential information and they will have to be understandable. Beyond that, however, different communications need different levels of polish. For example, a coworker who has asked you to provide a small piece of information is probably not expecting highly polished prose—and might be very surprised to receive it. If you provide the key fact clearly, your memo will probably succeed no matter how rough your prose is. In contrast, if you are writing a proposal to a prospective customer, you are probably writing to someone who expects your prose to be very polished. In fact, if your writing isn't polished, this reader may infer that you or your organization is not capable of doing high-quality work on the proposed project.

## Determining How Good a Communication Needs To Be

To determine how good something you are writing needs to be, you can do three things: think about your purpose, look at similar communications, and consider a few general expectations that people at work hold about writing.

### Think about Purpose

Most important, a communication needs to be good enough to achieve its purpose. Therefore, when determining how good your communication needs to be, begin by reviewing the information you developed about the communication's purpose when you were defining its objectives (see Part II). Perhaps you recorded that information in the Worksheet for Defining Objectives (Figure 5-4), which you could pull out again now.

When reviewing the communication's purpose, look first at the task you want it to enable your readers to perform while reading. For example, do you want your communication to enable your readers to compare facts, locate and apply advice, or reconsider a previous decision? Then, ask yourself, "How good does my writing have to be to enable my readers to perform that task?"

While considering your communication's purpose, think also about the way you want your communication to affect your readers' attitudes. For example, you may want to persuade your readers that a certain course of action is desirable. Or, you may want to reinforce your readers' good opinion of your abilities. Then, ask yourself, "How polished does my communication need to be to affect the readers' attitudes in that way?"

### Look at Similar Communications

In addition to thinking about your communication's purpose, looking at similar communications prepared within your organization can help you learn how good your communication needs to be. Expectations concerning quality are determined in part by custom. Your efforts to learn what is customary will give you a general sense of which kinds of communications you need to write most carefully.

*Consider General Expectations about Writing Quality*

Finally, you can learn how well you need to write by considering three factors that influence expectations about writing quality:

- When you are *trying to gain something,* you usually need to polish your writing more than you do when you are not trying to gain something. For example, you need to write proposals more carefully than reports.
- When you are *writing to readers outside your organization,* you usually need to polish your writing more than you do when writing to readers inside your organization.
- When you are *writing to readers higher than you in your organization,* you usually need to polish your writing more than you do when writing to those who are at or close to your own level.

These general expectations can be translated into a list that ranks some typical communications according to the level of polish they usually need. That list is shown in Figure 25-1.

*An Additional Consideration*

There is one last point that you should note when thinking about how good something you are writing needs to be. When you are new on the job, people who are not involved with your work from day to day may base their

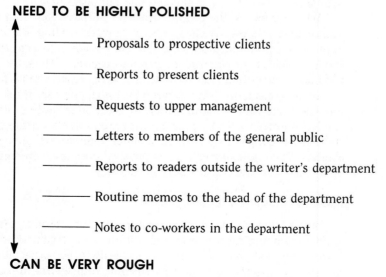

**NEED TO BE HIGHLY POLISHED**

Proposals to prospective clients

Reports to present clients

Requests to upper management

Letters to members of the general public

Reports to readers outside the writer's department

Routine memos to the head of the department

Notes to co-workers in the department

**CAN BE VERY ROUGH**

**Figure 25-1.** Amount of Polish Typically Needed by Several Kinds of Communication.

opinions of your ability solely upon your writing. For that reason, when you begin work you should be especially careful to write well.

## Determine How Good Your Communication Is Now, Before You Revise It

To plan your revising wisely, you must know not only how good you want your communication to be, but also how good it is now, before you revise it. Of course, two equally good drafts may require different amounts of revision—if they are for different communications that must exhibit different levels of writing quality in the final draft.

You will already have found out how good your communication is if you have followed the guidelines for evaluating your drafts. These guidelines, you will recall, tell you how to evaluate your drafts by checking the draft yourself (Chapter 22), by giving it to someone else to review (Chapter 23), and by giving it to members of your target audience to test (Chapter 24).

Sometimes your evaluation will show that your draft is *already* as good as it must be—so that it needs no revision at all.

## Determine Which Revisions Will Improve Your Communication Most

When you evaluate a draft, you will probably find many things that might be revised. To plan your revising wisely, you need to determine how much improvement each possible revision will bring and then rank them, either on paper or in your mind, from those that will bring the most improvement to those that will bring the least.

Figure 25-2 shows the ranked list that Mario developed after studying the many suggestions offered by the reviewers of his recommendation report (mentioned at the beginning of this chapter). Mario's list resembles very closely the ranked list of improvements that are developed by professional editors who, in some organizations, help other employees prepare their most important communications. As you look at Figure 25-2, keep three important points in mind:

- *The Importance of a Particular Revision Differs from One Communication to Another.* That's because the importance of any particular revision is relative to the importance of the other revisions you might make. In Mario's list of revisions (Figure 25-2), for example, consider the importance of supplying topic sentences. In comparison with the other items listed, supplying topic sentences is *relatively* unimportant. However, in a communication where Mario had handled effectively all of the other aspects of writing, providing topic sentences might bring more improvement than any other revision he could make.

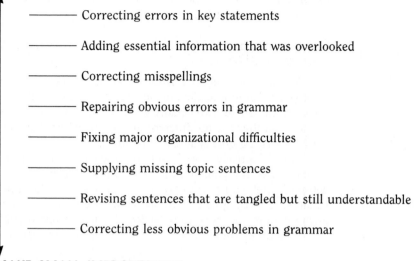

**MAKE GREAT IMPROVEMENT**

——— Correcting errors in key statements

——— Adding essential information that was overlooked

——— Correcting misspellings

——— Repairing obvious errors in grammar

——— Fixing major organizational difficulties

——— Supplying missing topic sentences

——— Revising sentences that are tangled but still understandable

——— Correcting less obvious problems in grammar

**MAKE SMALL IMPROVEMENT**

**Figure 25-2.**    Amount of Improvement that Various Revisions Would Make in a Particular Communication.

- *Revisions to Some Parts of a Communication Are More Important than Revisions to Other Parts.* That's because some parts (for example, a summary addressed to decision-makers) contribute more than others (for example, an appendix) to the overall effectiveness of the communication. The same principle applies within sections: some parts are more crucial than others. Generally, the beginning and the end are more important than the middle. Parts that make key points are more important than those that provide background information or explanations. Figure 25-3 shows the relative amounts of improvement that can be made by comparable revisions in various parts of a typical communication.

- *Obvious Errors in Spelling and Grammar Must Always Be Corrected.* The working world is very intolerant of technical errors in writing—particularly errors in spelling and basic grammar. In fact, the working world is much harsher in judging lapses in these areas than are most college teachers (including most college writing teachers). On the job, a typographical error can seriously affect the readers' attitudes toward the writing—even though the error doesn't prevent the readers from knowing what the writer meant. That's why, for example, Mario ranked correcting his spelling and grammar as even more important than fixing major organizational difficulties (Figure 25-2).

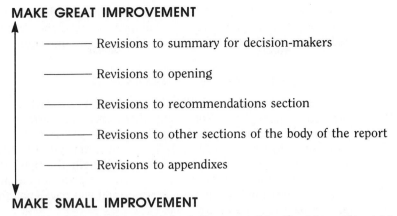

**MAKE GREAT IMPROVEMENT**

———— Revisions to summary for decision-makers

———— Revisions to opening

———— Revisions to recommendations section

———— Revisions to other sections of the body of the report

———— Revisions to appendixes

**MAKE SMALL IMPROVEMENT**

Figure 25-3.   Amount of Improvement that Comparable Revisions Would Make in Various Parts of a Communication.

# INTRODUCTION TO GUIDELINES 4 THROUGH 6

The three guidelines you have just read tell you what kinds of information you need to gather in order to make a good plan for revising a communication you write on the job. The next three guidelines describe some principles you should follow when using that information to create your revising plan.

Guideline

4

## Plan To Make the Most Important Revisions First

As you make your revising plan, keep in mind that you are deciding how much of your energy, talent, and time you are going to invest in polishing a particular communication rather than in working on one of your other responsibilities.

You want, as much as anything else, to avoid squandering your resources on revisions that don't make any difference to the outcome of your writing. Revisions make no difference if they involve minor matters that don't alter the communication's overall impact or if you make them after you had already raised the communication to the point where it is good enough to achieve its purpose.

Further, when revising you want to avoid spending your time making any revision—even one that brings about noticeable improvement—if you could spend the same time making a revision that would improve your communication even more.

Consequently, to create a plan that will enable you to invest your revising time productively, you need to do one simple thing: plan to make the most important revisions first. Begin with any revisions that are needed to bring the draft up to the minimum level of quality. Then, choose the revisions that will most increase the effectiveness of your communication.

### Example of a Plan That Is Usually Bad

Because the plan to make the most important revisions first seems so sensible, you might be wondering what a bad plan looks like—and whether anyone actually follows a bad plan. In fact, one particular bad plan is followed quite often by writers on the job. That plan is to begin with the first sentence on the first page and then to proceed sequentially through the draft, making as many revisions as time will permit. Such a plan can be satisfactory when you are preparing brief communications and you have enough time to make all the revisions you desire.

But when you are writing a longer communication or when you are facing a tight deadline, such a plan can be good only under certain rare circumstances. One of these rare circumstances is when all the revisions that will make the most difference are also those that are at the beginning of the communication. Another is when *every* revision *must* be made in order to make the communication good enough.

### Importance of Making the Most Important Revisions First

In sum, you will produce the greatest improvement with the least time and effort if you make the most important improvements first. Figure 25-4

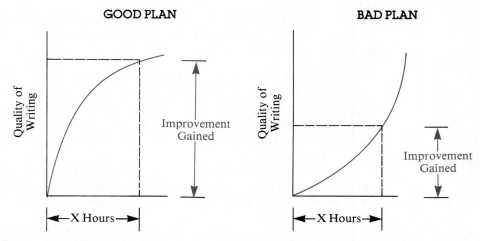

**Figure 25-4.**    Comparison of Improvement Gained in *X* Hours of Revising with a Good and a Bad Plan.

shows how much *more* improvement you can produce in a fixed amount of time by following such a good plan. Figure 25-5 shows how much more *quickly* you can achieve a given level of quality by following a good plan.

<table>
<tr><td>Guideline</td><td></td></tr>
<tr><td>5</td><td></td></tr>
</table>

### Plan When To Stop Revising

Plan to quit revising once your communication is good enough to do its job. Don't plan to do more revising than is necessary.

There is only one situation in which you should ignore this guideline: ignore it when you are following Guideline 6, discussed below.

<table>
<tr><td>Guideline</td><td></td></tr>
<tr><td>6</td><td></td></tr>
</table>

### Revise To Become a Better Writer

So far this chapter has focused on the ways that revising can help you improve the effectiveness of your writing. It has suggested that you usually should not spend more time revising than is absolutely necessary to make your communication good enough to achieve its purpose. That advice should be qualified, however. In appropriate situations, you should invest *extra* time in revising because revising is one of the chief ways of learning to write better.

When revising, you often work at applying writing skills that you have yet to master fully. By consciously practicing the skills you have yet to master, you can learn them. Then you can use them when you *draft*. For example, if you spend much of your revising time supplying topic sentences for

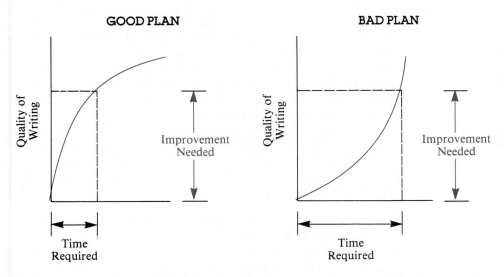

**Figure 25-5.**   Comparison of Time Required to Reach a Given Level of Quality with a Good and a Bad Plan.

your paragraphs, you are probably doing so because you have not yet learned to write those topic sentences when preparing your first drafts. Through the practice that revising can give you at performing this important writing activity, you can master the use of topic sentences so that you will supply them automatically in your first drafts of future communications.

Because you can learn so much from revising, you would be wise to devote a substantial effort to revising the assignments you prepare in the writing course for which you are using this book. Beyond doing the revising that your instructor requests, make additional revisions that focus on writing skills that you know you need to practice.

## A CAUTION

This chapter has focused largely on *efficiency*. It has suggested that you avoid spending more time than necessary at revising your drafts.

At the same time, you must also be sure to avoid spending *less* time than necessary. Although some of the communications you write at work won't need to be highly polished, others will. Be sure that you spend the time and effort necessary to make them good enough to succeed. If you stop revising a communication before you reach that point, you will have wasted your revising time by investing it in a communication that doesn't have a chance of achieving its purpose.

## SUMMARY

This chapter has presented six guidelines for revising your drafts. Underlying all of this advice is the assumption that at work you will be a busy person. Revising your communications will be only one of your many responsibilities. Furthermore, at work you will often be faced with deadlines that prevent you from spending as much time revising as you would otherwise spend. For both of these reasons, you will often have to abandon the goal of writing perfect communications—or even the more modest goal of writing as well as you are capable of writing.

Instead, you will have to concentrate on investing your energy, talent, and time wisely. You should aim to spend only as much time revising as will be beneficial. Furthermore, you will want every hour you spend revising to produce as much improvement as possible. The guidelines presented in this chapter will help you achieve those goals.

1. *Determine how good your communication needs to be.* Some communications need more polish than others. You can determine how much polish a particular communication needs by considering your purpose, looking at similar communications, and considering the general expectations about writing at work.

2. *Determine how good your communication is now, before you revise it.* Use the results of your efforts to evaluate your draft (Chapters 22–24). By comparing your communication's present quality with the quality that is needed, you can see how much improvement, if any, your revising needs to bring about.

3. *Determine which revisions will improve your communication most.* Some *kinds* of revisions bring about a greater improvement in the overall communication than do other revisions. Furthermore, revisions to some *parts* of a communication bring about greater improvement than do comparable revisions to other parts. Be sure to correct all errors in spelling and basic grammar.

4. *Plan to make the most important revisions first.* By making the most important revisions first, you spend your revising time as productively as possible.

5. *Plan when to stop revising.* Once your work at revising ceases to provide any additional benefit, you should stop revising—even if you can continue to improve the quality of the writing. The only exception is when you are following Guideline 6, below.

6. *Revise to become a better writer.* Revising gives you practice at skills that you have not completely mastered. Through this practice, you can learn how to write better.

# Using Conventional Formats

On the job, your readers will expect your communications to "look" right. They will want your letters to look like the letters they are accustomed to seeing, just as they will want your memos, formal reports, and other writing to have a familiar appearance.

To create communications that look right, you need to learn the conventional *formats* for them. In this section, you will study two kinds of formats. The first are for whole communications: letters, memos, and books. (The book format is used for formal reports, instruction manuals and other communications that have covers and bindings.) The second kind of format applies to the documentation—the reference lists, footnotes, and bibliographies—that you use to direct your readers' attention to other sources of information on your subject.

## CONTENTS

**Chapter 26**
**Formats for Letters, Memos, and Books**

**Chapter 27**
**Formats for References, Footnotes, and Bibliographies**

# Formats for Letters, Memos, and Books

**PREVIEW OF CHAPTER**

Importance of Using Conventional Formats
Formats and Employers' Style Guides
Choosing the Best Format for Your Communication
Preparing Neat, Legible Communications
Organization of the Rest of This Chapter
Using the Letter Format
Using the Memo Format
Using the Book Format

Formats are conventional packages for presenting messages. In this chapter, you will learn how to use the three formats most often employed at work: letter format, memo format, and book format.

Two of these three formats are familiar to you already, although you may never have written them yourself. You have surely read business letters sent to you by your college and by advertisers. And even if you've never read a memo addressed specifically to you, you have seen many sample memos in other chapters of this book. Perhaps much less familiar to you is the book format used at work. In many ways it resembles the format used for the novels and textbooks that you are accustomed to calling "books." For instance, it includes front and back covers, title page, table of contents, and separate chapters or sections. However, the book format includes some features not usually found in novels and textbooks, such as a summary placed at the beginning and appendixes placed at the end. And it is used not only with communications hundreds of pages long but also with ones as short as ten or twenty pages.

In this chapter, you will find detailed advice that will enable you to use all three of these formats successfully.

## IMPORTANCE OF USING CONVENTIONAL FORMATS

Why is it important for you to be able to use conventional formats well? As explained in the introduction to Part III, conventional formats can help you write effectively and efficiently. That's because the formats are *functional*—they have features designed to help readers read and use them—and because they are *familiar*—readers already know how and where to read to find the information they need.

Consider, for instance, the two special features of the book format that were just mentioned: a summary and appendixes. In a summary, you state the main points of your message in a very brief space. A summary is especially helpful to people who want to learn the gist of your message without searching through many pages of text to discover it. In appendixes, you include detailed information that your readers might want to examine but which they don't need in order to understand your overall message. For instance, in a report on an experiment you may discuss your results in the body of your communication but relegate to an appendix the lengthy calculations you used. Thus, you use the appendix to provide the detailed information that your readers may want to consult without hindering them from following your general discussion. Many other features of all three formats are similarly functional.

Conventional formats are also helpful to readers because readers are familiar with them. That means, for instance, that the readers of a letter know exactly where to look for *your* address, in case they want to write back to you. Similarly, the readers of a long report in the book format know exactly where to look for the summary: after the title page and before the table of contents. Thus, your readers' familiarity with the formats enables them to use your communications efficiently.

Besides helping you present your messages in reader-centered ways, conventional formats also help you present yourself as a creditable source of information. By definition, conventional formats are customary or usual ways of presenting messages. To people at work, they seem like natural and obvious ways of doing things. If you choose not to follow these customary practices, your readers may guess that you are someone who doesn't know how things are done on the job—not only in writing but in other areas as well. On the other hand, if you do use standard formats, you establish and maintain your credibility as a knowledgeable person.

## FORMATS AND EMPLOYERS' STYLE GUIDES

Because using standard formats can contribute so significantly to the success of writing, some employers create and distribute style guides or other instructions that tell employees how to use the formats. These instructions vary from employer to employer, but only in details. For instance, on the cover of a report, one employer may want the title placed in a certain location and another employer may want it placed somewhere else. In a letter, one employer indents the first line of a paragraph and another does not. Because of such variations, there is no universally accepted version of any of the three formats described in this chapter.

Despite these variations, however, most letters, memos, and books look very much alike. The versions of the formats described in this chapter reflect the common practices of the working world. You can use them with confidence in situations where no one has specified that you use some other version. Also, if you learn these versions, you will have little trouble adapting to other ones, because the differences involved will be minor.

## CHOOSING THE BEST FORMAT FOR YOUR COMMUNICATION

Conventional formats are somewhat *rigid*. They are rigid enough that, for example, most letters written at work look similar. From another point of view, however, these formats are very *flexible*. Each can convey any type of message. Imagine, for instance, that you want to report on a project that you have completed. You could do so in a letter, memo, or book—just as you could use any of these formats to propose a project or provide instructions.

"If that is so," you may be wondering, "how can I decide which format to use for some particular communication?" Often, you will not need to choose at all. Your boss will have said, "Send a letter to Kettering" or "Write a memo to Horton." But in situations where you must choose, the following notes will be helpful.

1. *Letters and memos are usually used for shorter communications, and books for longer communications.* The book format provides the protective cover, table of contents, summary, and other features that help readers use long communications (about ten or more pages). For shorter communications, those aids would be not needed. They would merely make the communications unnecessarily complicated to read and needlessly difficult to fit in file drawers.

2. *Letters are usually sent to people outside the writer's organization, and memos to people inside.* The format for letters includes your address so that your readers can find it readily. Each letter also contains its reader's address, so that you can find the address readily for follow-up communications. In addition, because the style of letters is customarily more gracious and formal than that of memos, the letter is a more appropriate format for addressing customers, clients, and people you don't know well. On the other hand, the memo is designed to be written quickly, so that it is well suited for internal communications, in which addresses are not necessary and formality is not expected.

3. *Format preferences vary from organization to organization.* Some organizations use letters for communications that in other organizations would be sent as memos. Some use memos to send messages that others would communicate in the book format. Let your choices concerning format be guided by the usual practices in the organization that employs you.

Notice that decisions about which format to use are based on the lengths of your messages, the location of your readers, and the preferences of your employer, not on the subjects or purposes of your messages.

## PREPARE NEAT, LEGIBLE COMMUNICATIONS

Before you study the individual formats described in this chapter, there is one important point that you should note. To use any of the formats well, you must prepare your communications neatly and legibly. If you fail to do so, you make communications that are hard work for your readers to read, and you risk offending your readers, who may feel insulted or slighted if they think that you have been careless when writing for them. In addition, if your written communications are careless, your readers may suspect that you are sloppy in other aspects of your work as well.

Here are a few brief pointers about the appearance of the communications you prepare at work (and in your writing class). They apply equally to things you prepare on typewriters and those you prepare on word processors.

- Be sure your typewriter or printer has clean, straight keys. In your communication, the letters should be sharp and evenly aligned.

- Use a dark ribbon. If your ribbon prints gray letters rather than black ones, replace it.
- Make all your corrections neatly. Before putting in corrections, carefully and completely erase the mistakes. Or cover them thoroughly with white correction fluid. If you can't make the corrections neatly, start the page again.
- Use ample, consistent margins. Margins at work usually are about 1 inch on the sides and 1 to 1½ inches on the top and bottom.

If you are using a word processor, be sure to use a printer that produces copies good enough to meet your readers' expectations and preferences. Some dot-matrix printers create communications that are difficult to read. (Dot-matrix printers create each letter not as a solid image but as a pattern of dots.) Keep in mind that some people feel that letters prepared on dot-matrix printers look unprofessional.

## ORGANIZATION OF THE REST OF THIS CHAPTER

The rest of this chapter is designed for you to use as a reference manual. When you are writing a communication in one of the three formats described here, you can turn directly to the appropriate discussion and read the advice given there without worrying about what is said about the other formats.

Because the discussion of each format provides so many detailed pieces of advice, the discussions are not organized around guidelines like those provided in much of the rest of this book. Instead, each format is discussed one part at a time. For instance, the discussion of the book format tells you how to write the cover, then the title page, and so on. Consequently, you can turn to the discussion of each part as you are writing it. You should note, however, that you should not necessarily write the parts in the order they are presented here. The parts are presented in the order in which they appear in a finished communication but you may have good reasons for writing them in another order. For instance, although a report in the book format usually begins with a summary, you will usually be wisest to write your summary last, so that you will have decided fully what you want to say before you try to summarize your message.

## USING THE LETTER FORMAT

Letters are widely used for communicating relatively short messages to customers, clients, government agencies, and other readers outside the writer's organization.

Three variations of the letter format are shown in Figures 26-1 through 26-3. All three variations have the same parts, which are described below. Letters written at work are almost always typed.

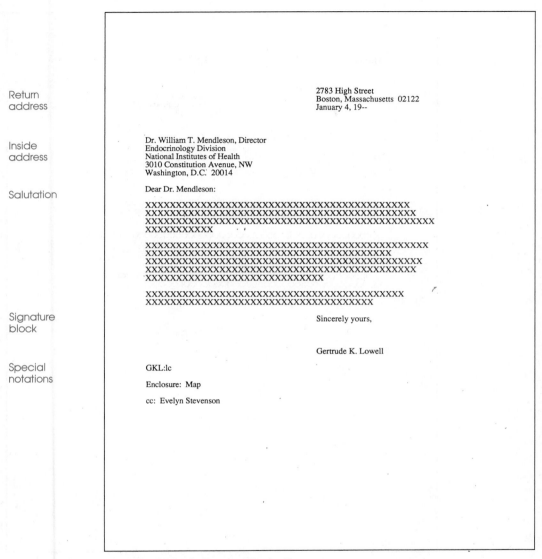

| Return address | 2783 High Street |
| | Boston, Massachusetts  02122 |
| | January 4, 19-- |

Inside address — Dr. William T. Mendleson, Director
Endocrinology Division
National Institutes of Health
3010 Constitution Avenue, NW
Washington, D.C.  20014

Salutation — Dear Dr. Mendleson:

XXXXXXXXXXXXXXXXXXXXXXXXXXXXXXXXXXXXXXXXXXXX
XXXXXXXXXXXXXXXXXXXXXXXXXXXXXXXXXXXXXXXXXXXX
XXXXXXXXXXXXXXXXXXXXXXXXXXXXXXXXXXXXXXXXXXXXXX
XXXXXXXXXXX

XXXXXXXXXXXXXXXXXXXXXXXXXXXXXXXXXXXXXXXXXXXX
XXXXXXXXXXXXXXXXXXXXXXXXXXXXXXXXXXXXXXXXXXXX
XXXXXXXXXXXXXXXXXXXXXXXXXXXXXXXXXXXXXXXXXXXX
XXXXXXXXXXXXXXXXXXXXXXXXXXXXXXXXXXXXXXXXXXXX
XXXXXXXXXXXXXXXXXXXXXXXXXXXXXXX

XXXXXXXXXXXXXXXXXXXXXXXXXXXXXXXXXXXXXXXXXXXX
XXXXXXXXXXXXXXXXXXXXXXXXXXXXXXXXXXXXXXX

Sincerely yours,

Gertrude K. Lowell

Signature block

Special notations — GKL:lc

Enclosure:  Map

cc:  Evelyn Stevenson

**Figure 26-1.**   Modified Block Format for Letters.

## Heading

The heading gives your address (but not your name) and the date. At work, people usually do not abbreviate words in the heading; they spell out *Street, Avenue,* and so on. There is one exception: people often use the Post Office's two-letter abbreviations for the states (for example, *NY* for *New York, TX* for *Texas).*

79 Contreras Avenue
San Diego, California 92108
March 12, 19--

Mr. James A. Waugh
Director of Engineering
Shippen Corporation
1634 Wespeiser Street
Kansas City, Kansas 66102

Dear Jim:

XXXXXXXXXXXXXXXXXXXXXXXXXXXXXXXXXXXXXXXXXX
XXXXXXXXXXXXXXXXXXXXXXXXXXXXXXXXXXXXXXXXXXXXXXXXX
XXXXXXXXXXXXXXXXXXXXXXXXXXXXXXXX

XXXXXXXXXXXXXXXXXXXXXXXXXXXXXXXXXXXXXXXXXXXXXXXXX
XXXXXXXXXXXXXXXXXXXXXXXXXXXXXXXXXXXXXXXXXXXX
XXXXXXXXXXXXXXXXXXXXXXXXXXXXXXXXXXXXXXXXXXXXXXX
XXXXXXXXXX

XXXXXXXXXXXXXXXXXXXXXXXXXXXXXXXXXXXXXXXXXXXX
XXXXXXXXXXXXXXXXXXXXXXXXXXXXXXXXXXXXXXXXXX

Best wishes,

Justin L. Kaputska

Enclosures (3)

**Figure 26-2.** Modified Block Format with Paragraph Indentations.

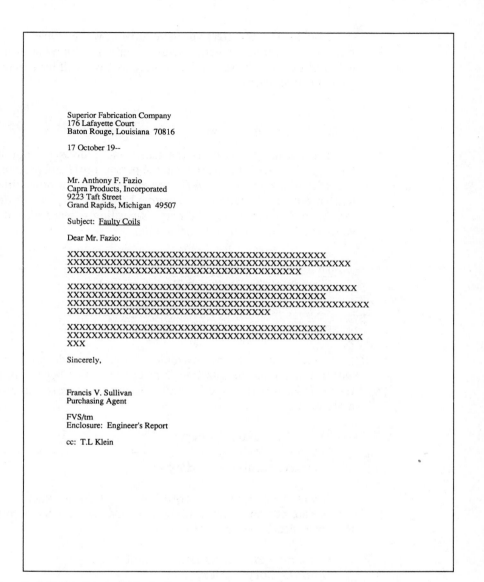

**Figure 26-3.**   Full Block Format for Letters.

Most likely, the letters you send at work will be typed on stationery that has your organization's name, address, and phone number printed on it. The date will be the only part of the heading that you will have to provide. Figure 26-4 shows an example.

## Inside Address

The inside address gives the name and address of your reader. When typing your reader's name, include the person's title and position. By custom, the titles *Mr., Mrs., Ms.,* and *Dr.* are abbreviated, but other titles are usually written out in full: *Professor, Senator.* Usually the reader's title is typed before his or her name, and the reader's position is typed after it. If you know your reader's middle initial, include it:

Reader's title ——————— Dr. Helen R. Reed, Manager ——————— Reader's position
Microbiology Research Division

Reader's title ——————— Professor John O. Jaworski, Chair ——————— Reader's position
Space Engineering Department

Notice that in the two examples above, the reader's department or unit within the overall organization is placed on a line below the person's title, name, and position. Sometimes the reader's position is also placed on a separate line:

Reader's title ——————— Dr. Harish Batterjhargee
Editor ——————— Reader's position
Journal of Molecular Physics

If you are writing to an organization but do not know the name of the appropriate person to contact, you may address your letter to the organization or a specific department in it:

Customer Relations Department
Spindex Corporation

## Salutation

The salutation is the letter's equivalent of "hello." By custom, a salutation includes the word *Dear,* usually followed by the reader's title, last name, and a colon:

Dear Professor Dobcheck:

**MIAMI UNIVERSITY**

Department of English
Bachelor Hall
Oxford, Ohio 45056
513 529-5221

August 12, 19--

Mr. Paul Ring
Box 143
Watson, Illinois   62473

Dear Mr. Ring:

I am delighted that you wish to learn about Miami University's
master's degree program in technical and scientific communication.
This professional, practice-oriented program prepares people for
careers in which they will help specialists in scientific, technical
and other fields communicate their knowledge in an understandable and
useful way.  The job market for our graduates is excellent.

People studying with us complete three semesters of course work and an
internship.  Within this framework, we strive to tailor each student's
course of study to his or her particular interests and career
objectives.  Consequently, our graduates work in a wide variety of
jobs and deal with many types of communication.  These types include
instruction manuals for computers and other high-tech equipment,
informational booklets given to cancer patients, technical
advertising, corporate procedure books, and technical reports and
proposals in many fields, such as chemistry, aerospace engineering,
pharmaceuticals, environmental protection, and health care.  A special
feature of our program is that it prepares people to advance rapidly
in the profession--to management, policy-making roles, or ownership of
their own communication companies.

We welcome applications from people with undergraduate degrees in many
different subjects, including English and the other humanities,
communication, natural and social sciences, engineering and other
technical fields, art, business, and education.  We strive to obtain
graduate assistantships or other financial aid for every student
accepted into the program who requests it.

I am enclosing a booklet describing our MTSC program in detail.  If
you wish to learn about similar programs at other schools, you may
want to purchase a copy of Academic Programs in Technical
Communication, which is sold for $12.00 by the Society for Technical
Communication, 815 Fifteenth Street, N.W., Washington, D.C. 20005.

If you have any questions about the MTSC program, please feel welcome
to write, call, or visit.

Sincerely,

C. Gilbert Storms, Acting Director
Master's Degree Program in Technical
and Scientific Communication

Enclosure:  Booklet

Excellence is Our Tradition

**Figure 26-4.** Letter Prepared on Preprinted Stationery.

If you know your reader well enough to use his or her first name in
conversation, then in your salutation you may use his or her first name
rather than last name and title. When writers use the reader's first name,
they sometimes follow the salutation with a comma rather than a colon. A
semicolon is never used:

Dear Leon,

If you do not know the name of the appropriate person to address, you may use the name of the department or organization:

Dear Customer Relations Department:

When you do not know the name of the person, you should avoid writing *Dear Sir* or *Gentlemen.* These salutations are now considered old-fashioned and even objectionable (because of the assumption they make about the sex of the person who will read and act on the letter).

## Subject Line

A subject line is like a title. It tells the reader what your letter is about and it helps your reader relocate the letter quickly in his or her files. Usually the subject line begins with the word *Subject* or *Re,* followed by a colon. You may place it above or below the salutation.

Subject:   Response to Your Letter of March 8, 19—

Re:   Continuing Problems with the SXD

Many letters are sent without subject lines. If you are unsure about whether to include one, consider the custom in your organization and the extent to which it will help the reader of the particular letter you are writing.

## Body

The body of a letter contains your message. Except in rare instances, the body is single-spaced with a double-space used between paragraphs. The paragraphs are usually short—ten lines long or less.

Customarily, the body of a letter has a three-part structure consisting of a beginning, middle, and end. In most letters, the beginning is one paragraph long. It announces the writer's reason for writing. In letters between people who have communicated frequently, it may include some personal news. The entire discussion of the writer's topic is usually contained in the middle section. The final paragraph usually includes a social gesture—thanking the other person for writing, expressing a willingness to be of further assistance, or the like.

## Complimentary Close

By custom, the complimentary close is the letter-writer's way of saying "goodbye." It begins with one of several familiar phrases, such as *Yours truly, Sincerely,* or *Cordially.* The first letter of the first word is always capitalized, but not the first letters of any other words. The complimentary close ends with a comma.

When choosing the phrase you will use in your complimentary close, consider your personal relationship with your reader. When writing to people you do not know, select one of the more formal phrases, such as *Sincerely* or *Sincerely yours*. When writing to acquaintances, use the more informal phrases, such as *Cordially* or *With best wishes*.

## Signature Block

In the signature block you give your name twice, once in handwriting (with a pen) and once typed. When you have a title, you may give it on the line following your typed name. When writing as a student, however, you should not give yourself a title.

Together, the complimentary close and signature block look like this.

| | | |
|---|---|---|
| Complimentary close | Sincerely yours, | Cordially, |
| Signature | *Raphael Goodman* | *Constance Idanopolis* |
| Typed name | Raphael Goodman | Constance Idanopolis |
| Title | Nutritionist | Head, Analytical Section |

## Special Notations

Following the signature block, you may include five kinds of notations:

- *Identification of Typist.* If the letter is typed by someone other than you, the typist may include your initials and his or hers, usually against the left-hand margin. Your initials appear first, in uppercase letters, followed by the typist's initials in lowercase letters. A colon or slash usually appears between the two:

    TLK:smc     TLK/smc

    Sometimes, only the typist's initials appear, always in lowercase letters.

- *Identification of Word-Processing or Computer File.* Letters typed on a word-processor or computer may include on a separate line the name or number of the file in which the letter is stored. Such a notation is especially common where printed copies of the letter will be distributed for review before the letter is sent to its intended reader and where the letter might, with some slight modification, also be sent to other readers. The file identification helps the writer or typist find the file again to make the revisions or changes.

- *Enclosure.* If you are enclosing materials with your letter, you may want to note that. You may also specify how many items you are enclosing or what the enclosures are:

  Enclosure

  Enclosures (2)

  Enclosure: Brochure

- *Distribution.* If you are going to distribute copies of the letter to other people, you may list those people in alphabetical order. The abbreviation *cc,* followed by a colon, appears before the names. The abbreviation *cc* stands for "carbon copy" but is used also when the copies are made on a photocopy machine.

  cc: T.K. Brandon
       F. Lassiter
       P.B. Waverly

  Sometimes it may be appropriate to include the titles, positions, and locations of the individuals in the distribution list. If you work in a large organization, some of your readers may not know who the other readers are unless you identify them in this way. Also, because people often change positions and employers, this information will help future readers know what departments were initially given copies of the communication:

  cc: T.K. Brandon, Vice President for Research
       Dr. F. Lassiter, Manager, Cryogenics Laboratory
       P.B. Waverly, Purchasing Department

- *Postscript.* A postscript is a note added to the end of a letter. Postscripts are useful for emphasis as in the following example from a technical sales letter:

  P.S. Our service contracts cost 15% less than those offered by our competitors.

  Postscripts are also useful for conveying personal notes in business correspondence addressed to friends:

  P.S. Perhaps after Thursday's meeting we could spend an hour in the Metropolitan Museum.

  If you find yourself writing a postscript because you *forgot* to make a key point in the body of your letter, you should probably draft the letter again.

## Placement of Text on the First Page

The customary margins for letters are 1 inch on the sides and 1½ inches at the top and bottom. Many letters are shorter than a full page, however. With them, you want to avoid making your letter look unusually high or low on the page. You can usually avoid those difficulties by placing your letter so that the middle line of your letter is a few lines above the middle of your page. Unless you are an experienced typist, you may find yourself typing some letters twice, once to see how many lines long it will be and once to place it pleasingly on the page.

## Format for Additional Pages

When you write a letter that is longer than one page, you should prepare the second and subsequent pages on plain paper (not on paper with your organization's name and address printed at the top). At the top of these additional pages, type a heading that includes the name of your reader, the page number, and the date of your letter. Here are two commonly used ways of arranging that heading:

Hasim K. Lederer                2           November 16, 19--

Katherine W. Hodges                                      Page 2
November 16, 19--

Begin your text with a double-space or triple-space below your heading. On additional pages, the heading is always placed ½ to 1 inch from the top of the sheet, no matter how much blank space might appear at the bottom of the page. If little except your complimentary close and signature block will appear on the final page, you might retype the preceding page with larger top and bottom margins so that you can carry over more material to the final page.

## Format for Envelope

The letters you write at work should be sent in business-sized envelopes (9½ by 4⅛ inches). Figure 26-5 shows where to place the reader's address and the return address on the envelope.

## Conventions about Style

You can use a letter to communicate any type of message. That means, also, that you can use a letter to communicate in any writing style that is appropriate at work. Nevertheless, there are some conventions about the style for letters that are widely—though not universally—observed.

Return address
down 1/2 inch,
over 1/2 inch

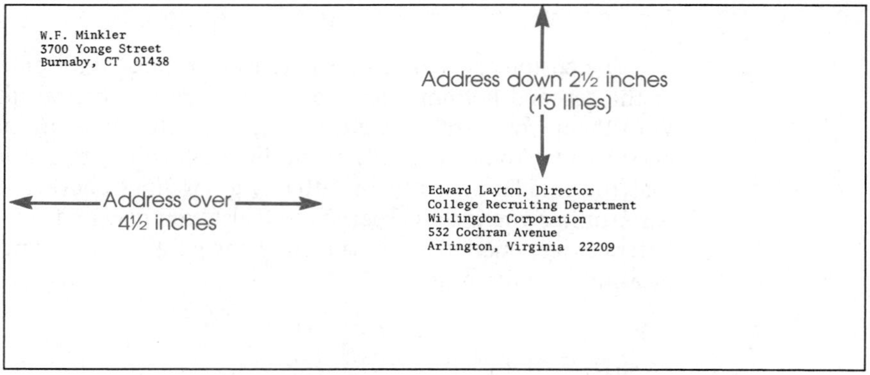

W.F. Minkler
3700 Yonge Street
Burnaby, CT  01438

Address down 2½ inches
(15 lines)

←————— Address over —————→
4½ inches

Edward Layton, Director
College Recruiting Department
Willingdon Corporation
532 Cochran Avenue
Arlington, Virginia  22209

**Figure 26-5.**   Format for Envelope.

Of the three formats discussed in this chapter, the letter usually uses the most gracious style. The polite phrasing of the salutation (*Dear* . . . ) and the complimentary close (*Sincerely yours*) are among the signs that people associate the letter with a polite conversation. In a similar fashion, the opening sentence of the body of a letter often contains the words *you* and *I*.

This style of direct, well-mannered exchange is often carried into the body of a letter, where writers often use a conversational style. They use *you* and *I*, employ relatively short sentences, and generally write as if they were talking in person with the reader. In a similar fashion, the final paragraph usually refers to the relationship between the writer and reader, for instance by expressing the writer's gratitude or eagerness to be of further assistance to the reader.

Of course, some conversations are much more formal than others—and so are some letters. Among the most formal are official letters that communicate policy or that serve as legal rulings or contracts. In the end, you have to decide upon the style of your letter in light of your purpose and reader.

## USING THE MEMO FORMAT

Memos are communications written on preprinted forms like the one shown in Figure 26-6. They are most often addressed to the writer's co-workers, managers, and other readers inside the writer's own organization.

By custom, memos may be typed or handwritten, with handwriting usually reserved for brief messages that will not be preserved in files for future reference.

### Heading

The memo's distinguishing characteristic is its heading, which has preprinted spots for you to enter your name, your reader's name, and the date.

# Rentscheller Company
Interoffice Memorandum

**FROM:**      Natalie Sebastian

**TO:**        Dwight Levy

**DATE:**      June 16, 19--

**SUBJECT:**   Questions about Peptides

    XXXXXXXXXXXXXXXXXXXXXXXXXXXXXXXXXXXXXXXXXX
XXXXXXXXXXXXXXXXXXXXXXXXXXXXXXXXXXXXXXXXXXXXXXXXX
XXXXXXXXXXXXXXXXXXXXXXXXXXXX

    1.  XXXXXXXXXXXXXXXXXXXXXXXXXXXXXXXXXXXXXXX
       XXXXXXXXXXXXXXXXXXXXXXXXXXXXXXXXXXXXXXXXX
       XXXXXXXXXXXXXXXXXXXXXXXXXXXXXXXXXX
       XXXXXXXXX

    2.  XXXXXXXXXXXXXXXXXXXXXXXXXXXXXXX

    3.  XXXXXXXXXXXXXXXXXXXXXXXXXXXXXXXXXXXXXXXX
       XXXXXXXXXXXXXXXXXXXXXXXXXXXXXXXXXX
       XXXXXXXXXXXXXXXXXXXXXXXXXXX
       XXXXXXXXXXXXXXXXXXXXXXXXXXXXXXXXXXXX

    XXXXXXXXXXXXXXXXXXXXXXXXXXXXXXXXXXXXXXXXXXXXXX
XXXXXXXXXXXXXXXXXXXXXXXXXXXXXXXXXXXXXXXXXXXX

cc: Marty Gonzoles

**Figure 26-6.**    Memo Prepared on Preprinted Stationery.

In some organizations it is customary to include the writer's and reader's positions and departments. If you do that, place them on the same lines as the names, or on the next lines. Do not include addresses.

With the increasing use of word-processing equipment that can store format material, more and more writers are using word processors to create memo forms as they write. If you are requested to write a memo in your writing course, your instructor may ask you to do the same thing on your typewriter or word processor. Figure 26-7 shows how you might do that.

```
          INTEROFFICE MEMORANDUM
          Mathematics Department


                                             March 6,  19--

                           TM
          FROM:  Trina May, Chair, Committee on Undergraduate Excellence

          TO:    Aldon Butler, English Department

          RE:    MEETING ON MARCH 18


               Thank you for agreeing to talk with the Committee on
          Undergraduate Excellence on March 18.  I have put you third on the
          agenda for that day's meeting.

               After reviewing the various topics you mentioned in your
          recent memo, the committee has decided that it would like you to
          address the following issues:

               1. Ways of giving recognition to excellent student
                  accomplishments in all aspects of university life.

               2. Strategies for encouraging students to develop their
                  academic strengths through activities not tied to their
                  classes.

               The committee will meet at 3 p.m. in Room 112 of Upham Hall.
          We all look forward to hearing your remarks.


          192A
```

**Figure 26-7.**   Memo Format Typed along with the Message.

## Subject Line

As in a letter, you can use a subject line to indicate briefly the topic of your communication. Subject lines are much more common in memos than in letters. In fact, many organizations use memo forms that provide as part of their headings a preprinted spot for a subject line.

## Body

The body of a memo closely resembles the body of a letter, with its three-part structure: beginning, middle, and end. As in a letter, the beginning states the purpose of the communication and the middle presents the gist of the writer's message.

With respect to the ending of memos, practice varies from organization to organization. In many, writers include final paragraphs just like those found at the end of most letters—ones that express thanks, indicate an eagerness to work further with reader, or the like. In many other organizations, those paragraphs are omitted on the grounds that they are redundant: everyone knows that fellow employees are willing to help one another and that they are grateful for the assistance they receive.

## Signature

There are several conventional ways of signing memos. In some organizations, writers sign only their initials, while in others they sign their full names. In some organizations they sign at the bottom of the memo, and in others they sign by the line in the heading that gives their name. In memos, writers rarely use the kind of complimentary close and signature block that are used in letters.

## Special Notations

Memos may include any of the five kinds of notations that letters sometimes have: identification of the typist, identification of the word-processing or computer file, enclosure list, distribution list, and postscript. See the discussion of those notations in the section on letters, above.

## Placement of Text on the First Page

The body of a memo begins two or three lines below the heading, regardless of the amount of blank space that might leave at the bottom of the page.

## Format for Additional Pages

When a memo extends beyond the first page, the additional pages use the same sort of heading that is used for the second and subsequent pages of letters. See the description of those headings in the section on letters, above.

## Conventions about Style

The memo format is designed to be efficient. The writer can write quickly, without needing to take time to spell out full addresses or even the words *To* and *From*. The writer usually does not write out a complimentary close—and may sign with initials instead of writing out his or her whole name.

This same concern with efficiency sometimes carries over into the writing style of the memo. Whereas letters are often (though not always) gracious, memos are often (not always) brisk. They often begin and end without the social amenities found in the beginnings and endings of many letters. Throughout, they are often crisp, though not to the point of intentional rudeness or of lack of clarity.

You must determine the best style for each memo you write in light of your purpose and your readers. Although memos are well suited to situations in which you want to communicate with your co-workers with as little fuss as possible, you can also use memos in situations where more sociability is desirable.

## USING THE BOOK FORMAT

The last of the three formats used most often at work is the book format. Whereas letters and memos are usually used for short communications, books are used for longer ones—ranging from about ten pages to many volumes.

The book format derives its name from the features it shares with novels, textbooks, and other publications we generally call books. These features include a strong binding, a protective and informative cover, a table of contents, and separate chapters or sections for each of the major blocks of material. However, the book format used at work also differs in many important ways from the format used for novels, textbooks, and similar publications. Often, communications prepared at work in the book format are typewritten on 8½ by 11 inch paper. Sometimes, the text is printed on only one side of the sheets. The binding might be a set of staples or a plastic spine with teeth that are inserted into perforations along one edge of the pages. Furthermore, the book format has several special elements such as summaries and appendixes that are not usually found in novels and textbooks. So, although the book format used at work resembles the format we usually associate with books, it differs from that format in significant ways also.

In fact, the term *book format* is a special one devised for the textbook you are now reading. In the workplace, there is no single term for this format. When reports and proposals are prepared in this format, they are often called *formal* reports and *formal* proposals to distinguish them from reports and proposals presented in letters and memos. When instructions are presented in this format, they are usually called *instruction manuals,* to distinguish them from instructions presented in letters, memos, or single-page instruction sheets.

Why, then, should the book format be given a special name in the textbook you are now reading? Giving the format its own name enables you to focus your attention on it, to study it separately from the conventional patterns for organizing reports, proposals, and instructions that are discussed in Part IX, "Using Conventional Superstructures." That separate study has two advantages.

First, it helps you learn efficiently. By studying the book format as a separate subject, you will learn the basic principles that apply to its use with all kinds of messages. Although you will have to pay attention to some minor ways in which its use in reports and proposals differs from its use in instructions, you won't have to study the format once to learn about formal reports, again to learn about formal proposals, and a third time to learn about instruction manuals.

Second, by studying formats separately from types of messages, you will learn to distinguish your decisions about the format you are using (for example, letter, memo, book) from your decisions about the conventional patterns for organizing the type of messages you are writing (for example, report, proposal, instructions). As a result, you will become a more creative and resourceful writer, able to see the full range of ways you can combine and adapt conventional formats with conventional superstructures to achieve your purposes.

The following discussion of the book format begins with a description of its three main elements: front matter, body, and supplementary materials. After that, you will find information about typing, page numbering, and style.

## Front Matter

Front matter consists of all the materials that precede the body of the communication: cover, title page, summary, table of contents, and list of figures and tables.

### Cover

Figures 26-8 and 26-9 show typical covers, one for a report and one for an instruction manual. Covers are usually printed on heavy but flexible paper.

On a cover you should give your communication's title, along with other information that will help people who file and later want to find your communication. For instance, if you have written a report for readers in some other organization, you might give both your organization's name (the name under which your readers might file it) and the name of your readers' organization (the name under which your organization might file it).

You can prepare your cover in a variety of ways. One simple method is to make a master copy on regular typing paper, then print the material onto the heavier cover paper using a photocopy machine. To make words in large type, you can use press-on letters or a Kroy lettering machine (both available in many copy centers). Some companies have covers typeset by a printer, in which case you will simply have to tell the printer what you want the cover to say.

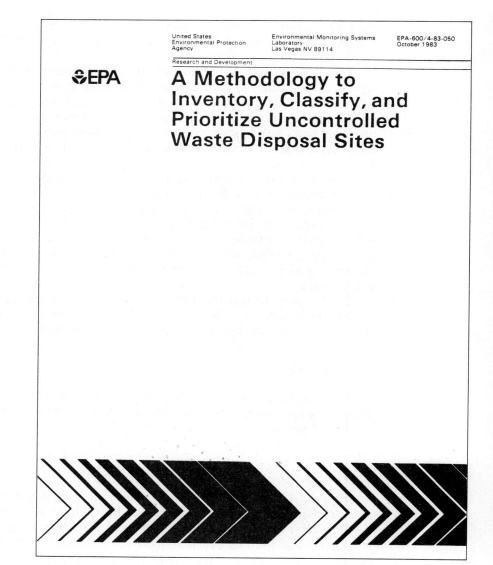

United States
Environmental Protection
Agency

Environmental Monitoring Systems
Laboratory
Las Vegas NV 89114

EPA-600/4-83-050
October 1983

Research and Development

♻EPA

# A Methodology to Inventory, Classify, and Prioritize Uncontrolled Waste Disposal Sites

**Figure 26-8.** Cover of a Report.

## Title Page

The title page of a formal report contains all of the information on the cover. You may also add the following items:

- *Names of the authors and other contributors.*

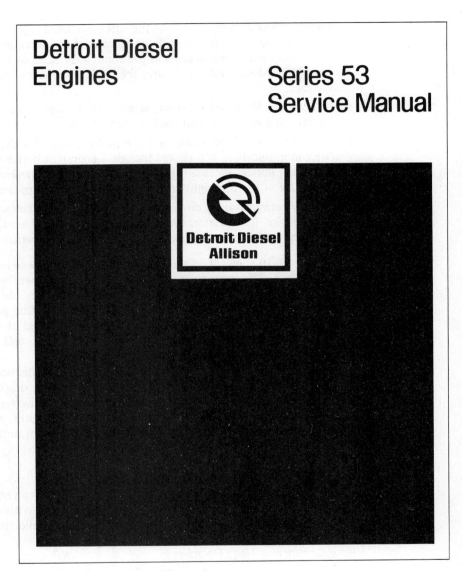

**Figure 26-9.** Cover of an Instruction Manual.

Courtesy of Detroit Diesel Allison Division, General Motors Corporation.

- *Addresses of the authors and of the organization for which they work.* If you prepared the communication for some organization other than the one that employs you, you might include that name and address also.
- *Cross-referencing material.* In a report, for example, you might include the name of the contract under which the work was done (for example,

Contract Number WEI-377-B), or the titles of related reports. In proposals, you might tell the title or number of the document issued by your readers when they asked you to submit your ideas to them. In an instruction manual, you might give the model numbers of the equipment described.

- *File number.* Some organizations assign such numbers to the reports, manuals, and other communications they produce.

- *Copyright notice.* If your employer claims a copyright on the material in your communication, you should include a copyright notice on the title page. Such notices include the word *Copyright* or the symbol ©, followed by the date of the copyright and the name of the organization.

- *Notice of restrictions on distribution or photocopying.* You would include such a notice, for instance, when writing a communication that contains information that your employer does not want its competitors to obtain.

Figure 26-10 shows the title page from the report whose cover is shown in Figure 26-8. That report was prepared by two researchers working under a contract with the U.S. Environmental Protection Agency. As you can see, in addition to the information shown on the cover, the title page gives the names and addresses of the authors, the name and address of the person in the EPA who was responsible for overseeing the report, and the contract number under which the authors prepared the report.

Figure 26-11 (page 574) shows another title page, this one for the instruction manual whose cover is shown in Figure 26-9. In this case, the title page adds the following information: the file number of the report, the date of the most recent revision, the address of the manufacturer, a note telling where readers can obtain additional copies, and a copyright notice.

## Summary

Summaries convey the essence of a communication in a very brief space, usually one page or less. Almost all reports and proposals written in the book format begin with them. Those that don't are usually less effective as a result.

### How Summaries Help Readers

By providing a summary, you help your readers in one or more of the following ways. First, your summary enables busy people (such as decision-makers) to learn the *essential contents* of your communication without reading more than a page or so. Having learned the essential message of the report, readers can then read selectively through the report for more information on the specific topics that are important to them.

Second, your summary provides your readers with a *preview* of the main points of your communication. Such a preview is especially useful to those readers who will read sequentially through the report from beginning to end.

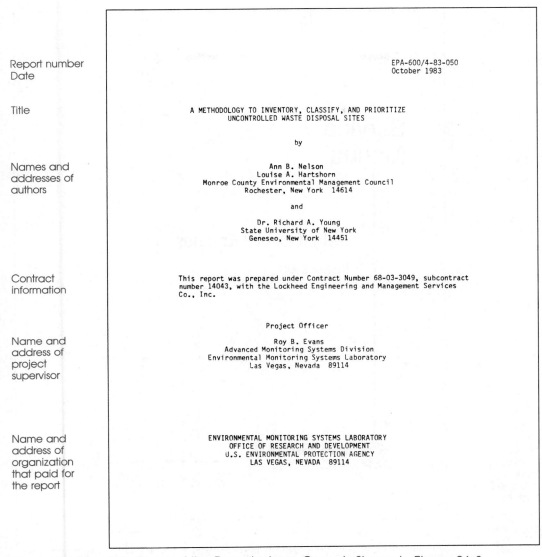

Report number
Date

Title

Names and
addresses of
authors

Contract
information

Name and
address of
project
supervisor

Name and
address of
organization
that paid for
the report

EPA-600/4-83-050
October 1983

A METHODOLOGY TO INVENTORY, CLASSIFY, AND PRIORITIZE
UNCONTROLLED WASTE DISPOSAL SITES

by

Ann B. Nelson
Louise A. Hartshorn
Monroe County Environmental Management Council
Rochester, New York  14614

and

Dr. Richard A. Young
State University of New York
Geneseo, New York  14451

This report was prepared under Contract Number 68-03-3049, subcontract
number 14043, with the Lockheed Engineering and Management Services
Co., Inc.

Project Officer

Roy B. Evans
Advanced Monitoring Systems Division
Environmental Monitoring Systems Laboratory
Las Vegas, Nevada  89114

ENVIRONMENTAL MONITORING SYSTEMS LABORATORY
OFFICE OF RESEARCH AND DEVELOPMENT
U.S. ENVIRONMENTAL PROTECTION AGENCY
LAS VEGAS, NEVADA  89114

**Figure 26-10.**   Title Page of the Report whose Cover Is Shown in Figure 26-8.

As explained in Chapter 11, previews help readers understand your message
readily by enabling them to build a mental framework for organizing and
understanding the detailed information you provide in the body of your communication.

Third, your summary can help people determine the *key results* reported
in your communication. This particular use of summaries occurs most often

File number
and date of
revision

Title

Name and
address of
company that
published the
manual

Copyright

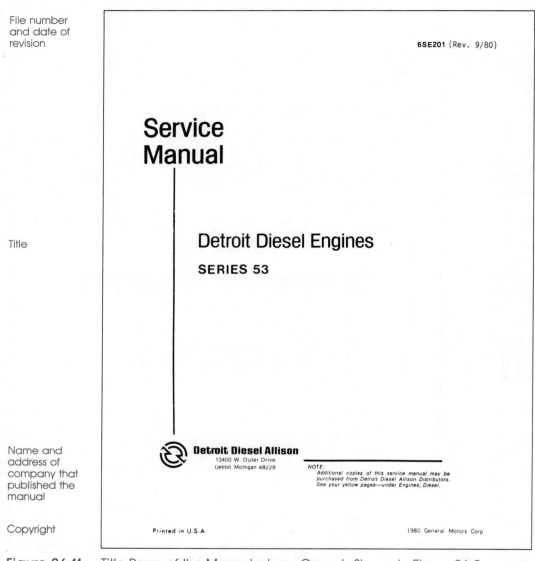

6SE201 (Rev. 9/80)

# Service Manual

### Detroit Diesel Engines

**SERIES 53**

**Detroit Diesel Allison**
13400 W. Outer Drive
Detroit, Michigan 48228

NOTE:
Additional copies of this service manual may be
purchased from Detroit Diesel Allison Distributors.
See your yellow pages—under Engines, Diesel.

Printed in U.S.A.

1980 General Motors Corp

**Figure 26-11.**   Title Page of the Manual whose Cover Is Shown in Figure 26-9.
Courtesy of Detroit Diesel Allison Division, General Motors Corporation.

among scientists, engineers, and other specialists who want to determine
what others have found when addressing problems like the ones that they
now face. Accordingly, they go to corporate, university, and public libraries
where research reports relevant to their field are stored. If you write reports
that will be placed in a library, your summary will probably be collected with

other summaries in card files, bound volumes, computerized data bases, or other research aids. By looking through these resources, researchers can quickly determine which reports have information they can use.

### Features of a Good Summary

What makes a good summary? No single pattern or formula is appropriate to all of them. You must shape each one according to your purpose and readers, asking yourself, "If I were allowed only 200 words (for example) to affect my readers in the way I desire, what would I say?"

Your challenge is particularly great because, by convention, summaries must be entirely self-contained. That means, for instance, that you must not only list your recommendations, but also introduce the problem, tell how you conducted your investigation, describe the significant results, and answer such questions about your recommendations as, How much will they cost? What will be the benefit? How will the work be managed? and, What will the schedule be? Those sample questions, as you can tell, concern the summary for a report addressed primarily to a decision-maker. In a report addressed to other researchers or designers, the emphasis would be very different. The summary would still have to introduce the problem, but would emphasize the methods used, the key results obtained, and the implications for future work.

Because summaries are self-contained, you must also present your material very clearly and precisely, so that your reader will not have to look at the body of your report to see what you mean.

### Sample Summaries

Figure 26-12 (page 576) shows the summary (called an *abstract* in this case) from the EPA report mentioned above. This summary identifies the contents of the report for readers who might be looking for the kind of tools that the report describes for identifying and evaluating waste disposal sites. It stresses the capabilities of the methodology described in the report.

Figure 26-13 (page 577) shows the kind of summary that is often written for reports addressed to decision-makers. It focuses on the practical results of the writer's study and on the action recommended by the writer.

Figure 26-14 (page 578) shows the kind of summary that is often written for other researchers. It focuses on the method of the research and on the new knowledge generated by it.

The variety of these summaries should help you appreciate the extent to which you must adapt each of the summaries you write to your particular purpose and readers. At the same time, all the summaries shown share some features. All provide background information about the purpose of the work undertaken, all describe in specific terms the nature of that work, and all tell the results obtained.

Purpose of
reported
method

Overview of
method

Significance of
method

Capabilities of
method

ABSTRACT

A comprehensive approach has been developed for use by local governments
to inventory active and inactive waste disposal sites for which little or no
information is available, and to establish priorities for further investigation.
This approach integrates all available historic, engineering, geologic, land
use, water supply, and public agency or private company records to develop as
complete a site profile as possible. Historic aerial photographs provide the
accuracy and documentation required to compile a precise record of site boundaries, points of access, and adjacent land use. Engineering borings for construction projects in the vicinity of suspected sites can be integrated with
geologic information to construct reasonable hydrogeologic models to evaluate
potential leachate impact on water wells or nearby inhabitants. Sites are
systematically ranked in terms of potential hazard based on current land use,
hydrogeology, and proximity to water wells. Greatest attention is given to
those sites which could impact public or private drinking water supplies.
This kind of evaluation is a necessary step in the prioritization of abandoned
dump sites where little is known about contents and where numbers of sites
preclude a comprehensive drilling or testing program. Case histories from Monroe County, New York, indicate that a well-designed study provides a conserva-
tive estimate of the number of large dump sites which deserve further consideration. The Monroe County study also provided a comprehensive, 50-year inventory of all potentially significant sites in a large urban area (Rochester,
N.Y.) in which at least 90 percent of initially identified targets were either
eliminated or were not classified as high priority sites.

This report was submitted in fulfillment of Contract No. 14043 under the
sponsorship of the U.S. Environmental Protection Agency. This report covers a
period from July 1981 to July 1982, and work was completed as of November 1,
1982.

- iii -

**Figure 26-12.**   Summary from the Report whose Cover is Shown in Figure 26-8.

## When To Write Your Summary

Generally, you will be able to write your summaries most easily if you
prepare them after you have completed a full draft of your communication—
perhaps even after you have completed the *final* draft of it. That's because
the summary should only make points made in the report itself. Even though

EXECUTIVE SUMMARY

*Main point of report*

The Accounting Department recommends that Columbus International Airport purchase a new operating system for its InfoMaxx Minicomputer. The airport purchased the InfoMaxx minicomputer in 1985 to replace an obsolete and failing Hutchins computer system. However, the new InfoMaxx computer has never successfully performed one of its key tasks: generating weekly accounting reports based on the expense and revenue data fed to it. When airport personnel attempt to run the computer program that should generate the reports, the computer issues a message stating that it does not have enough internal memory for the job.

*Background concerning the problem addressed*

*Source of problem*

Our department's analysis of this problem revealed that the InfoMaxx would have enough internal memory if the software used that space efficiently. Problems with the software are as follows:

1. The operating system, BT/Q-91 uses the computer's internal memory wastefully.

2. The SuperReport program, which is used to generate the accounting reports, is much too cumbersome to create reports this complex with the memory space available on the InfoMaxx computer.

*Possible solutions*

Consequently, we evaluated three possible solutions:

1. Buying a new operation system (BT/Q-101) at a cost of $350. It would double the amount of usable space and also speed calculations.

2. Writing a more compact program in BASIC, at a cost of $1000 in labor.

3. Revising SuperReport to prepare the overall report in small chunks, at a cost of $750. SuperReport now successfully runs small reports.

*Recommendation and reason for it*

We recommend the first alternative, buying a new operating system, because it will solve the problem for the least cost. The minor advantages of writing a new program in BASIC or of revising SuperReport are not sufficient to justify their cost.

**Figure 26-13.**   Summary of a Report Directed Primarily to Decision-Makers.

Description of
study: focus on
method

New
knowledge
produced by
study

Summary

Canadian respondents from the 1981 Edmonton Area
Study (mean age 35 years) and the 1981 Winnipeg Area
Study (mean age 43 years) completed indexes of loneliness
(e.g., UCLA Loneliness Scale), satisfaction, and fear of crime.
The LISREL program was used to construct a model
explaining the relationship between these endogenous
variables and age, gender, and the number of people living
in the respondent's household.

Findings indicate that males and females were so
different in their responses that separate models had to be
built. In the male model, number of people in the household
contributed to loneliness, and age had a direct effect on fear.
In the female model, age had a significant effect on
loneliness, and loneliness determined satisfaction. Younger
females expressed greater feelings of loneliness than older
females. For females, it was living alone rather than being
lonely that created fear. For males, living alone was related
to loneliness, but loneliness did not necessarily lead to fear.

**Figure 26-14.**   Summary from a Report Directed Primarily to Other Researchers.
From Robert A. Silverman and Leslie W. Kennedy, "Loneliness, Satisfaction and Fear of Crime: A
Test for Non-recursive Effects," *Canadian Journal of Criminology,* volume 27 (1985):1–13.

it is placed at the beginning of a communication, it is not an introduction
but a condensation of the body of the communication. Only after you have
worked out fully the contents of the body are you ready to write the summary.

### How Long Your Summary Should Be

As you sit down to write a summary, you may find yourself wondering
how long you should make it. In many situations, someone (like your boss)
will set your limit by specifying the maximum number of words or the maximum amount of space it can take. If you have no such guidance, you might
follow this rule of thumb: make your summary roughly 5 to 10% of the
length of your overall communication. Thus, if you are writing a 20-page
communication, make the summary between 1 and 2 pages long.

## Table of Contents

A table of contents serves two purposes. It helps readers find particular
parts of the communication—the description of the research method, the

proposed schedule, the project budget, and so on. In addition, like a summary, a table of contents provides readers with a preview of the communication that can help them read and understand its contents efficiently.

Figure 26-15 shows a sample table of contents, taken from the EPA report mentioned earlier. Notice how the writers have used blank lines to help indicate the organization of the communication: two blank lines set off the

TABLE OF CONTENTS

|                                                                          | Page |
|--------------------------------------------------------------------------|------|
| Abstract                                                                 | iii  |
| Preface                                                                  | iv   |
| List of Figures                                                          | vi   |
| List of Tables                                                           | vii  |
| List of Abbreviations                                                    | viii |
| Acknowledgements                                                         | ix   |

CHAPTER

| 1. | Introduction                                                        | 1    |
| 2. | Site Identification                                                 | 5    |
| 3. | Site Characterization                                               | 23   |
| 4. | Geologic Analysis                                                   | 33   |
| 5. | Hydrogeologic Hazard Analysis                                       | 48   |
| 6. | Application of Methodology to Rank Sites                            | 65   |

| References                                                               | 73   |
| Bibliography                                                             | 75   |
| Appendices                                                               |      |

| A. | Administrative Procedures and Pert Demonstration Model              | 76   |
| B. | Addresses of Project Resources                                      | 91   |
| C. | Call-In Campaign Form                                               | 94   |
| D. | Call-In Campaign Flyer                                              | 95   |
| E. | Site Activity Record and Guide for Completing the Form             | 96   |
| F. | Symbols for Notation on Aerial Photographs                         | 104  |
| G. | Waste Disposal Site Information Sheet                              | 105  |
| H. | Notification Letter to Individual Well Owner                       | 106  |
| I. | Notification Letter to Landlord                                    | 107  |
| J. | A Geologic Case Study for an Industrial Landfill                   | 108  |

| Glossary                                                                 | 118  |

- v -

**Figure 26-15.** Table of Contents of the Report whose Cover Is Shown in Figure 26-8.

front matter from the body of the report, and other blank lines set off other major parts. If the writers had included information about the major sections of each chapter, they might also have placed blank lines between the chapters. You will find other tables of contents in Figures 15-9 and 15-11.

For detailed advice about writing tables of contents, see Guideline 6 in Chapter 15.

## List of Figures and Tables

When your readers are looking for some part of your communication, they may be looking not for a certain paragraph, but for a particular table, drawing, or other graphic aid. You can help them by including in the front matter a list of figures and tables. In some communications, it might even be helpful to list tables separately from figures. By custom, the list of figures and tables (or the separate lists) follows the table of contents. Figure 26-16 shows the list of figures from the EPA report mentioned above. That report also contains a list of tables, which has the same format.

## Body

The body of your communication is the heart of your treatment of your subject. In communications written in the book format, the body consists of the book's chapters (often called sections).

What should these chapters be about and how should they be developed? The answer depends entirely on your purpose and readers. For advice about how to organize the body, see Part III, "Planning." Based upon the advice given there and upon your study of your particular purpose and readers, you might organize the body using one of the conventional superstructures described in Chapters 28 through 33. Or, you might find that you are best able to achieve your purpose with your readers by devising your own original structure.

### Writing the Introduction

Despite the many differences among communications written in the book format, all share one structural feature: an introduction. This introduction may appear under many titles: "Introduction," "Problem," "Need," "Background," and so on. The contents of introductions vary widely, depending on the writer's purpose and readers. For general guidelines about writing introductions, see Chapter 13 on "Writing the Beginning of a Communication." For suggestions about the contents of introductions for the specific kind of communication you are writing, see the chapters on conventional superstructures (Chapters 28 through 33).

When you begin writing an introduction, you may wonder whether to repeat the introductory material that will go into your summary. The answer

FIGURES

| Number | | Page |
|---|---|---|
| 2-1 | Site activity notation | 12 |
| 2-2 | Orthophotomap | 18 |
| 2-3 | Inventory of sites:<br>Rochester West Quadrangle | 19 |
| 4-1 | A portion of the Bedrock Surface Map,<br>Monroe County, N.Y. | 43 |
| 4-2 | A portion of the Groundwater Contour Map,<br>Monroe County, N.Y. | 44 |
| 4-3 | A portion of the Thickness of Overburden Map,<br>Monroe County, N.Y. | 45 |
| 5-1 | Geologic ranking sheet | 60 |
| 6-1 | Matrix for ranking waste disposal sites | 67 |
| 6-2 | Distribution of waste disposal sites for<br>Town of Greece, Monroe County, N.Y. | 71 |
| A-1 | PERT demonstration model | 80 |
| J-1 | Aerial photographs of Weiland Road<br>industrial landfill | 109 |
| J-2 | Weiland Road industrial landfill | 111 |
| J-3 | Preliminary Diagrammatic geologic cross-section:<br>Weiland Road industrial landfill | 113 |
| J-4 | Interpretative geologic cross-section:<br>Culver-Ridge Shopping Center site | 115 |

- vi -

**Figure 26-16.** List of Figures from the Report whose Cover Is Shown in Figure 26-8.

is *yes*. It's true that most readers will read your summary immediately before reading the introduction. Nevertheless, the universal custom is to write the introduction as if the readers were beginning there—even if it means that the first sentence of the summary and the first sentence of the introduction are exactly the same. To put it another way (and to repeat an observation made above in the discussion of summaries): a summary should not provide any information—including introductory information—that is not also provided in the body of the communication.

### Writing the Conclusion

Because you have just read that all communications written in the book format begin with an introduction, you may be wondering whether they all end with a conclusion. Many do, but not all. Instruction manuals, for instance, often end after conveying the last bit of information the reader needs to operate the equipment or perform the task being described. For general advice about how to conclude the communications you write in the book format, see Chapter 14 on "Writing the Ending of a Communication." For advice about ending the particular kind of communication you are writing, see the chapters on conventional superstructures (Chapters 28 through 33).

### Writing the Chapters

No matter how you structure the body of a communication you are writing in the book format, you should begin each of its chapters on a new page and give each its own chapter number. Usually, arabic numbers (1, 2, . . .) are used for chapters, although roman numbers are sometimes employed. Figure 26-17 shows the first page of a chapter in the sample EPA report, and Figure 26-18 shows the first page from a chapter in the instruction manual whose cover is shown in Figure 26-9. The chapters in this manual (which are called sections) are so long that each has its own table of contents.

Most likely your chapters will vary considerably in length. For instance, if you look again at the sample table of contents shown in Figure 26-15, you will see that the "Introduction" to that report is four pages long, the chapter on "Application of Methodology to Rank Sites" is eight pages long, and the chapter on "Hydrogeologic Hazard Analysis" is seventeen pages long.

If you find that chapters in your communications are similarly varied in length, do not worry—even if one or more of your chapters is less than a page long. Remember that the chapters are supposed to help the reader find information and understand the structure of your communication. The chapters should reflect the communication's logic, even if that means that they do not divide it into approximately equal parts.

## Supplementary Elements

Almost every communication written in the book format contains all of the elements you have read about so far—cover, title page, summary, table of contents, list of figures and tables, and body. The major exceptions are that shorter communications often omit the list of figures and tables, and that instruction manuals do not include summaries.

Many communications in book format also contain one or more of the following supplementary elements: appendixes; list of references, endnotes, or bibliography; glossary or list of symbols; index; and letter of transmittal.

```
                         CHAPTER 1

                         INTRODUCTION

         In the late 1970's the public became concerned about
    uncontrolled waste disposal sites that could pose a hazard to
    human health.  In order to determine the location and impact of
    these sites, accurate information is needed on site locations,
    boundaries,  contents,   subsurface hydrogeologic conditions, and
    proximal land uses.  Documentation of past waste disposal
    activities is, at best, incomplete and, in many instances,
    nonexistent. An accurate and inexpensive method is needed to
    develop  site  information based on existing data so that
    expensive drilling and testing programs can be focused on  those
    sites of greatest potential hazard to human health.

         This  report  describes a comprehensive approach that can be
    used by local governments, particularly counties and large
    municipalities, to inventory active and inactive sites for which
    little or no information is available, and to  establish
    priorities for further investigation.  The methods were designed
    by agencies in Monroe County, New York, in response to a 1978
    county legislature request to locate hazardous sites and a 1979
    New York State law requiring counties to identify suspected
    inactive hazardous waste sites and to report their locations to
    the New York State Department of Environmental Conservation
    (DEC).   The study has at various times been financially
    supported by the County of Monroe, the State of New York, and
    the United States Environmental Protection Agency.  This broad
    base of support illustrates the concern felt at  all  levels  of
    government that uncontrolled hazardous waste sites be identified
    and their impacts accurately assessed so that potential health
    hazards can  be identified and corrected.  It also reflects the
    fact that limited available resources must be committed to
    cleaning up the worst sites.

         The  study  was  conducted under the direction of the Monroe
    County Landfill Review Committee (LRC).  This committee, chaired
    by the county Director of Health, includes representatives from
    the county  departments  of  Health,  Planning,  and  the
    Environmental Management Council (EMC); the New York State
    Departments of Health and Environmental Conservation; the City
    of Rochester; and the local Industrial Management Council.  The
    participation of individuals from this broad range of interests
    facilitated  access  to  information  and provided  valuable

                             - 1 -
```

**Figure 26-17.** First Page of a Chapter in the Report whose Cover Is Shown in Figure 26-8.

## Appendixes

Appendixes can help you overcome one of the more common problems that vex people who are writing long communications. Sometimes you may find that you want to make certain information available to your reader in order to achieve your purpose, but that if you include the information in the

---

# SECTION 2

## FUEL SYSTEM AND GOVERNORS

### CONTENTS

| | |
|---|---|
| Fuel System ................................................................................ | **2** |
| Fuel Injector (Crown Valve) ........................................................ | **2.1** |
| Fuel Injector (Needle Valve) ....................................................... | **2.1.1** |
| Fuel Injector Tube ...................................................................... | **2.1.4** |
| Fuel Pump ................................................................................... | **2.2** |
| Fuel Pump Drive .......................................................................... | **2.2.1** |
| Fuel Strainer and Fuel Filter ....................................................... | **2.3** |
| Fuel Cooler ................................................................................. | **2.5.1** |
| Mechanical Governors ................................................................. | **2.7** |
| Limiting Speed Mechanical Governor (In-Line Engine) ................... | **2.7.1** |
| Limiting Speed Mechanical Governor (6V Engine) ......................... | **2.7.1.1** |
| Limiting Speed Mechanical Governor (8V Engine) ......................... | **2.7.1.2** |
| Limiting Speed Mechanical Governor (Variable Low-Speed) ........... | **2.7.1.3** |
| Limiting Speed Mechanical Governor (Fast Idle Cylinder) ............. | **2.7.1.4** |
| Limiting Speed Mechanical Governor (Variable High-Speed) .......... | **2.7.1.5** |
| Variable Speed Mechanical Governor (Pierce) (In-Line Engine) ...... | **2.7.2** |
| Variable Speed Mechanical Governor (6V Engine) ......................... | **2.7.2.1** |
| Variable Speed Mechanical Governor (Enclosed Linkage) (In-Line Engine).... | **2.7.2.2** |
| Variable Speed Mechanical Governor (Pierce) (In-Line Tractor)...... | **2.7.2.3** |
| Variable Speed Mechanical Governor (Exposed Linkage) (In-Line Engine) .... | **2.7.2.4** |
| Variable Speed Mechanical Governor (8V Engine) ......................... | **2.7.2.5** |
| Constant Speed Mechanical Governor (In-Line Engine) .................. | **2.7.3** |
| Hydraulic Governors .................................................................... | **2.8** |
| SG Hydraulic Governor ................................................................. | **2.8.1** |
| Hydraulic Governor Drive ............................................................ | **2.8.3** |
| Hydraulic Governor Synchronizing Motor ...................................... | **2.8.4** |
| Fuel Injector Control Tube .......................................................... | **2.9** |
| Shop Notes - Trouble Shooting - Specifications - Service Tools........ | **2.0** |

## FUEL SYSTEM

The fuel system (Figs. 1 and 2) includes the fuel injectors, fuel pipes (inlet and outlet), fuel manifolds (integral with the cylinder head), fuel pump, fuel strainer, fuel filter and fuel lines.

Fuel is drawn from the supply tank through the fuel strainer and enters the fuel pump at the inlet side. Leaving the pump under pressure, the fuel is forced through the fuel filter and into the inlet fuel manifold, then through fuel pipes into the inlet side of each injector.

The fuel manifolds are identified by the words "IN" (top passage) and "OUT" (bottom passage) which are cast in several places in the side of the cylinder head. This aids installation of the fuel lines.

Surplus fuel returns from the outlet side of the injectors to the fuel return manifold and then back to the supply tank.

All engines are equipped with a restrictive fitting in the fuel outlet manifold to maintain the fuel system pressure. On V-type engines, the restricted fitting is located at the rear of the left-bank cylinder head. Refer to Section 13.2 for the size fitting required.

**Figure 26-18.** First Page of a Chapter in the Instruction Manual whose Cover Is Shown in Figure 26-9.

Courtesy of Detroit Diesel Allison Division, General Motors Corporation.

body of your communication you will *reduce* your communication's effectiveness.

How can that happen? The following paragraphs describe three typical situations in which it might.

1. *When you must include detailed information that would otherwise interfere with your general message.* Imagine, for instance, that you are preparing a report on a research project. You realize that you need to include a two-page account of some calculations that you used to analyze your data. Some of your readers might want to check those calculations, and others might want to use them in another experiment. At the same time, you realize that your readers will have trouble following the general thread of your overall research strategy if they are diverted into studying the details of your calculations.

   You can solve this problem by placing the calculations in an appendix, where they are available but won't interfere with your presentation of the body of the report.

   In addition to lengthy calculations, other kinds of detailed information that are often placed in appendixes include the following:

   - Detailed data obtained in a research study.

   - Detailed drawings and illustrations.

   - Lengthy tables of values that a reader would refer to while operating some equipment or performing some procedure.

   - Detailed descriptions of the professional qualifications of the people who will work on a proposed project.

2. *When your readers are unlikely to read the body of your communication if it exceeds a certain length.* Sometimes, you may find yourself addressing a group of readers that includes some key readers who will not read the body of a communication if it appears to be too long. In fact, you may even find situations in which your readers state a limit to the length of the body of your communication.

   In situations like this, you can create a communication of the appropriate length by saying as much as you can in the number of pages you think your primary readers will read—and then including everything else of importance in appendixes. The result might be, for example, a report that has a one-page summary, a fifteen-page body, and sixty (or more) pages of appendixes. To signal your readers that the body of your communication is relatively short, you can print the appendixes on paper of a different color, so that your readers will know at a glance how much of the thick communication is devoted to its body.

3. *When you must write one communication that will meet the diverse needs of different readers.* For instance, when writing a proposal that will be read by both a decision-maker and a technical adviser, you may find that some information that will be largely irrelevant to the decision-maker may

be essential to the technical adviser. To use appendixes effectively in this situation, you first decide who your primary reader will be. Then, place material directed mainly at your primary reader in the body, and place material mainly for other readers in appendixes.

When planning appendixes, there is one other important consideration. Some writers find it tempting to use appendixes to present all kinds of material that *no* reader is likely to need. Avoid using appendixes to collect material that doesn't belong *anywhere* in your communication.

If you decide to use one or more appendixes, be sure to tell your readers that they exist. List them in your table of contents (see Figure 26-15). Give each appendix an informative title that, when listed in the table of contents, indicates clearly what the appendix contains. Also, mention each appendix in the body of your report at the point where your readers might want to refer to it:

Printouts from the electrocardiogram appear in Appendix II.

Appendix A contains the names and addresses of companies in our area that provide the type of service contract that we recommend.

In the book format, each appendix begins on its own page, just as each chapter does. It is customary to arrange and label the appendixes in the same order in which they are mentioned in the body of the communication. If you have only one appendix, it is sufficient to label it "Appendix." If you have more than one appendix, you may use roman numerals, arabic numerals, or capital letters to label them:

Appendix I, Appendix II . . .
Appendix 1, Appendix 2 . . .
Appendix A, Appendix B . . .

## List of References, Endnotes, or Bibliography

Whenever you are writing at work, you may want to direct your readers' attention to other sources of information on your subject. Perhaps you want to do so to acknowledge the sources of your information or to help your readers learn more about your topic. In the shorter formats (letters and memos), these references are often worked right into the body of the communication. In the book format, however, they are often gathered in one place. Depending on the format being used, this place might be a list of references, an endnotes section, or a bibliography.

In the book format, reference lists and bibliographies usually follow the body of the communication, or they go immediately after the appendixes.

Sometimes, however, a separate list or bibliography goes at the end of each chapter.

To learn what sources to mention and how to construct reference lists, endnotes, and bibliographies, see Chapter 27 on "Documentation."

## Glossary or List of Symbols

When you are writing at work you may sometimes use specialized terms or symbols that are not familiar to some of your readers. Sometimes, you will want to explain these terms and symbols in the body of your communication. At other times you may want to present your explanations in a separate glossary or list of symbols.

How can you decide which strategy to use? Think about your readers in the act of reading. If you are using a special term or symbol only once or in only one small part of your communication, your readers probably have no use for a glossary or list of symbols. If you include your explanation in the text, they will be able to remember it or relocate it quickly when the same term or symbol appears a few sentences later. For that reason, glossaries and lists of symbols are rarely used with short communications such as letters and memos.

On the other hand, if you are going to use the same term or symbol throughout a long communication you may be able to help your readers greatly by including a glossary or list of symbols. Imagine, for instance, that you are writing an instruction manual in which special terms appear on pages 3, 39, and 72, and twelve other places. You might define the term the first time that it appears, in the hope that your readers will remember the definition when they encounter it again. However, the readers may not be able to remember. Or they may go directly to one of the other pages without reading page 3 because the particular information they want is located elsewhere in your manual.

In either case, your readers will have to scan through pages of your text to find the definition you provided earlier. You could save your readers that search by defining the term in every place it appears, but such repetition could annoy readers once they have learned the term. By defining the term in a glossary, you provide the definition in a place that is easy to find but that keeps the definition out of the readers' way if they do not need it.

Because of the way they make explanations available without requiring readers to read them, glossaries and lists of symbols are also useful in communications where some of your readers need explanations and some don't.

Figure 26-19 shows the first page of the glossary from the EPA report previously mentioned. Notice how the author has underlined the terms being defined, so that readers can find them easily. You can use many other strategies for achieving this same goal, such as using boldface type, or placing the terms in a column of their own on the left-hand side of the page while putting the explanations of the terms on the right-hand side.

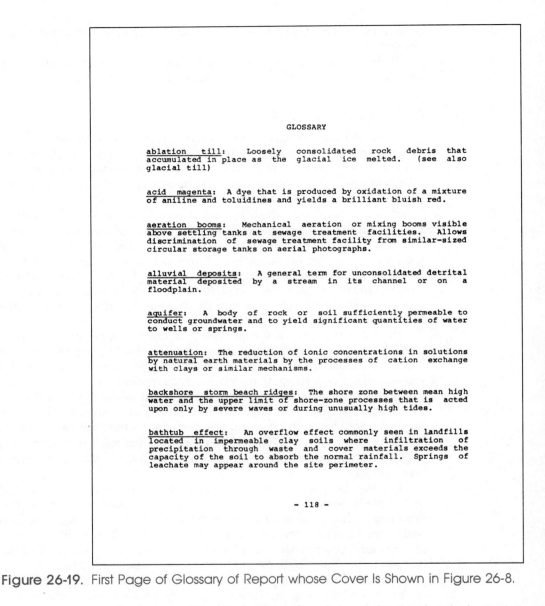

GLOSSARY

**ablation till:** Loosely consolidated rock debris that accumulated in place as the glacial ice melted. (see also glacial till)

**acid magenta:** A dye that is produced by oxidation of a mixture of aniline and toluidines and yields a brilliant bluish red.

**aeration booms:** Mechanical aeration or mixing booms visible above settling tanks at sewage treatment facilities. Allows discrimination of sewage treatment facility from similar-sized circular storage tanks on aerial photographs.

**alluvial deposits:** A general term for unconsolidated detrital material deposited by a stream in its channel or on a floodplain.

**aquifer:** A body of rock or soil sufficiently permeable to conduct groundwater and to yield significant quantities of water to wells or springs.

**attenuation:** The reduction of ionic concentrations in solutions by natural earth materials by the processes of cation exchange with clays or similar mechanisms.

**backshore storm beach ridges:** The shore zone between mean high water and the upper limit of shore-zone processes that is acted upon only by severe waves or during unusually high tides.

**bathtub effect:** An overflow effect commonly seen in landfills located in impermeable clay soils where infiltration of precipitation through waste and cover materials exceeds the capacity of the soil to absorb the normal rainfall. Springs of leachate may appear around the site perimeter.

- 118 -

**Figure 26-19.** First Page of Glossary of Report whose Cover Is Shown in Figure 26-8.

By custom, glossaries and lists of symbols may be placed at the beginning of a communication (for instance, directly before the introduction) or at the end. In either case, be sure to include them in your table of contents.

## Index

An index is one of the devices you can use to help your readers rapidly locate specific pieces of information in a long communication. At work, indexes are most common in instruction manuals. They appear much less often in reports and proposals, where readers can usually find what they want quickly enough by using the table of contents. Figure 26-20 shows the first page of the index of the instruction manual whose cover is shown in Figure 26-9. You can find another sample index in Figure 15-17.

For detailed advice about making indexes, see Guideline 7 in Chapter 15.

---

**DETROIT DIESEL 53**

### ALPHABETICAL INDEX

| Subject | Section | Subject | Section |
|---|---|---|---|
| | | Charging pump--Hydrostarter | 12.6.1 |
| **A** | | Charts: | |
| | | Engine coolant | 13.3 |
| Accessory drives | 1.7.7 | Engine operating conditions | 13.2 |
| Accumulator--hydrostarter | 12.6.1 | Injector timing gage | 14.2 |
| Adaptor--power take-off | 8.1.4 | Lubrication | 15.1 |
| Air box drains | 1.1.2 | Model description--engine | • |
| Air cleaner | 3.1 | Preventive maintenance | 15.1 |
| Air compressor | 12.4 | Cleaner--air | 3.1 |
| Air inlet restriction | 15.2 | Clearance--exhaust valve | 14.1 |
| Air intake system | 3 | Clutch adjustment | 8.1 |
| Air shutdown housing | 3.3 | Clutch pilot bearing | 1.4.1 |
| Air silencer | 3.2 | Cold weather operation--Hydrostarter | 12.6.1 |
| Alarm system | 7.4.2 | Cold weather starting | 12.6 |
| Alternator--battery-charging | 7.1 | Compression pressure | 15.2 |
| | | Compressor--air | 12.4 |
| **B** | | Connecting rod | 1.6.1 |
| | | Connecting rod bearings | 1.6.2 |
| Balance shaft | 1.7.2 | Converter--Torqmatic | 8 |
| Balance weights--front | 1.7 | Coolant--engine | 13.3 |
| Battery--storage | 7.2 | Coolant--filter | 5.7 |
| Battery-charging generator | 7.1 | Cooler--fuel | 2.5.1 |
| Battery-charging generator regulator | 7.1.1 | Cooler--oil (engine) | 4.4 |
| Bearings: | | Cooler--oil (marine gear) | 9.1.3 |
| Camshaft and balance shaft | 1.7.2 | Cooling system | 5 |
| Clutch pilot | 1.4.1 | Coupling--drive shaft | 1.4.2 |
| Connecting rod | 1.6.2 | Cover--engine front (lower) | 1.3.5 |
| Connecting rod (clearance) | 1.0 | Cover--engine front (upper) | 1.7.8 |
| Crankshaft main | 1.3.4 | Cover--valve rocker | 12.4 |
| Crankshaft main (clearance) | 1.0 | Crankshaft | 1.3 |
| Crankshaft outboard | 1.3.5.1 | Crankshaft oil seals | 1.3.2 |
| Fan hub | 5.4 | Crankshaft pulley | 1.3.7 |
| Idler gear--engine | 1.7.4 | Crankshaft timing gear | 1.7.5 |
| Belt adjustment--fan | 15.1 | Crankshaft vibration damper | 1.3.6 |
| Bilge pump | 12.2 | Cross section view of engines | • |
| Block--cylinder | 1.1 | Cylinder block | 1.1 |
| Blower (in-line and 6V) | 3.4 | Cylinder head | 1.2 |
| Blower (8V) | 3.4.1 | Cylinder liner | 1.6.3 |
| Blower drive gear | 1.7.6 | Cylinder--misfiring | 15.2 |
| Blower drive shaft | 1.7.6 | | |
| Blower end plates | 3.0 | **D** | |
| Bluing injector components | 2.0 | | |
| Breather--crankcase | 4.8 | Damper--vibration | 1.3.6 |
| By-pass valve--oil filter | 4.2 | Description--general | • |
| | | Diesel principle | • |
| **C** | | Dipstick--oil level | 4.6 |
| | | Drains--air box | 1.1.2 |
| Cam followers | 1.2.1 | Drive--accessory | 1.7.7 |
| Camshaft | 1.7.2 | Drive--fuel pump | 2.2.1 |
| Camshaft and balance shaft gears | 1.7.3 | Drive--hydraulic governor | 2.8.3 |
| Cap--coolant pressure control | 5.3.1 | | |

*General Information Section

© 1975 General Motors Corp.

February, 1975   **Page 1**

---

**Figure 26-20.** Index of Instruction Manual whose Cover Is Shown in Figure 26-9.

Courtesy of Detroit Diesel Allison Division, General Motors Corporation.

## Typing

Communications written in the book format are always typed or typeset. When typed, many are double-spaced, although many others are single-spaced. If you single-space, double-space between paragraphs. If you double-space, do not leave any additional blank lines between paragraphs, but be sure to indent the first line of each one.

## Page Numbering

You can handle the page numbering in the body of your report in either of two ways. First, you can give the first page of your introduction the number *1* and then number all your pages in one long sequence.

Alternatively, you can begin a new sequence at the first page of each chapter. If you do that, you would give each page a number that has two parts. The first part tells the chapter number, and the second part tells the number of the page within the chapter. Thus, page 1–2 is the second page of Chapter 1 and page 3–4 is the fourth page of the third chapter.

The advantage of this system is that it lets writers and typists prepare each chapter separately. Page numbers can be placed on the pages in Chapter 4, for instance, before the number of pages in the earlier chapters has been determined. In addition, if some material needs to be added or deleted from a chapter at the last minute, the only pages that would need to be renumbered are those in that chapter.

In the book format, pages that appear before the first page of the body either are given no numbers or are numbered in lowercase roman numerals (i, ii, iii, iv, and so on). Page numbering in the appendixes and other supplementary materials that follow the body depend partly upon the style of numbering used in the body of the report. If the chapters in the body are numbered separately, the supplementary materials usually are also. Thus, the second page of the first appendix might be A–2 or I–2. That same method may be used even if the pages in the body are numbered in one long sequence. Or the paging sequence of the body might be continued through all of the supplementary parts.

## Letter of Transmittal

When you prepare communications in the book format, you will often send them (rather than hand them) to your readers. In those cases, you will want to accompany them with a letter (or memo) of transmittal.

The exact contents of such a letter will depend greatly upon your purpose. Here, nevertheless, are some general observations about them:

- They usually begin by mentioning the enclosed communication. If appropriate, they will also explain or remind the reader why the communication was written:

To the director of marketing:

Today the printer delivered the instruction manuals for our new Model 100 voltage regulator. I am attaching a copy.

To a prospective client:

In response to your recent request for proposals, BioLabs is pleased to submit the enclosed plan for continuously monitoring effluent from your Eaton plant for various pollutants, including heavy metals.

To the director of manufacturing operations:

As you know, during the past six months our Number 1 machine has been shut down three times because of damaged bearings. The enclosed report details my study of this problem.

- The letter may then say something about the purpose, contents, or special features of the communication. What writers say about the enclosed communication depends very much on their purpose in the letter. For instance, the writer of an instruction manual may want the reader to be impressed with the good writing job he or she did. Consequently, the writer may point out special features of the manual and tell the readers what it will accomplish. The writers of a research report may want to be sure that the readers note the most important consequences of the reported findings. Accordingly, they may briefly explain the main findings and list the most important recommendations. (In this instance, the letter would repeat some of the information contained in the summary; such repetition is common in letters of transmittal.)

- The letter may acknowledge the assistance of key contributors to the communication.

- Like most letters of any kind, a letter of transmittal usually ends with a short paragraph (often one sentence) that does something like one of the following things: expresses the hope that the reader will find the communication helpful or satisfactory, or states his or her willingness to work further with the readers or to answer any questions the readers may have.

Figure 26-21 shows a letter of transmittal prepared in the workplace, and Figure 26-22 shows one prepared by a student.

## Conventions about Style

There are no conventions about writing style—tone of voice, level of formality, and so on—that apply generally to all communications prepared in the book format. Some are informal, others quite formal, almost stiff. There

ELECTRONICS CORPORATION OF AMERICA
MEMORANDUM

TO:     Myron Bronski, Vice-President, Research

FROM:   Margaret C. Barnett, Satellite Products Laboratory

DATE:   September 30, 19--

RE:     REPORT ON TRUCK-TO-SATELLITE TEST


On behalf of the entire research team, I am pleased to submit the
attached copy of the operational test of our truck-to-satellite
communication system.

The test shows that our system works fine.  More than 91% of our
data transmissions were successful, and more than 91% of our voice
transmissions were of commercial quality.  The test helped us
identify some sources of bad transmissions, including primarily
movement of a truck outside the "footprint" of the satellite's
strongest broadcast and the presence of objects (such as trees) in
the direct line between a truck and the satellite.

The research team believes that our next steps should be to develop
a new antenna for use on the trucks and to develop a configuration
of satellites that will place them at least 25° above the horizon
for trucks anywhere in our coverage area.

We're ready to begin work on these tasks as soon as we get the okay
to do so.  Let me know if you have any questions.



Encl: Report (2 copies)

**Figure 26-21.**   Letter of Transmittal Written at Work.

are, however, some conventions about the style used in some of the specific
kinds of communications prepared in the book format—such as the formal
report, the formal proposal, and the instruction manual. For information
about the style of the particular type of communication you are writing, see
the chapters on conventional superstructures (Chapters 28 through 33).

Box 114, Bishop Hall
Miami University
Oxford, Ohio  45056
December 10, 19--


Professor Thomas P. Weissman
Department of English
Miami University
Oxford, Ohio  45056

Dear Professor Weissman:

    I am enclosing my final project for your technical writing
course, a proposal for a crime prevention program directed to
elderly citizens of Oxford.  As you recall, I have developed this
proposal at the request of the City Manager, Tom Dority.

    While working on this project, I learned that persons 65 and
older comprise 18% of the non-student population of Oxford.
Experience nationwide suggests that these individuals are
particularly vulnerable to crime, but that much of that crime can
be prevented through simple precautions.  I propose that the City
of Oxford help protect its elderly by participating in two
nationwide programs, Whistle Alert and Operation Safe Return, and
that the city offer a series of nine presentations on crime
prevention for the elderly.  The entire effort could be supported
by donated supplies and services, with no cost to the City of
Oxford.

    Throughout my work on this project, I received much help from
the Oxford Police Department's Crime Prevention Officer, Dwight
Johnson.  I have also been assisted by the staff at the Oxford
Senior Citizens Center.

    I believe that there is a reasonable chance that the city will
accept my plan.  Officer Johnson has already said that he likes it.

    Thank you very much for your help and encouragement.

                    Sincerely,

                    *Tricia Daniels*

                    Tricia Daniels


Enclosure:  Final Project

**Figure 26-22.**   Letter of Transmittal Written by a Student.

# Formats for References, Footnotes, and Bibliographies

## PREVIEW OF CHAPTER

**Purposes of Documentation**
**Deciding What To Acknowledge**
**Choosing a Format for Documentation**
**Using Author–Year Citations Combined with a Reference List**
**Using Numbered Citations Combined with a Reference List**
**Using Footnotes**
**Using Bibliographies**

In many of the communications you write at work, you will want to tell your readers about other sources of information concerning your subject. Documentation consists of the references, footnotes, bibliographies, and other devices you use to do that.

This chapter will help you handle the many decisions and details involved with using documentation. First, you will learn when to provide documentation—and when not to. Then, you will learn how to select the most appropriate documentation format for a particular communication. You will also learn, in detail, how to use the three most popular formats for acknowledging the sources from which you have drawn the facts and ideas you present in the body of your communication. Finally, you will learn how to prepare bibliographies, through which you can provide your readers with a wider list of sources of information about your subject.

## PURPOSES OF DOCUMENTATION

At work, you may provide documentation for any of four reasons:

- *To acknowledge the people and sources that have provided you with ideas and information.* At work, as at school, such acknowledgement is more than a courtesy: it is an ethical obligation.

- *To help your readers find additional information about something that you have discussed.* In many situations, one of your readers' major questions will be: "Where can we learn more about that?" You can answer that question through documentation.

- *To persuade your readers to consider seriously a particular idea.* By showing that an idea was expressed by some respected person or in some respected publication, you are arguing that the idea merits acceptance.

- *To explain how your research relates to the development of new knowledge in your field.* In research proposals and in research reports published in professional journals, writers often include literature survey sections to help demonstrate how their research projects contribute to the knowledge and capabilities in their fields. For more information about how to use references in this way, see the discussion of literature survey sections in Chapter 25.

## DECIDING WHAT TO ACKNOWLEDGE

The first of the four reasons for providing documentation deserves your special attention. At work you will have the same sort of ethical obligation that you have at school to acknowledge the sources of your information. However, the standard used to determine what material you need to acknowledge is somewhat different at work than at school.

In both places, you must document material (1) that you have derived from someone else and (2) that is not "common knowledge." The difference between the standards at school and at work lies in the interpretation of what is to be considered common knowledge. At school common knowledge is knowledge that everyone possesses without doing any special reading. Thus, you must document *any* material you find in print. In contrast, at work common knowledge is the knowledge usually possessed by people *in your field*. Thus, you do not need to acknowledge material you obtained through your college classes, your textbooks, the standard reference works in your field, or similar sources.

## CHOOSING A FORMAT FOR DOCUMENTATION

Once you have decided which sources to acknowledge, you must determine where to place information about them and how to present that information to your readers. Your decisions about these matters are greatly simplified because there are standard formats for documentation. Unfortunately, however, there is no single format that is correct for all situations. There are many different formats, some very different from one another, some differing only in small details. Consequently, to document correctly you must do two things:

1. Find the particular format that is required or most appropriate for the communication you are writing.
2. Follow that format to the last detail.

How can you find the appropriate format? First, ask someone, "Is there a specific format I must use for documentation?" Your employer may specify a particular format. If you are writing to people in another organization, that organization may have its own preferences or requirements about documentation. And all professional journals tell authors exactly what format to use. To find information about the documentation format a journal prefers, look inside the front or back cover or on the page that gives the journal's address.

To describe the documentation format they want their employees to use, many employers have created and issued style guides that include rules for documentation together with sample citations. Many other employers—and almost all professional journals—ask writers to follow one of the popular published style guides. These include the *APA Style Guide* (published by the American Psychological Association), the *CBE Style Guide* (published by the Council of Biological Editors), and the *MLA Style Guide* (published by the Modern Language Association). Each of these style guides is used widely in particular fields: the APA in the social sciences, the CBE in the life sciences, and the MLA in the humanities.

In addition, there are some style guides that are not tied to any particular fields. Of these, the most widely used in the working world is *The Chicago Manual of Style* (published by the University of Chicago Press).

If it turns out that there is no specific style guide that you must follow, then find out what *type* of format your readers are accustomed to using. Overall, the hundreds of formats fall into three types:

- *Author–year citations combined with reference lists.* This type of format is used in the sciences, the social sciences, engineering, and many other fields. Overall, it is gradually becoming much more common than either of the others.
- *Numbered citations combined with reference lists.* This type of format is a variation of the author–year type. It is used, though not as widely, in the same fields.
- *Footnotes.* This type of format is used widely in the humanities and is preferred in many business and government organizations regardless of the field of their work.

Once you have found out which type of format to use, turn to some convenient source that describes it. One such source is this chapter, which treats all three formats in detail and also discusses the uses and formats for bibliographies.

# USING AUTHOR–YEAR CITATIONS COMBINED WITH A REFERENCE LIST

In the author-year format, you cite a source by putting the name of the author and the year of publication at the appropriate place in the body of your communication. This citation refers your readers to the full bibliographic information that you provide in an alphabetical list of sources at the end of your communication. The following sections tell you how to write the citation, where to put it in the body of your communication, and how to write the entry in your list of references.

The information provided below about the author–date format follows *The Chicago Manual of Style* almost exactly, varying only in a few places where simplification seems both possible and desirable for the writing you will do on the job.

## Writing Author–Year Citations

To write an author–year citation, enclose the author's last name and the year of publication in parentheses *inside* your normal sentence punctuation. Do not put any punctuation between the author's name and the year:

The first crab caught in the trap attracts others to it (Tanner 1985).

If you are referring to one particular part of your source, you can help your reader find that part by including its page numbers in your citation. Place a comma between the year and the page numbers:

(Angstrom 1982, 34–49)

If you use the author's name in part of your sentence, place the year of publication and page numbers immediately afterwards in parentheses. Do not repeat the author's name:

Angstrom (1982, 34–49) showed that the strength of these desires is inversely related to the person's level of self-confidence.

In some of your communications, you may cite two or more sources by the same author. If they were published in *different* years, your reader will have no trouble telling which work you are referring to. If the works were published in the *same* year, you can distinguish between them by placing lowercase letters after the publication dates in your citations and in your reference list:

(Burkehardt 1981a)

(Burkehardt 1981b)

If you are citing a work with two or three authors, give the names of all authors. Notice that no comma precedes the word *and:*

(Hoeflin and Bolsen 1986)

(Wilton, Nelson and Dutta 1978)

If you are citing a work with more than three authors, give the first author's name, followed by *et al.,* which is an abbreviation for the Latin phrase *et alii* ("and others"):

(Dutta et al. 1984)

If you are citing a source that does not name an individual or set of individuals as authors, give the name of the organization that publishes or sponsors the source:

(National Cancer Institute 1983)

If you need to name two or more sources in one place, enclose them all within a single pair of parentheses, separating them from one another with semicolons. Do not use the word *and* between sources:

(Justin 1984; Skol 1972; Weiss 1986)

In this example, the three sources are arranged alphabetically. They could also be arranged chronologically.

## Deciding Where To Place Citations

Your primary objective when deciding where to place your citations should be to make clear to your readers what part of your text is being referred to by each citation. That will be easy with citations that pertain to a single fact, sentence, or quotation. You simply place your citation immediately after the appropriate material:

According to D.W. Orley (1978, 37), "We cannot tell how to interpret these data without conducting further tests."

Researchers have shown that a person's self-esteem is based upon performance (Dore 1964), age (Latice 1981), and weight (Swallen and Ditka 1970).

If your note refers to material in several sentences, you can place your citation in what your readers will clearly see as a topic sentence for the affected material. Your readers will then understand that the citation covers all the material that relates to that topic sentence. To assist your readers even more, you may want to use the author's name (or a pronoun) in more than one sentence:

A much different account of the origin of oil in the earth's crust has been advanced by Thomas Gold (1983). He argues that . . . To critics of his views, Gold responds . . .

## Making the List of References

In the list of references you should describe each source in enough detail to help your readers find the source quickly in a library. If you are using author–year citations, arrange the list alphabetically by author. No matter how many times you cite the same work, place it in your list only once.

For your convenience, the following discussion groups three types of listings: those for books, those for journal articles, and those for other kinds of sources you are likely to cite on the job.

## References to Books

When describing a book, you should provide the following information:

- *Author.* Give the author's last name first, followed by the author's initials. Alternatively, you may use the author's given name and middle initials, copying exactly the way they appear in the book you are citing. Follow the author's name with a period.

    If there is more than one author, give the names of the second and additional authors in their natural order (first initial, middle initial, last name). Separate the names of the various authors with commas, and place *and* before the last author. Follow the last author's name with a period.

- *Year of publication.* Follow the year of publication with a period.
- *Title.* Capitalize only the first word of the title, the first word of the subtitle, if there is one, and proper nouns. Underline and follow with a period.
- *Edition.* If you are using the second or subsequent edition, tell which one you used. Abbreviate *edition* to *ed.* Use *2nd* and *3rd* rather than *second* and *third.*
- *Publisher.* Write the city of publication, a colon, and the name of the publisher. Follow with a period.

Below you will find examples that show how this information is presented for various commonly cited types of books. In these samples, the authors' first and middle names are represented only by their initials even if their names are given in full on the title page. As you study the examples, note carefully their capitalization and punctuation.

### One Author

Ayers, R.U. 1969. Technological forecasting and long-range planning. New York: McGraw-Hill.

### Two or Three Authors

Harris, J., and R. Kellermayer. 1970. The red cell: Production, metabolism, destruction, normal and abnormal. Cambridge, Massachusetts: Harvard University Press.

Middleditch, B.S., S.R. Missler, and H.B. Hines. 1981. Mass spectrometry of priority pollutants. New York: Plenum Press.

Be sure to place a comma between the first author's name and the second author's name.

*More Than Three Authors*

> Cooke, R.U., D. Brunsden, J.C. Doornkamp, and D.K.C. Jones. 1982. Urban geomorphology in drylands. New York: Oxford University Press.

In your reference list, be sure to name *all* the authors of a source, even though in the body of your communication you give only the first author's name followed by *et al.*

*Corporate Author or Sponsor*

> National Institutes of Health. 1976. Recombinant DNA research. Washington, D.C.: U.S. Government Printing Office.

This form is used to cite publications that do not name an individual or group of individuals as authors. If the sponsoring organization is also the publisher, you may give the organization's name twice in the listing, once as the author and once as the publisher.

*Editor of a Collection of Essays or Articles*

> Sabin, M.A., ed. 1974. Programming techniques in computer aided design. London: NCC Publications.

This form is used to cite books that give an editor's name, not an author's name, on the title page. Such books typically include essays or articles by many individuals. If you wish to cite only one essay in the book, use the format described below under the heading "References to Other Kinds of Sources." If the book has more than one editor, use the plural abbreviation *eds.* after the last editor's name.

*Second or Subsequent Edition*

> Hay, J.G. 1978. The biomechanics of sports techniques. 2nd ed. Englewood Cliffs, N.J.: Prentice-Hall.

## References to Journal Articles

When describing a journal article, you should provide the following information:

- *Author.* Give the names of authors of journal articles in the same way you give the names of authors of books (see above).
- *Year of Publication.* Follow with a period.
- *Article Title.* Capitalize only the first word of the title, the first word of

the subtitle, if any, and proper nouns. Do *not* underline or enclose in quotation marks. Follow with a period.

- *Journal Title.* Capitalize all major words in the title. Underline.
- *Volume.* Give the volume number in arabic numerals even if the journal uses roman numerals. Do not place any punctuation between the journal's title and the volume number. Follow the volume number with a colon.
- *Issue.* If the journal numbers its pages in a single sequence throughout the volume, omit the issue number. If it begins numbering pages anew with each issue, identify the issue by placing its date in parentheses directly after the volume number—and before the colon.
- *Page Numbers.* Give the numbers of the first and last pages of the article. Do not put any spaces between the first page number and the colon that follows the volume or issue number.

The following samples show how this information is presented. The form varies slightly according to the number of authors.

*One Author*

McNerney, W.J. 1980. Control of health care costs in the 1980's. New England Journal of Medicine 303:1088–95.

*Two Authors*

Lynn, S.J., and J.W. Rhue. 1986. The fantasy-prone person: Hypnosis, imagination, and creativity. Journal of Personality and Social Psychology 50:404–8.

*Three or More Authors*

Feldhamer, G.A., J.E. Gates, D.M. Harman, A.J. Loranger, and K.R. Dixon. 1986. Effects of interstate highway fencing on white-tailed deer activity. Journal of Wildlife Management 50:497–503.

Be sure to name all the authors in your reference list, even though in the body of your communication you give only the first author's name followed by *et al.* when citing articles with more than three authors.

*Journal That Begins Numbering Pages Anew in Each Issue*

Hoeflin, R., and N. Bolsen. 1986. Life goals and decision making: Educated women's patterns. Journal of Home Economics 78(Summer):33–5.

## References to Other Kinds of Sources

The following examples show how to describe some other kinds of sources you may use when writing on the job. The forms for several of these kinds of sources follow the patterns, just described, for listing books and journal articles.

### Pamphlet or Booklet

> U.S. Environmental Protection Agency. 1977. Is your drinking water safe? Washington, D.C.: U.S. Government Printing Office.

Treat pamphlets and booklets in the same way you treat books that do not identify any author or authors on the title page. Note that where the title ends with a punctuation mark (question mark or exclamation point), the period after the title is omitted.

### Essay in a Book

> Morris, M.G. 1976. Conservation and the collector. In Moths and butterflies of Great Britain and Ireland, ed. J. Heath, vol. 1, 107–16. London: Curwen.

Treat the title of the essay just as you would treat the title of a journal article. Place the word *In* before the title of the book. After the name of the book, place a comma, the abbreviation *ed.* (for "edited by"), and the name of the editor. This example shows the listing for an essay in a multivolume collection. For essays in books with only one volume, leave out the volume number. A comma would still follow the editor's name and a period would still follow the page numbers.

### Paper in a Proceedings

> Stover, E.L. 1982. Removal of volatile organics from contaminated ground water. In Proceedings of the second national symposium on aquifer restoration and ground water monitoring, ed. D.M. Nielsen, 77–84. Worthington, Ohio: National Well Water Association.

You should list a paper in a proceedings in the same way you list one essay in a collection of essays.

### Encyclopedia Article

> Aller, L.H. 1982. Astrophysics. McGraw-Hill encyclopedia of science and technology 5th edition.

Your listings for articles in encyclopedias should resemble your listings for essays in essay collections. Be sure to include the edition number and omit the city and year of publication, as well as the page numbers (because the work is arranged alphabetically). If the encyclopedia does not name an author, begin the listing with the article's title, as in the following example:

Asparagus. Encyclopaedia Britannica 15th edition, 1986.

### Article in a Popular Magazine

Davis, D. 1985. Furniture made fun. Newsweek 106(4 November):82–3.

Treat listings for articles in popular magazines in the same way you treat articles in journals that begin numbering pages anew in each issue: follow the title of the magazine with the volume number, the date of the issue (in parentheses), a colon, and the page numbers. If the magazine does not name an author for the article, begin the listing with the article's title.

### Newspaper Article

Lewis, P.H. 1986. UNIX and MS-DOS: Dueling for dominance in computers. New York Times, 13 May, sec. C, 9.

List articles in newspapers in much the same way you list articles in popular magazines. Give the section number as well as the page number. If the newspaper does not name an author, begin the listing with the article's title.

### Letter to You

Torgul, D., Director of Research and Development, Sterling Corporation. Letter to the author, 10 August 1986.

### Interview with You

Cawthorne, L., Attorney at Law. Interview with the author in New York City, 18 March 1986.

## Sample List of References

Figure 27-1 (pages 606 and 607) shows a sample list of references. Notice that if your list includes two entries by the same author, the author's name is given only for the first. The second and subsequent entries begin with a short, dashed line. (See the entries for the two books by R. Ayers.)

# USING NUMBERED CITATIONS COMBINED WITH A REFERENCE LIST

This second format is a variation of the author–year format just described. The only difference in your reference list is that you assign a number to each entry. However, the citations you make in the body of your communication are very different from author–year citations. Instead of giving the author and year of the source, you give the number you assigned the source in your reference list. Also, instead of placing your citations inside of the sentence punctuation, you place them outside:

> Since 1970, the number of industrial robots in the United States has increased 10,000 times. (4)

If you wish to include page numbers in your citation, put them after a colon. No space separates the colon from the page numbers:

> Since 1970, the number of industrial robots in the United States has increased 10,000 times. (4:522–3)

When you are using numbered citations, you may arrange your reference list alphabetically or place the sources in the order of their citation in your prose. If you cite a particular source more than once, use the same number for it in each citation.

# USING FOOTNOTES

The footnote format is especially common in the humanities, but is occasionally used in some other fields as well. In this format, you cite reference material with a number that is raised one-half line in your prose, like this: [1]. In communications written at work, the citations usually refer to notes gathered at the end of either the communication or sections within the communication. The section of gathered notes is entitled "Notes," "Footnotes," or "Endnotes." Less often, each note is printed at the bottom of the page on which its citation appears.

The information provided below about the footnote format follows the widely used *Chicago Manual of Style*.[1]

## Writing Footnote Citations

Footnote citations generally go outside sentence punctuation. Whenever possible, the citations should come at the end of a sentence:

> Thomas Gold has a new theory about the origins of oil.[1]

REFERENCES

Aller, L.H. 1982. Astrophysics. McGraw-Hill encyclopedia of science and technology 5th edition.

Asparagus. Encyclopaedia Britannica 15th edition, 1986.
Ayers, R.U. 1969. Technological forecasting and long-range planning. New York: McGraw-Hill.

——. 1984. The next industrial revolution: Reviving industry through innovation. Cambridge, Massachusetts: Ballinger.

Cawthorne, L., Attorney at Law. Interview with author in New York City, 18 March 1986.

Cooke, R.U., D. Brunsden, J.C. Doornkamp, and D.K.C. Jones. 1982. Urban geomorphology in drylands. New York: Oxford University Press.

Davis, D. 1985. Furniture made fun. Newsweek 106 (4 November): 82–3.

Feldhamer, G.A., J.E. Gates, D.M. Harman, A.J. Loranger, and K.R. Dixon. 1986. Effects of interstate highway fencing on white-tailed deer activity. Journal of Wildlife Management 50:497–503.

Harris, J., and R. Kellermayer. 1970. The red cell: Production, metabolism, destruction, normal and abnormal. Cambridge, Massachusetts: Harvard University Press.

Hay, J.G. 1978. The biomechanics of sports techniques. 2nd ed. Englewood Cliffs, N.J.: Prentice-Hall.

Hoeflin, R., and N. Bolsen. 1986. Life goals and decision making: Educated women's patterns. Journal of Home Economics 78(Summer): 33–5.

Figure 27-1. List of References.

If it would be confusing or misleading to place the footnote citation at the end of the sentence, the citation should still be placed outside sentence punctuation:

According to Haljmer Sunderstan,[1] whose truthfulness has been questioned in official proceedings,[2] the events were as follows:

Lewis, P.H. 1986. UNIX and MS-DOS: Dueling for dominance in computers. New York Times, 13 May, sec. C, 9.

Lynn, S.J., and J.W. Rhue. 1986. The fantasy-prone person: Hypnosis, imagination, and creativity. Journal of Personality and Social Psychology 50:404–8.

McNerney, W.J. 1980. Control of health care costs in the 1980's. New England Journal of Medicine 303:1088–95.

Middleditch, B.S., S. Missler, and H. Hines. 1981. Mass spectrometry of priority pollutants. New York: Plenum Press.

Morris, M.G. 1976. Conservation and the collector. In Moths and butterflies of Great Britain and Ireland, ed. J. Heath, vol. 1, 107–16. London: Curwen.

National Institutes of Health. 1976. Recombinant DNA research. Washington, D.C.: U.S. Government Printing Office.

Sabin, M.A., ed. 1974. Programming techniques in computer aided design. London: NCC Publications.

Stover, E.L. 1982. Removal of volatile organics from contaminated ground water. In Proceedings of the second national symposium on aquifer restoration and ground water monitoring, ed. D.M. Nielsen, 77–84. Worthington, Ohio: National Well Water Association.

Torgul, D., Director of Research and Development, Sterling Corporation. Letter to the author, 10 August 1986.

U.S. Environmental Protection Agency. 1977. Is your drinking water safe? Washington, D.C.: U.S. Government Printing Office.

**Figure 27-1.**    *Continued*

You should number the footnotes sequentially. If you are writing a communication with several chapters (often called "sections" in the writing done at work), you may use one continuous numbering sequence throughout your communication, or you may begin anew with the number 1 in each chapter.

## Writing the Individual Notes

Your readers need the same information to find your sources whether you use the footnote format, the author–year format, or the numbered citation format. Consequently, you should include the same information in your notes that you would include in entries in a reference list. By convention, however, notes present this information in the somewhat different manner described below.

An important difference between footnotes and entries in a reference list concerns the handling of sources that are cited more than once. In a reference list, an individual source is listed only once, no matter how many times it is cited in the body of the communication. With the footnote system, however, each citation in the text requires its own footnote. To reduce the amount of redundancy, the second and subsequent notes referring to a particular work differ from the first note. The following discussion tells you how to write first notes for books, journal articles, and other kinds of sources you are likely to cite on the job. It then tells you how to write additional citations of the same source.

### First Notes for Books

When citing a book for the first time, you should provide the following information:

- *Author.* Give the author's names in natural order: first initial (or name), middle initial, and last name. If your source has two authors, place *and* between them. If the source has three authors, separate the names with commas and place *and* before the last one. If your source has four or more authors, name the first, followed by *et al.* Place a comma after the names of the author or authors and before the title.
- *Title.* Capitalize all major words. Underline. Put no punctuation after the title.
- *Publication Information.* In parentheses, write the city of publication, colon, publisher's name, comma, and year of publication. Follow the parentheses with a period, unless you cite only part of the work rather than the entire work. When citing part of the work, follow the parentheses with a comma.
- *Pages.* If you are citing part of the work, rather than the entire work, write the page number or numbers. Follow with a period.

Here are some sample footnotes that show how this information is presented for various commonly cited types of books. In these examples, the authors' names are given exactly as they appear on the title page: first and middle names are spelled out where the title page spells them out but only initials are provided if the title page provides only initials.

*One Author*

1. Robert U. Ayers, <u>Technological Forecasting and Long-range Planning</u> (New York: McGraw-Hill, 1969), 216–34.

*Two Authors*

2. J. Harris and R. Kellermayer, <u>The Red Cell: Production, Metabolism, Destruction, Normal and Abnormal</u> (Cambridge, Massachusetts: Harvard University Press, 1970), 552–73.

*Three Authors*

3. Brian S. Middleditch, Stephen R. Missler, and Harry B. Hines, <u>Mass Spectrometry of Priority Pollutants</u> (New York: Plenum Press, 1981), 237.

*Four or More Authors*

4. R.U. Cooke et al., <u>Urban Geomorphology in Drylands</u> (New York: Oxford University Press, 1982), 86.

*Corporate Author or Sponsor*

5. National Institutes of Health, <u>Recombinant DNA Research</u> (Washington, D.C.: U.S. Government Printing Office, 1976), 42–7.

This form is used to cite publications that do not name an individual or group of individuals as authors. If the sponsoring organization is also the publisher, you may give the organization's name twice in the listing, once as the author and once as the publisher.

*Editor*

6. M.A. Sabin, ed., <u>Programming Techniques in Computer Aided Design</u> (London: NCC Publications, 1974), 110–42.

This form is used to cite books that give an editor's name, not an author's name, on the title page. Such books typically include essays or articles by many individuals. If you wish to cite only one essay in the book, use the format described below under the heading "References to Other Kinds of Sources." If the book has more than one editor, use the plural abbreviation *eds.* after the last editor's name.

*Second or Subsequent Edition*

> 7. James G. Hay, <u>The Biomechanics of Sports Techniques</u>, 2nd ed. (Englewood Cliffs, N.J.: Prentice-Hall., Inc, 1978), 382–405.

## First Notes to Journal Articles

When you are citing a journal article, you should supply the following information:

- *Author.* You should present the names of authors of journal articles in the same way you present the names of authors of books (see above).
- *Article Title.* Capitalize all major words. Follow the article title with a comma. Place the title and the comma in quotation marks.
- *Journal Title.* Capitalize all major words. Underline the title.
- *Volume Number.* Use arabic numerals even if the journal itself uses roman numerals. Do not place a comma between the journal title and the volume number.
- *Date of Publication.* Place the date in parentheses. If the journal numbers its pages in a single sequence throughout the volume, give the year only. If it begins numbering pages anew with each issue, give the date of the issue. Do not leave any spaces between the volume number and the parentheses that enclose the publication date. Follow the parentheses with a colon.
- *Page Number.* Unless you want to acknowledge the entire article, give the numbers of only the pages that contain the information you are using in your communication.

*Journal That Numbers Pages Consecutively Throughout Volume*

> 8. Walter J. McNerney, "Control of Health Care Costs in the 1980's," <u>New England Journal of Medicine</u> 303(1980):1093.

*Journal That Begins Numbering Pages Anew in Each Issue*

> 9. Ruth Hoeflin and Nancy Bolsen, "Life Goals and Decision Making: Educated Women's Patterns," <u>Journal of Home Economics</u> 78(Summer 1986):33–5.

## First Notes to Other Kinds of Sources

The following examples show how to describe some other sources you may use when writing on the job. The forms for several of these kinds of sources follow the patterns, just described, for listing books and journal articles.

### Pamphlet or Booklet

10. U.S. Environmental Protection Agency, <u>Is Your Drinking Water Safe?</u> (Washington, D.C.: U.S. Government Printing Office, 1977), 15.

Treat pamphlets and booklets in the same way you treat books that do not identify any author or authors on the title page. Note that when the title ends with a punctuation mark (question mark or exclamation point), the comma after the title is omitted.

### Essay in a Book

11. M.G. Morris, "Conservation and the Collector," in <u>Moths and Butterflies of Great Britain and Ireland</u>, ed. John Heath (London: Curwen, 1976), vol. 1, 107–16.

Present the title of the essay in much the same way you would present the title of a journal article. Add the word *in* before the title of the book. After the book title, place a comma, the abbreviation *ed.* (for "edited by"), and the name of the editor. Put no punctuation after the editor's name. Next, give the city of publication, colon, publisher's name, comma, and year of publication, all in parentheses, followed by a comma. After the publication facts, give volume and page numbers. For essays in books with only one volume, leave out the volume number.

By the way, this example is for a book that tells the first name of the editor *(John Heath)* but only the initials for the writer of the essay cited *(M.G. Morris)*.

### Paper in a Proceedings

12. Enos L. Stover, "Removal of Volatile Organics from Contaminated Ground Water," in <u>Proceedings of the Second National Symposium on Aquifer Restoration and Ground Water Monitoring</u>, ed. David M. Nielsen (Worthington, Ohio: National Well Water Association, 1982), 77–84.

Describe a paper in a proceedings in the same way you describe an essay in a collection of essays.

### Encyclopedia Article

13. Lawrence H. Aller, "Astrophysics," <u>McGraw-Hill Encyclopedia of Science and Technology</u> 5th ed.

Your listings for articles in an encyclopedia should resemble your listings for essays in essay collections. Be sure to include the edition number and omit the city and year of publication, as well as the page numbers (because the work is arranged alphabetically). If the encyclopedia does not name an author, begin the listing with the article's title, as in the following example:

14. "Asparagus," <u>Encyclopaedia Britannica</u> 15th edition, 1986.

### Article in a Popular Magazine

15. Douglas Davis, "Furniture Made Fun," <u>Newsweek</u> 106(4 November 1985), 82–3.

Treat listings for articles in popular magazines in the same way you treat articles in journals that begin numbering pages anew in each issue. If the magazine does not name an author for the article, begin the listing with the article's title.

### Newspaper Article

16. Peter H. Lewis, "UNIX and MS-DOS: Dueling for Dominance in Computers," <u>The New York Times</u>, 13 May 1986, sec. C, 9.

Treat articles in newspapers in much the same way you list articles in popular magazines. Notice, however, that a comma follows the newspaper's name, separating it from the date. Also, give the section number as well as the page number. If the newspaper does not name an author, begin the listing with the article's title.

### Letter to You

17. Dwight Torgul, Director of Research and Development, Sterling Corporation, letter to the author, 10 August 1983.

### Interview with You

18. Linda Cawthorne, Attorney at Law, interview with the author in New York City, 18 March 1981.

## Second Notes

If you cite the same source more than once, in the second and subsequent references you should use an abbreviated note that signals the reader to look to the earlier note for a full description of the source. These notes usually include only the author's last name and the page number.

19. Hay, 179.

If you cite two or more works by the same author, however, you must let your reader know which of them you are referring to in second and subsequent notes. Do that by adding a short version of the title. The following example is a second citation for R. Ayers' book *Technological Forecasting and Long-Range Planning:*

> 20. Ayers, <u>Technological Forecasting</u>, 207.

If you refer to the same source in two or more adjacent notes, then you have the option of replacing the author's name and the publication's short title (if needed) with the abbreviation *Ibid.* in the second and subsequent notes. For instance, if it refers to the Ayers book already cited in footnote 20, footnote 21 might look like this:

> 21. Ibid., 92.

## Sample Page of Endnotes

Figure 27–2 (page 614) shows a sample page of endnotes.

# USING BIBLIOGRAPHIES

A bibliography is an alphabetized list of sources. In the communications you write at work, you can use a bibliography for two purposes. The first is to provide your readers with a *longer* list of sources on your subject than you cited in the body of your communication. You might do that, for instance, to indicate general sources of information that you used but did not refer to specifically in the body.

The second purpose of a bibliography pertains to communications in which you use the footnote format. With footnotes, you present your sources out of alphabetical order and, if you place your notes on the bottom of the page, you scatter your notes throughout your communication. These circumstances make it difficult for your readers to quickly determine what sources you used. Readers might want to check your sources quickly in order to guide their own further reading or else to see whether or not you overlooked some particular source they know to be important. You can help your readers quickly find all your sources if you list all your sources alphabetically in a bibliography.

## The Form for Bibliography Entries Varies

The form of the entries in a bibliography will depend upon the format you use to cite sources in the body of your communication. If you are creating a bibliography for a communication in which you use one of the reference list formats (with either author–year citations or numbered citations), write the entries in the bibliography in the same way you write the entries in the

ENDNOTES

1. National Institutes of Health, Recombinant DNA
   Research (Washington, D.C.: U.S. Government Printing
   Office, 1976), 342–75.

2. J. Harris and R. Kellermayer, The Red Cell: Production,
   Metabolism, Destruction, Normal and Abnormal
   (Cambridge, Massachusetts: Harvard University Press,
   1970), 552–73.

3. Brian S. Middleditch, Stephen R. Missler, and Harry B.
   Hines, Mass Spectrometry of Priority Pollutants (New
   York: Plenum Press, 1981), 86–101.

4. Robert U. Ayers, Technological Forecasting and Long-
   range Planning (New York: McGraw-Hill, 1969),
   216–34.

5. Robert U. Ayers, The Next Industrial Revolution:
   Reviving Industry through Innovation (Cambridge,
   Massachusetts: Ballinger, 1984), 14.

6. Ayers, Technological Forecasting, 237.

7. Walter J. McNerney, "Control of Health Care Costs in
   the 1980's," New England Journal of Medicine
   303(1980):1093.

8. M.A. Sabin, ed., Programming Techniques in Computer
   Aided Design (London: NCC Publications, 1974), 110–
   42.

9. Dwight Torgul, Director of Research and Development,
   Sterling Corporation, letter to the author, 10 August
   1986.

10. James G. Hay, The Biomechanics of Sports Techniques,
    2nd ed. (Englewood Cliffs, N.J.: Prentice-Hall, Inc,
    1978), 382–405.

11. Harris and Kellermayer, 313.

12. Ibid., 310.

**Figure 27-2.**   Endnotes.

reference list. If you use the footnote format, you should write the entries in
your bibliography in the same way that you write the footnotes—with the
following three exceptions:

● Write the first author's name with the last name first.

● Use periods (not commas or parentheses) to separate the major parts of the citation. For a book, place these periods after the author's name, the book's title, and the publication information (city, publisher, and date). For an article, place the periods after the author's name, the article's title, and the publication information (journal name, volume, and page numbers).

● Include page numbers *only* when you are indicating the first and last pages of an article in a journal, book, proceedings or similar collection.

Because bibliography entries are so similar to footnotes, you may find it helpful to review discussions given above for the types of communications you are citing.

## Writing Bibliography Entries for Communications That Use Footnotes

The rest of this section presents sample entries for a bibliography appearing in a communication that uses footnotes.

### Bibliography Entries for Books

*One Author*

Ayers, Robert U. Technological Forecasting and Long-range Planning. New York: McGraw-Hill, 1969.

*Two or Three Authors*

Harris, J., and R. Kellermayer. The Red Cell: Production, Metabolism, Destruction, Normal and Abnormal. Cambridge, Massachusetts: Harvard University Press, 1970.

Middleditch, Brian S., Stephen R. Missler, and Harry B. Hines. Mass Spectrometry of Priority Pollutants. New York: Plenum Press, 1981.

*More Than Three Authors*

Cooke, R.U., D. Brunsden, J.C. Doornkamp, and D.K.C. Jones. Urban Geomorphology in Drylands. New York: Oxford University Press, 1982.

*Corporate Author*

National Institutes of Health. Recombinant DNA Research. Washington, D.C.: U.S. Government Printing Office, 1976.

*Editor*

> Sabin, M.A., ed. Programming Techniques in Computer Aided Design. London: NCC Publications, 1974.

*Second or Subsequent Edition*

> Hay, James G. The Biomechanics of Sports Techniques, 2nd ed. Englewood Cliffs, N.J.: Prentice-Hall, Inc., 1978.

## Bibliography Entries for Journal Articles

*One Author*

> McNerney, Walter J. "Control of Health Care Costs in the 1980's." New England Journal of Medicine 303(1980): 1088–95.

*Two Authors*

> Lynn, Steven J., and Judith W. Rhue. "The Fantasy-prone Person: Hypnosis, Imagination, and Creativity." Journal of Personality and Social Psychology 50(1986):404–8.

*Three or More Authors*

> Feldhamer, George A., J. Edward Gates, Dan M. Harman, Andre J. Loranger, and Kenneth R. Dixon. "Effects of Interstate Highway Fencing on White-tailed Deer Activity." Journal of Wildlife Management 50(1986):497–503.

Be sure to name all the authors in your bibliography, even though in a footnote for an article with more than three authors you give only the first author's name followed by *et al.*

*Journal That Begins Numbering Pages Anew in Each Issue.*

> Hoeflin, Ruth, and Nancy Bolsen. "Life Goals and Decision Making: Educated Women's Patterns." Journal of Home Economics 78(Summer 1986):33–5.

## Bibliography Entries to Other Sources

*Pamphlet or Booklet*

> U.S. Environmental Protection Agency. Is Your Drinking Water Safe? Washington, D.C.: U.S. Government Printing Office, 1977.

### Essay in a Book

Morris, M.G. "Conservation and the Collector." In <u>Moths and Butterflies of Great Britain and Ireland,</u> ed. John Heath, vol. 1, 107-16. London: Curwen, 1976.

### Paper in a Proceedings

Stover, Enos L. "Removal of Volatile Organics from Contaminated Ground Water." In <u>Proceedings of the Second National Symposium on Aquifer Restoration and Ground Water Monitoring,</u> ed. David M. Nielsen, 77–84. Worthington, Ohio: National Well Water Association, 1982.

### Encyclopedia Article

Aller, Lawrence H. "Astrophysics." <u>McGraw-Hill Encyclopedia of Science and Technology</u> 5th edition, 1982.

"Asparagus." <u>Encyclopaedia Britannica</u> 15th edition, 1986.

### Article in a Popular Magazine

Davis, Douglas. "Furniture Made Fun." <u>Newsweek</u> 106(4 November 1985):82–3.

### Newspaper Article

Lewis, Peter H. "UNIX and MS-DOS: Dueling for Dominance in Computers." <u>New York Times</u>, 13 May 1986, sec. C, 9.

### Letter to You

Torgul, Dwight, Director of Research and Development, Sterling Corporation. Letter to the author, 10 August 1986.

### Interview with You

Cawthorne, Linda, Attorney at Law. Interview with the author in New York City, 18 March 1986.

## Sample Bibliography

Figure 27-3 shows a sample bibliography for a communication that uses footnotes.

BIBLIOGRAPHY

Aller, Lawrence H. "Astrophysics." McGraw-Hill Encyclopedia of Science and Technology 5th edition, 1982.

"Asparagus." Encyclopaedia Britannica 15th edition, 1986.

Ayers, Robert U. Technological Forecasting and Long-range Planning. New York: McGraw-Hill, 1969.

——. The Next Industrial Revolution: Reviving Industry through Innovation. Cambridge, Massachusetts: Ballinger, 1984.

Cawthorne, Linda, Attorney at Law. Interview with the author in New York City, 18 March 1986.

Cooke, R.U., D. Brunsden, J.C. Doornkamp, and D.K.C. Jones. Urban Geomorphology in Drylands. New York: Oxford University Press, 1982.

Davis, Douglas. "Furniture Made Fun." Newsweek 106 (4 November 1985):82–3.

Feldhamer, George A., J. Edward Gates, Dan M. Harman, Andre J. Loranger, and Kenneth R. Dixon. "Effects of Interstate Highway Fencing on White-tailed Deer Activity." Journal of Wildlife Management 50(1986): 497–503.

Harris, J., and R. Kellermayer. The Red Cell: Production, Metabolism, Destruction, Normal and Abnormal. Cambridge, Massachusetts: Harvard University Press, 1970.

Hay, James G. The Biomechanics of Sports Techniques, 2nd ed. Englewood Cliffs, N.J.: Prentice-Hall, Inc., 1978.

Hoeflin, Ruth, and Nancy Bolsen. "Life Goals and Decision Making: Educated Women's Patterns." Journal of Home Economics 78(Summer 1986):33–5.

**Figure 27-3.**   Bibliography for Use with Footnotes.

## NOTE

1. The *MLA Handbook,* once widely used as a source of information about footnotes, now advises authors to use a format somewhat like the author–year format described earlier in this chapter. Although this new MLA style is widely used in the humanities, the use of footnotes has not been abandoned in business, industry, and government. See Joseph Gibaldi and Walter S. Achtert, *MLA Handbook for Writers of Research Papers* (New York: Modern Language Association, 1984).

Lewis, Peter H. "UNIX and MS-DOS: Dueling for Dominance in Computers." New York Times, 13 May 1986, sec. C, 9.

Lynn, Steven J., and Judith W. Rhue. "The Fantasy-prone Person: Hypnosis, Imagination, and Creativity." Journal of Personality and Social Psychology 50(1986):404–8.

McNerney, Walter J. "Control of Health Care Costs in the 1980's." New England Journal of Medicine 303(1980): 1088–95.

Middleditch, Brian S., Stephen R. Missler, and Harry B. Hines. Mass Spectrometry of Priority Pollutants. New York: Plenum Press, 1981.

Morris, M.G. "Conservation and the Collector." In Moths and Butterflies of Great Britain and Ireland, ed. John Heath, vol. 1, 107–16. London: Curwen, 1976.

National Institutes of Health. Recombinant DNA Research. Washington, D.C.: U.S. Government Printing Office, 1976.

Sabin, M.A., ed. Programming Techniques in Computer Aided Design. London: NCC Publications, 1974.

Stover, Enos L. "Removal of Volatile Organics from Contaminated Ground Water." In Proceedings of the Second National Symposium on Aquifer Restoration and Ground Water Monitoring, ed. David M. Nielsen, 77–84. Worthington, Ohio: National Well Water Association, 1982.

Torgul, Dwight, Director of Research and Development, Sterling Corporation. Letter to the author, 10 August 1986.

U.S. Environmental Protection Agency. Is Your Drinking Water Safe? Washington, D.C.: U.S. Government Printing Office, 1977.

**Figure 27-3.**    *Continued*

## EXERCISES

1. Pick a topic in your field and find six articles, books, and other printed sources of information about it. For each, do the following:
   - Write an entry for use in a list of references.
   - Write a footnote (cite some particular page or pages in each source).
   - Write a bibliography entry.

Alternatively, you may find six sources concerning the subject of an assignment you are preparing in your writing course.

2. Find a journal and a book concerning topics in your field. Compare the journal's documentation format with the format described in this chapter that is closest to it. Do the same for the book. Note similarities and differences in detail.

3. For each of the following sources do these three things:

- Write an entry for use in a list of references.
- Write a footnote (cite some particular page or pages in each source).
- Write a bibliography entry.

A. The third edition of a book entitled *Occupational Safety Management and Engineering,* published by Prentice-Hall, which has its headquarters in Englewood Cliffs, New Jersey. The author is Willie Hammer. This third edition was issued in 1976.

B. An article by Ida Brambilla, Alcide Bertani, and Remo Reggiani in volume 123, issue number 5, of the *Journal of Plant Physiology*. This issue was published in June 1986, and the article is entitled, "Effects of Inorganic Nitrogen Nutrition (Ammonium and Nitrate) on Aerobic and Anaerobic Metabolism in Excised Rice Roots." Pages 419–28.

C. An article entitled "Space Goals for 21st Century Depicted in Report to White House," which appears on pages 16 and 17 of the May 26, 1986, issue of *Aviation Week*. This issue is number 21 of volume 24. The journal does not identify the author of the article. *Aviation Week* is published by McGraw-Hill, which has its headquarters in New York City.

D. An article entitled "Genes Could Be the Enemy in Fighting the Battle of the Bulge," which was published on Tuesday, August 5, 1986, in the *Atlanta Constitution,* a newspaper. The article was written by Jane C. Allison, Staff Writer, and it appears on page A-20.

E. An article entitled "An Evaluation of the Sulfite-AQ Pulping Process," which appears on pages 102 through 105 of the August 1986 issue of the *Tappi Journal*. The journal identified the authors as I.B. Sanborn and K.D. Schwieger. The *Tappi Journal* is published by the Technical Association of the Pulp and Paper Industry, which has its headquarters in New York City. This is number 8 of volume 69.

F. An essay entitled "Identifying Potentially Active Faults and Unstable Slopes Offshore." It appeared in an essay collection, edited by J.I. Ziony. The title of the collection is *Evaluating Earthquake Hazards in the Los Angeles Region—An Earth Science Perspective*. The book was published by the U.S. Geological Survey in Washington, D.C. The essay appeared in pages 347 through 373. It was published in 1985. The authors are S. H. Clarke, H. G. Greene, and M. P. Kennedy.

# Using Conventional Superstructures

Superstructures are general frameworks for constructing communications.[1] You use them every day. When you talk with a friend, parent, or instructor, you probably follow the conventional superstructure for a conversation: you start by saying "Hi," you proceed to discuss one or more topics, and you conclude by saying "Bye."

At work, people employ many conventional superstructures. These patterns provide general outlines for communications. They indicate what a communication's major parts might be and what order those parts might have. For example, the conventional superstructure for a research report indicates that early in the report the writers should tell what they were trying to find out. Later, according to the superstructure, the writers should describe their research method, present their results, and explain their conclusions.

But conventional superstructures do much more than provide a list of topics and their order of presentation. The superstructures also indicate what the writers might include in each part and how they might develop that content and relate it to the content of the other parts. Consequently, conventional superstructures provide broad and extensive guidance to you as a writer.

Part IX describes conventional superstructures for three types of communication you will write often at work: reports, proposals, and instructions.

## RELATIONSHIP BETWEEN SUPERSTRUCTURES AND FORMATS

As you read about these conventional superstructures, keep in mind that superstructures are quite distinct from formats. Formats are packages into which you put your messages when you present them to your readers. They are ways of putting messages on paper. You will find information about three widely used formats—letter, memo, and book—in Part VIII. In contrast to formats, superstructures are ways of organizing messages. They are independent of formats. You can, for instance, present a research report in the memo, letter, or book format. In any of these formats, your proposal will have the same superstructure.

## HOW KNOWLEDGE OF SUPERSTRUCTURES HELPS YOU

Knowledge of conventional superstructures can help you in several stages of writing. When you are *planning* your communication, you can refer to the conventional superstructures described in the following chapters for suggestions about what to include and how to organize. Similarly, when you are *drafting,* you can refer to the superstructures for ideas about how to develop the various parts of your communication.

And when you are *evaluating* your completed draft, you can look again at the superstructures to see if they suggest ways you can improve your communication.

Because a knowledge of superstructures helps first when you are planning, you will find comments on superstructures in the introduction to Part III, "Planning." A few basic points from that discussion are worth repeating here.

First, superstructures embody the wisdom and experience of people who have tried to do the very thing you are trying to do. For example, one of the most common activities on the job is to propose that someone take a certain action, such as to change a policy or purchase a piece of equipment. When you write a proposal, you will have a great deal in common with all the other people who have written proposals. You can benefit from their experiences if you use the conventional superstructure for proposals, because you will be using an organizational pattern that many other writers have found to be successful. Superstructures are helpful to you because they are patterns that work successfully in such recurring situations.

Second, conventional superstructures help your readers. The superstructures remind you of the kinds of information that your readers want in such recurring situations, and they suggest patterns for organizing that information in ways your readers will find helpful.

Third, you should avoid imitating superstructures blindly. To use conventional superstructures successfully, you must employ them thoughtfully and creatively. Though most communication situations closely resemble many other situations, no two situations are exactly alike: purposes differ, readers differ, and circumstances differ. Consequently, the conventional superstructures are general—too general merely

to be mimicked. You will need to shape them to your own situation.

The following chapters present various conventional superstructures in ways designed to help you do that shaping. Besides explaining each superstructure in detail, the chapters also describe the types of situations for which each is best suited. By comparing these typical situations with the situation in which you are writing, you can learn how to adapt the superstructure to your purpose and readers.

## CONTENTS OF THIS SECTION

**Chapter 28**
**Reports: General Superstructure**

**Chapter 29**
**Empirical Research Reports**

**Chapter 30**
**Feasibility Reports**

**Chapter 31**
**Progress Reports**

**Chapter 32**
**Proposals**

**Chapter 33**
**Instructions**

## NOTE

1. There is no generally accepted term for the overall structural patterns conventionally used for communications. For this book, I have borrowed the term *superstructures* from cognitive psychologists, who study how these conventional patterns help people comprehend the things they read. See, for example, the following: (1) Teun A. van Dijk, "Semantic Macrostructures and Knowledge Frames in Discourse Comprehension." In *Cognitive Processes in Comprehension*, ed. Marcel Adam Just and Patricia A. Carpenter, 3–32. Hillsdale, New Jersey: Erlbaum, 1977; (2) Walter Kintsch and Teun A. van Dijk, "Toward a Model of Text Comprehension and Production." *Psychological Review* 85(1978):363–94; and (3) Teun A. van Dijk, *Macrostructures*. Hillsdale, New Jersey: Erlbaum, 1980.

# CHAPTER 28

# Reports: General Superstructure

## PREVIEW OF CHAPTER

How to Use the Advice in Chapters 28 through 31
Varieties of Report-Writing Situations
Your Readers Want to Use the Information You Provide
The Questions that Readers Ask Most Often
General Superstructure for Reports
    Introduction
    Method of Obtaining Facts
    Facts
    Discussion
    Conclusions
    Recommendations
A Note about Summaries
Sample Reports

At work you will often be called upon to convey and interpret information for other people. That's because you will possess a great deal of important knowledge that others will want to use. Sometimes, this will be knowledge you have gained in your college courses. At other times, it will be knowledge you have gained through special research, perhaps in the library or the laboratory. And at still other times, it will be knowledge you have gained through general, thoughtful observation of the things that go on around you on the job. Whatever the source of that knowledge, you will often respond to people's requests for it by writing a report.

This chapter is the first of four that will help you write effective reports on the job. In it, you will learn about the basic superstructure for reports. The advice given here will be useful almost every time you prepare a report at work.

The following three chapters describe three variations of the basic superstructure. Each represents an adaptation of the basic superstructure to meet the needs of readers in a particular type of reporting situation. The superstructures described in those chapters are for empirical research reports (Chapter 29), feasibility reports (Chapter 30), and progress reports (Chapter 31). Whenever you write one of those types of reports, you will find those chapters to be helpful *supplements* to this one. You should not, however, use them as *substitutes* for this chapter because this one contains much basic advice that is not repeated in the others.

## HOW TO USE THE ADVICE IN CHAPTERS 28 THROUGH 31

Whether you read only about the basic superstructure or also about the variations of it, keep in mind that no two reporting situations are exactly alike. You will have to use your imagination and creativity to adapt the superstructures you find here to the particular reports you are writing. If you have not already read the general advice about using conventional superstructures that appears in the introduction to Part IX, you might find it helpful to do so now.

Remember also that whatever superstructure you use for a report, you will have to present the report in some format, usually one of the three discussed in Chapter 26: letter, memo, or book. Therefore, as you prepare your report, look also at the discussion of the appropriate format.

## VARIETIES OF REPORT-WRITING SITUATIONS

Reports come in many varieties. The following examples will give you some idea of their diversity:

- A one-hundred-page report on a seven-month project to test a special method of venting high-speed engines for use in space vehicles.

- A twelve-page report based on library research to determine which long-distance telephone company provides the most reliable service.
- A two-paragraph report based upon a manufacturing engineer's visit to a new plant that is about to be put into service.
- A two-hundred-page report addressed to the general public concerning the environmental impact of mining certain portions of public land in Utah.

As these examples suggest, there are many ways in which the reports you write may differ from one another and from the reports written by other people:

- *Sources of Your Information.* You may base your report on information gathered from one or more of a wide variety of sources, including your own research, reading, and interviews.
- *Amount of Time You Spent Gathering Your Information.* This may vary from a few minutes to many years.
- *Number of Readers.* Your report may have many readers or only a few. Rarely will it have only one.
- *Kinds of Readers.* Your readers may be people employed in your own organization or they may be employed in other organizations. In some situations, you may write to the general public.

## YOUR READERS WANT TO USE THE INFORMATION YOU PROVIDE

Despite these and the many other differences among them, almost all report-writing situations have one factor in common: your readers will want to put the information you provide to some professional or practical use. The precise kind of use will vary, of course, from situation to situation. For example, your readers may want to use your information to solve an organizational problem (where typical goals are to increase efficiency and profit), a social problem (where typical goals are to improve the general health and welfare of groups of people), or a personal problem (where typical goals are to satisfy individual preferences and values). Regardless of these differences, however, the readers' desire to use your information will almost always be a key factor when you write reports, a factor that you should take into account when planning and writing every part of your communication.

Many other times in this book, you have read about the importance of considering the way your readers will use your communications. The point deserves repetition—and special emphasis—here because so many writers forget their readers when they write reports. They persuade themselves that their purpose is to "tell what I know" or "tell what I have done" rather than to provide their readers with information they can use.

## THE QUESTIONS THAT READERS ASK MOST OFTEN

When trying to use the information they find in reports, readers usually ask the same basic questions. The general superstructure for reports is a pattern that writers and readers have found to be successful for answering those basic questions. Therefore, by thinking about the readers' questions, you will prepare yourself to understand the superstructure and to use it effectively. The readers' six basic questions are as follows:

- *What will we gain by reading your report?* Most people at work want to read only these communications that are directly useful to them.

- *Are your facts reliable?* Readers want to be certain that the facts you supply will provide a sound basis for their decisions or actions.

- *What do you know that is useful to us?* Readers don't want you to tell them everything you know about your subject; they want you to tell them only those facts they must know to do the job that lies before them. (*Example*: "The most important sales figures for this quarter are as follows: . . . ")

- *How do you interpret those facts from our point of view?* Facts alone are meaningless. To give facts meaning, people must interpret them by pointing out relationships or patterns among them. (*Example:* "The sales figures show a rising demand for two products but not for two others.") Usually, your readers will want you to make those interpretations rather than leave that work to them.

- *How are those facts significant to us?* Readers generally want you to go beyond an interpretation of the facts to explain what the facts mean in terms of the readers' responsibilities, interests, or goals. (*Example:* "The demand for one product falls during this season every year, though not quite this sharply. The falling demand for the other may signal that the product is no longer competitive.")

- *What do you think we should do?* Because you have studied the facts in detail, readers often want you to tell them what action you think they should take. (*Example:* "You should continue to produce the first product, but monitor its future sales closely. You should find a way to improve the second product or else quit producing it.")

Of course, those six questions are very general. For large reports, writers need to take hundreds, even thousands, of pages to answer them. That's because readers seek answers to these basic questions by asking a multitude of more specific, subsidiary questions.

## GENERAL SUPERSTRUCTURE FOR REPORTS

The general superstructure for reports contains six elements, one for each of the six basic questions you just read about: introduction, method of obtaining facts, facts, discussion, conclusions, and recommendations.

Figure 28-1 shows how the six elements relate to the readers' basic questions. Of course, each element of the superstructure may serve important purposes in addition to answering the general question identified with it. Also, the six elements may be arranged in many ways, and one or more of them may be omitted if circumstances warrant. In some brief reports, for example, the writers begin with a recommendation, move to a paragraph in which the facts and conclusions are treated together, and state the sources of their facts in a concluding, single-sentence paragraph. Also, writers sometimes present two or more of the six elements under a single heading. For instance, they may include in their introduction information about how they obtained their facts, and they frequently present and interpret their facts in a single segment of their report.

The following paragraphs briefly describe each of the six elements of the general superstructure for reports.

## Introduction

In the introduction of a report, you answer your readers' question, "What will we gain by reading your report?" In some reports, you can answer the question in a sentence or less. Consider, for instance, the first sentence of a report written by Lisa, an employee of a university's fund-raising office, who was asked to investigate the university's facilities and programs in horseback

| Report Element | Readers' Question |
| --- | --- |
| Introduction | What will we gain by reading your report? |
| Method of obtaining facts | Are your facts reliable? |
| Facts | What do you know that is useful to us? |
| Discussion | How do you interpret those facts from our point of view? |
| Conclusions | How are those facts significant to us? |
| Recommendations | What do you think we should do? |

**Figure 28-1.**    Elements of a Report and Their Relationship to the Basic Questions Readers Ask.

riding. Because her reader, Matt, had assigned her to prepare the report in the first place, she could tell him what he would gain from it simply by reminding him why he had requested it:

> In this report I present the information you wanted to have before deciding whether to place new university stables on next year's list of major funding drives.

In longer reports, your explanation of the relevance of your report to your readers may take many pages, in which you tell your readers such things as (1) what problem your report will help solve, (2) what activities you performed toward solving that problem, and (3) how your readers can apply your information in their own efforts toward solving the problem. In the discussion of Guideline 2 in Chapter 13, you will find detailed advice about how to tell readers why they should read your report.

Besides telling your readers what your communication offers them, your introduction may serve many other functions. The most important of these is to tell your main points. In most reports, your main point will be your major conclusions and recommendations. Although you should save a full discussion of these topics for the sections devoted to them at the end of your report, your readers will usually appreciate a brief summary of them—perhaps in a sentence or two—in your introduction. Lisa provided such a summary in the second, third, and fourth sentences of her horseback-riding report:

Summary of conclusions ———

> Overall, it seems that the stables would make a good fund-raising project because of the strength of the current programs offered there, the condition of the current facilities, and the existence of a loyal core of alumni who used the facilities while undergraduates. The fund-raising should focus on the

Summary of recommendations ———

> construction of a new barn, costing $125,000. An additional $150,000 could be sought for a much-needed arena and classroom, but I recommend that this construction be saved for a future fund-raising drive.

In brief reports (for example, one-page memos), a statement of your main points may even replace the conclusions and recommendations that would otherwise appear at the end. For an additional discussion of the value of putting your main points in your introduction, see Guideline 5 in Chapter 13.

The other functions of the introduction to a report are also described in Chapter 13. These include telling your readers how your report is organized, telling its scope, and providing background information your readers will need in order to understand the rest of your report.

## Method of Obtaining Facts

In a report, your discussion of your method of obtaining your facts can serve a wide variety of purposes. Report readers want to assess the reliability of the facts you present: your segment on method tells them how and where you got your facts. It also suggests to your readers how they can gain additional information on the same subject. If you obtained your information through reading, for example, you direct your reader to those sources. If you obtained your information through an experiment, survey, or other special technique, your account of your method may help readers design similar projects.

In her investigation of the university stables, Lisa gathered her information through interviews. She reported her method in the following way:

> I obtained the information given below from Peter Troivinen, Stable Manager. Also, at last month's Alumni Weekend, I spoke with a half-dozen alumni interested in the riding programs. Information about construction costs comes from Roland Taberski, whose construction firm is experienced in the kind of facility that would be involved.

## Facts

Your facts are the individual pieces of evidence that underlie and support your conclusions and recommendations. If your report, like Lisa's, is based upon interviews, your facts are the things people told you. If your report is based upon laboratory, field, or library research, your facts are the verifiable pieces of information that you gathered: the data you obtained, the survey responses you recorded, or the knowledge you assembled from printed sources. If your report is based upon your efforts to design a new product, procedure or system, your facts are the various aspects of the thing you designed or created. In sum, your facts are the separate pieces of information you present as objectively verifiable.

You may present your facts in a section of their own or you may combine your presentation of your facts with your discussion of them, as explained below.

## Discussion

Taken alone, facts mean nothing. They are a table of data, a series of isolated observations or pieces of information without meaning. Therefore, an essential element of every report you write will be your discussion of the facts, in which you interpret the facts in a way significant to your readers.

Sometimes, writers have trouble distinguishing between a presentation of the facts and a discussion of them. The following example may help to make the distinction clear. Imagine that you observed that when the temperature on the floor of your factory is 65°F, workers produce 3% rejected parts; when it is 70°F, they produce 3% rejected parts; when it is 75°F, they produce 4.5% rejected parts; and when it is 80°F, they produce 7% rejected parts. Those would be your facts. If you were to say, "As the temperature rises above

70°F, so too does the percentage of rejected parts," you would be interpreting those facts. Of course, in many reports you will be dealing with much larger and more complicated sets of facts that require much more sophisticated and extended interpretation. But the basic point remains the same: when you begin to make general statements based upon your facts, you are interpreting them for your readers. You are discussing them.

In many of the communications you write, you will weave your discussion of the facts together with your presentation of them. In such situations, the interpretations often serve as the topic sentences for paragraphs. Here, for example, is a paragraph in which Lisa mixes facts and discussion:

Interpretation ——

Facts ——

> The university's horseback riding courses have grown substantially in the recent years, due largely to the enthusiastic and effective leadership of Mr. Troivinen, who took over as Stable Manager five years ago. When Mr. Troivinen arrived, the university offered three courses: beginning, intermediate, and advanced riding. Since then two new courses have been added, one in mounted instruction and one in the training of horses.

Whether you integrate your presentation and discussion of the facts or treat the two separately, it is important for you to remember that your readers count upon you not only to select the facts that are relevant to them, but also to discuss those facts in a way that is meaningful to them.

## Conclusions

Like interpretations, conclusions are general statements based on your facts. However, they focus not simply on interpreting the facts but on answering the readers' question, "How are those facts significant to us?" In her report, for instance, Lisa provided many paragraphs of information about the university riding programs, the state of the current stable facilities, and the likely interest among alumni in contributing money for new stable facilities. After reading her presentation and discussion of those facts, Lisa's reader, Matt, might ask, "But what, exactly, does all that mean in terms of my decision about whether to start a fund-raising project for the stables?" To anticipate and answer this question, Lisa provided the following conclusions:

> In conclusion, my investigation indicates that the university's riding programs could benefit substantially from a fund-raising effort. However, the appeal of such a program will be limited primarily to the very supportive alumni who used the university stables while students.

Such brief, explicit statements of conclusions are almost always desired and welcomed by report readers.

## Recommendations

Just as conclusions grow out of interpretations of the facts, recommendations grow out of conclusions. They answer the readers' question, "If your conclusions are valid, what should we do?" Depending upon many factors, including the number and complexity of the things you are recommending, you may state your recommendations in a single sentence or in many pages.

As mentioned above, you can help your readers immensely by stating your major recommendations at the beginning of your report. In short reports where you can state your recommendations in a few words or sentences, that may be the only place you need to present them. On the other hand, if your communication is long or if a full discussion of your recommendations requires much space, you can summarize your recommendations generally at the beginning of your report and then treat them more extensively at the end. That's what Lisa did when she summarized her recommendations in two sentences in the first paragraph of her introduction and presented and explained the recommendations in three paragraphs at the end of her report. To be sure that her readers could readily find this fuller discussion, she placed it under the heading "Conclusions and Recommendations," and she began the first paragraph of her recommendations with the words, *I recommend*.

Although readers usually want recommendations in reports, you may encounter some situations in which you will not want to include recommendations. That might happen, for instance, in either of the two following situations:

- The decision being made is clearly beyond your competence and you have been asked to provide only a small part of the information your readers need to make the decision.
- You are working in a situation where the responsibility for making recommendations belongs to your boss or other readers.

Nevertheless, in the usual situation your recommendations will be expected, or at least welcomed. If you are uncertain about whether or not to provide them, ask your boss or the person who asked you to write. Don't omit recommendations out of shyness or because you are guessing about what is wanted.

# A NOTE ABOUT SUMMARIES

The preceding discussion concentrates on the elements found in most reports written on the job. Many longer reports share another feature: they are preceded by a separate summary of the report overall. (This summary is distinct from the summary of conclusions and recommendations that appears in the introduction.) Such summaries are often called *executive summaries* because they usually are addressed to decision-makers. Executive summaries devote special attention to conclusions and recommendations, but they also provide a general overview of the report, introducing it briefly, summarizing the writer's method of obtaining facts, and citing and discussing (perhaps only in a sentence or two) the key findings and their interpretation.

Although such summaries are often associated with reports in the book format, they are also appropriate at the beginning of reports in the letter and memo formats that are more than a few pages long. You will find detailed advice about how to write report summaries in Chapter 26, "Formats for Letters, Memos, and Books," where summaries are treated as one of the elements of the book format.

## SAMPLE REPORTS

Figure 28-2 shows Lisa's full report. The next three chapters also contain samples that will help you understand the general superstructure for reports. While those samples show specific varieties of report (empirical research, feasibility, progress), those varieties are really just specialized versions of the general report superstructure. You might be interested in comparing the sample empirical research report (Chapter 29), which uses the book format, with Lisa's report shown in Figure 28-2, which uses the memo format.

## WRITING ASSIGNMENTS

Assignments that involve writing reports are included in Appendix I.

## MEMORANDUM

### Central University
### Development Office

**FROM:** Lisa Beech                          February 27, 19--

**TO:**    Matt Fordyce, Director of Funding Drives

**SUBJECT:** POSSIBLE FUNDING DRIVE FOR UNIVERSITY
STABLES

In this report I present the information you wanted to
have before deciding whether to place new university
stables on next year's list of major funding drives. Overall,
it seems that the stables would make a good fund-raising
project because of the strength of the current programs
offered there, the condition of the current facilities, and the
existence of a loyal core of alumni who used the facilities
while undergraduates. The fund-raising should focus on the
construction of a new barn, costing $125,000. An additional
$150,000 could be sought for a much-needed arena and
classroom, but I recommend that this construction be saved
for a future fund-raising drive.

I obtained the information given below from Peter
Troivinen, Stable Manager. Also, at last month's Alumni
Weekend, I spoke with a half-dozen alumni interested in the
riding programs. Information about construction costs
comes from Roland Taberski, whose construction firm is
experienced in building the kind of facility that would be
involved.

### RIDING PROGRAMS

Begun in 1936, Central University's riding programs
fall into two categories: regular university courses offered
through the stable, and the university's horse show team,
which competes nationally through the Intercollegiate
Horse Show Association.

Introduction
explains
significance of
report to reader
and
summarizes
conclusions
and
recommendations.

Writer explains
method of
obtaining facts.

Writer begins
first of four
sections that
present and
discuss her
findings.

**Figure 28-2.**    Report that Uses the General Superstructure.

February 26, 19--                                    Page 2

### University Courses

The university's horseback riding courses have grown substantially in recent years, due largely to the enthusiastic and effective leadership of Mr. Troivinen, who took over as Stable Manager five years ago. When Mr. Troivinen arrived, the university offered three courses: beginning, intermediate, and advanced riding. Since then two new courses have been added, one in mounted instruction and one in the training of horses. In all riding courses, students may choose either English or Western style.

In the past five years, the number of students taking a horseback riding course for college credit has more than doubled from 310 to 725. That has placed a tremendous burden on the both the stable and the horses.

### Intercollegiate Show Team

When Mr. Troivinen began working at the stable, he helped students form an intercollegiate show team, which originally had six members. It now has over one hundred, making it one of the largest in the country. Furthermore, the team has become competitive both regionally and nationally. Three years ago it was first in the region; two years ago it was third in the nation in English; and last year it was first in the nation in Western.

### NEED

Mr. Troivinen and the alumni identify three major problems facing the equestrian programs: poor barn facilities, lack of a classroom and meeting room, and lack of an all-weather riding area.

### Barn

Clearly, the most pressing problem facing the equestrian programs is the size of the barn. The barn does not provide adequate stabling for the increased number of horses Mr. Troivinen has purchased in response to the rising number of riders in classes and on the horse show

*Writer begins second section that reports and discusses her findings.*

**Figure 28-2.**   *Continued*

February 26, 19--                                           Page 3

team. To provide at least some space for these horses, he
has made some modifications to the barn. When originally
built, the barn had six box stalls and two tie stalls. Box
stalls are the best housing for most horses because they
allow the horses to move around. Tie stalls have many uses,
including providing a place for horses between classes and
while they are being curried. However, tie stalls are not
desirable for permanent housing because the horses' heads
are tied at all times. Mr. Troivinen has had to convert all
but one of the box stalls to tie stalls in order to make room
for the additional horses. Also, some horses are now kept in
the aisles.

The barn is now full, so that no additional horses can
be purchased to meet the student demand for riding
courses. As a result, many of the horses work seven hours
a day instead of the more reasonable four or five that is
standard for the type of horse used for riding. With a seven-
hour workload, the chances of health problems increase--
and the workload of a sick horse must be shifted to other
horses. Also, with this workload, the horses grow tired by
the afternoon classes and they are no longer willing to
cooperate with their riders. Consequently, the riders
(especially beginning riders) have trouble getting their
horses to perform. However, the only way to increase the
number of horses now would be to tie them up in
enclosures along a fence behind the barn. Mr. Troivinen is
reluctant to do that because horses kept there would have
no shelter from the weather.

Other problems with the barn include the following:

- Storage space for feed is too small. Hay, grain, and
  the special feeds required for some of the horses are
  stored in the aisle, at one end of the barn. Horses
  that get loose during the night sometimes eat the
  feed until they get sick.

- Lack of storage place for equipment. Saddles and
  bridles are kept along the aisles, where they are
  exposed to damage from bites or kicks of passing
  horses. Moreover, other necessary equipment can't
  be purchased because there is no place to store it.

**Figure 28-2.**   *Continued*

February 26, 19--                                          Page 4

- <u>Lack of a wash area</u>. Washing has to be done in the courtyard, where drainage is poor and there is no place to tie the horses. As a result, the horses rarely get a bath, rendering them more susceptible to fungus and skin disease.

### Classroom and Meeting Room

At present, there is no classroom for riding classes. In good weather the classes meet outdoors and in bad weather they meet in the aisles of the barn. For the basic and intermediate riding classes, that is not a great problem. It is a problem, however, for the advanced riding class and the elementary training class, where lectures, films, and other indoor instruction are appropriate.

Also, there is no meeting room for the riding club, which in addition to its regular business meetings sometimes has guest speakers and films. The nearest room they can use for these purposes are classrooms in Haddock Hall, a quarter mile from the barn.

### Riding Tracks

Riding is done on four outdoor cinder tracks. Therefore, student interest is very low during winter sessions, when the weather is cold and the riding tracks are often muddy or even covered with snow and ice. Snow and ice create additional problems. They greatly increase the chances that a horse will slip, injuring itself and its rider. Also, snow builds up on the horses' feet, impairing their movements. When that happens, riding is restricted to the slower gaits of walk and trot, so students don't receive the full content of the course they are taking. Finally, the cold air of winter is hard on the horses' lungs.

### POSSIBLE DEVELOPMENT PROJECTS

Mr. Troivinen has identified two subjects for a possible fund drive: a new barn to replace the old one and a riding arena with a classroom and viewing area attached.

Writer begins third section on her findings

**Figure 28-2.** *Continued*

<u>Barn</u>

According to both Mr. Troivinen and the alumni I spoke with, an adequate barn would be a 70 × 170-foot steel building. The building would contain 32 box stalls, 27 tie stalls, 3 storage rooms for equipment, another storage area for feed, and a washing area for bathing the horses. Like the present barn, it would also include an office for the Stable Manager.

The stall area and aisles would have dirt floors to provide good drainage. The other areas would have regular concrete floors, except the washing area, which would need a special nonslip concrete floor. The building could be faced with wood siding and roofed with wood shingles to help it blend in with its setting.

The cost of such a barn would be about $125,000. Small savings could be realized by replacing some of the box stalls with tie stalls, which cost about one-third as much, but as explained above the tie stalls are not as desirable.

<u>Arena and Classroom</u>

An indoor riding facility could be provided by building an 80 × 200-foot arena. It would be used for riding instruction, team practice, and horse shows. Like the barn, this could be a steel building covered on the outside with appropriate siding and roofing. It would have a dirt floor.

The arena would have no seating area, under the assumption that a classroom would be attached to its side. The classroom would extend for 40 feet along one wall of the arena, and would be 30 feet deep. Along the common wall, it would have a 30-foot window through which students could watch demonstrations and spectators could watch horse shows. The room would be equipped with a movie screen and restroom facilities. It would also provide a meeting room for the horse show club. The arena and classroom together would cost about $150,000.

**Figure 28-2.**   *Continued*

February 26, 19--                                              Page 6

### PROSPECTS FOR SUCCESSFUL FUND RAISING

This is the last
section in which
the writer
presents and
discusses her
findings.

When I spoke with alumni at last month's alumni
weekend, I found that a fund-raising drive for the stables
would not be widely appealing to Central's general alumni.
Many alumni perceive that contributions to such a drive
would benefit only a relatively small number of students,
primarily those on the riding team.

However, I also found that alumni who have been
involved in the riding programs over the years are very
supportive, especially alumni from more than ten years
ago, when students who owned their own horses could
board them at the barn. These alumni are particularly
impressed with the success of the intercollegiate show
team. In the past few months, one alumnus has pledged
$4000 toward the new barn, and the parents of a current
member of the riding club have donated a new treadmill,
valued at over $5000, which will be used to train young
horses. Mr. Troivinen has a long list of alumni who have
expressed support for the program.

### CONCLUSIONS AND RECOMMENDATIONS

Writer explicitly
states her
conclusions.

In conclusion, my investigation indicates that the
university's riding programs could benefit substantially
from a fund-raising effort. However, the appeal of such a
program will be limited primarily to the very supportive
alumni who used the university stables while students.

Writer presents
her detailed
recommendations.

Therefore, I recommend that we conduct a fund-raising
drive focused on the barn alone, leaving the arena and
classroom for a future program. The drive should be
announced to all alumni in one of the brief sketches in the
brochure we send to all alumni each year to describe our
development plans. The description should emphasize the
classes taught and the need to provide better housing for
the horses.

A more extensive description of the project should be
prepared for alumni known to be interested in the riding
programs. In addition to the information provided to all

**Figure 28-2.**   *Continued*

February 26, 19-- Page 7

alumni, the description should emphasize the success of the horse show team and appeal to the desire of the alumni to have a top-quality equestrian program at their alma mater.

Though Mr. Troivinen will be disappointed with a decision not to seek funds for the arena and classroom next year, I am sure that he will work with us enthusiastically and effectively in the fund-raising drive I have outlined.

**Figure 28-2.** *Continued*

# Empirical Research Reports

## PREVIEW OF CHAPTER

Typical Writing Situations
The Questions Readers Ask Most Often
Superstructure for Empirical Research Reports
    Introduction
    Objectives of the Research
    Method
    Results
    Discussion
    Conclusions
    Recommendations
An Important Note about Headings
Sample Research Report

Vernon has just completed a major research project—not library research but an experiment. He has tested two new, lightweight metal alloys to see which one will work best in the blades of jet engines. Working with a team of engineers, he placed fan blades made of each alloy in engines, which he ran through the same fifty-hour-long sequence of operating conditions. Throughout that time, he and the others monitored the performance of the blades, and afterwards they analyzed the blades to determine how well they were holding up.

Elaine is just finishing another research project, a survey to learn what kinds of outdoor recreation activities elderly people prefer and what factors influence how much they participate in those activities. With the help of six assistants, she has gathered responses to 80 carefully written questions from a scientifically selected group of 200 elderly individuals, each interviewed at home. Then, she spent weeks performing statistical analyses of the responses.

Research of the kinds performed by Vernon and Elaine is called *empirical research*. In it, the researcher gathers information by means of carefully planned, systematic observations or measurements. Although your major field may be much different from those of Vernon and Elaine, you will probably conduct some kind of empirical research regularly in your career. In this chapter, you will learn how to use the conventional superstructure for reporting empirical research.

## TYPICAL WRITING SITUATIONS

You will be able to use the superstructure for empirical research reports most successfully if you understand the purposes of the research discussed in them. Basically, there are two distinct purposes for empirical research. Most of it aims to help people make practical decisions. For instance, the results of Vernon's experiment will be used by the engineers who design engines for his employer. The results of Elaine's survey will be used by the state agency in charge of outdoor recreation as it decides what sorts of services and facilities it must provide to meet the needs of older citizens. Similarly, when you write about your empirical research, you may be writing to people who want to use your information to make practical decisions; these people may work in your own organization or in a client's.

A smaller amount of the empirical research has a different purpose: to extend general human knowledge. The researchers set out to learn how fish remember, what the molten core of the earth is like, or why people fall in love. This research is carried out even though it has no immediate practical value, and it is usually reported in scholarly journals, such as the *Journal of Chemical Thermodynamics*, the *Journal of Cell Biology*, and the *Journal of*

*Social Psychology*, whose readers are concerned not so much with making practical, business decisions as with extending the frontiers of human understanding and knowledge.

In some situations, these two aims of research overlap. Some organizations sponsor basic research, usually in the hope that what is learned can later be turned to some practical use. Likewise some practical research turns up results that are of interest to those who desire to learn more about the world in general.

## THE QUESTIONS READERS ASK MOST OFTEN

Whether it aims to support practical decisions, extend human knowledge, or achieve some combination of the two purposes, almost all empirical research is customarily reported in the same superstructure. That's largely because the readers of all types of empirical research have the same seven general questions about it:

- *Why is your research important to us?* Readers concerned with solving specific practical problems want to know what problems your research will help them address. Readers concerned with extending human knowledge want to know how you think your research contributes to what humans know.
- *What were you trying to find out?* A key part of an empirical research project is the careful formulation of the research questions that the project will try to answer. Readers want to know what those questions are so they can determine whether they are significant questions.
- *Was your research method sound?* Unless your method is appropriate to your research questions and unless it is intellectually sound, your readers will not place any faith in your results or in the conclusions and recommendations you base upon them.
- *What results did your research produce?* Naturally, your readers will want to find out what results you obtained.
- *How do you interpret those results?* Your readers will want you to interpret your results in ways that are meaningful to them.
- *What is the significance of those results?* What answers do your results imply for your research questions, and how do your results relate to the problems your research was to help solve or to the area of knowledge your research set out to expand?
- *What do you think we should do?* Readers concerned with practical problems want to know what you advise them to do. Readers concerned with extending human knowledge want to know what you think your results imply for future research.

## SUPERSTRUCTURE FOR EMPIRICAL RESEARCH REPORTS

To answer the readers' typical questions about empirical research projects, writers use a superstructure that has the following elements: introduction, objectives of the research, method, results, discussion, conclusions, and recommendations. As Figure 29-1 shows, each of these elements addresses one of the readers' seven questions.

The rest of this chapter explains how you can develop each element of an empirical research report. If you have not already done so, you should also read Chapter 28, which describes the *general* superstructure for reports, and the introduction to Part IX, which gives a practical introduction to the nature and use of superstructures. Both provide information not repeated here that will help you write empirical research reports effectively.

Much of the advice provided in this chapter is illustrated through the use of two sample reports. The first is presented in full at the end of this chapter. Its aim is practical. It was written by engineers who are developing a satellite communication system that will permit companies with large fleets of trucks to communicate directly with their drivers at any time.[1] In the report, the writers tell decision-makers and other engineers in their organization about the first operational test of the system, in which they sought to answer several practical engineering questions.

In contrast, the aim of the second sample report is to extend human knowledge. The researcher tells about his study of the ways that people develop friendships. Selected passages from this report are quoted below; if you

| Report Element | Readers' Question |
|---|---|
| Introduction | Why is your research important to us? |
| Objectives of the research | What were you trying to find out? |
| Method | Was your research method sound? |
| Results | What results did your research produce? |
| Discussion | How do you interpret those results? |
| Conclusions | What is the significance of those results? |
| Recommendations | What do you think we should do? |

**Figure 29-1.** Elements of an Empirical Research Report and Their Relationship to the Questions Readers Ask.

wish to read the entire report you will find it in the April 1985 issue of the *Journal of Personality and Social Psychology*.[2]

## Introduction

In the introduction to an empirical research report, you should seek to answer the readers' question, "Why is this research important to us?" Typically, writers answer that question in two steps: they announce the topic of their research and then explain the importance of the topic to their readers.

### Announcing the Topic

You can often announce the topic of your research simply by including that topic as the key phrase in the opening sentence of your report. For example, consider the first sentence of the report on the satellite communication system:

Topic of report

For the past eighteen months, the Satellite Products Laboratory has been developing a system that will permit companies with large, nationwide fleets of trucks to communicate directly to their drivers at any time through a satellite link.

Here is the first sentence from the report on the way that people develop friendships:

Topic of report

Social psychologists know very little about the way real friendships develop in their natural settings.

### Explaining the Importance of the Research

To explain the importance of your research to your readers, you can use either or both of the following methods: state the relevance of the research to your organization's goals, and review the previously published literature on the subject.

#### Relevance to Organization Goals

In reports written to readers in organizations (whether your own or a client's), you can explain the relevance of your research by relating it to some organizational goal or problem. Sometimes, in fact, the importance of your research to the organization's needs will be so obvious to your readers that merely naming your topic will be sufficient. At other times, you will need to discuss at length the relevance of your research to the organization. In the first paragraph of the satellite report, for instance, the writers mention the potential market for the satellite communication system they are developing.

That is they explain the importance of their research by saying that it can lead to a profit. For detailed advice about how to explain the importance of your research to readers in organizations, see the advice given in Guideline 2 of Chapter 13, "Writing the Beginning of a Communication."

## Literature Reviews

A second way to establish the importance of your research is to review the existing knowledge on your subject. Writers usually do that by reviewing the previously published literature. Generally, you can arrange a literature review in two parts. First, present the main pieces of knowledge communicated in the literature. Then, identify some significant gap in this knowledge—the very gap your own research will fill. In this way, you establish the special contribution that your research will make.

The following passage from the opening of the report on the friendship study shows this two-stage pattern of development. Notice that the writer uses author–year citations, as described in Chapter 27:

The writer tells what is known on his topic.

The writer identifies the gaps in knowledge that his research will fill.

> A great deal of research in social psychology has focused on variables influencing an individual's attraction to another at an initial encounter, usually in laboratory settings (Bergscheid and Walster, 1978; Bryne, 1971; Huston and Levinger, 1978), yet very little data exists on the processes by which individuals in the real world move beyond initial attraction to develop a friendship; even less is known about the way developing friendships are maintained and how they evolve over time (Huston and Burgess, 1979; Levinger, 1980).

The writer continues this discussion of previous research for three paragraphs. Each follows the same pattern: it identifies an area of research, tells what is known about that area, and identifies gaps in the knowledge—gaps that will be filled by the very research that the writer has conducted. These paragraphs serve an important additional function also performed by many literature reviews: they introduce the established facts and theories that are relevant to the writers' work and necessary to the understanding of the report.

Writers almost always include literature reviews in the reports they write for professional journals. In contrast, they often omit reviews when writing to readers inside an organization. That's because such reviews are often unnecessary when addressing organizational readers. Organizational readers judge the importance of a report in terms of its relevance to the organization's goals and problems, not in terms of its relation to the general pool of human knowledge. For example, the typical readers of the truck-and-satellite communication report were interested in the report because they wanted to learn how well their company's system would work. To them, a general survey of the literature on statellite communication would have seemed irrelevant—and perhaps even annoying.

A second reason that writers often omit literature reviews when addressing readers in organizations is that such reviews rarely help such readers understand the reports. That's because the research projects undertaken within organizations usually focus so sharply on a particular, local question that published literature on the subject is beside the point. For example, a review of previously published literature on satellite communications would not have helped readers understand the truck-and-satellite report.

Sometimes, of course, literature reviews do appear in reports written to organizational readers. Often, they say something like this: "In a published article, one of our competitors claims to have saved large amounts of money by trying a new technique. The purpose of the research described in this report is to determine whether or not we could enjoy similar results." Of course, the final standard for judging whether or not you should include a literature review in your report is your understanding of your purpose and readers. In some way or another, however, the introductions to all your empirical research reports should answer your readers' question, "Why is this research important to us?"

## Objectives of the Research

Every empirical research project has carefully constructed objectives. These objectives define the focus of your project, influence the choice of research method, and shape the way you interpret your results. Thus, readers of empirical research reports want and need to know what the objectives are.

The following example from the satellite report shows one way you can tell your readers about your objectives:

> In particular, we wanted to test whether we could achieve accurate data transmissions and good-quality voice transmissions in the variety of terrains typically encountered in long-haul trucking. We wanted also to see what factors might affect the quality of transmissions.

When reporting on research that involves the use of statistics, you can usually state your objectives by stating the hypotheses you tested. Where appropriate, you can explain these hypotheses in terms of existing theory, again citing previous publications on the subject. The following passage shows how the writer who studied friendship explains some of his hypotheses. Notice how the author begins with a statement of the overall goal of the research:

Overall goal —

First objective (hypothesis) —

Second objective (hypothesis) —

The goal of the study was to identify characteristic behavioral and attitudinal changes that occurred within interpersonal relationships as they progressed from initial acquaintance to close friendship. With regard to relationship benefits and costs, it was predicted that both benefits and costs would increase as the friendship developed, and that the ratings of both costs and benefits would be positively correlated with ratings of friendship intensity. In addition,

Third objective
(hypothesis) ———— the types of benefits listed by the subjects were expected to change as the friendships developed. In accord with Levinger and Snoek's (1972) model of dyadic relatedness, benefits listed at initial stages of friendship were hypothesized to be more activity centered and to reflect individual self-interest (e.g., companionship, information) than benefits at later stages, which were expected to be more personal and reciprocal (e.g., emotional support, self-esteem).

## Method

When reading reports of your empirical research, people will look for precise details concerning your method. Those details serve three purposes. First, they let your readers assess the soundness of your research design and its appropriateness for the problems you are investigating. Second, the details enable your readers to determine the limitations that your method might place upon the conclusions you can draw. Third, the description of your method provides information that will help your readers repeat your experiment if they wish to verify your results or conduct similar research projects of their own.

The nature of the information you should provide about your method depends upon the nature of your research. For instance, the writer studying friendship began his description of his research methods in this way:

> At the beginning of their first term at the university, college freshmen selected two individuals whom they had just met and completed a series of questionnaires regarding their relationships with those individuals at 3-week intervals through the school term.

In the rest of that paragraph, the writer explains that the questionnaires asked the freshmen to tell about such things as their attitudes toward each of the other two individuals and the specific things they did with each of the other two. However, that paragraph is just a small part of the researcher's account of his method. He then provides a 1200-word discussion of the students he studied and of the questionnaires and procedures he used.

The writers of the satellite report likewise provided detailed information about their procedures: three paragraphs and two tables explaining their equipment (truck radios and satellite), two paragraphs and one map describing the eleven-state region covered by the trucks, and two paragraphs describing their data analysis.

How can you decide what details to include? The most obvious way is to follow the general reporting practices in your field. Find some research reports that use a method similar to yours and see what they report. You can check the scope of your description in the following ways:

● List every aspect of your procedure that you made a decision about when planning your research.

- Identify *every* aspect of your method that your readers might ask about.
- Ask yourself what aspects of your procedure might limit the conclusions you can draw from your results.
- Identify *every* procedure that other researchers would need to understand in order to design a similar study.

## Results

The results of empirical research are the data you obtain. Although your results are the heart of your empirical research report, they may take up a very small portion of it. Generally, results are presented in one of two ways:

- *Tables and Graphs.* The satellite report, for instance, uses two tables. The report on friendship uses four tables and eleven graphs. For information on using tables, graphs, and other visual aids, see Part V, "Drafting Visual Elements."
- *Sentences.* When placed in sentences, results are often woven into a discussion that combines data and interpretation, as is explained in the next paragraphs.

## Discussion

Sometimes writers briefly present all their results in one section and then discuss them in a separate section. Sometimes they combine the two in a single, integrated section. Whichever method you use, your discussion must link your interpretative comments with the specific results you are interpreting.

One useful way of making that link is to refer to the key results shown in a table or other visual aid and then comment on them as appropriate. The following passage shows how the writers of the satellite report did that for some of the results that they presented in one of their tables:

Writers emphasize a key result shown in a table.

Writers draw attention to other important results.

Writers interpret those results.

As Table 3 shows, 91% of the data transmissions were successful. These data are reported according to the region in which the trucks were driving at the time of transmission. The most important difference to note is the one between the rate of successful transmissions in the Southern Piedmont region and the rates in all the other regions. In the Southern Piedmont area, we had the truck drive slightly outside the ATS-6 footprint so that we could see if successful transmissions could be made there. When the truck left the footprint, the percentage of successful data transmissions dropped abruptly to 43%.

When you present your results in prose only (rather than in tables and graphs), you can weave those results together with your discussion by beginning your paragraphs with general statements that are really interpretations of your data. Then, cite the relevant results as evidence in support of the interpretation. Here is an example from the friendship report:

General interpretation.

Specific results presented as support for the interpretation.

> Intercorrelations among the subjects' friendship intensity ratings at the various assessment points showed that friendship attitudes became increasingly stable over time. For example, the correlation between friendship intensity ratings at 3 weeks and 6 weeks was .55; between 6 weeks and 9 weeks, .78; between 9 weeks and 12 weeks, .88 (all $p < .001$).

In a single report, you may use both of these methods of combining the presentation and discussion of your results.

## Conclusions

Besides interpreting the results of your research, you need to explain what your results mean in terms of the original research questions and the general problem you set out to investigate. Your explanations of these matters are your conclusions.

If your research project is sharply focused on (for example) only a single hypothesis, your conclusion can be very brief, perhaps only a restatement of your chief results. However, if your research has many threads, your conclusion should draw those strands together.

In either case, the presentation of your conclusions should correspond very closely to the objectives for your research that you identified toward the beginning of your report. Consider, for instance, the correspondence between objectives and conclusions in the satellite study. The first objective was to determine whether accurate data transmissions and good-quality voice transmissions could be obtained in the variety of terrains typically encountered in long-haul trucking. The first of the conclusions addresses that objective:

> The Satellite Product Laboratory's system produces good-quality data and voice transmissions throughout the eleven-state region covered by the satellite's broadcast footprint.

The second objective was to determine what factors affect the quality of transmissions, and the second and third conclusions relate to it:

> The most important factor limiting the success of transmissions is movement outside the satellite's broadcast footprint, which accurately defines the satellite's area of effective coverage.

> The system is sensitive to interference from certain kinds of objects in the line of sight between the satellite and the truck. These include trees, mountains and hills, overpasses, and buildings.

The satellite research concerns a practical question. Hence its objectives and conclusions address practical concerns of particular individuals—in this case, the engineers and managers in the company that is developing the satellite system. In contrast, research that aims primarily to extend human knowledge often has both objectives and conclusions that focus on theoretical issues.

For example, at the beginning of the friendship report, the researcher identifies several questions that his research investigated, and he tells what answers he predicted his research would produce. In his conclusion, then, he systematically addresses those same questions, talking about them in terms of the results his research produced. Here is a summary of some of his objectives and conclusions. (Notice how he uses the technical terminology commonly employed by his readers.)

| Objective | Conclusion |
|---|---|
| As they develop friendships, do people follow the kind of pattern theorized by Guttman, in which initial contacts are relatively superficial and later contacts are more intimate? | Yes. "The initial interactions of friends . . . correspond to a Guttman-like progression from superficial interaction to increasingly intimate levels of behavior." |
| Do *both* the costs (or unpleasant aspects) and the benefits of personal relationships increase as friendships develop? | Yes. "The findings show that personal dissatisfactions are inescapable aspects of personal relationships and so, to some degree, may become immaterial. The critical factor in friendships appears to be the amount of benefits received. If a relationship offers enough desirable benefits, individuals seem willing to put up with the accompanying costs." |
| Are there substantial differences between the friendships women develop with one another and the kind men develop with one another? | Apparently not. "These findings suggest that—at least for this sample of friendships—the sex differences were more stylistic than substantial. The bonds of male friendship and female friendship may be equally strong, but the sexes may differ in their manner of expressing that bond. Females may be more inclined to express close friendships through physical or verbal affection; males may express their closeness through the types of companionate activities they share with their friends." |

Typically, in a discussion of the conclusions of an empirical research project directed primarily at increasing human knowledge, writers will also discuss the relationship of their findings to the findings of other researchers and to various theories that have been advanced concerning their subject. The writer of the article on friendship did that. The table you have just read presents only a few snippets from his overall discussion, which is several thousand words long and is full of thoughts about the relationship of his results to the results and theories of others.

In the discussion sections of their empirical research reports, writers sometimes discuss any flaws in their research method or limitations on the generalizability of their conclusions. For example, the writer on friendship points out that his subjects were all college students and most lived in dormitories. It is possible, he cautions, that the things he found while studying this group may not be true for other groups.

## Recommendations

The readers of some empirical research reports want to know what, based on the research, the writer thinks should be done. That is especially true in situations where the research is directed at solving a practical problem. Consequently, research reports usually include recommendations. For example, the satellite report contains three. The first is the general recommendation that work on the project be continued. The other two involve specific actions that the writers think should be taken: design a special antenna for the trucks and develop a plan that tells what satellites would be needed to provide coverage throughout the 48 contiguous states, Alaska, and Southern Canada. As is common in research addressed to readers in organizations, these recommendations concern practical business and engineering decisions.

Even in reports directed toward extending human knowledge, writers sometimes have recommendations. These usually concern their ideas about future studies that should be made, adjustments in methodology that seem to be called for, and the like. In the last paragraph of the friendship report, for instance, the writer suggests that additional research be conducted that studies different groups (not just students) in different settings (not just college) to establish a more comprehensive understanding of how friendships develop.

# AN IMPORTANT NOTE ABOUT HEADINGS

In the discussion you have just read, you learned about the seven elements of the conventional superstructure for reports on empirical research. Such reports may divided into seven sections that correspond directly to those seven elements. That is especially likely to happen in long reports written to readers in organizations (as distinct from the readers of professional journals). However, in some reports some of the elements are combined under a single heading. For example, in the satellite report, the objectives are included at the end of the section entitled "Introduction," and the results and discussion are combined under the single heading "Results and Discussion." Even when only a few major headings are used, almost all empirical research reports include all seven of the elements described above. You should determine which headings to use in your reports by considering your purpose and your readers' expectations.

## SAMPLE RESEARCH REPORT

Figure 29-2 shows the full report on the truck-and-satellite communication system, which is addressed to readers within the writers' own organization. To see examples of empirical research reports presented as journal articles, consult journals in your field.

## WRITING ASSIGNMENT

An assignment that involves writing an empirical research report appears in Appendix I.

## NOTES

1. This report is an adaptation of material from Roy E. Anderson, Richard L. Frey, and James R. Lewis, *Satellite-Aided Mobile Communications Limited Operational Test in the Trucking Industry* (Schenectady, New York: General Electric Company, 1980).

2. Robert B. Hays, "A Longitudinal Study of Friendship Development," *Journal of Personality and Social Psychology* 48(1985):909–24. Copyright 1985 by the American Psychological Association.

Cover

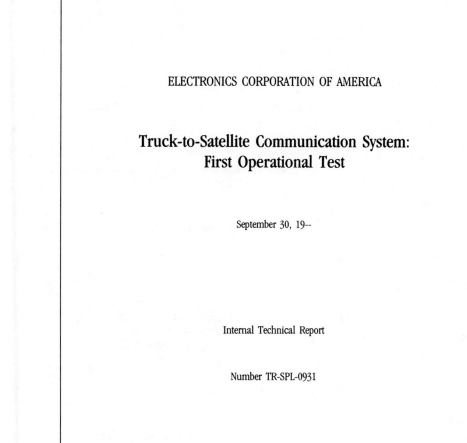

ELECTRONICS CORPORATION OF AMERICA


**Truck-to-Satellite Communication System:**
**First Operational Test**


September 30, 19--


Internal Technical Report


Number TR-SPL-0931

**Figure 29-2.**   Empirical Research Report.

Title Page

ELECTRONICS CORPORATION OF AMERICA

Truck-to-Satellite Communication System:
First Operational Test

September 30, 19--

<u>Research Team</u>
Margaret C. Barnett
Erin Sanderson
L. Victor Sorrentino
Raymond E. Wu

Internal Technical Report
Number TR-SPL-0931

Read and Approved:

_____     _____
Laboratory Director                      Date

**Figure 29-2.** *Continued*

EXECUTIVE SUMMARY

Topic of report

For the past eighteen months, the Satellite Products Laboratory has been developing a system that will permit companies with large, nationwide fleets of trucks to communicate directly to their drivers at any time through a satellite link. During the week of May 18, we tested our concepts for the first time, using the ATS-6 satellite and five trucks that were driven over an eleven-state region with our prototype mobile radios.

Method

Results and discussion

More than 91% of the 2500 data transmissions were successful and more than 91% of the voice transmissions were judged to be of commercial quality. The most important factor limiting the success of transmissions was movement outside the satellite's broadcast footprint. Other factors include the obstruction of the line of sight between the truck and the satellite by highway overpasses, mountains and hills, trees, and buildings.

Conclusions

Overall, the test demonstrated the soundness of the prototype design. Work on it should continue as rapidly as possible. We recommend the following actions:

Recommendations

- Develop a new antenna designed specifically for use in communications between satellites and mobile radios.

- Explore the configuration of satellites needed to provide thorough footprint coverage for the 48 contiguous states, Alaska, and Southern Canada at an elevation of 25° or more.

i

**Figure 29-2.**   *Continued*

FOREWORD

Cross-reference
to related
reports

Acknowledg-
ments

Previous technical reports describing work on this
project are: TR-SPL-0785, TR-SPL-0795 through TR-SPL-
0798, TR-SPL-0823, and TR-SPL-0862.

We could not have conducted this test without the
gratifying cooperation of Smith Moving Company and, in
particular, the drivers and observers in the five trucks that
participated. We are also grateful for the cooperation of the
United States National Aeronautics and Space Administration in the use of the ATS-6 satellite.

The test described in this report was supported through
internal product development funds.

ii

**Figure 29-2.**   *Continued*

Table of
contents

TABLE OF CONTENTS

1. Introduction                                          1

2. Method                                                2

    Equipment                                           2
    Region Covered                                      5
    Data Analysis                                       6

3. Results and Discussion                                7

    Data Transmissions                                  7
    Voice Transmissions                                 7
    Factors Affecting the Quality of Transmissions     10

4. Conclusions                                          12

5. Recommendations                                      13

iii

**Figure 29-2.** *Continued*

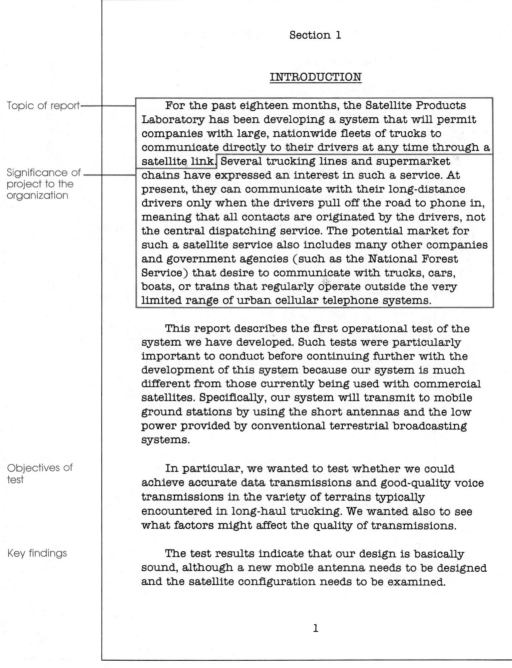

Section 1

## INTRODUCTION

Topic of report—

Significance of project to the organization—

For the past eighteen months, the Satellite Products Laboratory has been developing a system that will permit companies with large, nationwide fleets of trucks to communicate directly to their drivers at any time through a satellite link. Several trucking lines and supermarket chains have expressed an interest in such a service. At present, they can communicate with their long-distance drivers only when the drivers pull off the road to phone in, meaning that all contacts are originated by the drivers, not the central dispatching service. The potential market for such a satellite service also includes many other companies and government agencies (such as the National Forest Service) that desire to communicate with trucks, cars, boats, or trains that regularly operate outside the very limited range of urban cellular telephone systems.

This report describes the first operational test of the system we have developed. Such tests were particularly important to conduct before continuing further with the development of this system because our system is much different from those currently being used with commercial satellites. Specifically, our system will transmit to mobile ground stations by using the short antennas and the low power provided by conventional terrestrial broadcasting systems.

Objectives of test

In particular, we wanted to test whether we could achieve accurate data transmissions and good-quality voice transmissions in the variety of terrains typically encountered in long-haul trucking. We wanted also to see what factors might affect the quality of transmissions.

Key findings

The test results indicate that our design is basically sound, although a new mobile antenna needs to be designed and the satellite configuration needs to be examined.

1

**Figure 29-2.**  *Continued*

Section 2

## METHOD

In this experiment, we tested a full-scale system in which five trailer trucks communicated with an earth ground station via a satellite in geostationary orbit 23,200 miles above the earth. This section describes the equipment we used, the area covered by the test, and the data analyses we performed.

### Equipment

The five trucks, operated by Smithson Moving Company, were each equipped with a prototype of our newly developed 806 megahertz (MHz) two-way mobile radio equipment. Each radio had a speaker and microphone for voice communications, along with a ten-key keyboard and digital display for data communications. Equipped with dipole antennas, the radios broadcast at 1650 MHz with 12 to 15 watts of power. They received signals at 1550 MHz and had an equivalent antenna temperature of 800°K, including feedline losses. Technical specifications for the receivers and transmitters of these radios are given in Table 1.

The satellite used for this test was the ATS-6, which has a larger antenna than most commercial communication satellites, making it more sensitive to the low-power signals sent from the mobile stations. Technical specifications for its receiver and transmitter are given in Table 2.

Through the ATS-6 satellite, the five trucks communicated with the Earth Ground Station in King of Prussia, Pennsylvania. This facility is a relatively large station, but not larger than is planned for a fully operational commercial system.

2

Overview of method

Forecasting statement for section

Paragraph organized from general to particular (as are all others)

Reference to technical data in table

Use of a large station is justified

Figure 29-2.   *Continued*

Table 1
Specifications for Satellite-Aided Mobile Radio

<u>Transmitter</u>

| | |
|---|---|
| Frequency | 1655.050 MHz |
| Power Output | 16 watts nominal<br>12 watts minimum |
| Frequency Stability | ± 0.0002% ($-30°$ to $+60°C$) |
| Modulation | $16F_3$ Adjustable from 0 to<br>± 5 kHz swing FM with<br>instantaneous modulation<br>limiting |
| Audio Frequency Response | Within +1 dB and $-3$ dB of<br>a 6 dB/octave pre-emphasis<br>from 300 to 3000 HZ per<br>EIA standards |
| Duty Cycle | EIA 20% Intermittent |
| Maximum Frequency Spread | ± 6 MHz with center tuning |
| RF Output Impedance | 50 ohms |

<u>Receiver</u>

| | |
|---|---|
| Frequency | 1552.000 MHz |
| Frequency Stability | ± 0.0002% ($-30°$ to $+60°C$) |
| Noise Figure | 2.6 dB referenced to<br>transceiver antenna jack |
| Equivalent Receiver Noise<br>Temperature | 238° Kelvin |
| Selectivity | $-75$ dB by EIA Two-Signal<br>method |
| Audio Output | 5 watts at less than 5%<br>distortion |
| Frequency Response | Within +1 and $-8$ dB of a<br>standard 6 dB per octave de-<br>emphasis curve from 300 to<br>3000 Hz |
| Modulation Acceptance | ± 7 kHz |
| RF Input Impedance | 50 ohms |

3

Figure 29-2. *Continued*

Table 2
Performance of ATS-6 Spacecraft L-Band Frequency
Translation Mode


### Receive

| | |
|---|---|
| Receiver Noise Figure (dB) | 6.5 |
| Equivalent Receiver Noise Temperature (°K) | 1005 |
| Antenna Temperature Pointed at Earth (°K) | 290 |
| Receive System Temperature (°K) | 1295 |
| Antenna Gain, peak (dB) | 38.4 |
| Spacecraft G/T, peak (dB/°K) | 7.3 |
| Half Power Beamwidth (degrees) | 1.3 |
| Gain over Field of View (dB) | 35.4 |
| Spacecraft G/T over Field of View (dB/°K) | 4.3 |


### Transmit

| | |
|---|---|
| Transmit Power (dBw) | 15.3 |
| Antenna Gain, peak (dB) | 37.7 |
| Effective Radiated Power, peak (dBw) | 53.0 |
| Half Power Beamwidth (degrees) | 1.4 |
| Gain over Field of View (dB) | 34.8 |
| Effective Radiated Power over Field of View (dBw) | 50.1 |

4

**Figure 29-2.**   *Continued*

### Region Covered

The five trucks drove throughout the region covered by the "footprint" of ATS-6. The footprint is shown as the area within the oval in Figure 1. It is defined as the area in which the broadcast signals received are within at least 3 dB of the signal received at the center of the beam. In all, the trucks covered eleven states: Georgia, South Carolina, North Carolina, Tennessee, Virginia, West Virginia, Ohio, Indiana, Illinois, Iowa, and Nebraska.

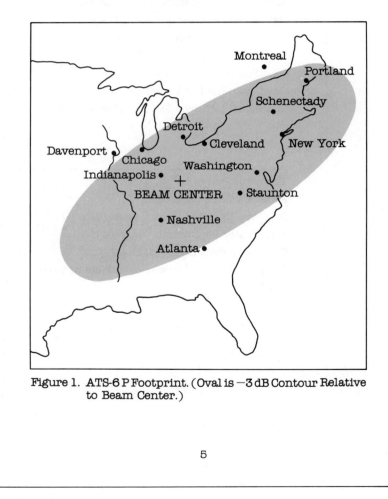

Figure 1. ATS-6 P Footprint. (Oval is −3 dB Contour Relative to Beam Center.)

5

Figure 29-2. *Continued*

Within this region, the trucks drove through the kinds of terrain usually encountered in long-haul trucking, including both urban and rural areas in open plains, foothills, and mountains.

Data Analysis

All test transmissions were recorded at the earth station on a high-quality reel-to-reel tape recorder. The strength of the signals received from the trucks via the satellite was recorded on a chart recorder that had a frequency response of approximately 100 hertz. In addition, observers in the trucks recorded all data signals received and all data codes sent. They also recorded information about the terrain during all data and voice transmissions.

We analyzed the data collected in several ways:

- To determine the accuracy of the data transmissions, we compared the information recorded by the observers with the signals recorded on the tapes of all transmissions.

- To determine the quality of the voice transmissions, we had an evaluator listen to the tape using high-quality earphones. For each transmission, the evaluator rated the signal quality on the standard scale for the subjective evaluation of broadcast quality. On it, Q5 is excellent and Q1 is unintelligible.

- To determine what factors influenced the quality of the transmissions, we examined the descriptions of the terrain that the observers recorded for all data transmissions that were inaccurate and all the voice communications that were rated 3 or less by the evaluator. We also looked for relationships between the accuracy and quality of the transmissions and the distance of the trucks from the edge of the broadcast footprint of the ATS-6 satellite.

Presentation of data analyses parallels list of objectives in the introduction

6

**Figure 29-2.** *Continued*

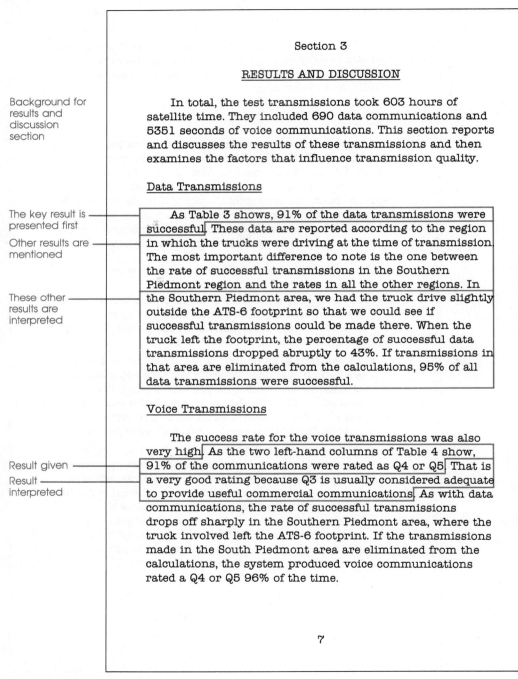

Background for results and discussion section

Section 3

RESULTS AND DISCUSSION

In total, the test transmissions took 603 hours of satellite time. They included 690 data communications and 5351 seconds of voice communications. This section reports and discusses the results of these transmissions and then examines the factors that influence transmission quality.

Data Transmissions

The key result is presented first

Other results are mentioned

These other results are interpreted

As Table 3 shows, 91% of the data transmissions were successful. These data are reported according to the region in which the trucks were driving at the time of transmission. The most important difference to note is the one between the rate of successful transmissions in the Southern Piedmont region and the rates in all the other regions. In the Southern Piedmont area, we had the truck drive slightly outside the ATS-6 footprint so that we could see if successful transmissions could be made there. When the truck left the footprint, the percentage of successful data transmissions dropped abruptly to 43%. If transmissions in that area are eliminated from the calculations, 95% of all data transmissions were successful.

Voice Transmissions

Result given

Result interpreted

The success rate for the voice transmissions was also very high. As the two left-hand columns of Table 4 show, 91% of the communications were rated as Q4 or Q5. That is a very good rating because Q3 is usually considered adequate to provide useful commercial communications. As with data communications, the rate of successful transmissions drops off sharply in the Southern Piedmont area, where the truck involved left the ATS-6 footprint. If the transmissions made in the South Piedmont area are eliminated from the calculations, the system produced voice communications rated a Q4 or Q5 96% of the time.

7

Figure 29-2. *Continued*

Table 3
Success in Decoding DTMF Automatic Transmitter

| REGION | STATES | ATIS TRANSMISSIONS | | ELEVATION |
| --- | --- | --- | --- | --- |
| | | Sent by Vehicle | Received and Decoded Correctly | Angle to Satellite (°) |
| Open Plains | Indiana, Ohio, Nebraska, Illinois, Iowa | 284 | 283 (100%) | 17–26 |
| Western Appalachian Foothills | Ohio, Tennessee | 55 | 53 (96%) | 15–19 |
| Appalachian Mountains | West Virginia | 112 | 93 (83%) | 15–17 |
| Piedmont | Virginia, North Carolina | 190 | 178 (97%) | 11–16 |
| Southern Piedmont | Georgia, South Carolina | 49 | 21 (43%) | 17–18 |
| TOTAL | | 690 | 628 (91%) | |

8

**Figure 29-2.**  *Continued*

Table 4
Quality of Voice Communication Signal[1]

| Area | Transmission Time (Seconds) | No Blockage Time (Secs) | | Trees | | | Mountains and Hills | | | Overpasses (Momentary Dropouts) | | | Buildings | | |
|---|---|---|---|---|---|---|---|---|---|---|---|---|---|---|---|
| | | Q5 | Q4 | Q3 | Q2 | Q1 | Q3 | Q2 | Q1 | Q3 | Q2 | Q1 | Q3 | Q2 | Q1 |
| Open Plains | 2481 | 2334 | 73 | 1 | 1 | 0 | 4 | 1 | 3 | 13 | 2 | 20 | 15 | 10 | 4 |
| Western Foothills | 344 | 322 | 2 | 0 | 0 | 0 | 6 | 12 | 13 | 0 | 0 | 0 | 0 | 0 | 0 |
| Appalachian Mountains | 1037 | 614 | 267 | 42 | 17 | 0 | 20 | 31 | 37 | 0 | 2 | 7 | 0 | 0 | 0 |
| Piedmont | 1219 | 481 | 623 | 5 | 15 | 19 | 25 | 16 | 10 | 4 | 1 | 4 | 8 | 4 | 1 |
| Southern Piedmont | 270 | 0 | 149 | 109 | 3 | 12 | 0 | 0 | 0 | 0 | 0 | 2 | 0 | 0 | 0 |
| TOTAL TIME (seconds) | 5351 | 3751 | 1114 | 157 | 36 | 31 | 55 | 60 | 63 | 17 | 5 | 33 | 23 | 14 | 5 |

[1]Total times in seconds for each quality of received signals. Q5 is excellent; Q1 is unintelligible.

9

**Figure 29-2.**   *Continued*

Thus, within the footprint of the satellite, 95% or more of the transmissions of both data and voice were of commercial quality. That success rate is very good: the specifications for the mobile radio systems used by police and fire departments usually require only 90% effectiveness for the area covered.

Factors Affecting the Quality of Transmissions

The factor having the largest effect on transmission quality is the location of the truck within the footprint of ATS-6. The quality of transmissions is even and uniformly good throughout the footprint, but almost immediately outside of it the quality drops well below acceptable levels.

Several factors were found to disrupt transmissions even when the trucks were in the satellite's footprint. The four right-hand columns in Table 4 show what these factors are for the 4% of the transmissions in the footprint that were not of commercial quality. In all cases the cause is some object passing in the line-of-sight between the satellite and the truck.

Trees caused 45% of the disruptions, more than any other source. At the frequencies used for broadcasting in this system, trees and other foliage have a high and very sharp absorption. Of course, the tree must be immediately beside the road and also tall enough to intrude between the truck and the satellite, which is at an average elevation of 17° above the horizon. And the disruption caused by a single tree will create only a very brief and usually insignificant dropout of one second or less. Only driving past a group of trees will cause a significant loss of signal. Yet, this happened often in the terrain of the Appalachian Mountains and the South Piedmont.

We believe we could eliminate many of the disruptions caused by trees if we developed an antenna specifically for use in communications between satellites and mobile radios. In the test, we used a standard dipole antenna. Instead, we might devise a Wheeler-type antenna that is

10

**Figure 29-2.**    *Continued*

omnidirectional in azimuth and with gain in the vertical direction to minimize ground reflections.

Mountains and hills caused 36% of the disruptions. That happened mostly in areas where the satellite's elevation above the horizon was very low. Otherwise a hill or mountain would have to be very steep to block out a signal. For example, if a satellite were only 17° above the horizon, the hill would have to rise over 1500 feet per mile to interfere with a transmission--and the elevation would have to be precisely in a line between the truck and the satellite.

Most of the disruptions caused by mountains and hills can be eliminated by using satellites that have an elevation of at least 25°. That would place them above all but the very steepest slopes.

Highway overpasses accounted for 11% of the disruptions, but these disruptions had little effect on the overall quality of the broadcasts. As one of the test trucks drove on an open stretch of interstate highway, the signal was strong and steady, with fading less than 2 decibels peak-to-peak. About two seconds before the truck entered the overpass, there was detectable but not severe multipath interference. The only serious disruption was a one-second dropout while the truck was directly under the overpass. This one-second dropout was so brief that it did not cause a significant loss of intelligibility in voice communications. Only a series of overpasses, such as those found where interstate highways pass through some cities, cause a significant problem.

Finally, buildings and similar structures accounted for about 8% of the disruptions. These were experienced mainly in large cities, and isolated buildings usually caused only brief disruptions. However, when the trucks were driving down city streets lined with tall buildings, they were unable to obtain satisfactory communications until they were driven to other streets.

11

**Figure 29-2.** *Continued*

Presentation of
conclusions
parallels list of
objectives in
the introduction

Section 4

<u>CONCLUSIONS</u>

This test supports three important conclusions:

- The Satellite Products Laboratory's system produces good-quality data and voice transmissions through the eleven-state region covered by the satellite's broadcast footprint.

- The most important factor limiting the success of transmissions is movement outside the satellite's broadcast footprint, which accurately defines the satellite's area of effective coverage.

- The system is sensitive to interference from certain kinds of objects in the line of sight between the satellite and the truck. These include trees, mountains and hills, overpasses, and buildings.

12

**Figure 29-2.**   *Continued*

Section 5

RECOMMENDATIONS

Overall recom-
mendation
given first

    Based upon this test, we believe that work should
proceed as rapidly as possible to complete an operational
system. In that work, the Satellite Products Laboratory
should do the following things:

1. Develop a new antenna designed specifically for use in
   communications between satellites and mobile radios.
   Such an antenna would probably eliminate many of the
   disruptions caused by trees, the most common cause of
   poor transmissions.

2. Define the configuration of satellites needed to provide
   service throughout our planned service area. We are now
   ready to determine the number and placement in orbit of
   the satellites we will need to launch in order to provide
   service to our planned service area (48 contiguous
   states, Alaska, Southern Canada). Because locations
   outside of the broadcast footprint of a satellite probably
   cannot be given satisfactory service, our satellites will
   have to provide thorough footprint coverage throughout
   all of this territory. Also, we should plan the satellites so
   that each will be at least 25° above the horizon
   throughout the area it serves; in that way we can almost
   entirely eliminate poor transmissions due to
   interference from mountains and hills.

13

**Figure 29-2.**   *Continued*

# Feasibility Reports

## PREVIEW OF CHAPTER

Typical Writing Situation
The Questions Readers Ask Most Often
Superstructure for Feasibility Reports
    Introduction
    Criteria
    Method of Obtaining Facts
    Overview of Alternatives
    Evaluation
    Conclusions
    Recommendations
Sample Feasibility Report

W hen you perform a feasibility study, you evaluate the practicality and desirability of pursuing some course of action. No matter what kind of employer you work for, you are likely to be asked to conduct many feasibility studies in your career. Imagine, for instance, that you work for a company that designs and builds sailboats. The company thinks it might be able to reduce manufacturing costs without hurting sales if it uses new, high-strength plastics to manufacture some parts traditionally made from metal. Before making such a change, however, the company wants you to answer many questions. Would plastic parts be as strong, durable, and attractive as metal ones? Is there a supplier who might make the parts for the boat company? Would the parts be less expensive than the metal ones? Would boat buyers accept this change? The information and analyses you provide in your report will be the company's primary basis for deciding whether to pursue this course of action. In this chapter, you will learn how to write effective reports on feasibility studies.

## TYPICAL WRITING SITUATION

All feasibility reports share one essential characteristic: they are written to help the readers choose between two or more courses of action. In some cases, it might at first appear that your readers will be considering only one course of action. For instance, it might seem that the readers of the feasibility report on sailboats will be evaluating only the single course of action of using plastic parts. However, even when a feasibility report focuses primarily on one course of action, the readers are always considering a second course: to leave things the way they are—for example, to continue to use metal parts. In other writing situations, your readers will already have decided that some change is necessary and will be choosing between two or more alternatives to the status quo.

## THE QUESTIONS READERS ASK MOST OFTEN

As they think about the choice they must make, readers ask many questions. From situation to situation, these questions remain the same. That's what makes it possible for one superstructure to be useful across nearly the full range of situations in which people write feasibility reports.

The questions most often asked by readers of feasibility reports are the following:

- *Why is it important for us to consider these alternatives?* Readers ask this question because they want to know why they have to make any choice in the first place. If you have been studying a course of action that addresses a problem that is well-known to your readers, you may be able to explain the importance of considering the alternatives simply by reminding your readers of the problem. In other situations, your readers may include people who need long explanations to appreciate the importance of considering the alternative courses of action.

- *Are your criteria reasonable and appropriate?* To help your readers .choose between the alternative courses of action, you must evaluate the alternatives in terms of specific criteria. At work, readers want these criteria to reflect the needs and aims of their organization. And they want you to tell what the criteria are so that they can judge them.

- *Are your facts reliable?* Readers want to be sure that your facts are reliable before they make any decisions based on those facts.

- *What are the important features of the alternatives?* So that they can understand your detailed discussion of the alternatives, readers want you to highlight the key features of each one.

- *How do the alternatives stack up against your criteria?* The heart of a feasibility study is your evaluation of the alternatives in terms of your criteria. Readers of your report want to know the results.

- *What overall conclusions do you draw about the alternatives?* Based upon your detailed evaluation of the alternatives, you will reach some general conclusions about their merits. Because these conclusions can form the basis for decision-making, readers want to learn what they are.

- *What do you think we should do?* In the end, your readers must choose one of the alternative courses of action. Because of your expertise on the subject, they want you to help them by telling what you recommend.

## SUPERSTRUCTURE FOR FEASIBILITY REPORTS

To answer your readers' questions about your feasibility studies, you can use a superstructure that has the following seven elements: introduction, criteria, method of obtaining facts, overview of alternatives, evaluation, conclusions, and recommendation. As Figure 30-1 shows, each element corresponds to one of the readers' questions described above. Of course, you may combine the elements in many different ways, depending upon your situation. For instance, you may integrate your conclusions into your evaluation, or you may omit a separate discussion of criteria if they need no special explanation and if you organize your evaluation around them. But when writing any feasibility report, you should consciously determine how to include each of the seven elements, based upon your understanding of your purpose, readers, and situation.

The rest of this chapter explains how you can develop each of the elements to create an effective feasibility report. If you have not already done so, you should also read Chapter 28, which describes the general superstructure for reports, and the introduction to Part IX, which gives a practical

| Report Element | Readers' Questions |
| --- | --- |
| Introduction | Why is it important for us to consider these alternatives? |
| Criteria | Are your criteria reasonable and appropriate? |
| Method of Obtaining Facts | Are your facts reliable? |
| Overview of Alternatives | What are the important features of the alternatives? |
| Evaluation | How do the alternatives stack up against your criteria? |
| Conclusions | What overall conclusions do you draw about the alternatives? |
| Recommendations | What do you think we should do? |

**Figure 30-1**   Elements of a Feasibility Study and Their Relationship to the Questions Readers Ask.

introduction to the nature and use of superstructures. Both provide information not repeated here that will help you write effective reports about your feasibility studies.

## Introduction

In the introduction to a feasibility report you should answer your readers' question, "Why is it important for us to consider these alternatives?" The most persuasive way to answer this question is to identify the problem your feasibility study will help your readers solve or the goal it will help them achieve. Be sure to identify a problem or goal that is significant from the point of view of your employer's or client's organization: to reduce the number of rejected parts, to increase productivity, and so on. Beyond identifying the problem or goal that your study addresses, your introduction should also make clear what alternative courses of action you studied and what you did to investigate them.

Consider, for example, the way Ellen wrote the introduction of a feasibility report she prepared. (Ellen's entire report is presented at the end of this chapter.) A process engineer in a paper mill, Ellen was asked to evaluate the feasibility of substituting one ingredient for another in the furnish for one of the papers it produces (*furnish* is the combination of ingredients used to make the pulp for paper):

> At present we rely on the titanium dioxide ($TiO_2$) in our furnish to provide the high brightness and opacity we desire in our paper. However . . . the price of $TiO_2$ has been rising steadily and rapidly for several years. We now pay roughly $1400 per ton for $TiO_2$, or about 70¢ per pound.
>
> Some mills are now replacing some of the $TiO_2$ in their furnish with silicate extenders. Because the average price for silicate extenders is only $500 per ton, well under half the cost of $TiO_2$, the savings are very great.
>
> To determine whether we could enjoy a similar savings for our 30-pound book paper, I have studied the physical properties, material handling requirements, and cost of two silicate extenders, Tri-Sil 606 and Zenolux 26 T.

Problem — *(marginal annotation)*

Possible solution — *(marginal annotation)*

What writer did to investigate possible solution — *(marginal annotation)*

Besides telling your readers why they should consider the alternative or alternatives you studied, your introduction can also perform several other important functions. Foremost among these is to summarize your main recommendation or conclusion. When reading a feasibility report, your readers will want, more than anything else, to know what they should do. Although you will explain your conclusions and recommendations more fully elsewhere in your report, you should give your readers a clear indication of the outcome of your study right away. Ellen did that by adding the following sentence to the end of the last of the paragraphs quoted above:

> I conclude that one of the silicate extenders, Zenolux 26 T, looks promising enough to be tested in a mill run.

Another important function of your introduction is to provide information that will enable your readers to understand and evaluate information you will provide later in your report. Consider, for example, a feasibility report written by Phil, who earned a degree in home economics with an emphasis in marketing. He was asked by his employer, a department store chain, to investigate the feasibility of opening a men's clothing store in a college town. In the introduction to his report, Phil explained the subject of his report very briefly:

> This report discusses the feasibility of opening a men's clothing store in Oxford, Ohio, a college town about 35 miles northwest of Cincinnati.

However, Phil needed several pages to explain the problem that his report would help his readers solve. In the past few years, the chain has had mixed success in opening stores in college towns. A few stores prospered, but the majority were closed within a year at a considerable financial loss to the chain. To explain this problem fully, Phil had to provide details concerning this past experience and analyze why such stores succeed or fail. This detailed analysis helped his readers understand the criteria he used to evaluate the feasibility of opening the store in Oxford. It also helped explain how he interpreted the information he gathered during his study. Furthermore, through this discussion, he helped persuade his readers that he was a knowledgeable person who had conducted a thoughtful, thorough, and reliable study.

In your introduction, you should also tell your readers about the scope and organization of your report. For further advice about writing your introduction, see Chapter 13, "Writing the Beginning of a Communication."

## Criteria

Criteria are the standards that you apply in a feasibility study to evaluate the course (or courses) of action that you are assessing. For instance, when evaluating the feasibility of opening a store in the college town, Phil used many criteria, including the following: the presence of various kinds of potential customers, the availability of store space of a certain size in particular kinds of locations, and the availability of certain means of advertising. If any of these criteria could not have been met, Phil would have determined that opening a successful store in the town would not be feasible. Likewise, Ellen evaluated the two silicate extenders in terms of several specific properties.

### Two Ways of Presenting Criteria

There are two common ways of telling your readers what your criteria are. One is to devote a separate section to identifying and explaining them. Writers often do that in long reports or in reports where the criteria themselves require extended explanation.

The second way to tell your readers about your criteria is to integrate your presentation of them into other elements of the report. Ellen did that in the following sentence from the third paragraph of her introduction:

> To determine whether we could enjoy a similar savings for our 30-pound book paper, I have studied the physical properties, material handling requirements, and cost of two silicate extenders, Tri-Sil 606 and Zenolux 26 T.

Criteria named

For each of the general criteria named in the quoted sentence, Ellen had some more specific criteria. These she described when she discussed her methods and results. For instance, at the beginning of her discussion of the physical properties of the two extenders, she named the three properties she evaluated.

### Present Your Criteria Early in Your Feasibility Report

Whether or not you present your criteria in a separate section, you should introduce them early in your report. There are three good reasons for doing that. First, your readers may want to evaluate the criteria themselves. They will ask, for instance, "Did you take into account all of the considerations relevant to this decision?" and "Are the standards that you are applying reasonable in this circumstance?" Readers are very concerned about your criteria because they know that the validity of your conclusions will depend upon the soundness of the standards you use to evaluate the alternatives.

Second, your discussion of the criteria tells your readers a great deal about the scope of your study. Did you restrict yourself to technical questions, for instance, or did you also consider relevant organizational issues such as profitability and management strategies? Thus, Ellen's description of her general criteria in the third paragraph of her report indicates the scope of her study.

The third reason for presenting your criteria early in your report is that your discussion of the alternatives will make much more sense to your readers if they know in advance the criteria you applied when evaluating the alternatives.

## Sources of Your Criteria

You may be wondering how you will come up with the criteria you will use in your study and report. Often, when someone asks you to undertake a study, that person will tell you what criteria to apply. In other situations, particularly when you are conducting a feasibility study that requires technical knowledge that you have but your readers don't, your readers may expect you to identify the relevant criteria for them.

In either case, you are likely to refine your sense of the criteria as you conduct your study. Writing, too, can help you refine your criteria because when you write you must think in detail about the information you have obtained and decide how best to evaluate it.

## Four Common Types of Criteria

As you write your report (and conduct your study) you may find it helpful to keep in mind that the criteria people use to evaluate various courses of action are often designed to answer one or more of the following general questions:

- *Will the course of action work?* Will it really do what's wanted? This question is especially common when the problem is a technical one: Will the new design really make the flange strong enough? Will the new type of programming really reduce the computer time?

- *Can we implement it?* Even though a particular course of action may work technically, it may not be practical. Maybe it requires that existing operations be changed too much or maybe it requires equipment or materials that are not readily available. Maybe it requires special skills that present employees do not possess—and there is no chance of hiring additional personnel.

● *Can we afford it?* Cost can be treated in several ways. You may be looking for a course of action that costs less than some fixed amount, or that will save enough to pay for itself in a fixed period (for example, two years). You may simply be asked to determine whether the costs are "reasonable."

● *Is it desirable?* Sometimes, a solution must be more than effective, implementable, and affordable. Many otherwise feasible courses of action are rejected on the grounds that they create undesirable side effects. For example, a plan for increasing productivity may be rejected because it would have the undesirable side effect of decreasing employee morale. Similarly, the citizens of a region might feel that although a new power plant is technically feasible and would bring many benefits, the land it would occupy should be used for other purposes, such as farming.

Of course, the criteria you apply in a particular feasibility study depend upon the problem you address and the professional responsibilities, goals, and values of the people who will use your report. In some instances, you will need to deal only with criteria related to the question, Does it work? At other times, you might need to deal with all the criteria mentioned above, plus others. No matter what your criteria, however, you should tell your readers what they are before you discuss your evaluation.

## Method of Obtaining Facts

In your explanation of your method of obtaining your facts, you answer your readers' question, "Are your facts reliable?" By showing that you obtained your facts through reliable methods, you assure your readers that your facts form a sound basis for decision-making.

The source of your facts will depend upon the nature of your study—library research, calls to companies that sell the things your readers are thinking of buying, interviews, meetings with other experts in your organization, surveys, laboratory research, and the like.

How much detail do you need to provide about your method of gathering information? The answer is: enough to assure your readers that your information is trustworthy. How much you need to say in order to provide that assurance depends upon your readers and situation. For example, Ellen used some highly technical procedures in her analysis of the physical properties of the silicate extenders. However, those procedures are standard in the paper industry and are well-known to her readers. For that reason, she did not need to provide a detailed explanation of them.

In contrast, Phil used a wide variety of methods of gathering information while studying the feasibility of opening the men's clothing store. Some of these methods were standard and required no special explanation. For instance, Phil talked to real estate agents and building owners to find where the best locations were in town and to find out which might be available. However, he also created and used a nonstandard market survey. He needed to do this because the standard market surveys are designed for use in areas where most of the potential customers live permanently near the shopping area. In this college town, however, most of the potential customers for the store are students who have their permanent residences in other cities. Because his survey was unique, Phil provided a detailed discussion of the strategy he used in it, together with a copy of his questionnaire, so that his readers would know exactly what he asked and why.

The best place to describe your methods depends largely upon how many different techniques you used. If you used one or two techniques—say reading and interviews—you might explain each in a separate paragraph or section near the beginning of your report. You could name the books read and individuals interviewed there or, if the list is long, in an appendix.

On the other hand, if you used several different techniques, each pertaining to a different part of your analysis, you should mention each of them at the point at which you discuss the results you obtained. For instance, Ellen used three different methods of studying the physical properties of the extenders. Instead of describing all three methods in a single section of her report, she explained the first method along with the results it produced, the second along with its results, and so on.

## Overview of Alternatives

Your readers will be able to follow your detailed evaluation of the alternatives most easily if they first have an overall understanding of the alternatives you are investigating. Sometimes, you will write to people who fully understand the alternatives before reading your report. For instance, readers of Phil's report will certainly need no special explanation to understand what Phil means when he talks about opening a men's clothing store. And they are already familiar with the alternative to opening a new one in the college town, which is to continue without a branch store there.

However, you may sometimes find yourself writing in situations where your readers do need to have the alternatives explained to them. For example, you may be evaluating two strategies for extracting medicinal chemicals from a particular species of plant. If those procedures are not familiar to your readers, you should explain them overall before evaluating them in detail. You may need a few sentences or many pages to describe those processes, depending upon their complexity.

## Evaluation

The heart of a feasibility report is the detailed evaluation of the course or courses of action you studied. In most feasibility studies, writers organize their evaluation sections around their criteria. For instance, Phil arranged his evaluation section around the six criteria he used: existing competition, potential market, advertising channels, location, logistics for supplying merchandise, and budgets.

In reports in which you treat two or more alternatives, you can use either the alternating or divided pattern of organization. These patterns, treated briefly below, are explained in greater detail in the discussion of comparisons in Chapter 12.

## Alternating Pattern

In the alternating pattern, you organize your report around the criteria rather than around the alternatives. To give you a clearer idea about how this pattern fits into your overall discussion, the following outline includes the statement of the criteria and the overview of the alternatives:

---

Statement of Criteria

Overview of the Two Alternatives

Evaluation of the Alternatives
    Criterion 1
        Alternative A
        Alternative B
    Criterion 2
        Alternative A
        Alternative B
    Criterion 3
        Alternative A
        Alternative B

---

With this pattern, because all the pertinent information about both alternatives is presented together, you enable your reader to make a point-by-point comparison without flipping back and forth through the communication. It is the pattern Ellen used to compare the two silicate extenders she studied.

## Divided Pattern

In the divided pattern, you organize not around the criteria but around the alternatives themselves. Only after you have described each thoroughly do you present your evaluation:

---

Statement of Criteria

Overview of the Two Alternatives

Detailed Description of Alternatives
    Alternative A
        Criterion 1
        Criterion 2
        Criterion 3
    Alternative B
        Criterion 1
        Criterion 2
        Criterion 3
Evaluation of Alternatives

---

The divided pattern is well suited to situations where both the general nature and the details of each alternative can be described in a short space, say one page or so. For instance, it is the pattern used by an employee of an insurance company to investigate the feasibility of using a variety of competing computer programs to create graphs showing sales figures and other data of interest. He described the key features of each system on a single page, using a standard organization for all of them. Then, he made an overall comparison at the end. This enabled his readers to find the details about each program in one place, while still providing the readers with an overall sense of how they stacked up against one another.

## Handling Obviously Unsuitable Alternatives

Sometimes you will want to mention several alternatives but treat only one or a few thoroughly. Perhaps your investigation showed that the other alternatives fail to meet one or more of the critical criteria so that they should not be considered seriously. Perhaps you suspect that your readers have a prejudice in favor of a particular alternative that you know to be completely impractical.

Usually, it makes no sense to discuss obviously unqualified solutions at length. One good way to handle them is to make a slight adjustment to the basic superstructure for feasibility reports: explain briefly the alternatives that you have dismissed and tell why you have dismissed them. This entire discussion might take only a sentence or a paragraph. You might include it in the introduction (when you are talking about the scope of your report) or in your overview of the alternatives.

It is usually not wise to postpone the discussion of unqualified solutions until the end of a report. If you do that, throughout the time they are reading the earlier parts of your report the readers may keep asking, "But why didn't the writer consider so and so?" That question may distract their attention from important material that you are presenting.

## Putting Your Most Important Points First

As explained in Chapter 12, "Writing Segments," you should generally begin each segment of your communications with the most important information. That is the information your readers want most. This advice applies to the segments in which you evaluate the alternatives you are considering. For instance, Ellen begins one part of her evaluation of the silicate extenders in this way:

> With respect to material handling, I found no basis for choosing between Zenolux and Tri-Sil.

Ellen then spends two paragraphs reporting the facts she has gathered about the physical handling of the two extenders.

Similarly, Phil begins his section on the potential market for the men's store in this way:

> Through a specially designed market survey, I found that there is a sufficient market for the prospective store, provided that it concentrates on casual wear in a price range suitable to students. Students are not likely to buy more expensive items, like suits, away from home. And townspeople will usually drive to shopping centers for these items because of the larger selection.

He then explains this survey and the detailed results he obtained.

This advice to begin with generalizations pertains to the entire evaluation section as well as to its parts. You will help your readers by briefly presenting at the beginning of the evaluation section any broad generalizations you can make about your results.

## Conclusions

Your conclusions are your overall assessment of the feasibility of the courses of action you studied. You might present them in two or three places in your report. You should certainly mention them in summary form near the beginning. If your report is long (say several pages), you should remind your readers of your overall conclusion at the beginning of your evaluation section. Finally, you should provide your detailed discussion of your conclusions in a separate section that follows your evaluation of your alternatives.

## Recommendations

It is customary to end a feasibility report by answering the readers' question, "What do you think we should do?" Because you have investigated and thought about the alternatives so thoroughly, your readers will place special value on your recommendations. Depending upon the situation, you might need to take only a single sentence or paragraph, or else many pages to present your recommendations.

Sometimes your recommendations will pertain directly to the course of action you studied: "Do this" or "Don't do it." At other times you may perform *preliminary* feasibility studies to determine whether the course of action you are studying is promising enough to warrant a more thorough study. In that case, your recommendation would not focus on whether or not to take that course of action, but whether or not to continue investigating it. Phil's report about opening the men's clothing store is of that type. He determined that there is a substantial possibility of making a profit with the store, and so he recommends that a more detailed study be developed.

Sometimes, too, you may discover that you are unable to gather all the information you need to make a firm recommendation about the course of action you are studying. Perhaps your deadline is too short or your funds too small. Perhaps you uncovered an unexpected question that needs further investigation. In all these situations, you should point out the limitations of your report and let your readers know what else they should find out so that they can be confident that they are making a well-informed decision.

## SAMPLE FEASIBILITY REPORT

Figure 30-2 shows the feasibility report written by Ellen. If you would like to see another feasibility report, look at Figure 2-4.

## WRITING ASSIGNMENT

A writing assignment that involves a feasibility report is included in Appendix I.

# Rᴇɢᴇɴᴄʏ ɪɴᴛᴇʀɴᴀᴛɪᴏɴᴀʟ ᴘᴀᴘᴇʀ ᴄᴏᴍᴘᴀɴʏ

## Memorandum

FROM:   Ellen Hines, Process Engineer

TO:   Jim Shulmann, Senior Engineer

DATE:   December 13, 19--

SUBJECT: FEASIBILITY OF USING SILICATE EXTENDERS
FOR 30-POUND BOOK PAPER

### Summary

I have investigated the feasibility of using a silicate extender to replace some of the $TiO_2$ in the furnish for our 30-pound book paper. Because the cost of the extenders is less than half the cost of $TiO_2$, we could enjoy a considerable savings through such a substitution.

The tests show that either one of the two extenders tested can save us money. In terms of retention, opacity, and brightness, Zenolux is more effective than Tri-Sil. Consequently, it can be used in smaller amounts to achieve a given opacity or brightness. Furthermore, because of its better retention, it will place less of a burden on our water system. With respect to handling and cost, the two are roughly the same.

I recommend a trial run with Zenolux.

### Introduction

At present we rely on the titanium dioxide ($TiO_2$) in our furnish to provide the high brightness and opacity we desire in our paper. However, as Figure 1 shows (at the end of this report), the price of $TiO_2$ has been rising steadily and rapidly for several years. We now pay roughly $1400 per ton for $TiO_2$, or about 70¢ per pound.

Some mills are now replacing some of the $TiO_2$ in their furnish with silicate extenders. Because the average price

Entire report summarized in 125 words

Summary emphasizes conclusions and recommendations

Problem

Possible solution

**Figure 30-2.**   Sample Feasibility Report.

Silicate Extenders                                    Page 2

*What writer did to investigate possible solution (general criteria)*

for silicate extenders is only $500 per ton, well under half the cost of $TiO_2$, the savings are very great.

To determine whether we could enjoy a similar savings for our 30-pound book paper, I have studied the physical properties, material handling requirements, and cost of two silicate extenders, Tri-Sil 606 and Zenolux 26 T. I conclude that one of the silicate extenders, Zenolux 26 T, looks promising enough to be tested in a mill run.

*Major conclusion and reccomendation*

### Tests of Physical Properties

*Criteria for evaluating physical properties*

*General result*

The three physical properties I tested are retention, opacity, and brightness. In all three areas, Zenolux is superior.

### Retention

*Reason for testing retention*

As with any ingredient in our furnish, we must be concerned with the proportions of a silicate extender that will be retained in the paper and the proportion that will be left in the water, where it is wasted and may cause problems in our water system. To test retention of the two silicate extenders, I made two dozen handsheets, each containing the equivalent of three grams of oven-dried pulp and 2 grams of oven-dried extender. By weighing the finished handsheets, I determined how much silicate extender had been lost from each.

*Method*

*Evaluation (results and discussion)*

The results showed that the average retention for Zenolux was 75% whereas the average retention for Tri-Sil was 51%. Higher retention should result in higher opacity and brightness because more particles remain in the furnish to prevent clumping of the $TiO_2$.

### Opacity

*Method of testing opacity*

To determine the effectiveness of each extender in preventing light from passing through the paper, I conducted a two-stage test of opacity. First, I investigated the opacity of $TiO_2$, Tri-Sil, and Zenolux when each is used alone. To do that, I made the following sets of handsheets:

- 6 sheets containing each of the following loadings of $TiO_2$: 2%, 4%, 6%, 8%, 10%, 12%, 14%, and 16%.

**Figure 30-2.**   *Continued*

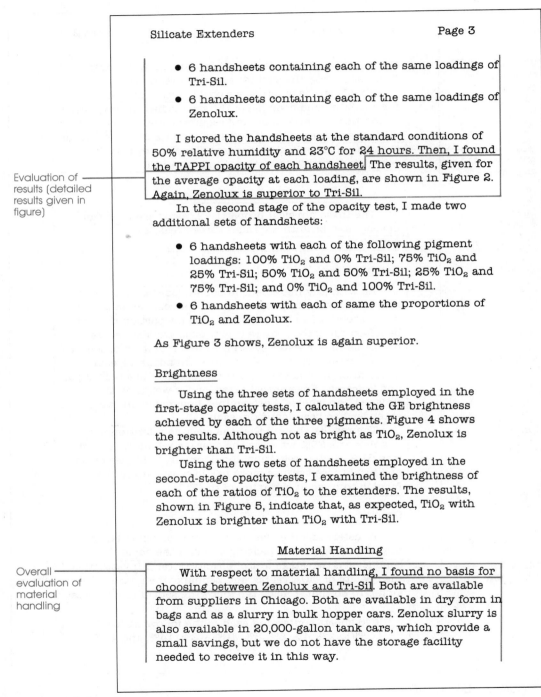

Silicate Extenders                                          Page 3

- 6 handsheets containing each of the same loadings of Tri-Sil.
- 6 handsheets containing each of the same loadings of Zenolux.

I stored the handsheets at the standard conditions of 50% relative humidity and 23°C for 24 hours. Then, I found the TAPPI opacity of each handsheet. The results, given for the average opacity at each loading, are shown in Figure 2. Again, Zenolux is superior to Tri-Sil.

In the second stage of the opacity test, I made two additional sets of handsheets:

- 6 handsheets with each of the following pigment loadings: 100% $TiO_2$ and 0% Tri-Sil; 75% $TiO_2$ and 25% Tri-Sil; 50% $TiO_2$ and 50% Tri-Sil; 25% $TiO_2$ and 75% Tri-Sil; and 0% $TiO_2$ and 100% Tri-Sil.
- 6 handsheets with each of same the proportions of $TiO_2$ and Zenolux.

As Figure 3 shows, Zenolux is again superior.

Brightness

Using the three sets of handsheets employed in the first-stage opacity tests, I calculated the GE brightness achieved by each of the three pigments. Figure 4 shows the results. Although not as bright as $TiO_2$, Zenolux is brighter than Tri-Sil.

Using the two sets of handsheets employed in the second-stage opacity tests, I examined the brightness of each of the ratios of $TiO_2$ to the extenders. The results, shown in Figure 5, indicate that, as expected, $TiO_2$ with Zenolux is brighter than $TiO_2$ with Tri-Sil.

Material Handling

With respect to material handling, I found no basis for choosing between Zenolux and Tri-Sil. Both are available from suppliers in Chicago. Both are available in dry form in bags and as a slurry in bulk hopper cars. Zenolux slurry is also available in 20,000-gallon tank cars, which provide a small savings, but we do not have the storage facility needed to receive it in this way.

Evaluation of results (detailed results given in figure)

Overall evaluation of material handling

**Figure 30-2.**   *Continued*

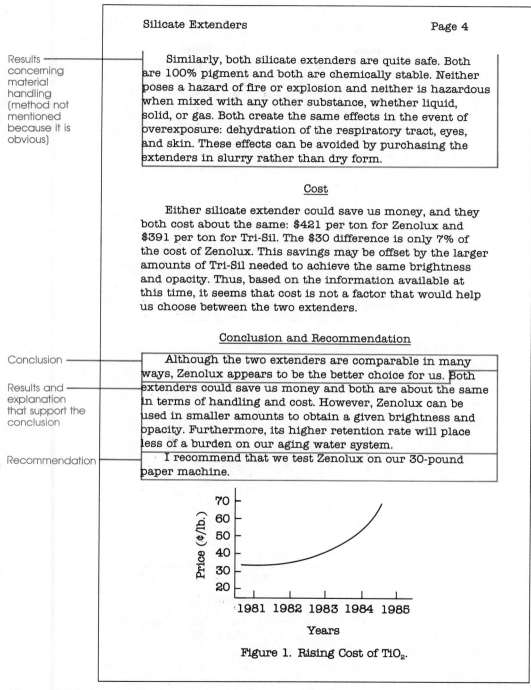

Silicate Extenders                                    Page 4

Results concerning material handling (method not mentioned because it is obvious)

Similarly, both silicate extenders are quite safe. Both are 100% pigment and both are chemically stable. Neither poses a hazard of fire or explosion and neither is hazardous when mixed with any other substance, whether liquid, solid, or gas. Both create the same effects in the event of overexposure: dehydration of the respiratory tract, eyes, and skin. These effects can be avoided by purchasing the extenders in slurry rather than dry form.

Cost

Either silicate extender could save us money, and they both cost about the same: $421 per ton for Zenolux and $391 per ton for Tri-Sil. The $30 difference is only 7% of the cost of Zenolux. This savings may be offset by the larger amounts of Tri-Sil needed to achieve the same brightness and opacity. Thus, based on the information available at this time, it seems that cost is not a factor that would help us choose between the two extenders.

Conclusion and Recommendation

Conclusion

Results and explanation that support the conclusion

Recommendation

Although the two extenders are comparable in many ways, Zenolux appears to be the better choice for us. Both extenders could save us money and both are about the same in terms of handling and cost. However, Zenolux can be used in smaller amounts to obtain a given brightness and opacity. Furthermore, its higher retention rate will place less of a burden on our aging water system.

I recommend that we test Zenolux on our 30-pound paper machine.

Figure 1. Rising Cost of $TiO_2$.

**Figure 30-2.** *Continued*

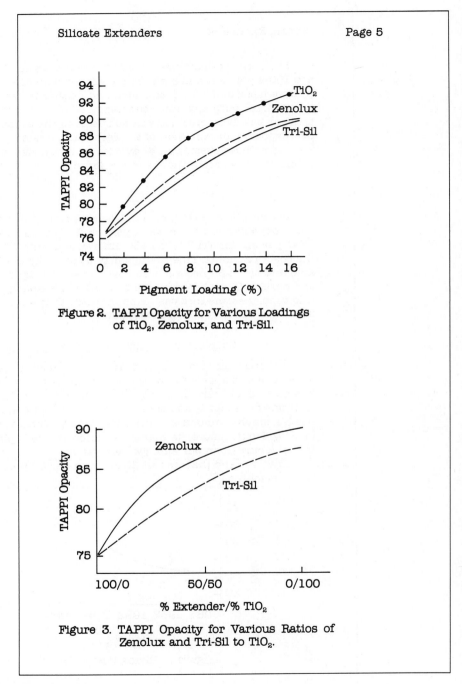

Figure 2.  TAPPI Opacity for Various Loadings
of TiO$_2$, Zenolux, and Tri-Sil.

Figure 3.  TAPPI Opacity for Various Ratios of
Zenolux and Tri-Sil to TiO$_2$.

**Figure 30-2.**   *Continued*

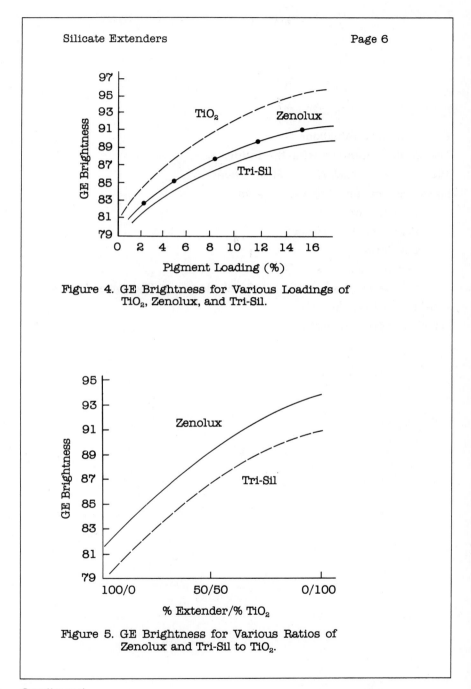

Silicate Extenders                                    Page 6

Figure 4. GE Brightness for Various Loadings of
TiO$_2$, Zenolux, and Tri-Sil.

Figure 5. GE Brightness for Various Ratios of
Zenolux and Tri-Sil to TiO$_2$.

**Figure 30-2.**  *Continued*

# Progress Reports

PREVIEW OF CHAPTER

**Typical Writing Situations**
**The Readers' Concern with the Future**
**The Questions Readers Ask Most Often**
**Superstructure for Progress Reports**
     **Introduction**
     **Facts and Discussion**
     **Conclusions**
     **Recommendations**
     **A Note on the Location of Conclusions and Recommendations**
**Tone in Progress Reports**
**Sample Progress Report**

I n a progress report, you tell about work that you have begun but not yet completed. The typical progress report is one of a series submitted at regular intervals, such as every week or month.

## TYPICAL WRITING SITUATIONS

Progress reports are written in two types of situations. In the first, you tell your readers about your progress on one *particular* project. As a geologist employed by an engineering consulting firm, Lee must do that. His employer has assigned him to study the site that a large city would like to use for a civic center and large office building. The city is worried that the site might not be geologically suited for such construction. Every two weeks, Lee must submit a progress report to his supervisor and to the city engineer. Lee's supervisor uses the progress report to be sure that Lee is conducting the study in a rapid and technically sound manner. The city engineer uses the report to see that Lee's study is proceeding according to the tight schedule planned for it. She also uses it to look for preliminary indications about the likely outcome of the study. Other work could be speeded up or halted as a result of these preliminary findings.

In the second type of situation, you write progress reports that tell about your work on *all* your projects. Many employers require their workers to report on their activities at regular intervals all year round, year in and year out. Jacqueline is a person who must write such progress reports (often called periodic reports). She works in the research division of a large manufacturer of consumer products, where she manages a department that is responsible for improving the formulas for the company's laundry detergents—making them clean and smell better, making them less expensive to manufacture, and making them safer for the environment. At any one time, Jacqueline's staff is working on between ten and twenty different projects.

As part of her regular responsibilities, Jacqueline must write a report every two weeks to summarize the progress on each of the projects. These reports have many readers, including the following people: her immediate superiors, who want to be sure that her department's work is proceeding satisfactorily; researchers in other departments, who want to see whether her staff has made discoveries they can use in the products they are responsible for (for example, dishwashing detergents); and corporate planners, who want to anticipate changes in formulas that will require alterations in production lines, advertising, and so on.

As the examples of Lee and Jacqueline indicate, progress reports can vary from one another in many ways: they may cover one project or many; they may be written to people inside the writer's organization or outside it; and they may be used by people with a variety of reasons for reading them, such as learning things they need to know to manage your work and to make decisions.

## THE READERS' CONCERN WITH THE FUTURE

Despite their diversity, however, almost all progress reports have this in common: their readers are primarily concerned with the *future*. That is, even though most progress reports talk primarily about what has happened in the past, their readers usually want that information so that they can plan for and act in the future.

Why? Consider the responsibilities that your readers will be fulfilling while reading your progress reports. From your report, they may be trying to learn the things they need to know to manage *your* project. They will want to know, for instance, what they should do (if anything) to keep your project going smoothly or to get it back on track. The reports written by Lee and Jacqueline are used for this purpose by some of their readers.

From your report, some readers may also be trying to learn things they need to know to manage *other* projects. That's because in an organization almost all projects are interdependent with other projects. For example, other people and departments may need the results of your project as they work on their own projects. Maybe you are running a test whose results they need so that they can design a new part, or maybe you are building a piece of equipment that must be installed before they can install other equipment. If your project is going to be late, the schedules of those other projects will have to be adjusted accordingly. Similarly, if your project costs more than expected, money and resources will have to be taken away from other activities to compensate. Because of interdependencies like these, your readers need information about the past accomplishments and problems in your project so that they can make plans for the future. The progress reports written by Lee and Jacqueline are used for this purpose.

Similarly, your readers will be interested in learning the preliminary results of your work. Suppose, for instance, that you complete one part of a research project before you complete the others. Your readers may very well be able to use the result of that part immediately. The city engineer who reads Lee's reports about the possible building site is especially interested in making this use of the information Lee provides.

In sum, the readers of your progress reports will be mainly concerned about the future. The information you provide about your progress will help them make decisions about the management of your work in the future. And the results you report now may be information they can use in their future work on other projects.

## THE QUESTIONS READERS ASK MOST OFTEN

The readers' concern with the implications of your progress for their future work and decisions leads them to want you to answer the following questions in your progress reports. If your report describes more than one project, your readers will ask these questions about each of them.

- *What work does your report cover?* To be able to understand anything else in a progress report, readers must know what project or projects and what time period the report covers.

- *What is the purpose of the work?* Readers need to know the purpose of your work to see how your work relates to their responsibilities and to the other work, present and future, of the organization.

- *Is your work progressing as planned or expected?* Your readers will want to determine if adjustments are needed in the schedule, budget, number of people assigned to the project or projects you are working on.

- *What results did you produce?* The results you produce in one reporting period may influence the shape of work in future periods. Also, even when you are still in the midst of a project, readers will want to know about any results they can use in other projects now, before you finish your overall work.

- *What progress do you expect during the next reporting period?* Again, your readers' interests will focus on such management concerns as schedule and budget and on the kinds of results they can expect.

- *How do things stand overall?* This question arises especially in long reports. Readers want to know what the overall status of your work is, something they may not be able to tell readily from all the details you provide.

- *What do you think we should do?* If you are experiencing or expecting problems, your readers will want your recommendations about what should be done. If you have other ideas about how the project could be improved, they too will probably be welcomed.

## SUPERSTRUCTURE FOR PROGRESS REPORTS

To answer your readers' questions, you can use the conventional superstructure for writing progress reports, which has the following elements: introduction, facts and discussion, conclusions, and recommendations.

Figure 31-1 shows the relationship between these elements and the general questions readers ask. The rest of this chapter explains how you can develop each of those four elements to create an effective progress report. If you have not already done so, you should also read Chapter 28, which describes the general superstructure for reports, and the introduction to Part IX, which gives a practical introduction to the nature and use of superstructures. Both provide information not repeated here that will help you write progress reports effectively.

### Introduction

In the introduction to a progress report, you should answer the first two questions shown in Figure 31-1. You can usually answer the question, "What

| Report Element | Readers' Question |
|---|---|
| Introduction | What work does your report cover? What is the purpose of the work? |
| Facts and Discussion Past work | Is your work progressing as planned or expected? What results did you produce? |
| Future work | What progress do you expect during the next reporting period? |
| Conclusion | (In long reports: How do things stand overall?) |
| Recommendation | What do you think we should do? |

**Figure 31-1.** Elements of a Progress Report and Their Relationship to the General Questions Readers Ask.

work does your report cover?" by opening with a sentence that tells what project or projects your report concerns and what time period it covers.

Sometimes you will not need to answer the second question—"What is the purpose of the work?"—because all your readers will already be quite familiar with your work's purpose. At other times, however, it will be crucial for you to tell your work's purpose because your readers will include people who don't know or may have forgotten it. You are especially likely to have such readers when your report will be widely circulated in your own organization and when you are writing to another organization that has hired your employer to do the work you describe. You can usually explain purpose most helpfully by describing the problem that your project will help your readers solve.

The following sentences show how one group of writers answered the readers' first two questions:

Project and
period covered

Purpose of
project

> This report covers the work done on the Focus Project from July 1 through September 1. Sponsored by the U.S. Department of Energy, the aim of the Focus Project is to overcome the technical difficulties encountered in manufacturing photovoltaic cells that can be used to generate commercial amounts of electricity.

Of course, in your introduction you should also provide your readers with any background information they will need in order to understand the rest of your report.

## Facts and Discussion

In the discussion section of your progress report, you should answer these questions from your readers: "Is your work progressing as planned or expected?" "What results did you produce?" and, "What progress do you expect during the next reporting period?"

### Answering Your Readers' Questions

In many situations, the work for each reporting period is planned in advance. In such cases you can easily tell about your progress by comparing what happened with what was planned. Where there are significant discrepancies between the two, your readers will want to know why. The information you provide about the causes of problems will help your readers decide how to remedy them. It will also help you explain any recommendations you make later in your report.

When you are discussing preliminary results that your readers might use, be sure to explain them in terms that allow your readers to see their significance. In research projects, preliminary results are often tentative. If that is the case for you, let your readers know how certain—or uncertain—the results are. That information will help your readers decide how to use the results.

### Providing the Appropriate Amount of Information

When preparing progress reports, writers often wonder how much information they should include. Generally, progress reports are brief because readers want them that way. While you need to provide your readers with specific information about your work, don't include details except when the details will help your readers decide how to manage your project or when you believe that your readers can make some immediate use of them. In many projects, you will learn lots of little things and you will have lots of little setbacks and triumphs along the way. Avoid talking about these matters. No matter how interesting they may be to you, they are not likely to be interesting to your readers. Stick to the information your readers can use.

### Organizing the Discussion

You can organize your discussion section in many ways. One is to arrange your material around time periods:

I. What happened during the most recent time period.
II. What's expected to happen during the next time period.

You will find that this organization is especially well suited for reports in which you discuss a single project that has distinct and separate stages, so that you work on only one task at a time. However, you can also expand this structure for reports that cover either several projects or one project in which several tasks are performed simultaneously:

I. What happened during the most recent time period.
   A. Project A (or Task A)
   B. Project B

II. What's expected to happen in the next time period.
   A. Project A
   B. Project B

When you are writing reports that cover more than one project or more than one task, you might also consider organizing around those projects or tasks:

I. Work on Project A (or Task A)
   A. What happened during the last time period.
   B. What's expected to happen during the next time period.

II. Work on Project B
   A. What happened during the last time period.
   B. What's expected to happen during the next time period.

This organization works very well in reports that are more than a few paragraphs long because it keeps all the information on each project together, making the report easy for readers to follow.

### Emphasizing Important Findings and Problems

Your findings and problems are important to your readers. Your findings are important because they may involve information that can be used right away by others. The problems you encounter are important because they may require your readers to change their plans.

Because your findings and problems can be so important to your readers, be sure that you devote enough discussion to them to satisfy your readers' needs and desires for information. Also, place these discussions prominently and mark them with headings or other appropriate devices so that they are easy to find.

### Conclusions

Your conclusions are your overall views on the progress of your work. In short progress reports, there may be no need to include them, but if your report covers many projects or tasks, a conclusion may help the reader understand the general state of your progress.

## Recommendations

If you have any ideas about how to improve the project or increase the value of its results, your readers will probably want you to include them. Your recommendations might be directed at overcoming some difficulty that you have experienced or anticipate. Or they might be directed at refocusing or otherwise altering the project.

## A Note on the Location of Conclusions and Recommendations

For most of your readers, your conclusions and recommendations are the most important information in your progress report. Therefore, you should generally include them at the beginning, either in the introduction or at the head of your discussion section. That may be the only place you need to state them if your conclusions and recommendations are brief. If they are long or if your readers will be able to understand them only after reading your discussion section, you can present your conclusions and recommendations at the end of your report, while still including a summary of them at the beginning.

# TONE IN PROGRESS REPORTS

You may wonder what tone to use in the progress reports you write. In them, you will generally aim to persuade your readers that you are doing a good job. The pressure you feel to make your readers feel satisfied can be especially great when you are new on the job and when your readers might discontinue a project if they feel that it isn't progressing satisfactorily.

Because of this strong persuasive element in progress reports, some writers are tempted to use an inflated or highly optimistic tone. That sort of tone, however, can lead to difficulties. It might lead you to make statements that sounds unbusinesslike—more like advertising copy than a professional communication. Such a tone is more likely to make your readers suspicious than agreeable. Also, if you present overly optimistic accounts of what can be expected, you risk creating an unnecessary disappointment if things don't turn out the way you seem to be promising. And if you consistently turn in overly optimistic progress reports, your credibility with your readers will quickly diminish.

In progress reports, it's best to be straightforward about problems so that the readers can take appropriate measures to overcome them and so that they can adjust their expectations realistically. It is best to sound pleased and proud of your accomplishments without seeming to puff them up.

## SAMPLE PROGRESS REPORT

Figure 31-2 (page 700) shows a sample progress report prepared by a student who needed to tell her instructor about her work on a project in her technical writing class.

## EXERCISE

1. A. Recommend ways that the writer could improve the progress report printed below. Focus on his selection and organization of information. The report concerns work done at a company that is developing a product that construction companies can use to make large numbers of styrofoam panels at their building sites. The product is to consist of a liquid and a powder that are poured into heated metal molds, where they will foam up to fill the mold and take the mold's shape.

   B. Rewrite the memo by following the recommendations you developed for the writer.

---

FROM:    Buddy McCormick
TO:       Craig Skelton
DATE:    January 29, 19--
RE:       PROGRESS REPORT ON INSTANT STYROFOAM
            MOLDS

Over the past month, we have been experimenting with designs for the molds for the new "Instant Styrofoam" product being developed by the Chemical Products Division. We started by looking for devices that could clamp our standard mold shut once these new ingredients were poured into it and began to foam up. The only appropriate devices found to clamp the standard mold and hold it closed during the foaming process were C-clamps (12) around all the edges. Obviously, this type of clamping required considerable time for closing and opening, but a variety of quick-action type clamping devices (all designed to take the temperatures and pressures expected) proved unsatisfactory in earlier experiments.

Whereas the clamps did hold the mold closed, this resulted in some secondary damage to the molds, specifically warpage of the aluminum plates. This was due to the internally generated pressure during the foaming of the panels. This pressure reached a maximum about 10 minutes after placing the mold into the oven. This resulted in cleaning problems of both the mold and oven, and produced an unusable part.

Having determined that the aluminum could not withstand the pressure without deflection, it was decided to build a steel mold using 1.25 cm (1/2 inch) thick steel and to bolt it together. To reduce the time required to close the mold, the lid was split two-thirds and one-third, allowing two-thirds of the mold to be bolted shut prior to the adding of the foam ingredients. Once the foam ingredients were added, only about a dozen bolts needed to be inserted. Such a mold was fabricated and the first attempt at molding a part resulted in blowing the hinges off the end of the mold. Replacing the hinges and doubling up on the bolts with the hinges on, a second molding was made which actually caused a deflection of the 1.25 cm (1/2 inch) thick steel.

Thus, we have been unsuccessful in our efforts to design a mold that can accommodate the high pressure created during foaming by the higher density foams. I recommend that during the next month you schedule us to work exclusively on this important problem.

## WRITING ASSIGNMENT

An assignment that involves writing a progress report is given in Appendix I.

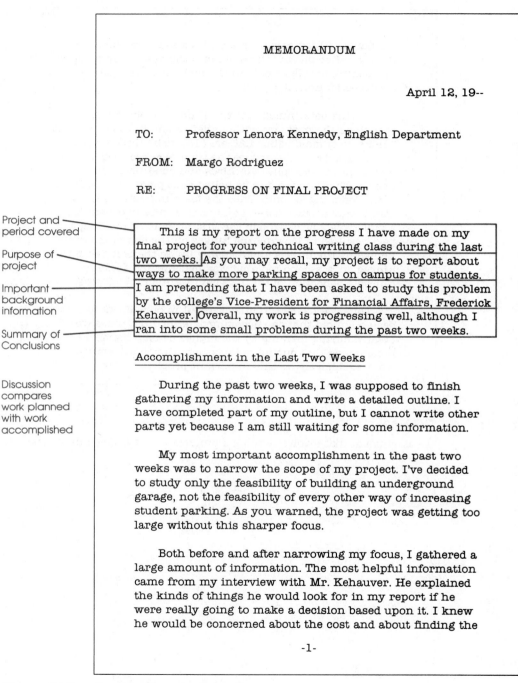

Project and period covered

Purpose of project

Important background information

Summary of Conclusions

Discussion compares work planned with work accomplished

MEMORANDUM

April 12, 19--

TO:     Professor Lenora Kennedy, English Department

FROM:   Margo Rodriguez

RE:     PROGRESS ON FINAL PROJECT

This is my report on the progress I have made on my final project for your technical writing class during the last two weeks. As you may recall, my project is to report about ways to make more parking spaces on campus for students. I am pretending that I have been asked to study this problem by the college's Vice-President for Financial Affairs, Frederick Kehauver. Overall, my work is progressing well, although I ran into some small problems during the past two weeks.

Accomplishment in the Last Two Weeks

During the past two weeks, I was supposed to finish gathering my information and write a detailed outline. I have completed part of my outline, but I cannot write other parts yet because I am still waiting for some information.

My most important accomplishment in the past two weeks was to narrow the scope of my project. I've decided to study only the feasibility of building an underground garage, not the feasibility of every other way of increasing student parking. As you warned, the project was getting too large without this sharper focus.

Both before and after narrowing my focus, I gathered a large amount of information. The most helpful information came from my interview with Mr. Kehauver. He explained the kinds of things he would look for in my report if he were really going to make a decision based upon it. I knew he would be concerned about the cost and about finding the

-1-

**Figure 31-2.** Sample Progress Report.

Professor Kennedy                              April 12, 19--
Page 2

money to pay for it. But I was surprised at all the other
concerns he would have, such as the effect on the beauty of
the campus and the possibility of traffic jams.

My second source of information was Kelly Horn, an
engineer for Hamilton Construction, which has put up
several of the newer buildings on campus. He pointed out
that many sites on campus have underground springs or
unstable soil that make them unsuitable for building an
underground garage. He's promised to send me a map that
identifies these spots. Best of all, though, he's going to send
me some cost figures--so many dollars per parking space
and the like. He will include different sets of figures for the
different kinds of locations that I might choose.

Third, I've talked with many students to get their views
on a parking garage. They all like it, but some say that they
wouldn't pay to park there, and some are very concerned
about where it might be put. They don't want it to be too far
from their classes.

Finally, I have talked on the phone with several
administrators at Green University, which has recently
built an underground garage similar to the one I am
investigating. From them I learned a lot about how to
evaluate alternative plans for such a structure and I
learned some of the arguments people can make for (and
against) building an underground garage here.

Problems Encountered in the Last Two Weeks

I have run into a couple of problems getting the
information I need to complete my outline. The biggest one
is that Cecilia Norton, the Assistant City Planner, has been
on vacation and won't be back until next week. City Hall
has told me she's the person I must depend upon for
information about the city's regulations and concerns.

Also, I found out from Ron Thiemann, who is a special
aide to the university president, that the university's board
of trustees almost certainly wouldn't approve an addition to

Important result emphasized

Separate section devoted to problems

**Figure 31-2.** *Continued*

Professor Kennedy                                    April 12, 19--
Page 3

student tuition to pay for the lot. Also, the Alumni Office
says that it would be impossible to give money for a
parking garage. Alumni want to pay for classroom
buildings and scholarships, not garages. A tuition
surcharge and alumni contributions were my best ideas
about funding the lot, so I will have to see if there are other
ways to pay the bill.

Overall, during the past two weeks I've gotten lots of
information and started my detailed outline, but I still need
more information before I can finish it.

Plan for the Next Two Weeks

Expected
progress during
next reporting
period

Like everyone else in the class, I am rushing so that I
can complete my rough draft April 29, the day we will have
our drafts reviewed by other students. I think I can make it.
Ms. Norton will be back in three days, and if the maps and
prices from Mr. Horn haven't arrived by then, I'll call him
up again or drive to his office to get the things he's
promised. So, by the middle of next week, I should be able to
know what locations are really practical sites for such a
garage. Then I'll be able to complete my outline and show it
to you.

In the meantime, I will begin writing my introduction
and my statement of the criteria for evaluating the
feasibility of the underground garage. I have all the
information I need to write those sections. They should be
done before the weekend, and I don't have any tests for
papers due next week, so I will have plenty of time to work
on the rest of the draft.

Conclusion

Conclusion tells
how the project
stands overall

Although I'm a little behind schedule, I feel good about
what I've accomplished so far. In fact, I've started to think
about actually giving my report to Mr. Kehauver, if it turns
out well. He said he'd be interested in reading it. I am
confident I will be able to complete both my rough draft
and my entire project on time.

**Figure 31-2.**   *Continued*

# Proposals

## PREVIEW OF CHAPTER

Variety of Proposal-Writing Situations
    Example Situation 1
    Example Situation 2
Proposal Readers Are Investors
The Questions Readers Ask Most Often
Strategy of the Conventional Superstructure for Proposals
    Answering Your Readers' Questions Persuasively
    Various Lengths of Proposals
Conventional Superstructure for Proposals
    Introduction
    Problem
    Objectives
    Product
    A Note on the Relationships among
      Problem, Objectives, and Product
    Method
    Resources
    Schedule
    Qualifications
    Management
    Costs
Sample Proposal

n a proposal, you make an offer. And you try to persuade your readers to accept that offer. You say that, in exchange for money or time or some other form of support from your readers, you will give them something they want, make something they desire, or do something they wish to have done.

Throughout your career you will have many occasions to make such offers. You may think up a new product that you could develop—if your employer will give you the time and funds to do so. Or, you may devise a plan for increasing your employer's profits, but you must obtain your employer's authorization to put it into effect. Perhaps, you will work for one of the many companies that sell their products and services through proposals (rather than through advertising), in which case you may offer to provide those products and services to your employer's prospective clients and customers.

In this chapter, you will learn about the conventional superstructure for writing proposals, one that will help you persuade people to accept your offer and invest their money, time, and trust in your ideas and in your employer's products and services.

## VARIETY OF PROPOSAL-WRITING SITUATIONS

As the second paragraph of this chapter suggests, you may write proposals in a wide variety of situations. Some of the major ways in which these situations vary from one another are as follows.

- Your readers may be employed in your own organization, or they may be employed in other organizations.
- Your readers may have asked you to submit a proposal, or you may submit it to them on your own initiative.
- Your proposal may be in competition against others, or it may be considered on its own merits alone.
- Your proposal may need to be approved by various people in your organization before you submit it to your readers, or you may submit it directly yourself.
- You may have to follow regulations concerning the content, structure, and format of your proposal, or you may be free to write your proposal entirely as you think best.
- Once you have delivered your proposal to your readers, the readers may follow any of a wide variety of methods for evaluating it.

To illustrate these variables in actual proposal-writing situations, the following paragraphs describe the circumstances in which two successful proposals were written. The information provided about these two proposals will also be useful to you later in this chapter, where several pieces of advice are explained through the example of these proposals.

## Example Situation 1

Helen wanted permission to undertake a special project. She thought that her employer should develop a computer program that employees could use to reserve conference rooms. She concluded that her company needs such a program after several instances in which she arrived for a meeting at a conference room she had reserved only to find that someone else had reserved it also. Because she is employed to write computer programs, she is well qualified to write this one. However, her work is assigned to her by her boss, and she cannot write the scheduling program without permission from him and from his boss. Consequently, she wrote a proposal to them.

As she wrote, Helen had to think about only two readers, because her boss and her boss's boss would decide about her proposed project without consulting other people. Because her employer has no specific guidelines for the way such internal proposals should be written, she could have used whatever content, structure, and format she thought would be most effective. Furthermore, she did not need anyone's approval to submit this proposal to people within her own department, although she would need approval before sending a proposal to another department.

Finally, Helen did not have to worry about competition from other proposals, because hers would be considered on its own merits. However, her readers would approve her project only if they were persuaded that the time she would spend writing the scheduling program would not be better spent on her regular duties. (Helen's entire proposal is presented at the end of this chapter.)

## Example Situation 2

The second proposal was written under much different circumstances than was Helen's. To begin with, three people, not just one, wrote it. The writers are college instructors who wanted funds from the U.S. Department of Education (DOE) to produce an environmental education program. The instructors, who teach in a college's Institute of Environmental Sciences, began designing the environmental education project when they learned that the DOE was inviting proposals for projects in environmental education. To learn more about what the DOE wanted, the instructors obtained copies of the DOE's "request for proposals" (RFP). After studying the RFP, the group decided to propose to develop materials that high school teachers and community leaders could use to teach about hazardous wastes.

In their proposal, the instructors addressed an audience much different from Helen's. The DOE receives about four proposals for every one it can fund. To evaluate these proposals, it follows a procedure widely used by government agencies. It sends batches of the proposals to appropriate experts around the country. These experts, called reviewers, rate and comment on each proposal. Every proposal is seen by several reviewers. Then, the reviews for each proposal are gathered and interpreted by staff members at the DOE.

Those proposals that have received the best response from the reviewers are funded. To secure funding, then, the instructors needed to persuade their reviewers that their proposed project came closer to meeting the announced objectives of the DOE than did at least three-quarters of the other proposed projects.

Before the instructors could even mail their proposal to the DOE, they had to obtain approval for it from several administrators at their college. That's because the proposal, if accepted, will become a contract between the college and the DOE. Through these approvals, the college assures itself that all the contracts it makes are beneficial to the school.

## PROPOSAL READERS ARE INVESTORS

The descriptions of the proposals written by Helen and by the three instructors illustrate some of the many differences that exist in various proposal-writing situations. Despite these differences, however, almost all proposal-writing situations have two important features in common—features that profoundly affect the way you should write your proposals:

- In your proposals, you ask your readers to invest some resource, such as time or money, so that the thing you propose can be done.

- Your readers will make their investment decisions *cautiously*. They will be acutely aware that their resources are limited, that if they decide to invest in the purchase or project you propose, those resources will not be available for other uses. For example, to let Helen spend two weeks creating the new scheduling program, her bosses had to decide that she would not spend that time on the other projects the department had. To spend money on the education project proposed by the three instructors, the DOE had to turn down some other projects that were seeking those same dollars. These readers, like all proposal readers, wanted to be sure they invested their resources wisely.

## THE QUESTIONS READERS ASK MOST OFTEN

As cautious investors, readers ask many questions about the purchases, projects, and other things proposed to them. From situation to situation, these questions remain basically the same. Furthermore, the kinds of answers that readers find satisfying and persuasive are—at a very basic level—the same also. That's what makes it possible for one superstructure to be useful across nearly the full range of proposal-writing situations.

The questions asked by proposal readers generally concern the following three topics:

- *Problem.* Readers of your proposals will want to know why you are making your proposal and why they should be interested in it. What problem, need, or goal does your proposal address—and why is that problem, need, or goal important to them?

- *Solution.* Your readers will want to know exactly what you propose to make or do and how it relates to the problem you described. Therefore, they will ask, "What kinds of things will a successful solution to this problem have to do?" and "How do you propose to do those things?" They will examine carefully your responses, trying to determine whether it is likely that your overall strategy and your specific plans will work.

- *Costs.* What will your proposed product or activity cost your readers—and is it worth the cost to them?

In addition, if you are proposing to perform some work (rather than supply a ready-made product), your readers will want the answer to this question:

- *Capability.* If your readers pay or authorize you to perform this work, how do they know they can depend upon you to deliver what you promise?

## STRATEGY OF THE CONVENTIONAL SUPERSTRUCTURE FOR PROPOSALS

The conventional superstructure for proposals is a framework for answering those questions—one found successful in repeated use in the kinds of situations you will encounter on the job.

When you follow this superstructure, you provide information on up to ten topics, which are listed in Figure 32-1. In some cases, you will include information on all ten topics, but in others, you will cover only some of them. Even in the briefest of proposals, however, you will probably need to treat the following four topics if you are to succeed: introduction, problem, solution, costs.

### Answering Your Readers' Questions Persuasively

Figure 32-1 links each of the ten possible topics of a proposal to the question by readers that you will be answering when you write on the topic. When you are providing information on these topics, however, you should do much more than supply data. You should also try to make a persuasive point. In its right-hand column, Figure 32-1 identifies the overall persuasive point you should try to make when discussing each of the ten topics of the conventional superstructure.

As you write, you will find it helpful to see the relationships among the ten topics. Think of them as a sequence in which you lead your readers through the following progression of thoughts:

| Topic | Your Readers' Questions | Your Persuasive Point |
|-------|-------------------------|------------------------|
| * Introduction | What is this communication about? | Briefly, I propose to do the following. |
| * Problem | Why is the proposed project needed? | The proposed action addresses a problem that is important to you. |
| Objectives | What kinds of things will a solution to this problem have to do in order to be successful? | A successful solution must achieve the following objectives. |
| * Product | How do you propose to do those things? | Here's what I plan to produce and how it will work effectively at achieving the objectives. |
| Method Resources Schedule Qualifications Management | Are you going to be able to deliver what you describe here? | Yes, because I have a good plan of action (method); the necessary facilities, equipment, and other resources; a workable schedule; appropriate qualifications; and a sound management plan. |
| * Costs | What will it cost? | The cost is reasonable. |

*Topics marked with an asterisk are important in almost every proposal, whereas the others are needed only in certain ones.

**Figure 32-1.**   Relationship of the Standard Topics in a Proposal to Your Readers' Questions and the Persuasive Points You Need To Make.

1. The readers learn generally what you want to do. (Introduction)
2. The readers are persuaded that there is a problem, need, or goal that is important to them. (Problem)
3. The readers are persuaded that the proposed action will be effective in solving the problem, meeting the need, or achieving the goal that the readers now agree is important. (Objectives, Product)

4. The readers are persuaded that you are capable of planning and managing the proposed solution. (Method, Resources, Schedule, Qualifications, Management)

5. The readers are persuaded that the cost of the proposed action is reasonable in light of the benefits the action will bring. (Costs)

There is no guarantee, of course, that your readers will actually read your proposal from front to back. Consider how readers often approach long proposals. These proposals are usually written in the book format, which means that they include a summary or abstract at the beginning. Instead of reading these long proposals straight through, many readers will read the summary, perhaps the first few pages of the body, and then skip around through the other sections.

In fact, in some competitive situations, reviewers of proposals are *prohibited* from reading the entire proposal. For instance, when companies compete for huge contracts to build major parts of space shuttles for the National Aeronautics and Space Administration (NASA), they submit proposals in three volumes: one explaining the problem and their proposed solution, one detailing their management plan, and one analyzing their costs. Each volume is evaluated by a separate set of experts: technical experts for the first volume, management experts for the second, and budget experts for the third.

Even when readers will not read your proposal straight through, the account given above of the relationships among the parts can help you write a tightly focused proposal in which all the parts support one another effectively.

## Various Lengths for Proposals

The preceding discussion mentions proposals that are several volumes long. Such proposals can run to hundreds or even thousands of pages. On the other hand, some proposals are less than one page. How will you know how long your proposals should be? There is no simple answer. In each case you need to determine how much you must say to win your point with your readers.

Sometimes you can be brief and still be persuasive. Often you will need to touch upon only a few of the ten topics listed in Figure 32-1. For instance, Helen's proposed project involved only one person: Helen. Consequently, she didn't need any management plan. Similarly, because her readers were already familiar with her qualifications as a writer of programs, she didn't have to say anything about them, except perhaps to point out the experience she had had in preparing the particular kind of program she proposed to write. And because she was asking only for two weeks of time to spend on the project she proposed, she didn't have to present a detailed budget, though she did need to justify her proposed schedule.

In other situations, like those involving the proposals to NASA mentioned above, you may need to write very lengthy proposals. Those will be long because you will need to address all ten topics, and your discussion of each of those topics must answer fully all the questions your readers will have. In the end, then, to decide how long a proposal needs to be, you must think about your readers, anticipating their questions and their reactions to what you are writing.

# CONVENTIONAL SUPERSTRUCTURE FOR PROPOSALS

The rest of this chapter describes in detail each of the ten topics that form the conventional superstructure for proposals. As you read this information, keep in mind that the conventional superstructure represents only a general plan. You must use your imagination and creativity to adapt it effectively to your particular situation. If you have not already read the general advice about using conventional superstructures that appears in the introduction to Part IX, you might find it helpful to do so now.

As you plan and write your proposal, remember that the ten topics identify kinds of information you need to provide, not necessarily the titles of the sections you will include. In brief proposals, some parts take only a sentence or a paragraph, so that several are grouped together. For instance, writers often combine their announcement of their proposal, their discussion of the problem, and their explanation of their objectives under a single heading, which might be "Introduction," "Problem," or "Need."

Also, remember that the conventional superstructure may be used with any of the three common formats: letter, memo, and book. While writing your proposal, you should consult Chapter 26 for information about the particular format you are going to use.

The sections that follow take up the ten topics in the order in which they appear in Figure 32-1: introduction, problem, objectives, product, method, resources, schedule, qualifications, management, and costs.

## Introduction

At the beginning of a proposal you want to do the same thing that you do at the beginning of anything else you write on the job: tell your reader what you are writing about. In a proposal that means announcing what you are proposing.

How long and detailed should this introductory announcement be? In proposals, introductions vary considerably in length but they are almost always relatively brief. By custom, writers reserve the full description of what they propose until later, after they have discussed the problem that their proposal will help to solve.

You may be able to introduce your proposal in a single sentence. Helen did that in her proposal:

I request permission to spend two weeks writing, testing, and implementing a program for scheduling conference rooms.

When you propose something more complex than a two-week project, you may need more words to introduce it. In addition, sometimes you may need to provide background information to help your readers understand what you have in mind. Here, for example, is the introduction from the instructors' proposal to the DOE:

> Chemicals are used to protect, prolong, and enhance our lives in numerous ways. Recently, however, society has discovered that some chemicals also present serious hazards to human health and the environment. In the coming years, citizens will have to make many difficult decisions to solve the problems created by these hazardous substances and to prevent future problems from occurring.
>
> To provide citizens with the information and skills they will need to decide wisely, the Institute of Environmental Sciences proposes to develop two educational packages entitled "Hazardous Substances: Handle with Care!" One package, for high school students, will include five fifteen-minute videotape programs and a teacher's guide. The other package, for communities, will consist of a thirty-minute color film and a discussion-leader's guide.

## Problem

Once you've announced to your readers what you are proposing, you must persuade them that your proposed action will address some problem significant *to them*. Your description of the problem is crucial to the success of your proposal. Although you might persuade your readers that your proposed project will achieve its objectives and that your project's costs are reasonable, you cannot hope to win approval unless you show that the project is worth doing in the first place—from your *readers'* point of view. You must not only identify a problem, but make that problem seem important to your readers; not only describe a need, but make it seem significant to your readers; not only define a goal, but make its achievement seem worthwhile to your readers.

To do that requires both creativity and research. However, the precise nature of your effort to describe the problem addressed by your proposed project will depend upon your proposal-writing situation. The following paragraphs offer advice that applies to each of three situations you are likely to encounter on the job: when your readers define the problem for you, when your readers provide you with a general statement of the problem, and when you must define the problem yourself.

### When Your Readers Define the Problem for You

You need to do the least research about the problem when your readers define it for you. That can happen when you are writing a proposal that your readers have asked you to submit. For instance, the readers might issue an RFP that explains in complete detail some technical problem that they would like your engineering firm (or one of your competitors) to solve. In such situations, your primary objective in describing the problem will be to show your readers that you thoroughly understand what they want.

### When Your Readers Provide a General Statement of the Problem

At other times, even when you have an RFP, you will need to devote considerable research and creativity to describing the problem. That happened to the three instructors who wrote the proposal concerning environmental education. In its RFP, the DOE provided only the general statement that it wanted to support projects that would "develop educational practices and resources dealing with the relation of various aspects of the natural and man-made environment to the total human environment." Proposal writers were then left to identify the particular kind of practice or resource they wished to develop and to devise their own arguments that those practices or resources were worthy of financial support from the DOE.

When you are in a similar situation, you should find out what sort of problem your readers will consider important. For instance, the instructors discovered that the DOE made the following statement elsewhere in the RFP: "Thus the environmental education process is multifaceted, multidisciplinary, and issue- or problem-oriented." This statement suggested to them that they should describe the problem they would address as an issue- or problem-oriented one that would require multidisciplinary knowledge to solve.

Accordingly, they decided to say that the materials they would prepare on hazardous substances would help citizens make decisions about the handling and regulating of hazardous substances in their own communities. Further, they described the citizens' problem in making these decisions as one that required them to think about hazardous wastes from many points of view: health, economics, technology, and so on. Once the instructors framed their general approach, they found appropriate facts and articles to explain and support this need. They also investigated to prove that materials addressed to this need did not already exist. Finally, they identified two groups of people in particular need of this information: high school students, who would soon have to use these skills as adult, voting citizens, and residents who already face decisions about hazardous wastes because industries in their communities produce them or because such wastes are disposed of in their region. In the end, the instructors included in their proposal a four-page discussion of the need for the particular kind of educational materials they would develop—a need they presented in terms they knew would be persuasive to their readers.

### When You Must Define the Problem Yourself

In other situations, you may not have the aid of explicit statements from your readers to help you formulate the problem. That is most likely to happen when you are writing a proposal on your own initiative, without being asked by someone else to submit it. The challenge of describing the problem in such situations can be particularly great because the argument that will be persuasive to your readers might be entirely distinct from your own reasons for writing the proposal.

Think about Helen's situation, for instance. She originally came up with the idea of writing the program for scheduling conference rooms because she felt frustrated and angry on the many occasions that she went to a meeting room that she had reserved only to find that someone else had reserved it too. Her bosses, however, are not likely to approve the scheduling project simply to help Helen avoid frustrations. For the project to appeal to them, it must be couched in different terms, terms that are allied to their own responsibilities and professional interests.

In such situations, you can pursue two strategies to define the problem. The first is to think about how you can make your proposed project important to your readers. What goals or responsibilities do your readers have that your proposal will help them achieve? What kinds of concerns do they typically express that your proposal could help them address? A good place to begin is to think about some of the standard concerns of organizations: efficiency and profit.

When she did that, Helen realized that from her employer's point of view, the problems involved with scheduling conference rooms were creating great inefficiencies. Time was wasted as people looked around for another place to meet. Further, the time wasted was not just of one person, but of all the people involved with the displaced meeting.

The second strategy for defining the problem is to speak with one or more of the people to whom you will send your proposal. This conversation can have two advantages. It will let you know whether or not your proposal has at least some chance of succeeding. If it doesn't, you might as well find out before you invest a large amount of time writing it. Second, by talking with someone who will later be a reader of your proposal, you can find out how the problem looks to him or her.

When Helen spoke to her boss, she discovered another aspect of the conference room scheduling that she had not thought of. Sometimes the rooms are used for meetings with customers. When customers see confusions arising over something as simple as meeting rooms, they are not only annoyed but also prompted to wonder if the company has similar problems with other parts of its operation—problems that would affect the company's products and services.

## Objectives

When using the conventional superstructure for proposals, you discuss in two stages your ideas for solving the problem you have identified:

1. You answer your readers' question, "What will a solution to this problem need in order to be successful?" Your answer will be a statement of the objectives of your proposed solution.

2. You answer your readers' question, "How do you propose to do those things?" In your answer, you describe the product of your proposed action—the thing you are asking your readers to authorize or support. You must do so in a way that persuades your readers that your proposed action will in fact achieve the objectives and thereby solve the problem.

As you can see, your statement of objectives plays a crucial role in the logical development of your proposal: it links your proposed action to the problem by telling how the action will solve the problem. To make that link tight, you must formulate each of your objectives so that it grows directly out of some aspect of the problem you describe. Here, for example, are three of the objectives that the instructors devised for their proposed environmental education program. To help you see how these objectives grew out of the instructors' statement of the problem, each objective is followed by the point from their problem statement that serves as the basis for it. (The instructors did not include the bracketed sentences in the objectives section of their proposal.)

1. To teach high school students the definitions, facts, and concepts necessary to understand both the benefits and risks of society's heavy reliance upon hazardous substances. [This objective is based upon the evidence presented in the problem section that high school students do not have that knowledge.]

2. To employ an interdisciplinary approach that will allow students to use the information, concepts, and skills from their science, economics, government, and other courses to understand the complex issues involved with our use of hazardous substances. [This objective is based upon the instructors' argument in the problem section that people need to understand the use of hazardous substances from many points of view to be able to make wise decisions about them.]

3. To teach a technique for identifying and weighing the risks and benefits of various possible courses of action. [This objective is based upon the instructors' argument in the problem section that people must be able to weigh the risks and benefits of the use of hazardous substances if they are to make sound decisions about that use.]

The instructors created similar lists of objectives for the other parts of their proposed project, each likewise based upon specific points made in their discussion of the problem.

Like the instructors, Helen carefully based her objectives upon her description of the problem. The exception is that she based her second objective not upon the problem that existed when she wrote the proposal but upon a problem that would have been created if her employer sought to solve the scheduling problem through another strategy, which Helen rejected. Here are two of Helen's three objectives (which she presented in paragraph form):

1. To maintain a completely up-to-date schedule of reservations that can be viewed by salaried personnel in every department. [This objective corresponds to her argument that the problem arises partly because there is no convenient way for people to find out what reservations have been made.]

2. To allow designated persons in every department to add, change, or cancel reservations through their terminals. [This objective relates to Helen's observation that making everyone follow the present regulations will create too much work for the people who would have to take everybody's calls about reservations.]

In proposals, writers usually describe the objectives of their proposed solution without describing the solution itself at all. Consequently, their readers can imagine many solutions that might achieve those objectives. The instructors, for example, wrote their objectives so that their readers could imagine achieving them through the creation of a textbook, an educational movie, a series of lectures, or a slide-tape show—as well as through the creation of the videotape programs the instructors proposed. Similarly, Helen described objectives that might be achieved by many kinds of computer programs. She withheld her ideas about the structure and strategies of her program's internal design until the next section of her proposal.

The purpose of separating the writer's objectives from the description of the solution is not to keep readers in suspense. In a proposal, writers generally tell in the first paragraph what they are proposing so readers know at least generally what is being proposed before reading the objectives. Rather, the separation of objectives from solution enables readers to evaluate the aims of the project separately from the writers' particular strategies for achieving those aims. Consequently, you must write your objectives so they will seem positive and desirable to your readers. Unless your readers feel that your objectives are desirable, they are unlikely to feel that your proposed way of achieving those objectives is worth their support. The most important way to make them seem desirable, of course, is to base them upon your description of the problem that your project will solve.

As the examples quoted from the instructors' proposal suggest, writers often present objectives in lists. Furthermore, even when they don't, they usually describe their objectives very briefly. For instance, Helen used only a paragraph to present her objectives. The instructors presented all of their objectives in three pages of their 98-page proposal. The challenge presented to you as a proposal writer is to make your objectives clear and appealing while still making them brief.

## Product

When you describe the product that your proposed project will produce, you explain your plan for achieving the objectives you told your readers about. For example, Helen described the various parts of the computer program she would write, talking both about how they would be built and how they would be used. The instructors described in detail each of the four components of their environmental education package. Scientists seeking money for cancer research would describe the experiments they wish to conduct.

To describe your product persuasively, you need to do three things. First, be sure to let your readers know how you will achieve each of your objectives. For instance, to be sure that the description of their proposed educational program matched their objectives, the instructors included detailed descriptions of the following: the definitions, facts, and concepts their videotape program would teach (see objective 1, above), the strategy they would use to help students take an interdisciplinary approach to the question of hazardous waste (objective 2), and the technique for weighing costs and benefits that they would help students learn (objective 3).

The second thing you must do when describing your product is provide enough detail to satisfy your readers that you have planned it carefully and thoughtfully. To do that, you will have to begin actual work on the project you are proposing. For instance, to provide enough detail to her boss about the computer program she would write, Helen had to design most of its aspects. Similarly, the instructors had to design many aspects of the proposed educational materials. For instance, they developed an outline for each of the five videotape programs and another outline for the teacher's guide that would accompany them.

The third thing you must do is to explain, where appropriate, the desirability of the product of your project. For example, the instructors planned to use a case-study approach in their materials, and in their proposal they offer this explanation of the advantages of that approach:

> Because case studies represent real-world problems and solutions, they are effective tools for illustrating the way that politics, economics, and social and environmental interests all play parts in hazardous substance problems. In addition, they can show the outcome of the ways that various problems have been solved—successfully or unsuccessfully—in the past. In this way, case studies provide students and communities with an opportunity to learn from past mistakes and successes.

Of course, you should include such statements only where they won't be perfectly obvious to your readers. In her proposal, Helen did not include any because she planned to use standard practices whose advantages would be perfectly evident to her readers, both of whom had been promoted to their present positions after several years of doing exactly the kind of work Helen is doing now.

## A Note on the Relationships among Problem, Objectives, and Product

The first three elements of a proposal are closely related to one another. In fact, you can increase the likelihood of succeeding in your proposal by ensuring that the three parts fit tightly together. Be sure that the objectives grow directly out of your statement of the problem, and that the project's product will address those objectives.

As a result of the close interrelationship among these parts, your work on any one of them has implications for the others. For example, your definition of the problem will shape the product you propose. When the instructors discovered the importance the DOE placed on issue-oriented and multidisciplinary approaches, they knew they would have to describe a problem, define an objective, and design their product (instructional materials) in issue-oriented, multidisciplinary terms. In a similar fashion, Helen discovered that, from the company's point of view, the impression that the present system creates on customers is a major problem. Therefore, she emphasized that her proposed computer program would include some method of creating priorities among potential users of the conference rooms so that users meeting with customers would be assured of having conference rooms to use even if other groups wanted them also.

It's crucial that you realize that the parts of your proposal must be well integrated and that your work on one topic will naturally affect what you will say about the other topics.

## Method

Readers of proposals often want to be assured that you can, in fact, produce the results that you promise. That happens especially in situations where you are offering something that takes special expertise—something to be customized or created only if your proposal is approved.

To assure themselves that you can produce what you promise, readers look for information about several aspects of your project: your method or plan of action for producing the result; the facilities, equipment, and other resources you plan to use; your schedule; your qualifications; and your plan for managing the project. Method is described in this section; the other aspects are discussed in the sections that follow.

To determine how to explain your proposed method, imagine that your readers have asked you, "How will you bring about the result you have described?"

In some cases, you will not need to answer that question. For example, Helen did not talk at all about the programming techniques she planned to use because her readers were already familiar with them. On the other hand, her readers did not know before reading her proposal how she planned to train people to use the program she would create. Therefore, in her proposal she explained her plans for training.

In contrast, the instructors had an elaborate plan for creating their educational materials. An important part of their plan, for instance, was to use three review teams, one to assess the accuracy of the materials they drafted, another to advise about the effectiveness of the videotapes, and a third to advise about the effectiveness of the community film and discussion leader's guide. In their proposal, the instructors described these review teams in great detail.

In addition, the instructors described each phase of their project to show that they would conduct all phases in a way that would lead to success. These phases include research, scripting, review, revision, production, field-testing, revision, final production, and distribution. In a similar fashion, when you write proposals you may have to supply detailed descriptions of your method of creating your product.

## Resources

By discussing the facilities, equipment, and other resources to be used for your proposed project, you assure your readers that you will use whatever special equipment is required to do the job properly. If part of your proposal is to request that equipment, tell your readers what you need to acquire and why.

Of course, when you propose something that requires no special resources, you do not need to include such a section. Helen did not need to include one. In contrast, the instructors needed many kinds of resources. In their proposal, they described the excellent library facilities that were available for their research. Similarly, to persuade their readers that they could produce high-quality programs, they described the videotaping facilities they would use.

## Schedule

Readers have several reasons for wanting to know the schedule for your proposed project. First, they want to know when they can enjoy the final result. Second, they want to know how the work will be structured so they can be sure the schedule is reasonable and is a sound way of organizing and scheduling the work. In addition, they may want to plan other work around the project: When will this project have to coordinate with others? When will it take people's attention from other work? When will other work be disrupted and for how long? Finally, readers want a schedule they can use once the project has begun so they can determine if the project is proceeding according to plan.

The most common way to provide a schedule is to use a Gannt chart. You can find detailed instructions for creating schedule charts in Chapter 20, "Twelve Types of Visual Aids." Figure 20-34 shows a sample Gannt chart.

## Qualifications

When they are thinking about investing in a project, readers want to be sure that the proposers have the experience and capabilities to carry out the project successfully. For that reason, a discussion of the qualifications of the personnel involved with a project is a standard part of most proposals. For example, in their proposal, the instructors discussed their qualifications in two places. First, in a section entitled "Qualifications," they presented the chief qualifications of each of the eight key people who will work on the project. In addition, they included a detailed curriculum vitae of each in an appendix.

In other situations, much less information might be needed. For instance, Helen's qualifications as a programmer were evident to her readers because they were employing her as one. If that experience alone were enough to persuade her readers that she could carry out the project successfully, Helen would not have needed to include any section on qualifications. However, her readers might have wondered whether she was qualified to undertake the particular program she proposed, because different kinds of programs require different knowledge and skills. Therefore, Helen wrote the following:

> As you know, although I usually work with our IBM system, I am also familiar with the Hewlett-Packard computer on which the schedule will be placed. In addition, as an undergraduate I took a course in scheduling and transportation problems, which will help me here.

In some situations, your readers will want to know not only about the qualifications of the personnel who will work on the proposed project, but also the qualifications of the organization for which you all work. Although the DOE did not specifically request it, the instructors included a five-page discussion of the qualifications of each of the three groups involved: the college, the Institute of Environmental Sciences, and the Ohio Environmental Protection Agency, which cooperated in the project.

## Management

When you propose a project that will involve more than about four people, you increase the persuasiveness of your proposal by describing the management structure of your group. That's because readers of proposals know that even well-qualified people cannot work successfully on a project if their activities aren't coordinated and overseen in an effective manner. In projects with relatively few people, you can describe the management structure by first identifying the person or persons who will have management responsibilities and then telling what their duties will be. In larger projects, you might need to provide a full organizational chart for the project (see Chapter 20 for information about creating organizational charts) and a detailed description of the management techniques and tools that will be used.

Because her project involved only one person, Helen did not establish or describe any special management structure. However, the instructors did. Because they had a complex project involving many parts and several workers, they set up a project planning and development committee to oversee the activities of the principal workers. In their proposal, they described the make-up and functions of this committee, and in the section on qualifications, they also described the credentials of the committee members.

## Costs

As emphasized throughout this chapter, when you propose something, you are asking your readers to invest resources, usually money and time. Naturally, then, you need to tell them how much your proposed project will cost.

One way to discuss costs is to include a budget statement. Sometimes, a budget statement needs to be accompanied by a prose explanation that persuades your readers that any unusual expenses are justified and that each of the items in the budget is calculated in a reasonable fashion. (See Chapter 20 for information about presenting budget statements.)

In proposals where dollars are not involved, information about the costs of required resources may be provided elsewhere. For instance, in her discussion of the schedule for her project, Helen explained the number of hours she would spend, the time that others would spend, and so on.

In some proposals, you may demonstrate the reasonableness of the costs of your proposal by also calculating the savings that will result from your project.

# SAMPLE PROPOSAL

Figure 32-2 shows Helen's proposal.

# WRITING ASSIGNMENT

An assignment that involves writing a proposal is included in Appendix I.

# PARKER MANUFACTURING COMPANY
Memorandum

TO:      Floyd Mohr and Marcia Valdez

FROM:   Helen Constantino

DATE:    July 14, 19--

RE:      Proposal to Write a Program for Scheduling
         Conference Rooms

*Writer tells what
is proposed
and what it will
cost*

     I request permission to spend two weeks writing, testing, and implementing a program for scheduling conference rooms in the plant. This program will eliminate several problems with conference room schedules that have become acute in the last six months.

### Present System

*Background
information*

     At present, the chief means of coordinating room reservations is the monthly "Reservations Calendar" distributed by Peter Svenson of the Personnel Department. Throughout each month, Peter collects notes and phone messages from people who plan to use one of the conference rooms sometime in the next month. He stores these notes in a folder until the fourth week of the month, when he takes them out to create the next month's calendar. If he notices two meetings scheduled for the same room, he contacts the people who made the reservations so they can decide which of them will use one of the other seven conference rooms in the new and old buildings. He then prints the calendar and distributes it to the heads of all seventeen departments. The department heads usually give the calendars to their secretaries.

     Someone who wants to schedule a meeting during the current month usually checks with the department secretary to see if a particular room has been reserved on the monthly calendar. If not, the person asks the secretary to note his or her reservation on the department's copy of the calendar. The secretary is supposed to call the

**Figure 32-2.**    Sample Proposal in the Memo Format.

July 14, 19--                                                    Page 2

reservation in to Peter, who will see if anyone else has called about using that room at that time.

Problems with the Present System

The present system worked adequately until about six months ago, when two important changes occurred:

* The new building was opened, bringing nine departments here from the old Knoll Boulevard plant.

* The Marketing Department began using a new sales strategy of bringing major customers here to the plant.

Problem

These two changes have greatly complicated the work of scheduling rooms. In the past, though secretaries rarely called Peter Svenson with reservations for the current month, few problems resulted. Now, with such greatly increased use of the conference rooms, employees often schedule more than one meeting for the same time in the same room. That problem has always created some loss of otherwise productive work time. Now, if one of the meetings involves customers brought here by the Marketing Department, we end up giving a bad impression of our ability to manage our business.

Consequences of problem that are important to the readers

A related problem is that even when concurrent meetings are scheduled for different rooms, both may be planning to use the same audiovisual equipment.

Possible Solutions

There are three main ways to solve the problem:

Writer describes unsatisfactory solutions first, telling why each is rejected

1. Insist that all the secretaries follow the existing procedures more diligently. However, slips will still occur. Furthermore, this will require Peter Svenson to devote more time to maintaining the schedule,

**Figure 32-2.** *Continued*

something the Personnel Department does not want him to do.

2.  Establish a separate reservation system in each building, and require departments to use only the conference rooms in their own building. This solution would significantly reduce the burden on Peter Svenson, but it would mean someone in another department would need to spend time managing the schedule. Furthermore, each building has conference rooms developed especially for certain kinds of meetings. It seems foolish to restrict a department from using the most appropriate room for a particular meeting just because that room is in another building.

3.  Maintain an easily up-dated and easily checked computer program for room scheduling that is available to every department. For that purpose, we could use the HP2632A, for which every department has terminals. This is the program that I propose to develop.

Objectives of Proposed Solution

> **Objectives of proposed project**

The program I propose to develop will replace the monthly Reservations Calendar with a central reservation system. The system's first objective will be to maintain a completely up-to-date schedule of reservations that can be viewed by salaried personnel in every department. Second, the system will allow designated persons in every department to add, change, or cancel reservations through their terminals. Third, it will show the reservation priority of each meeting so that people with higher priority (such as people hosting visits from potential customers) will be able to see which scheduled meetings they can ask to move to free up a room.

Details of Proposed Solution

> **Overview of proposed project**

Implementation of this program involves three steps: writing it, testing it, and training people in its use.

**Figure 32-2.**   *Continued*

Detailed
description of
product of
proposed work

**Writing the Program.** The program will have three routines. The first will display the reservations that have been made. When users access the program, the system will prompt them to tell which day's schedule they want to see and whether they want to see the schedule organized by room or by the hour. The calendar will display a name for the meeting, the name of the person responsible for organizing it, and the audiovisual equipment that will be needed.

The second routine will handle entries and modifications to the schedule. When users call up this program, they will be asked for their company identification number. To prevent tampering with the calendar, only people whose identification number is on a list given to the computer will be able to proceed. To make, change, or cancel reservations, users will simply follow prompts given by the system. Once a user completes his or her request, the system will instantly update the calendar that everyone can view. In this way, the calendar will always be absolutely up to date.

The third routine is for administration of the system. It will be used only by someone in the Personnel Department. Through it, this person can add and drop people from the list of authorized users. The person will also be able to see who made each addition, change, or deletion from the schedule. That information can be helpful if someone tampers with the calendar.

**Testing the Program.** I will test the program by having secretaries in four departments use it to create an imaginary schedule for one month. The secretaries will be told to schedule more meetings than they usually do, to be sure that conflicts arise. They will then be asked to reschedule some meetings and cancel some others.

**Training.** Training in the use of the program will involve preparing a user's manual and conducting training sessions. I will write the user's manual, and I will work with Joseph Raab in the Personnel Department to design and conduct the first training session. After that, he will conduct the remaining training sessions on his own.

**Figure 32-2.** *Continued*

July 14, 19--                                          Page 5

**Writer has already planned the testing**

Resources Needed. To write this program I will need no special resources. Testing and training will require the cooperation of other departments. I have already contacted four people to test the program, and Vicki Truman, head of the Personnel Department, has said that Joe Raab can work on it because that department is so eager to see Peter relieved of the work he is now having to do under the current system.

Schedule

**Schedule also tells the cost (in hours of work)**

I can write, test, and train in eight eight-hour days.

| Task | Hours |
|------|-------|
| Designing Program | 12 |
| Coding | 24 |
| Testing | 8 |
| Writing User's Manual | 12 |
| Training First Group of Users | 8 |
| | 64 |

The eight hours estimated for training includes the time needed both to prepare the session and to conduct it one time.

Qualifications

**Qualifications**

As you know, although I usually work with our IBM system, I am also familiar with the Hewlett-Packard computer on which the schedule will be placed. In addition, as an undergraduate I took a course in scheduling and transportation problems, which will help me here.

Conclusion

I am enthusiastic about the possibility of creating this much-needed program for scheduling conference rooms and hope that you are able to let me work on it.

**Figure 32-2.**   *Continued*

# Instructions

## PREVIEW OF CHAPTER

The Variety of Instructions
Three Important Points to Remember
    Instructions Shape Attitudes
    Good Visual Design Is Essential
    Testing Is Often Indispensable
Conventional Superstructure for Instructions
    Introduction
    Description of the Equipment
    Theory of Operation
    List of Materials and Equipment
    Directions
    Troubleshooting

You may often write instructions on the job. When doing so, you will act as your readers' guide and coach. Perhaps your readers will be people who have just purchased a product made by your employer. Through your instructions, you will tell them how to use that product. Perhaps your readers will be new employees in your organization. Through your instructions, you will teach them how to perform their basic duties. Perhaps your readers will be experienced engineers or scientists. Through your instructions, you will tell them how to perform a special procedure that you have developed.

Whoever your readers are, they will be counting on you to guide them quickly and safely toward the successful completion of their task. In this chapter, you will learn the conventional superstructure for writing instructions, which will help you guide your readers effectively.

## THE VARIETY OF INSTRUCTIONS

If you were to look at a sampling of the various kinds of instructions written at work, you would see that instructions vary greatly in length and complexity. The simplest and shortest are only a few sentences long. Consider, for example, the instructions that the state of Ohio prints on the back of the $1 \times 1$-inch registration stickers that Ohio citizens must buy and affix to their automobile license plates each year:

Application Instructions

1. Position sticker on clean, dry surface in lower right-hand corner of rear plate (truck tractor front plate). If plate has a previous sticker, place new sticker to cover old sticker.

2. Rub edges down firmly.

Note: Do not moisten or apply at temperatures less than 0° F.

Other instructions are hundreds—or even thousands—of pages long. Examples of these long and highly complex instructions are those written by General Electric, Rolls Royce, and McDonnell Douglas for servicing the airplane engines they manufacture. Other examples are the manuals that IBM, Burroughs, and NCR write to accompany their large mainframe computers.

This chapter describes the superstructure for instructions in a way that will enable you to use the patterns for any instructions you write at work, whether long or short.

## THREE IMPORTANT POINTS TO REMEMBER

When writing instructions, you should keep in mind three points: instructions shape attitudes, good visual design is essential, and testing is often indispensable. Each of these points is discussed briefly in the following paragraphs.

## Instructions Shape Attitudes

All the communications you write at work have a double aim: to help your readers perform some task and to affect your readers' attitudes in some way. However, many writers of instructions focus their attention so sharply on the task they want to help their readers perform that they forget about their readers' attitudes. To write effective instructions, you must not commit this oversight.

The most important attitude with which you should concern yourself is that of your readers toward the instructions themselves. Most people dislike using instructions. When faced with the work of reading, interpreting, and following a set of instructions, they are often tempted to toss the instructions aside and try to do the job using common sense. However, you and your employer will often have good reasons for wanting people to use the instructions you write. Maybe the job you are describing is dangerous if it isn't done a certain way, or maybe the product or equipment involved can be damaged. Maybe you know that failure to follow instructions will lead many readers to an unsatisfactory outcome, which they might then blame on your employer. For these reasons, it is often very important for you to persuade your readers that they should use your instructions.

In addition, as an instruction writer, you may want to shape your readers' attitudes toward your company and its products. If your readers feel that the product is reliable and that the company thoroughly backs it with complete support (including good instructions), they will be more likely to buy other products from your employer and to recommend those products to other people.

## Good Visual Design Is Essential

To create instructions that will help your readers and also shape their attitudes in the ways you want, you must pay special attention to the instructions' visual design, including both the page design and the design of the drawings, charts, flow diagrams, and other visual aids you might use.

### Page Design

In instructions, good page design is important for several reasons. First, readers almost invariably use instructions by alternating between reading and acting. They read a step and then do the step, read the next step and then do that step. By designing your pages effectively, you can help your readers easily find the instructions for the next step each time they turn their eyes back to your page. This may seem a trivial concern, but readers quickly become frustrated if they have to search through a page or a paragraph to find their places. When readers are frustrated by a set of instructions, they may quit trying to use them.

Second, through good page design you can help your readers quickly grasp the connections between related blocks of material in your instructions, such as the connection between an instruction and the drawing or other visual aid that accompanies it.

Third, the appearance of instructions influences readers to use or not use them. If the instructions look dense and difficult to follow, or if they look careless and unattractive, readers may decide not even to try them.

For advice about creating effective page designs, see Chapter 21, "Designing Pages."

## Visual Aids

You can increase the effectiveness of most instructions by including visual aids. Well-designed visual aids are much more economical than words in showing readers where the parts of a machine are located or what the result of a procedure should look like. On the other hand, visual aids that are poorly planned and prepared can be just as confusing and frustrating for readers as poorly written prose.

For general advice about creating effective visual aids, see Chapter 19, "Using Visual Aids." For specific advice about preparing twelve of the most commonly used types of visual aids, see Chapter 20, "Twelve Types of Visual Aids."

## Testing Is Often Indispensable

It may seem that instructions are among the easiest of all communications to write and therefore among those that least need to be tested. After all, when you write instructions, you usually describe a procedure you know very well—and your objective is simply to tell your readers as clearly and directly as possible what to do, one little step at a time. Actually, instructions present a considerable challenge to you as a writer. You will find that it is often difficult to find the words that will tell your readers what to do in a way that they will understand quickly and clearly. Also, because you know the procedure so well, it will be easy for you to accidentally leave out some critical information because you don't realize that your readers may need to be told it.

The consequences of even relatively small slips in writing—even in only a few of the directions in a set of instructions—can be very great. Every step contributes to the successful completion of the task, and the difficulties the readers have with any step can prevent them from completing the task satisfactorily. Even if the readers can eventually figure out how to perform all the steps, their initial confusion with one or two can greatly increase the time it takes them to complete the procedure. Furthermore, in steps that are potentially dangerous, one little mistake can create tremendous problems.

For these reasons, it's often absolutely necessary to determine for certain if your instructions will work for your intended audience. And the only way to find that out for sure is to give a draft to representatives of your audience and ask them to try the instructions out. For detailed advice about designing and creating tests of your instructions, see Chapter 24, "Testing."

## CONVENTIONAL SUPERSTRUCTURE FOR INSTRUCTIONS

The conventional superstructure for instructions contains six elements:

- Introduction
- Description of the equipment (if the instructions are for running a piece of equipment)
- Theory of operation
- List of materials and equipment
- Directions
- Guide to troubleshooting

The simplest instructions contain only the directions. More complex instructions contain some or all of the other five elements, the selection depending upon the aims of the writer and the needs of the readers.

Many instructions also include elements often found in longer communications such as reports and proposals. Among these elements are a cover, title page, table of contents, appendixes, list of references, glossary, list of symbols, and index. Because these elements are not peculiar to instructions, they are not described here but rather in the discussion of the book format in Chapter 26.

One good way to use the chapter you are now reading is to do the following. After you have carefully defined your purpose and studied your readers (see Chapters 4 and 5), read through the following sections, determining which elements of the conventional pattern for writing instructions will help you write effectively. Then, reread the sections for the elements that do seem relevant, thinking about ways to apply the advice given there to your particular situation.

That last step is important. No two writing situations are exactly alike. You cannot write successful instructions if you blindly follow the conventional superstructure described here. You must adapt that pattern to your particular readers, purpose, and circumstances by using your imagination and creativity and by following the guidelines given throughout this book. If you have not already read the general advice about using conventional superstructures that appears in the introduction to Part IX, you may find it helpful to so now.

## Introduction

As mentioned above, some instructions contain only directions, and no introduction. Often, however, readers find an introduction to be helpful—or even necessary. You will find general guidelines for writing introductions in Chapter 13, "Writing the Beginning of a Communication." The following paragraphs will help you see how to apply that general advice when you are writing instructions.

In the conventional superstructure for instructions, an introduction tells some or all of the following things about the instructions:

- Subject
- Aim (purpose or outcome of the procedure described)
- Intended readers
- Scope
- Organization
- Usage (advice about how to use the instructions most effectively)
- Motivation (reasons why readers should use rather than ignore the instructions)
- Background (information the readers will find helpful or necessary)

The following paragraphs discuss ways you can handle each of these six topics in your introduction.

### Subject

Writers usually announce the subject of their instructions in the first sentence. Here is the first sentence from the operator's manual for a ten-ton machine used at the end of assembly lines that make automobile and truck tires:

> This manual tells you how to operate the Tire Uniformity Optimizer (TUO).

Here is the first sentence from the owner's manual for a small, lightweight personal computer:

> This manual introduces you to the Apple Macintosh™ Computer.

### Aim

From the beginning, readers want to know the answer to the question, "What can we achieve by doing the things this communication instructs us to do?"

With some of the instructions you write, the purpose or outcome of the procedure described will be obvious. For example, most people who buy computers know many of the things that can be done with them. For that reason, a statement about what computers can do would be unnecessary in the Macintosh instructions, which in fact contain none.

However, other instructions do have to answer the readers' questions about the aim of the instructions. In instructions for operating pieces of equipment, for example, writers often answer the readers' inquiry about what the procedure will achieve by telling the capabilities of the equipment. Here, for instance, is the second sentence of the manual for the Tire Uniformity Optimizer:

> Depending upon the options on your machine, it may do any or all of the following jobs:
> - Test tires
> - Find irregularities in tires
> - Grind to correct the irregularities, if possible
> - Grade tires
> - Mark tires according to grade
> - Sort tires by grade

## Intended Readers

Many readers will ask themselves, "Are these instructions written for us—or for people who differ from us in interests, responsibilities, level of knowledge, and so on?"

Often, readers will know the answer to that question without being told explicitly. For instance, the operator's manual for the Tire Uniformity Optimizer is obviously addressed to people hired to operate that machine.

In contrast, people who pick up a computer manual often wonder whether the manual will assume that they know more (or less) about computers than they do. In such situations, it is most appropriate for you to answer that question. Here is the third sentence of the Macintosh manual:

> You don't need to know anything about Macintosh or any other computer to use this manual.

## Scope

Information about the scope of the instructions answers the readers' question, "What kinds of things will we learn to do in these instructions—and what things won't we learn?" The writers of the manual for the Tire Uniformity Optimizer answer that question in its third and fourth sentences:

This manual explains all the tasks you are likely to perform in a normal shift. It covers all of the options your machine might have.

The writers of the Macintosh manual answer the same question in this way:

This manual tells you how to:
- use the mouse and keyboard to control your Macintosh (Chapter 1)
- get started with your own work, make changes to it, and save it (Chapter 1)
- find out more about Macintosh concepts and how to use your new techniques to establish a daily working routine (Chapter 2)
- organize your documents on the Macintosh (Chapters 2 and 3)
- get the most out of your Macintosh system by adding other products to it (Chapter 5)
- care for your Macintosh (Chapter 6)
- do simple troubleshooting and find further help (Chapter 6)

## Organization

By describing the organization of their instructions, writers answer the readers' question, "How is the information given here put together?" Your readers may want to know the answer so they can look for specific pieces of information. Or, they may want to know about the overall organization simply because they can then understand the instructions more rapidly and thoroughly than they could without such an overview.

The writers of the Macintosh manual announce its organization at the same time that they tell the manual's scope. They do that by citing the appropriate chapter number when describing the manual's scope (see above).

The writers of the manual for the Tire Uniformity Optimizer (TUO) explain its contents in a different way (notice that this information fills out the readers' understanding of the scope of the manual):

The rest of this chapter introduces you to the major parts of the TUO and its basic operation. Chapter 2 tells you step-by-step how to prepare the TUO when you change the type or size of tire you are testing. Chapter 3 tells you how to perform routine servicing, and Chapter 4 tells you how to troubleshoot problems you can probably handle without needing to ask for help from someone else. Chapter 5 contains a convenient checklist of the tasks described in Chapters 3 and 4.

## Usage

As they begin to use a set of instructions, readers often ask themselves, "How can we get the information we need as quickly and easily as possible?"

Sometimes, the answer is obvious. If the readers' job is simply to follow the instructions from front to back or to look for a certain set of steps and then follow them, you don't need to say anything about how to use the instructions. The manual for the Tire Uniformity Optimizer is used in just such a straightforward way, so it contains no special advice about how readers should use it.

In contrast, in some of the instructions you write, you may be able to help your readers considerably by providing advice about how to use your communication. For that reason, the writers of the Macintosh manual provide this advice on the first page, under the heading "How to Use This Manual":

> Read Chapter 1 to learn the basics and to get started using one of the application programs you probably purchased with your Macintosh. Then continue on with this manual or go to the manual that came with the application you're going to use. Return to Chapter 3 of this manual when you want to know more about organizing your work. Use Chapter 4 for reference. Read Chapter 6 soon after you get your Macintosh to learn how to care for it.

## Motivation

As pointed out above, when people are faced with the work of using a set of instructions, they often are tempted to toss the instructions aside and try to do the job using common sense. There are several things you could do to persuade your readers not to ignore your instructions. For instance, you can use an inviting and supportive tone and an attractive appearance, such as are used in the Macintosh manual. You may also include sentences that tell the readers directly why it is important for them to use the instructions. The three examples that follow illustrate some of the kinds of statements that writers sometimes provide.

### From the Operating Instructions for a Typewriter

> To take advantage of the automatic features of the IBM 60 you need to take the time to do the training exercises offered in this manual.

### From a Service Manual for Electric Motors

> If, through proper installation and maintenance, we can keep our customers' motors in trouble free operation, we have satisfied customers. Everyone needs "satisfied customers" because . . .
>
> "OUR CUSTOMERS ARE OUR EMPLOYERS"

### From an Operating Manual for an Office Photocopy Machine

> Please read this manual thoroughly to ensure correct operation.

## Background

You may recall that Chapter 13 on "Writing the Beginning of a Communication" advised you to include in your introduction any background information that would help your readers understand and use the rest of your communication. That advice applies to instructions. The particular pieces of background information your readers need vary from one set of instructions to the next. However, for instructions that involve machines or other equipment, two kinds of background information—a description of the equipment and an explanation of the theory of operation of the equipment—are so often helpful that they are discussed separately in the next two sections.

So that you can see what their parts look like when put together, look at the introductions to the Tire Uniformity Optimizer and the Macintosh manuals shown in Figures 33-1 and 33-2. You will find the introduction to another manual in Figure 13-2.

As you look at these figures, notice that only the manual for the Tire Uniformity Optimizer uses the word *Introduction*. The introduction to the Macintosh manual is called "About This Manual" and the introduction to the Detroit Diesel 53 (Figure 13-2) is called "General Information." Indeed, the material that this chapter refers to as the introduction is called many other names in other instructions. Do not be distracted by the variety of names used for it. The material itself, whatever it is called, is a customary part of the conventional pattern for instructions, and it can help your readers greatly when you include it in appropriate situations.

## Description of the Equipment

Many sets of instructions concern the operation or repair of equipment: cars, computers, manufacturing machines, and laboratory instruments, for example. To be able to operate or repair the equipment, readers need to know the location and function of its parts. For that reason, a description of the equipment is an important section of many sets of instructions. The manual for the Tire Uniformity Optimizer, for instance, includes on the first page a photograph of the overall machine, with its major parts labelled. In many instructions, the drawings are accompanied by written explanations of the equipment and its parts.

## Theory of Operation

An explanation of the way a piece of equipment operates can be extremely useful to readers when they must do something more than merely follow a step-by-step procedure. If your readers will use the material in your instructions to design something (such as a computer program or an experiment), or if they must figure out what is wrong with a malfunctioning piece of equipment, they will be able to work much more effectively if they understand the theory of operation of the tools or equipment they use.

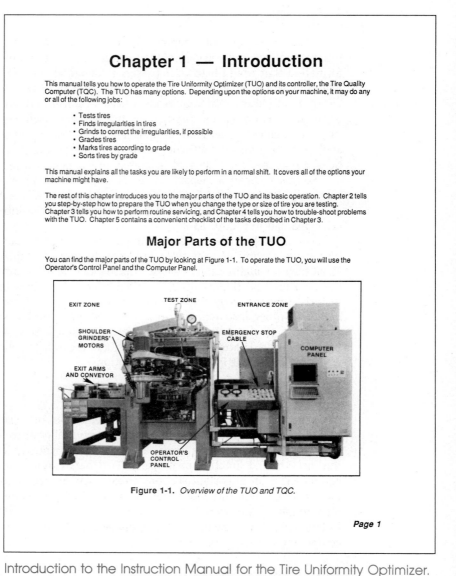

## Chapter 1 — Introduction

This manual tells you how to operate the Tire Uniformity Optimizer (TUO) and its controller, the Tire Quality Computer (TQC). The TUO has many options. Depending upon the options on your machine, it may do any or all of the following jobs:

- Tests tires
- Finds irregularities in tires
- Grinds to correct the irregularities, if possible
- Grades tires
- Marks tires according to grade
- Sorts tires by grade

This manual explains all the tasks you are likely to perform in a normal shift. It covers all of the options your machine might have.

The rest of this chapter introduces you to the major parts of the TUO and its basic operation. Chapter 2 tells you step-by-step how to prepare the TUO when you change the type or size of tire you are testing. Chapter 3 tells you how to perform routine servicing, and Chapter 4 tells you how to trouble-shoot problems with the TUO. Chapter 5 contains a convenient checklist of the tasks described in Chapter 3.

### Major Parts of the TUO

You can find the major parts of the TUO by looking at Figure 1-1. To operate the TUO, you will use the Operator's Control Panel and the Computer Panel.

**Figure 1-1.** *Overview of the TUO and TQC.*

*Page 1*

**Figure 33-1.**   Introduction to the Instruction Manual for the Tire Uniformity Optimizer.

From Akron Standard, *Operator's Manual for the Tire Uniformity Optimizer* (Akron, Ohio: Akron Standard, 1986), 1. Used with permission of Eagle-Picher Industries, Inc., Akron Standard Division.

Writers often place such information near the beginning of their instructions, especially if the instructions are for servicing or operating a piece of equipment. No such information is provided in the operator's manual for the Tire Uniformity Optimizer, probably because the operators are supposed to stick strictly to the jobs explicitly laid out for them. However, much general background information is provided at the beginnings of other manuals for

**About This Manual**

This manual introduces you to the Apple® Macintosh™ computer. Use it now to learn the basic Macintosh skills, and pick it up again later to use as a reference. You don't need to know anything about Macintosh or any other computer to use this manual. And you won't have to keep learning new ways of doing things. Once you've mastered a few new techniques, you'll use them whenever you use your Macintosh.

You can also take a guided tour of Macintosh by listening to the cassette tape (use it in any cassette player). In the guided tour, your Macintosh demonstrates itself, introducing—in a different way—the same skills this manual teaches.

This manual tells you how to:

☐ use the mouse and keyboard to control your Macintosh (Chapter 1)

☐ get started with your own work, make changes to it, and save it (Chapter 1)

☐ find out more about Macintosh concepts and how to use your new techniques to establish a daily working routine (Chapter 2)

☐ organize your documents on the Macintosh (Chapters 2 and 3)

☐ get the most out of your Macintosh system by adding other products to it (Chapter 5)

☐ care for your Macintosh (Chapter 6)

☐ do simple troubleshooting and find further help (Chapter 6)

**How to Use This Manual**

Read Chapter 1 to learn the basics and to get started using one of the **application programs** you probably purchased along with your Macintosh. Then continue on with this manual or go to the manual that came with the application you're going to use. Return to Chapter 3 of this manual when you want to know more about organizing your work. Use Chapter 4 for reference. Read Chapter 6 soon after you get your Macintosh to learn how to care for it.

The appendixes contain technical information. A glossary of Macintosh terms and an index are also included.

Now turn to the first chapter and get started.

5   ABOUT THIS MANUAL

**Figure 33-2.**   Introduction to the Manual for the Macintosh Personal Computer.

From Apple Computer, *Macintosh* (Cupertino, California: Apple Computer, 1986), 5. Reprinted by permission of Apple Computer, Inc.

this same machine. For example, the manual for setting up the machine provides an overview of the basic features of the special computer language used to program the machine. Without that overview, computer programmers would have a very difficult time writing programs to instruct the machine to do what they wanted done.

In the Macintosh manual, much of the overview material is presented in the chapter on "Finding Out More About Your Macintosh." There, for instance, readers find the answer to the question, "Where Does Your Information Go?" Knowledge of how information is stored by the computer enables the readers to figure out and remember some of the sequences of steps that are critical to their effective use of the computer.

## List of Materials and Equipment

Some instructions describe processes for which the readers will need to use materials or equipment that they wouldn't normally have at hand. When you are writing in such a situation, you can help your readers greatly by inserting a list of the things needed *before* you give the step-by-step directions. By doing so, you save the readers from the unpleasant surprise that they cannot go on to the next step until they have gone to the shop, supply room, or store to get something they didn't realize they would need. The instructions shown in Figure 33-3 include such a list.

## Directions

The heart of a set of instructions is the directions for performing the procedure they describe. The conventional pattern for writing effective directions is undoubtedly familiar to you. Its most obvious characteristics are the use of numbered steps and the use of illustrations. In the "How-To" section of the instructions in Figure 33-3, you will see many features of this pattern illustrated. Here are guidelines for writing directions that use that pattern:

### Present the Steps in a List

As explained above, people usually use instructions by reading about one step, performing that step, reading the next step, performing it, and so on. If you run your directions together within a paragraph, you make it difficult for your readers to find their place each time they look back at your instructions. In contrast, by presenting your directions in a list, you help your readers find their place each time. You can help your reader even more by placing some distinctive mark at the beginning of each item in your list, perhaps a number or a solid circle (called a *bullet*).

### In Your List, Give One Step Per Entry

If you present more than one step per number, you are clumping your directions into the kinds of paragraphs that the list is intended to avoid. There may be times when you want to present substeps. You can do that by using short indented lists within your larger list:

14. Drain the cannister.
   - Release the latch that locks the cannister's drain cap.
   - Unscrew the cap.

# Cracks in Concrete Sidewalks

**Your Problem**
- Small cracks in sidewalks are becoming larger.
- Uneven surfaces are dangerous.

**What You Need**
- Packaged ready-mixed mortar
- Epoxy concrete ("clear" type for narrow cracks and "gray" type for wide cracks and concrete breaks)
- Wire brush
- Pointing trowel and wood float
- Heavy-duty paint brush

**How-To: Repairing Cracks**

1. Caution! Repair only when concrete is dry.

2. Chisel out the crack or hole wider under the surface (fig. 1).

3. Clean the concrete surface thoroughly with the wire brush (fig. 2).

4. Mix a batch of mortar according to the directions on the package. Mix in the epoxy concrete with the mortar according to the direction on the epoxy container.

5. Using the trowel, put mixture into the crack (fig. 3).

6. Using the wood float, smooth the mixture even with concrete surface (fig. 4).

7. Clean the tools immediately with paint thinner.

8. Note: Work fast! Most epoxies will harden in an hour. If the patch should harden before the operation is completed, apply a second coat and smooth the surface again.

For big cracks, spread the mixture over the full width of the crack until the level of mortar is slightly above the concrete surface. If repairing a full break in the concrete, use the trowel to force the mortar mixture to the bottom of the break (fig. 5).

**Your Benefits**
- A more attractive sidewalk
- Prevention of further damage
- A safer walking surface

Fig. 1

Fig. 2

Fig. 3

Fig. 4

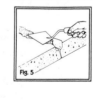

Fig. 5

37

**Figure 33-3.**   Typical Set of Brief Instructions.

From United States Department of Agriculture Extension Service, *Simple House Repairs . . . Outside* (Washington, D.C.: U.S. Government Printing Office, 1978).

## Use Headings and Titles to Indicate the Overall Structure of the Task

For instance, if you were writing a manual for operating a 35-mm camera, you might use the following headings and subheadings:

Preparing for use
    Installing the batteries
    Checking battery power
    Loading film
    Setting the film speed

Operating the camera
    Setting the shutter speed
    Setting the aperture
    Focusing
        Regular photography
        Infrared photography
    Making the exposure
    Advancing the film
    Unloading the film

By using headings, subheadings, and (in the book format) chapter titles to show the overall structure of the procedure you are describing, you help your readers *learn* the procedure so that they will be able to perform it without instructions in the future. You also help them find the directions for the specific parts of the procedure that they need assistance with.

## Use the Active Voice and the Imperative Mood

Active, imperative verbs give commands: "*Stop* the engine." They allow you to speak directly to your readers, telling them as briefly as possible what to do:

*Set* the dial to seven. (Much simpler than, "The operator then sets the dial to seven.")

*Clean* the parts with oil. (Much simpler than, "The parts should be cleaned in oil.")

## Use Illustrations

Drawings, photographs, and similar illustrations often provide the clearest and simplest means of telling your reader such important things as:

- *Where things are.* For instance, Figure 33-4 tells the readers of an instruction manual where to find a certain control lever.

Power Switch

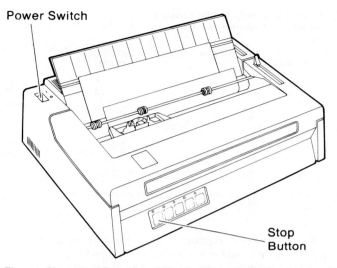

Stop
Button

**Figure 33-4.**   Figure Showing Readers Where To Locate Parts of a Machine.

From International Business Machines, *Wheelprinter E: Guide to Operations* (Lexington, Kentucky:
International Business Machines, 1985), 3–26. © 1985 by International Business Machines
Corporation.

- *How to perform steps.* For instance, by showing someone's hand
  performing a step, you provide your readers with a model to follow as
  they interpret the words that tell them what to do (see Figure 33-5).
- *What should result.* By showing readers what should result from
  performing a step, you help them understand what they are trying to
  accomplish. You also help them determine whether or not they have
  performed a step correctly.

### Place Warnings Where Readers Will See Them before Performing the Steps to Which They Apply

Your readers will depend upon you to warn them about actions that
would

- Endanger them or others
- Damage equipment they are using
- Ruin their results

You can ensure that your readers will see these warnings in time to
benefit from them by placing the warnings *before* the steps to which they
apply. Otherwise, your readers may look away to perform the step before
reading the warning:

**(b) If the paper is misfed in the entrance area of the fuse unit, remove it as shown.**
- Be careful as the fuser unit may be Hot.

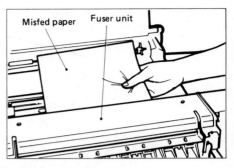

**Figure 33-5.**    Illustration Showing Readers How To Perform a Step.

From Toshiba Corporation, *Operator's Manual: Plain Paper Copier BD-7816* (Tokyo, Japan: Toshiba Corporation, 1984), 27. Courtesy of Toshiba Corporation.

WARNING: Before performing any of the following calibrations, follow the initial setup procedures described in sections 6.1 and 6.2. If you fail to do so, you could damage the chisel blades or the devices upon which they are mounted.

If the sentence describing the step is short, you can also place the warning immediately *after* that sentence. If you do that, you may want to use capital letters or some other device to ensure that your readers read the sentence containing the warning before performing the step described in the preceding sentence:

8. Rinse the reservoir. CAUTION: Do not use detergents to clean it. They contain chemicals that will damage the seals.

Both of these sample warnings explain why readers should follow them. Sometimes the reasons will be obvious. When the reasons aren't, readers may ignore or forget the warnings unless you explain the consequences of doing so.

### Tell Your Readers What To Do in the Case of a Mistake or Unexpected Result

You should anticipate the places where readers might make mistakes in the procedure. If it will not be obvious to them how to correct or compensate for a mistake, you should tell them how to do so. Similarly, you should tell your readers what to do in places where a correct action by them might not produce the expected result:

5. Depress and release the START, RESET, and RUN switches on the operator's panel. NOTE: If the machine stops immediately and the FAULT light illuminates, reposition the second reel and repeat step 5.

## Where Alternative Steps May Be Taken, Help Your Readers Quickly Find the One They Want

Sometimes you may write instructions where your readers may choose alternative courses of action, depending upon such things as the equipment or other resources they have or the results they desire. In these situations, be sure to make clear to your readers that they have a choice and then arrange your material so that they can quickly locate the alternative they want. Here is a sample from a manual written by IBM.[1]

You're ready to duplicate your new DPPX/SP system throughout the network. If you plan to:

First alternative
- Install remote sites with skilled personnel using SLU, repeat the central site installation procedures (Chapters 2 through 4). However, consider the note on CFE/IPF catalogs in Chapter 5 under the heading "Preparing the Distribution Sites for the SLU."

Second alternative
- Install the DPPX/SP Service Level Update again at each distributed site and customize each system *from the central site,* follow the steps in Chapter 5.

Third alternative
- Duplicate the central system on tape or diskettes, and restore the contents of the tape or diskettes at the distributed sites, follow the steps in Chapter 6.

Fourth alternative
- Send the customized central system to distributed sites with DSX, follow the steps in Chapter 7.

## Provide Enough Detail for Your Readers To Do Everything They Must Do

One of the questions you ask yourself as you prepare a set of directions is, "How much detail should I give when telling my readers what to do?" The answer, of course, is that you must tell your readers how to do everything your readers don't already know how to do.

To identify places where you may not have included enough detail, you can use the following strategy, which should sound familiar to you: think about your readers in the act of reading. Specifically, imagine whether or not, as they read each of the steps in your procedure, your readers will ask, "How do we do that?"

When you come to some steps, you will know immediately that your readers will not ask such a question. Here is an example:

Set the toggle switch to the ON position.

Because you know that your reader will not ask for detailed information about how to work a toggle switch, you do *not* need to include the following details about the substeps involved:

Extend the index finger of one of your hands.

Place the tip of your finger under the end of the toggle switch.

Push up on the switch until it snaps to the ON position.

With other steps, however, you may discover that your readers are likely to need additional information. Here are examples:

Calíbrate the scales. ("But *how* do I calibrate the scales?")

Set the potentiometers to the proper setting. ("*How* do I adjust the potentiometers?" "How can I find out what the proper setting *is*?")

In both of these examples, the writer should provide further information to answer the readers' questions.

### Troubleshooting

When they read instructions, people want to learn what to do if things don't work out as they expect—if the equipment they are using fails to work properly or if they don't get the result they want, for instance. You can provide that information in a troubleshooting section.

Usually, you will be able to provide troubleshooting information most helpfully if you use a table format. Figure 33-6 shows the format used in the troubleshooting section of the manual for the Tire Uniformity Optimizer, and Figure 33-7 (page 746) shows the format used in the troubleshooting section of the Macintosh manual. Other formats are possible, but they share these characteristics: the left-hand column lists the problems that might arise and the right-hand column lists the action to be taken. Information about the probable cause of the problem is often provided either in a middle column or in the right-hand column.

## CONCLUSION

In this chapter, you have learned about a conventional pattern for writing instructions. This pattern is very flexible, so you will be able to adapt it readily to many—if not all—the instructions you write on the job. Figure 33-8 (following the Exercises) shows a full set of instructions written by a student.

## NOTE

1. International Business Machines (IBM), *DPPX/SP Migration Guide* (Kingston, New York: International Business Machines, 1983), 24. Copyright © 1983 by International Business Machines Corporation. Reprinted by permission.

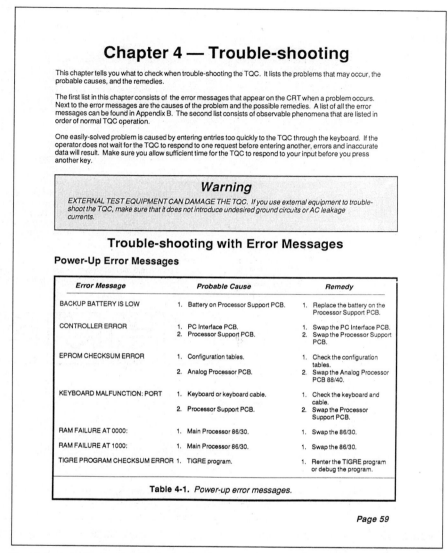

# Chapter 4 — Trouble-shooting

This chapter tells you what to check when trouble-shooting the TQC. It lists the problems that may occur, the probable causes, and the remedies.

The first list in this chapter consists of the error messages that appear on the CRT when a problem occurs. Next to the error messages are the causes of the problem and the possible remedies. A list of all the error messages can be found in Appendix B. The second list consists of observable phenomena that are listed in order of normal TQC operation.

One easily-solved problem is caused by entering entries too quickly to the TQC through the keyboard. If the operator does not wait for the TQC to respond to one request before entering another, errors and inaccurate data will result. Make sure you allow sufficient time for the TQC to respond to your input before you press another key.

---

### *Warning*

*EXTERNAL TEST EQUIPMENT CAN DAMAGE THE TQC. If you use external equipment to trouble-shoot the TQC, make sure that it does not introduce undesired ground circuits or AC leakage currents.*

---

## Trouble-shooting with Error Messages

### Power-Up Error Messages

| Error Message | Probable Cause | Remedy |
|---|---|---|
| BACKUP BATTERY IS LOW | 1. Battery on Processor Support PCB. | 1. Replace the battery on the Processor Support PCB. |
| CONTROLLER ERROR | 1. PC Interface PCB.<br>2. Processor Support PCB. | 1. Swap the PC Interface PCB.<br>2. Swap the Processor Support PCB. |
| EPROM CHECKSUM ERROR | 1. Configuration tables.<br><br>2. Analog Processor PCB. | 1. Check the configuration tables.<br>2. Swap the Analog Processor PCB 88/40. |
| KEYBOARD MALFUNCTION: PORT | 1. Keyboard or keyboard cable.<br><br>2. Processor Support PCB. | 1. Check the keyboard and cable.<br>2. Swap the Processor Support PCB. |
| RAM FAILURE AT 0000: | 1. Main Processor 86/30. | 1. Swap the 86/30. |
| RAM FAILURE AT 1000: | 1. Main Processor 86/30. | 1. Swap the 86/30. |
| TIGRE PROGRAM CHECKSUM ERROR | 1. TIGRE program. | 1. Renter the TIGRE program or debug the program. |

**Table 4-1.** *Power-up error messages.*

*Page 59*

**Figure 33-6.**   Troubleshooting Section from the Manual for the Tire Uniformity Optimizer.

From Akron Standard, *Operator's Manual for the Tire Uniformity Optimizer,* (Akron, Ohio: Akron Standard, 1986), 57. Used with permission of Eagle-Picher Industries, Inc., Akron Standard Division.

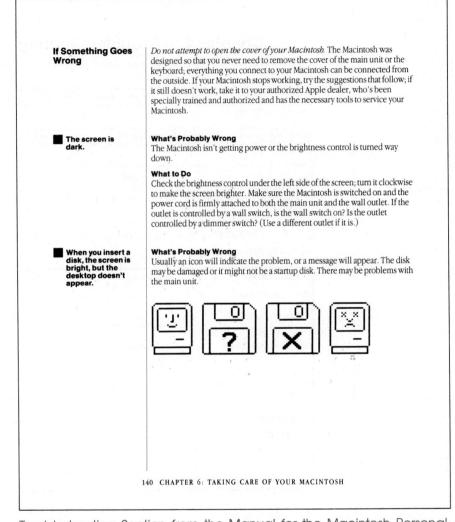

**If Something Goes Wrong**

*Do not attempt to open the cover of your Macintosh.* The Macintosh was designed so that you never need to remove the cover of the main unit or the keyboard; everything you connect to your Macintosh can be connected from the outside. If your Macintosh stops working, try the suggestions that follow; if it still doesn't work, take it to your authorized Apple dealer, who's been specially trained and authorized and has the necessary tools to service your Macintosh.

**The screen is dark.**

**What's Probably Wrong**
The Macintosh isn't getting power or the brightness control is turned way down.

**What to Do**
Check the brightness control under the left side of the screen; turn it clockwise to make the screen brighter. Make sure the Macintosh is switched on and the power cord is firmly attached to both the main unit and the wall outlet. If the outlet is controlled by a wall switch, is the wall switch on? Is the outlet controlled by a dimmer switch? (Use a different outlet if it is.)

**When you insert a disk, the screen is bright, but the desktop doesn't appear.**

**What's Probably Wrong**
Usually an icon will indicate the problem, or a message will appear. The disk may be damaged or it might not be a startup disk. There may be problems with the main unit.

140   CHAPTER 6: TAKING CARE OF YOUR MACINTOSH

**Figure 33-7.**   Troubleshooting Section from the Manual for the Macintosh Personal Computer.

From Apple Computer, *Macintosh* (Cupertino, California: Apple Computer, 1986), 140. Reprinted by permission of Apple Computer, Inc.

## EXERCISES

1. Find and photocopy a short set of instructions (five pages or less). Attach to the photocopy your analysis of the instructions, noting how the writers have handled each of the elements of the conventional superstructure. If the writers have omitted certain elements, explain why you think they did that. Be sure to comment on the page design and visual aids (if any) used in the instructions.

   Then evaluate the instructions. Tell what you think works best about them, and identify ways you think they can be improved.

2. Imagine that you have been asked to review the set of instructions shown in Figure 33-9, which are intended for people who have purchased a certain brand of waterbed. What are the strengths of the instructions? How could they be improved? Consider each element of the conventional superstructure for instructions. Be sure to comment on the page design and visual aids.

Writer uses Croy
lettering for
larger letters in
title and
headings

# Determining the Percentages of Hardwood and Softwood Fiber in a Paper Sample

Importance of
the procedure
explained

These instructions tell you how to analyze a paper sample to determine what percentage of its fibers are from hardwood and what from softwood. This information is important because the ratio of hardwood to softwood affects the paper's physical properties. The long softwood fibers provide strength but bunch up into flocks which give the paper an uneven formation. The short hardwood fibers provide an even formation but little strength. Consequently, two kinds of fibers are needed in most papers, the exact ratio depending upon the type of paper being made.

Overview of
procedure and
important
background
information

To determine the percentages of hardwood and softwood fiber, you perform the following major steps: preparing the slide, preparing the sample slurry, placing the slurry on the slide, staining the fibers, placing the slide cover, counting the fibers, and calculating the percentages. The procedure described in these instructions is an alternative to the test approved by the Technical Association of the Pulp and Paper Industry (TAPPI). The TAPPI test involves counting fibers in only one area of the sample slide. Because the fibers can be distributed unevenly on the slide, that procedure can give inaccurate results. The procedure given below produces more accurate results because it involves counting all the fibers on the slide.

## Equipment

All equipment
listed before
the directions

| | |
|---|---|
| Microscope | Clean cloth |
| Microscope slide | Hot plate |
| Microscope slide | Paper sample |
| cover | Blender |
| Microscope slide | Beaker |
| marking pen | Eyedropper |
| Acetone solvent | Graff "c" stain |

Figure 33-8. Instructions Written by a Student.

2

## Preparing the Slide

1. <u>Clean slide</u>. Using acetone solvent and a clean cloth, remove all dirt and fingerprints. NOTE: Do not use paper towel because it will deposit fibers on the slide.

2. <u>Mark slide</u>. With a marking pen, draw two lines approximately 1.5 inches apart across the width of the slide.

3. <u>Label slide</u>. At one end, label the slide with an identifying number. Your slide should now look like the one shown in Figure A.

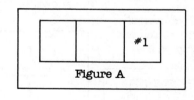

**Figure A**

4. <u>Turn on hot plate</u>. Set the temperature at warm. NOTE: Higher temperatures will "boil" off the softwood fibers that you will later place on the slide.

5. <u>Place slide on hot plate</u>. Leave the slide there until it dries completely, which will take approximately 5 minutes.

6. <u>Remove slide from hot plate</u>. Leave the hot plate on. You will use it again shortly.

## Preparing the Sample Slurry

1. <u>Pour 2 cups of water in blender</u>. This measurement can be approximate.

2. <u>Obtain paper sample</u>. The sample should be about the size of a dime.

3. <u>Tear sample into fine pieces</u>.

4. <u>Place sample in blender</u>.

5. <u>Turn blender on</u>. Set blender on high and run it for about 1 minute.

6. <u>Check slurry</u>. After turning the blender off, see if any paper clumps remain. If so, turn the blender on for another 30 seconds. Repeat until no clumps remain.

7. <u>Pour slurry into beaker</u>.

Explanation of reason for the caution

Figure placed immediately after its mention in text

**Figure 33-8.**    *Continued*

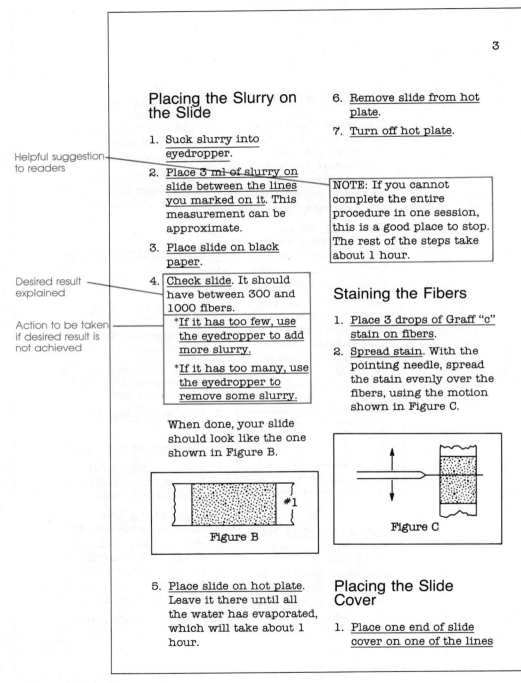

3

## Placing the Slurry on the Slide

1. Suck slurry into eyedropper.

2. Place 3 ml of slurry on slide between the lines you marked on it. This measurement can be approximate.

3. Place slide on black paper.

4. Check slide. It should have between 300 and 1000 fibers.

   *If it has too few, use the eyedropper to add more slurry.

   *If it has too many, use the eyedropper to remove some slurry.

When done, your slide should look like the one shown in Figure B.

*Helpful suggestion to readers*

*Desired result explained*

*Action to be taken if desired result is not achieved*

#1

**Figure B**

5. Place slide on hot plate. Leave it there until all the water has evaporated, which will take about 1 hour.

6. Remove slide from hot plate.

7. Turn off hot plate.

NOTE: If you cannot complete the entire procedure in one session, this is a good place to stop. The rest of the steps take about 1 hour.

## Staining the Fibers

1. Place 3 drops of Graff "c" stain on fibers.

2. Spread stain. With the pointing needle, spread the stain evenly over the fibers, using the motion shown in Figure C.

**Figure C**

## Placing the Slide Cover

1. Place one end of slide cover on one of the lines

**Figure 33-8.** *Continued*

Figure used to show a step not easily described in words

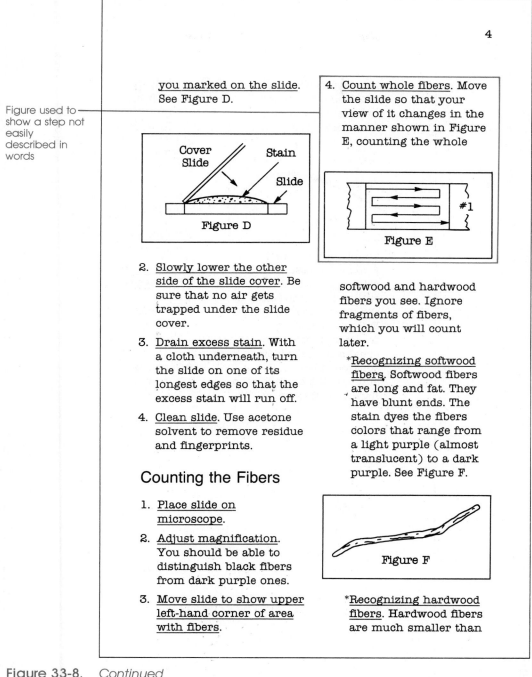

4

you marked on the slide. See Figure D.

Cover Slide

Stain

Slide

Figure D

2. Slowly lower the other side of the slide cover. Be sure that no air gets trapped under the slide cover.

3. Drain excess stain. With a cloth underneath, turn the slide on one of its longest edges so that the excess stain will run off.

4. Clean slide. Use acetone solvent to remove residue and fingerprints.

## Counting the Fibers

1. Place slide on microscope.

2. Adjust magnification. You should be able to distinguish black fibers from dark purple ones.

3. Move slide to show upper left-hand corner of area with fibers.

4. Count whole fibers. Move the slide so that your view of it changes in the manner shown in Figure E, counting the whole

#1

Figure E

softwood and hardwood fibers you see. Ignore fragments of fibers, which you will count later.

*Recognizing softwood fibers. Softwood fibers are long and fat. They have blunt ends. The stain dyes the fibers colors that range from a light purple (almost translucent) to a dark purple. See Figure F.

Figure F

*Recognizing hardwood fibers. Hardwood fibers are much smaller than

**Figure 33-8.**   *Continued*

5

softwood fibers. Their ends come to a point, and the stain dyes them deep black. See Figure G.

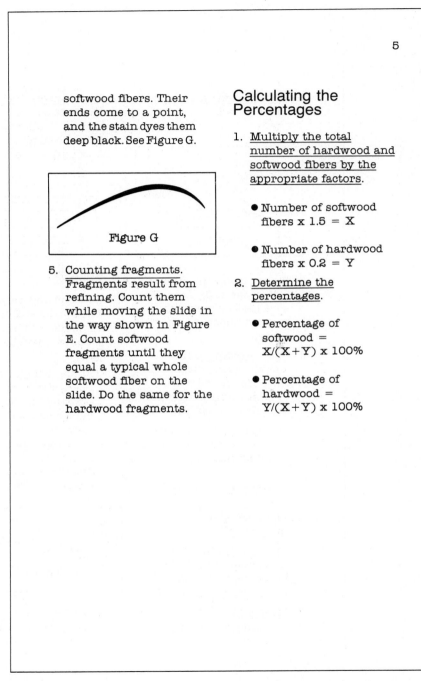

Figure G

5. Counting fragments. Fragments result from refining. Count them while moving the slide in the way shown in Figure E. Count softwood fragments until they equal a typical whole softwood fiber on the slide. Do the same for the hardwood fragments.

## Calculating the Percentages

1. Multiply the total number of hardwood and softwood fibers by the appropriate factors.

   - Number of softwood fibers x 1.5 = X

   - Number of hardwood fibers x 0.2 = Y

2. Determine the percentages.

   - Percentage of softwood = X/(X+Y) x 100%

   - Percentage of hardwood = Y/(X+Y) x 100%

**Figure 33-8.**   *Continued*

# HOW TO ASSEMBLE
# YOUR OWN WATERBED

Welcome to the wonderful world of waterbeds! If properly assembled, your new waterbed will provide you with years of comfort and enjoyment. This manual will guide you through the assembly, even if you have no previous experience with waterbeds. The manual is divided into the following six sections:

> *Frame Assembly
> *Pedestal Assembly
> *Decking Assembly
> *Headboard Assembly
> *Heater, Liner, and Mattress Placement
> *Filling Your Waterbed

Each section consists of several easy-to-follow steps. By reading each step carefully and completing it before moving on to the next step, you should be able to completely assemble your waterbed in about two hours. But, before beginning the assembly, there are a few things you should consider.

First, determine if your floor can handle the weight of a waterbed. (Generally, a floor is able to support a waterbed if the rafters are at most 16 inches apart.) Most buildings are adequately constructed to support this weight; however, you may wish to consult your waterbed dealer just to make sure. Also, if you live in an apartment or a condominium, it is a good idea to notify the building manager of your plans. Secondly, check the contents of your waterbed kit and compare them with the list on page two. If you are missing any necessary parts, you should contact your waterbed dealer. Now examine the Exploded-View Diagram on page two. Many steps in the assembly will refer back to this diagram.

You are now ready to begin the assembly with the aid of the following items:

| | |
|---|---|
| * screwdriver | NOTE: |
| * tape measure | Some steps of this assembly |
| * garden hose | may require two people. |
| * broomstick | |

Please proceed to page three.

1

**Figure 33-9.**   Instructions For Use with Exercise 2.

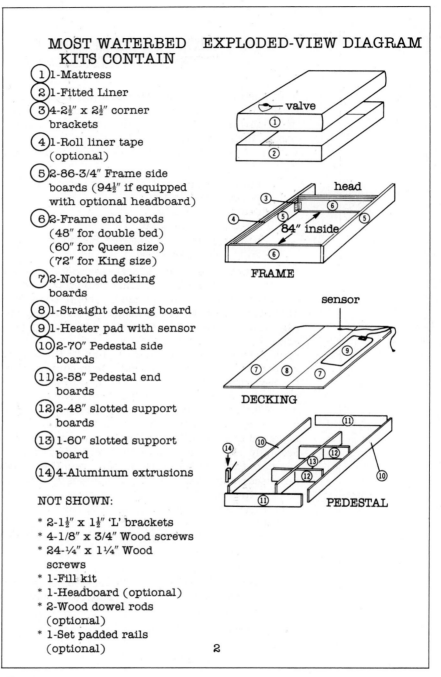

## MOST WATERBED KITS CONTAIN

1. 1-Mattress
2. 1-Fitted Liner
3. 4-$2\frac{1}{2}$" x $2\frac{1}{2}$" corner brackets
4. 1-Roll liner tape (optional)
5. 2-86-3/4" Frame side boards ($94\frac{1}{2}$" if equipped with optional headboard)
6. 2-Frame end boards (48" for double bed) (60" for Queen size) (72" for King size)
7. 2-Notched decking boards
8. 1-Straight decking board
9. 1-Heater pad with sensor
10. 2-70" Pedestal side boards
11. 2-58" Pedestal end boards
12. 2-48" slotted support boards
13. 1-60" slotted support board
14. 4-Aluminum extrusions

### NOT SHOWN:

* 2-$1\frac{1}{2}$" x $1\frac{1}{2}$" 'L' brackets
* 4-1/8" x 3/4" Wood screws
* 24-¼" x 1¼" Wood screws
* 1-Fill kit
* 1-Headboard (optional)
* 2-Wood dowel rods (optional)
* 1-Set padded rails (optional)

### EXPLODED-VIEW DIAGRAM

valve

FRAME

head

84" inside

DECKING

sensor

PEDESTAL

2

Figure 33-9. *Continued*

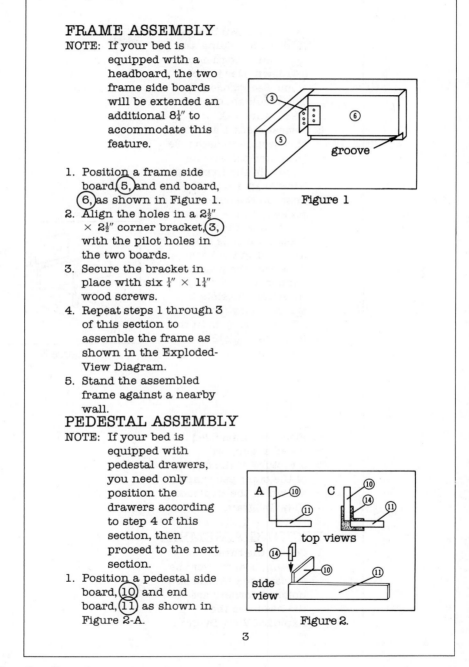

## FRAME ASSEMBLY

NOTE: If your bed is equipped with a headboard, the two frame side boards will be extended an additional $8\frac{1}{4}''$ to accommodate this feature.

1. Position a frame side board, 5, and end board, 6, as shown in Figure 1.
2. Align the holes in a $2\frac{1}{2}'' \times 2\frac{1}{2}''$ corner bracket, 3, with the pilot holes in the two boards.
3. Secure the bracket in place with six $\frac{1}{4}'' \times 1\frac{1}{4}''$ wood screws.
4. Repeat steps 1 through 3 of this section to assemble the frame as shown in the Exploded-View Diagram.
5. Stand the assembled frame against a nearby wall.

Figure 1

## PEDESTAL ASSEMBLY

NOTE: If your bed is equipped with pedestal drawers, you need only position the drawers according to step 4 of this section, then proceed to the next section.

1. Position a pedestal side board, 10 and end board, 11 as shown in Figure 2-A.

Figure 2.

3

Figure 33-9. *Continued*

2. Secure the two boards in place by sliding an aluminum extrusion, (14) down over top of them. See Figure 2-B. The result should look like Figure 2-C.

3. Repeat steps 1 and 2 of this section to assemble the outer pedestal as shown in the Exploded-View Diagram.

4. Position the outer pedestal in the desired location so that it is at least six inches from any wall in the room.

5. Assemble the pedestal support structure by inserting the slots in the two short boards, (12) into the slots in the long board, (13) See Figure 3.

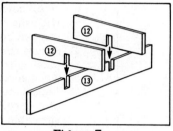

**Figure 3**

6. Place the assembled pedestal support structure in the center of the outer pedestal as shown in the Exploded-View Diagram.

## DECKING ASSEMBLY

1. Center the three decking boards over the pedestal so that the two notched corners are at the head. See the Exploded-View Diagram.

4

Figure 33-9.   *Continued*

2. Set the frame on top of the decking so that the grooves in the end boards hook onto the decking.

3. Place a $1\frac{1}{2}''$ x $1\frac{1}{2}''$ 'L' bracket midway along the side board and decking as shown in Figure 4.

4. Secure the bracket in place with two 1/8" x 3/4" wood screws.

5. Repeat steps 3 and 4 of this section to secure a bracket to the other side board.

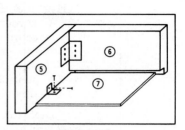

**Figure 4**

## HEADBOARD ASSEMBLY

NOTE: If your bed is not equipped with a headboard, you may proceed to the next section.

1. Insert a wood dowel rod in each of the two holes which appear on the top edges of the two sideboards, 5. See Figure 5.

2. Carefully position the headboard in place, making sure that the wood dowel rods are aligned with the holes on the bottom edge of the headboard. See Figure 5.

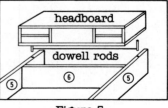

**Figure 5**

5

Figure 33-9.  *Continued*

## HEATER, LINER, AND MATTRESS PLACEMENT

NOTE: If your waterbed is equipped with the new self-shaping liner, you may skip steps 2 and 4 of this section.

1. Install the heater pad, ⑨ and sensor as shown in the Exploded-View Diagram. Do not <u>plug the heater pad into an outlet until after the bed has been filled.</u>

2. Run liner tape ④ along the inside of the frame, about two inches from the top edge and peel off the backing. See the Exploded-View Diagram.

3. Position the liner ② evenly inside the frame.

4. Fold the side of the liner ② up to the tape, ④ and fold the excess inward. See Figure 6.

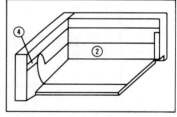

Figure 6

5. Place the mattress ① evenly inside the frame. See the Exploded-View Diagram.

6. If your bed is equipped with padded rails, slip them over the sides of the frame.

6

**Figure 33-9.** *Continued*

## FILLING YOUR WATERBED

1. Get out your fill kit and find the faucet adapter, hose adapter, and water conditioner.
2. Remove the airscreen on your faucet by unscrewing it counterclockwise. See Figure 7-A.
3. Screw the faucet adapter into your faucet and connect it to the female end of a garden hose. See Figure 7-B.
4. Remove the valve cap.
5. Screw the hose adapter onto the male end of the garden hose and slip the adapter through the mattress valve. See Figure 7-C.
6. Turn on both hot and cold water to desired temperature and fill the mattress until it is level with the top edge of the frame.

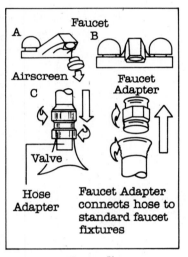

Figure 7

7. Pour the bottle of water conditioner into the mattress.
8. Roll a broomstick over the mattress towards the open valve to bleed out air.
9. Replace the valve cap.
10. If heat is desired, plug the heater pad into a nearby outlet.

7

Figure 33-9.   *Continued*

Congratulations! You have just assembled your first waterbed. Now it's time to lay back, relax, and float in fluid ecstasy. But don't fall asleep until you've read the notes below.

## IMPORTANT

*For proper comfort, your mattress should not be overfilled--you should float on the mattress.

*All mattresses require a periodic bleeding to dispel excessive air. Refer to step 8 on page 5.

*Be sure to add a new bottle of water conditioner to your mattress every six months.

8

**Figure 33-9.** *Continued*

# Making Oral Presentations

In your career, you will probably make hundreds—even thousands—of oral presentations. For these presentations, you will have many subjects and many audiences.

You may tell upper management about a project you are proposing, or report your progress to your co-workers on a project team. You may describe your company's products to prospective customers, or conduct a training course after they have made a purchase. If you become a manager, you may address the people in your department frequently and on a variety of topics.

Even if you give only a few talks in your career, they are likely to be very important to you. Through these talks you may seek funds or authorization for an important project or purchase. At work, talks are often made for such persuasive purposes. In these situations, your audience may make its decisions about your request based primarily on what you say—without reading anything you have written. Furthermore, your listeners—whether bosses, co-workers, or customers—will judge your professional abilities based in part upon the impressions you make as a speaker.

## HOW THIS SECTION RELATES TO THE REST OF THIS BOOK

The two chapters that follow are carefully coordinated with the other chapters in this book.

## A Listener-Centered Approach

Most important, these chapters teach the same basic strategy that you have been learning in the rest of the book: throughout all your work on a communication, think primarily about the people you are addressing. Don't think mainly about your subject matter, yourself, or anything else except the people whose needs you are trying to satisfy, whose opinions you are trying to shape, and whose actions you are trying to influence.

When you speak, of course, those people are your listeners. Accordingly, the chapters on oral presentations teach a *listener*-centered strategy rather than the *reader*-centered strategy you use in written communications. At bottom, though, the two strategies are the same.

## Coordinated Guidelines

Not only is the basic strategy for making oral presentations the same as it is for writing, but much of the more specific advice is identical also. To define the purpose of an oral presentation, to organize it effectively, to write its segments and sentences well, you do many of the same things you would do when preparing a written communication. For that reason, most of the guidelines for preparing written communications apply equally to preparing talks.

There are some differences, however. When speaking, you need to think about some things that don't come up in writing—like the best way to use your voice.

Also, some of the guidelines for writing need to be modified slightly when applied to speaking. For instance, when you find out how long your communication should be, you need to think in terms of minutes, not pages.

To allow you to build upon what you already know and to avoid unproductive repetition, the next two chapters focus on the additional things you need to know when preparing and presenting a talk. When reading them, though, keep in mind that most of what you know about how to write well will also help you speak well.

## COMPARISON OF WRITTEN AND ORAL COMMUNICATION

To help you use your knowledge about good writing as effectively as possible, the following paragraphs describe the similarities and differences between speaking and writing that are most important to you as you prepare and deliver a talk.

### Similarities

Oral and written communications are similar in the following crucial ways:

- *At work, the general aim of speaking and writing is the same: to make something happen.* You will try to persuade the board of directors to accept your proposal, to teach your company's customers to run the machine you've sold them, or to help your co-workers understand a new manufacturing process. Thus, speaking and writing are both forms of *action* in which you try to bring about some particular result by sharing information and ideas with others.
- *In both speaking and writing, you succeed or fail according to the way you affect the individual members of your audience.* Although you will deliver your talks to groups of people, each individual in the group sees through his own eyes, hears through her own ears, reacts in his or her own mind and heart. If your talk is to succeed with a group of people, it must succeed with the individuals who make up the group.
- *In both speaking and writing, the individual members of your audience react moment-by-moment as they hear or read your communication.* They respond to each statement as they come to it. Their response at one moment influences their responses in subsequent moments.
- *To prepare spoken and written messages, you follow the same general process.* You begin by defining your objectives (by thinking about your purpose and audience), then you plan, draft, test, and revise your communication.

### Differences

The following differences between listening and reading are important to you as you prepare and deliver talks at work:

- *Listeners can have more difficulty concentrating.* Have you noticed how easy it is for your mind to wander when you are listening to a lecture, even if the lecture is good? When speaking, you must work carefully at keeping your audience's attention.
- *Listeners can have more difficulty following the organization of your communication.* On the printed page, your readers find many clues to organization, including paragraph breaks, headings, indentations, and so

on. Listeners do not have a page to look at. When speaking, you must use your voice, visual aids, and other devices to help your listeners follow the organization of your message.

- *Listeners can't set their own course through your communication.* When you read, you can blaze your own trail through the material. You can flip around through the communication, looking for the information you most want. If a page contains material you already know, you can skim or skip it. If you want to review a certain point, you can turn back to it. Listeners, however, must follow the speaker through the material, or else they must quit listening to the speaker so they can review their memories of what he or she said a few moments ago. When you speak, you must be careful to set a course your readers will find satisfying.

- *Listeners can't turn off your talk the way a reader can put down your pages.* When you decide that you won't gain any further value from reading something, you can set it down. Listeners cannot do the same thing. Social convention prohibits them from telling the speaker to stop talking or from leaving a meeting until the speaker is done. In a sense, listeners are prisoners of the speaker. You must be careful to make every moment of your talks worthwhile to your listeners. If you don't, you may irritate them greatly. When that happens, you are unlikely to succeed in achieving your objectives.

## SUMMARY

The following two chapters are coordinated with the rest of this book to tell how to apply the advice given in other sections when your audience will hear rather than read your communication. The chapters also provide guidelines concerning the things you need to think about when speaking but not when writing—like using your voice effectively.

These two chapters are organized around the two major stages of making oral presentations.

## CONTENTS OF PART X

**Chapter 34**
**Preparing Your Talk**
**Chapter 35**
**Presenting Your Talk**

# Preparing
# Your Talk

## PREVIEW OF GUIDELINES

1. Define your objectives.
2. Select the form of oral delivery best suited to your purpose and audience.
3. Integrate visual aids into your presentation.
4. Select the visual medium best suited to your purpose, audience, and situation.
5. In your presentation, talk *with* your listeners.
6. Strongly emphasize your main points.
7. Make the structure of your talk evident.
8. Prepare for interruptions and questions.
9. Rehearse.

Tony is thrilled—and terrified. Today, he learned that a group of vice presidents wants to hear him talk about a project that he and three other people recently completed. The vice presidents will give him ten minutes to explain the aims, methods, results, and significance of the project. And then he should be ready to field questions.

"How nervous I'll be," he says, as he imagines himself standing before that audience, alone, with all eyes on him.

Then he thinks of the pride he will feel as he presents his ideas to such an audience. "What an opportunity to shine," he thinks.

Besides the opportunity to shine, Tony has some other pleasures to look forward to—pleasures that are not available when he communicates in writing. These include the enjoyment of talking *personally* to people who have come to hear what he has to say and the satisfaction of feeling *immediately* his audience's appreciation of his efforts.

If you are like most people, when you look forward to giving a talk at work you will experience much the same mixture of emotions that Tony is experiencing: great eagerness and anticipation, with some nervousness.

You can calm your nerves and increase the effectiveness of your talks by following the guidelines given below. They tell you how to prepare successful talks. In the next chapter, you will find nine guidelines that will help you present those talks effectively.

## ORGANIZATION OF THIS CHAPTER

The first seven guidelines in this chapter follow the sequence of activities involved with preparing a talk, activities comparable to those involved with preparing a written communication:

- Defining your objectives (Guideline 1)
- Planning (Guidelines 2 through 4)
- Drafting (Guidelines 5 through 7)

Guidelines 8 and 9 concern preparations that apply specifically to speaking, not to writing.

Guideline

1

### Define Your Objectives

You should begin working on an oral presentation in the same way you begin working on written communications: by figuring out exactly what you want to accomplish. To do that, you should follow much the same procedure that you follow when defining the objectives of a written communication. There are a few crucial differences, however.

## Purpose and Audience

First, think about who your listeners are and how you want to affect them. You can do that in exactly the same ways you think about the audience and purpose of a written communication. If you wish to review those procedures, turn to Chapters 4 and 5.

## Expectations

Next, consider people's expectations about what you will say and how you will say it. To do that you should consider the same factors that you consider when preparing a written communication (see Chapter 6), plus one other.

That additional factor is the length of time you are expected to speak. Often, you will be given a time limit. Even when you are not, your listeners are likely to have some firm expectations about the time you will take. Keep in mind that your talk will be only one part of your listeners' schedules. Do not encroach upon the time they have committed to other speakers or other business. You can ruin a good presentation by talking past your time limit.

## Scene

Finally, as you are defining objectives, consider the circumstances, or *scene*, of your talk. Three aspects of the scene are especially important:

- *Size of audience.* The smaller your audience, the smaller your visual aids can be, the more likely your audience is to expect informality from you, and the more likely your listeners will be to interrupt you to ask questions.
- *Location of your talk.* If the room has fixed seats, you will have to plan to use visual aids in spots where they are visible to everyone. If you can move the seats, you will be able to choose the seating arrangement best suited to your particular presentation.
- *Equipment available.* The kinds of equipment that are available determine the types of visual aids you can use. You can't use a blackboard or overhead projector if you can't obtain one for the room.

## Worksheet for Defining Objectives

Figure 34-1 shows a special worksheet for defining the objectives of the talks you give at work. It is a modified version of the worksheet for defining the objectives of written communications that is given on Figure 5-4. If all the members of your audience will have essentially the same reasons for listening to you and if they will all use the information you provide in essentially the same way, you can fill out the worksheet for one typical listener. On the other hand, if various members of your audience will have different

---

**Worksheet**
**DEFINING THE OBJECTIVES FOR AN ORAL PRESENTATION**

**TOPIC:** _____ **PRESENTATION DATE:** _____

**AUDIENCE:** _____ **TIME LIMIT:** _____

**LOCATION:** _____ **EQUIPMENT AVAILABLE:** _____

**FINAL RESULT** What final result do you want to achieve?

**LISTENERS' TASK** What will your listeners try to do while listening?

_____ **Compare Facts.** What criteria will your listeners use?

_____ **Understand.** What are the key points for your listeners to understand?

_____ **Apply Information.** In what situations will your listeners apply it?

_____ **Follow Step-by-Step Instructions.**

_____ **Other.** What are the important parts of this task?

**LISTENERS' ATTITUDES** How do you want to change your listeners' attitudes? (Tell what your listeners' attitudes are now and what you want them to be.)

**About you:**

**About your subject:**

**About themselves:**

**LISTENERS' CHARACTERISTICS**

**Role (decision-makers, advisers, implementers):**

**Familiarity with your specialty:**

**Special characteristics you should keep in mind:**

**Figure 34-1.**   Worksheet for Defining the Objectives for an Oral Presentation.

purposes and uses for your information, you will benefit from filling out worksheets not just for one listener but for several.

## INTRODUCTION TO GUIDELINES 2 THROUGH 4

Guidelines 2 through 4 provide you with advice about how to use your definition of your objectives to plan an effective talk.

<table>
<tr>
<td>Guideline<br>2</td>
<td>

### Select the Form of Oral Delivery Best Suited to Your Purpose and Audience

When people make presentations at work, they generally select from three major forms of oral delivery: the scripted talk, the outlined talk, and the impromptu talk. To choose among them, you should consider first your audience's expectations. In most of the situations in which you will speak at work, your listeners will have some fairly firm expectations concerning the form of talk you will give. You may reduce your chance of achieving your purpose if you disappoint these expectations.

In some situations, however, you will have complete freedom to select the form of delivery you desire. The following paragraphs provide information that will help you choose among the types.
</td>
</tr>
</table>

### Scripted Talk

With a scripted talk, you write out your entire talk, word for word, in advance. Then you deliver your talk by reading the script or by reciting it from memory.

The scripted talk is well suited for situations where you want to be very precise, because it lets you work out ahead of time exactly the phrasing you will use. This is a great advantage when you must present complex or detailed information clearly and when small slips in phrasing could be embarrassing or damaging. Government and corporate officials often use scripted talks when making formal statements, and speakers at professional conferences use scripted talks to ensure that they communicate clearly and concisely.

Besides precision, a script offers a great deal of security. It ensures that you will be able to say exactly what you had planned to say. You won't leave something out, digress, speak in confusing sentences, or forget to show your visual aids. And you won't exceed your time limit because you will know from your rehearsals precisely how long your talk will take.

Because of the security it offers, the scripted talk is a good choice when you expect to be nervous during your delivery. That might happen, for instance, when you give your first few talks on the job, before you are comfortable speaking in front of your co-workers and bosses. Even if you become too nervous to think straight, you will have your words spelled out in front of you.

A major disadvantage of a scripted talk is that you need to spend a long time preparing it. Another disadvantage is that when you are delivering your talk, you cannot easily adjust it in light of the reactions you see from your audience. Being thoroughly planned, the scripted talk is also rigid.

When delivering a scripted talk, you must be careful to avoid keeping your eyes riveted to your page. That will prevent you from establishing a good rapport with your audience. You should rehearse your script enough times so that you can deliver your talk smoothly by looking down at your notes only occasionally.

One way to avoid the temptation to keep your eyes on your script rather than your audience is to deliver your talk from memory. By doing so, you can create an impression of spontaneity while still retaining great precision. When you speak from memory, however, you risk forgetting your script. Therefore, you may want to keep a copy of your script (or an outline of your talk) at your side for safety's sake.

For many people, the greatest challenge of writing a scripted talk is that of achieving a style that sounds natural when read aloud. Most of us have a casual, informal speaking voice, and another voice—more formal, less lively—that we use when writing. Even when we have written a script that uses a natural style, some of us have trouble reading or reciting the script aloud in a manner that sounds natural. When rehearsing and delivering a scripted talk, you should concentrate on using your voice in the ways you usually do in conversation.

In sum, the scripted talk provides precision and security, but it also requires a great deal of time to prepare and a special effort to sound natural. If you haven't given scripted talks before, you will be surprised to learn how quickly you can improve your skill at doing so through practice.

## Outlined Talk

To prepare an outlined talk you do what the name implies: prepare an outline, perhaps very detailed, of the things you plan to say. You should, in addition, practice your talk, unless it is on a subject you have treated in oral presentations several times before. When practicing, you can work out your general treatment for each portion of your talk, decide how to emphasize your main points, and develop transitions that are clear and concise. You can also use this practice to judge how much you can say on each point and still keep within your time limit.

At work, outlined talks are much more common than scripted talks. They can be created much more quickly, and the presenter speaks with a "speaking voice," which helps increase interest and appeal. Furthermore, outlined talks are very flexible. Based upon your listeners' reactions, you can speed up, slow down, eliminate unnecessary material, or add something you discover is needed. Outlined talks are ideal for situations in which you speak on familiar topics to small groups, as when you are meeting with other people in your department.

The chief weakness of the outlined talk is that it is so flexible that unskilled speakers can easily run over their time limit, leave out crucial information, or encounter difficulty finding the phrasing that will explain their meaning clearly. For all these reasons, you may want to avoid giving outlined talks on unfamiliar material or to audiences you don't know well. Also, you may want to avoid them in situations where you might be unusually nervous and therefore likely to become tongue-tied or to forget your message.

## Impromptu Talk

An impromptu talk is one you give on the spur of the moment with little or no preparation. At most, you might jot down a few notes about the points you want to cover.

The chief advantage of the impromptu talk is that it takes little preparation time. It is well suited for situations in which you are speaking on subjects so familiar that you can express yourself clearly and forcefully with little or no forethought.

The chief disadvantage of the impromptu talk is that you prepare so little that you risk treating your subject in a disorganized, unclear, or incomplete manner. You may even miss the mark entirely by simply failing to address your audience's concerns at all.

For these reasons, the impromptu talks given at work are usually short, and they are usually used in informal meetings where the listeners can interrupt to ask for additional information and clarification. The impromptu talk is most common in meetings among people who work together regularly or who are at the same level within an organization. It is used much less often for formal presentations to people at higher levels or to large groups.

Of the three types of talks, the scripted and outlined talks are the most useful to you when you are trying to learn to give effective oral presentations. They require you to pay conscious attention to all the elements of a good presentation. In addition, they enable others to give you fairly precise advice about your plans before you give your talk, because your plans are evident in your written script or outline. If your instructor asks you to make an oral presentation in your class, it is likely to be a scripted or outlined talk.

## Guideline 3   Integrate Visual Aids into Your Presentation

Your talks at work will be much more effective if you present your message to your audience's eyes as well as to their ears. You can do that by making visual aids an integral part of your presentation.

## Benefits of Using Visual Aids

Visual aids can increase the effectiveness of your presentations in several ways:

- *Visual aids help you attract and hold your audience's attention.* Listeners are more likely than readers to let their minds wander. When they look away from a speaker, they may also be letting their thoughts drift off. By using visual aids, you provide listeners with a second place to gaze that is directly related to your message. Also, when you switch from one poster or slide to the next, your activity can draw wandering eyes back to you.

- *Visual aids help your listeners follow the organization of your talk.* You can use visual aids to provide headings that map your talk. For instance, you can display a poster that names the four topics you will cover, or the three recommendations you will make. If you leave such a poster up throughout the appropriate portion of your talk and point to it when you shift from one part of your talk to the next, you can help your listeners know at all times where you are in your presentation.

- *Visual aids help you emphasize key points.* By displaying your main points on a visual aid, you make them more prominent than the many points that you do not place on a poster.

- *Visual aids help your listeners remember what you have said.* Because visual aids communicate so succinctly and forcefully, the information they present is easy to recollect.

- *Visual aids help you explain your material.* Just as in a written communication, you can use drawings, graphs, charts, and other visual aids to communicate your material with an economy and effect that words alone cannot achieve.

- *Visual aids help you remember what you want to say.* Visual aids that are carefully integrated into your talk can serve as your notes while you speak. If you need to be reminded about what you intended to say next, you need only glance at the visual aid you are displaying or the one that will be displayed next.

## Using a Storyboard To Integrate Visual Aids into Your Presentation

For your visual aids to contribute as effectively as possible to your presentation, you must coordinate them with the words you will speak. You can do that by using a *storyboard,* a planning tool developed by people who write movie and television scripts.

Storyboards have two columns. One shows the words and other sounds that an audience will hear. The other describes (in words or sketches) the things the audience will see as those words are spoken. By reading a story-board, someone can tell what an audience will hear and see at each moment.

Figure 34-2 shows one kind of storyboard you can create for planning your talks. However, a storyboard doesn't have to be that elaborate to be effective. If you are going to use an outline rather than a script, you can make a few notes in the margins to help you combine the audio and visual aspects of your presentation into a single, effective unit.

Guideline

4

## Select the Visual Medium Best Suited to Your Purpose, Audience, and Situation

The preceding guideline suggests that you integrate visual aids into your talks. The guideline you are now reading concerns your decision about what kind of visual aid to use.

A *visual aid* is anything you give your audience to look at during your presentation. It might be something they touch and hold, such as a sample of your company's product. It might be a table, graph, drawing, or photograph. Whenever you use visual aids that can be printed or projected—such as tables, graphs, and drawings—you can choose from among a variety of media for presenting them to your listeners. At work, the media used most often are handouts, blackboards, overhead transparencies, posters, and 35-mm slides.

As you choose among these alternatives, you should think first about your purpose and audience. For some purposes, such as presentations at sales meetings, you may need the high polish that 35-mm slides can give. For other purposes, such as teaching, you may need to be able to make sponta-neous drawings—so that 35-mm slides would not be appropriate, although a blackboard, overhead transparencies, or posters might work well.

You should also consider your audience's expectations. If your audience expects you to draw lines on a blackboard but you present a slickly packaged slide show, your choice of visual medium may distract your audience and seriously reduce the effectiveness of your presentation. Conversely, if the members of your audience expect you to arrive with your visual aids all pre-pared, they may become impatient if you pause periodically in your talk to draw diagrams on the blackboard.

Finally, you should consider the situation in which you are to prepare and deliver your talk. How many days or weeks are there until your talk? How much of that time can you devote to preparing your visual aids? How big is the room in which you will speak? Will special equipment be available to you? How big is your budget?

The following paragraphs provide information about the five most com-monly used visual media. This information will help you select the medium that is best suited to your purpose, your audience's expectations, and the situation in which you will be speaking.

## Handouts

At work, handouts can be an effective medium for presenting information visually. You can prepare handouts inexpensively at the last minute. They are especially good for presenting detailed information (such as blueprints and large tables) that you could not display effectively on a screen or poster. You can also use them to provide listeners with an outline of your talk, one they can follow as they listen to you and look at your other visual aids, such as posters or slides. Many listeners like handouts because they can easily take notes on them and because they can file handouts as a record of your talk.

On the other side, handouts can disrupt your talk. They can be troublesome to distribute, particularly to large groups. Also, handouts can interfere with your talk if your listeners decide to read through your handouts when you want them to pay attention to something else.

## Blackboards

Blackboards are useful in situations where you want to create drawings during your talk. For instance, you could use them to gradually develop an illustration of a lengthy process, adding each step as you come to it. Blackboards also enable you to create visual materials (drawings, lists, and so on) spontaneously in response to questions.

On the other hand, the materials you draw on a blackboard are usually fairly rough, too unattractive for some purposes. Also, in order to draw on the blackboard, you must turn your back to your audience, which may make it difficult for them to hear you. The blackboard is not well suited for talks in which you will have lots of visual material. You will need to keep erasing, and you cannot easily refer to an earlier diagram after you have erased it.

## Overhead Transparencies

Overhead transparencies have most of the advantages of a blackboard—and some others. With them, you can create visual materials spontaneously without turning your back to your listeners. In addition, you can prepare some or all of your visuals in advance. These may be done in a casual, handwritten style, or they may be produced from carefully crafted, colorful artwork prepared by you or a graphic artist.

Overhead projection is good with large or small audiences because you can project overhead materials on a large screen, so that they are much bigger than material on a blackboard. An overhead projector can be used with the room lights on, so that your listeners can take notes, see you, and see one another.

On the other hand, when an overhead projector is placed on a table, it can block your listeners' view of your visual aids, you, or one another. A second disadvantage is that some overhead projectors have loud fans.

TALK TO COMPANY STEERING
COMMITTEE ON INTEGRATED
WORK STATION

Good morning. I want to thank Mr. Chin for inviting me to speak to you this morning about a project that my staff in the Computing Services Department think will increase our company's productivity considerably.

Show poster entitled "Uses of Computers." List: Management Functions, Research Analysis, Marketing Analysis, Design, Production, Word Processing.

As you know, people in our company use computers for many different purposes: our managers use computers to create budgets and schedules, our researchers use them to analyze their data, our marketing staff uses them to perform their duties, our development teams use them to design products, our manufacturing people use them to control production lines, and everybody uses them for word processing. In addition, people throughout the organization have expressed a desire for electronic mail.

Show poster entitled "Computer Barriers." List: Program to Program, Location to Location.

We estimate, however, that at present the company is realizing only about two-thirds of the labor-saving capability of present-day computer technology. To a large extent, that's because of *barriers* that prevent the easy flow of data from one computer application to another and from one location to another. For example, the data prepared by a researcher must be retyped by a secretary to be included in the research report. When our corporate planners want to use information developed by marketing analysts, they must obtain printouts of that data and then retype it into their own computers. And memos prepared through word processing that we could send almost instantly over telephone lines are instead printed out and sent via the Post Office or similar services.

Figure 34-2. Portion of a Storyboard for a Talk to a Group of Decision-Makers.

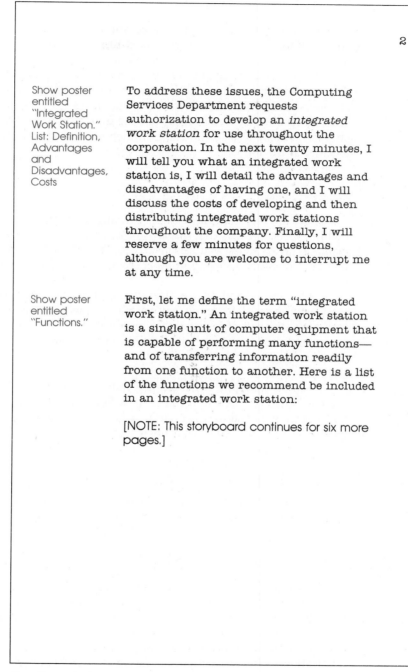

2

Show poster entitled "Integrated Work Station." List: Definition, Advantages and Disadvantages, Costs

To address these issues, the Computing Services Department requests authorization to develop an *integrated work station* for use throughout the corporation. In the next twenty minutes, I will tell you what an integrated work station is, I will detail the advantages and disadvantages of having one, and I will discuss the costs of developing and then distributing integrated work stations throughout the company. Finally, I will reserve a few minutes for questions, although you are welcome to interrupt me at any time.

Show poster entitled "Functions."

First, let me define the term "integrated work station." An integrated work station is a single unit of computer equipment that is capable of performing many functions— and of transferring information readily from one function to another. Here is a list of the functions we recommend be included in an integrated work station:

[NOTE: This storyboard continues for six more pages.]

**Figure 34-2.**   *Continued*

## Posters

You can create fancy posters in advance, casual ones at the last minute, and spontaneous ones during your talk. Beyond an easel or other means of holding them up, posters require no special equipment.

On the other hand, posters of the normal size hold only a very small amount of information, much less than you could display with a blackboard or overhead transparency. Posters are not suited for presentations to large groups because they cannot easily be read from a distance. (These comments on posters apply also to flip charts, which are like stacks of posters bound together at the top.)

## 35-mm Slides

Slides are very versatile. Besides showing photographs, you can generate highly polished visuals that display words, graphs, and many other kinds of materials. With slides, you can make excellent use of color and show things that are not easily displayed in other ways, such as color pictures of stained scientific specimens.

On the other hand, to use slides, you need special equipment and a darkened room. Also, slides aren't very flexible. You will have difficulty showing a particular slide out of sequence (for instance, in response to a question), and you cannot modify the content of a slide during your talk. And slides require considerable preparation time.

When planning the visual aids for your talks, keep in mind that you can use more than one type in the same talk. For instance, you may find it helpful to distribute handouts that outline your talk and to use one or more other types of visual aid to show your other materials.

# INTRODUCTION TO GUIDELINES 5 THROUGH 7

Guidelines 5 through 7 concern your work at drafting your talk. This advice applies whether your talk will be scripted, outlined, or impromptu.

Guideline
## 5

## In Your Presentation, Talk *with* Your Listeners

This guideline should sound familiar to you. It is a variation of the very first guideline in this book: "When writing, talk with your readers." Both guidelines emphasize the importance of responding to your audience's moment-by-moment reactions to what you are saying. You must respond effectively to those reactions because your audience's final reaction to your communication—whether written or oral—will be determined by these smaller reactions to one statement, then the next, and so on.

Of course, when applied to *writing,* this advice to talk with your readers isn't meant to be taken literally. To follow it, you create an imaginary portrait of your readers, then use that portrait to picture your reader's responses. When you *speak,* your listeners are present and it therefore might seem to be easy to talk with them. However, such is not necessarily so.

## Giving a Talk Differs from Conversing with People

Why not? Although your audience is present when you *deliver* a talk, they are not present when you *prepare* it. You prepare your talk in isolation, away from your listeners, who are then no closer to you than your readers are when you write.

The more detailed your preparations, the more difficult it will be for you to respond to your listeners' reactions. Thus, you will have the most difficulty talking with your listeners in scripted talks, where you work out beforehand even the phrasing of each individual sentence. Consequently, when you are preparing a scripted talk, you can benefit greatly from creating imaginary portraits of your listeners in the same way that you create such portraits of your readers (see Chapter 5).

You will encounter less difficulty responding to your listeners' reactions when you make outlined talks; you will have the least difficulty with impromptu talks. Even in an impromptu talk, however, speakers sometimes fail to respond to their listeners' reactions, either because the speakers are too nervous to observe those reactions or because they focus their attention on themselves or their subject matter, to the exclusion of their audience.

Thus, regardless of the form of oral delivery you choose, you should talk with your listeners by anticipating (or observing) their reactions and then responding accordingly. In this respect, preparing a talk is very much like drafting a written communication.

## The Importance of Building Rapport

When applied to oral presentations, the advice to talk with your listeners means something more than it does when applied to writing. When you give a talk, you and your listeners are right there in the same room, looking at each other. They expect you not merely to present information and ideas, but also to talk *to them,* to show an interest *in them.* Accordingly, to talk effectively with your listeners you must do more than respond to their reactions in a formal way. You must build some personal connection, a rapport, with your audience.

The following paragraphs list some specific suggestions for building rapport with your listeners:

● *While preparing your talk, imagine your listeners in the act of listening.* Your imaginary portrait of a typical listener can help you build rapport.

By imagining that you are *speaking* to this person, you will increase your ability to express your message in a way that sounds like one person speaking (rather than writing) to another.

- *Use the word* you *or* your *in the first sentence.* With those words, you signal that you intend to speak directly to your audience. You will also set a pattern of phrasing that you can follow in the rest of your preparations.

    There are many ways you can include *you* or *your* in the first sentence: thanking your listeners for coming to hear you, praising something you know about them (preferably something related to the subject of your talk), stating the reason they asked you to address them, or talking about the particular goals of theirs that you want to help them achieve.

- *Use a conversational style throughout.* Use shorter, less complex sentences than you might use in writing. Use personal pronouns *(I, we, you)* and the active voice.

- *Stick to points that are directly and clearly relevant to your listeners.* When you give a talk, you are making an implicit promise to provide information that your listeners will find useful or interesting. If you break that promise, you have broken the social connection between you and them. This point is especially important because it is tempting to digress in oral presentations, especially in impromptu and outlined talks. To maintain a good relationship with your listeners during your talk, give them what they want and can use.

- *Emphasize the implications of your points for your listeners.* In this way, you can assure your listeners that you have prepared remarks that are, in fact, for them.

- *Be very clear.* Nothing is more likely to make your listeners feel ill-treated than to speak in a way they cannot understand.

## Guideline 6    Strongly Emphasize Your Main Points

By emphasizing your main points, you improve your talk in two ways. First, your main points—not your incidental thoughts, not good stories—are what your *listeners* most want to hear. By emphasizing your main points, you draw your listeners' attention to the material they most desire.

Second, your main points are what *you* most want your listeners to take away from your talk. By emphasizing those points, you increase the chance that your listeners will notice and remember them.

In many situations at work, you will prepare talks that complement written reports. For instance, if you are proposing a major project, you might prepare an eighty-page written communication that you would introduce to your readers by making a fifteen-minute oral presentation. In your presentation, you would focus on the key points about your project, relying upon the written communication to fill in the details.

Here are some ways you can emphasize your main points:

- *Tell the listeners in advance that the main points are coming.* "This process has three major parts." "I will make three recommendations."
- *Announce each main point as you come to it.* "The second major step is as follows . . ." "My second recommendation is. . . ."
- *State your main points in your visuals.* In this way, your listeners will see as well as hear them.

When presenting your main points, remember to make clear their significance to your listeners.

<sub>Guideline</sub>
<sub>7</sub>

## Make the Structure of Your Talk Evident

To understand and remember your presentation fully, your listeners must be able to organize your points in their own minds. They stand the best chance of doing that if they are able to follow the structure of your talk.

*Listeners* must figure out your talk's structure without several of the aids that *readers* have. Listeners have no paragraph breaks, headings, or indentations to help them. Further, if they forget the overall structure or where they are in it, they cannot flip back through the pages to reorient themselves.

Here are two strategies you can use to help your listeners follow your organization.

- *Forecast the organization.* Tell your listeners in advance what your organization is. Your forecast will be especially effective if you both *tell* and *show* your organization. You can show it, for instance, by using a poster that contains a key word for each of the parts of your talk, or a handout that contains an outline of your talk. For detailed advice about forecasting the organization of your talk, see Chapter 15.
- *Clearly indicate the transitions from one topic to another.* Listeners can have real difficulty determining when you shift from one topic to another. The problem is especially critical because if they miss a transition, they will have trouble figuring out what your current topic is. Some ways you can make your shifts obvious are as follows:

  1. *Announce the shift.* "Now I would like to turn to my second topic."
  2. *Pause before beginning the next topic.* The pause in the flow of your talk will signal a shift.
  3. *Change visuals.* Take down one poster and put up another. Move to the next slide. Refer back to a visual that displays the overall organization of your talk.

4. *Move.* If you are giving a talk in a setting where you can move about, signal a shift from one topic to another by moving from one spot to another.

## INTRODUCTION TO GUIDELINES 8 and 9

The final two guidelines concern some special kinds of preparations that you should make for a talk but that are irrelevant when you are creating a written communication.

Guideline

8

### Prepare for Interruptions and Questions

At work, your talks will often be followed by a period for questions and discussion. That period provides your listeners with an opportunity to ask you for further information, to think through the significance of what you have said, and even to argue with you when they disagree. Such interchanges are a normal part of talks at work, and you should prepare for them.

To do that, think of the kinds of questions that might arise from your audience. In a sense, you do that when you plan the presentation itself, at least if you follow the guidelines given in Chapter 7, "Deciding What to Say." Usually, however, you will not have time in your talk to answer all the questions you expect that your listeners might ask. The questions you can't answer in your talk are ones your listeners may raise in a question period. Prepare for them by planning your responses.

Guideline

9

### Rehearse

All of your other good preparations can go for naught if you are unable to deliver your message in a clear and convincing manner. The next chapter provides you with guidelines for delivering your talk, but you should be *practicing* those guidelines well before you stand in front of your intended audience to give your talk.

For your rehearsal to be as valuable as possible to you, you should do the following:

- *Rehearse before other people.*
- *Pay special attention to your delivery of the key points.* Those are the points where stumbling can cause the greatest problems.
- *Rehearse with your visuals.* You need to practice coordinating your visual aids with your talk.
- *Time your rehearsal.* In this way, you can ensure that you keep within your time limit. Be sure to speak at a normal pace in rehearsal. If you race through your talk when you actually give it, you will make it difficult for your audience to follow. If you slow down when you deliver it, then you will take longer than you planned, perhaps exceeding your time limit.

# SUMMARY

The guidelines presented in this chapter are intended to help you prepare effective, listener-centered oral presentations.

1. *Define your objectives.* As when writing, describe the purpose and audience of your talk. Also learn what's expected of you, especially concerning the length of your talk. Finally, consider the scene of your presentation, including the audience size, room size, seating arrangement, and availability of visual equipment.

2. *Select the form of oral delivery best suited to your purpose and audience.* The *scripted talk* requires much preparation and may sound stiff and unnatural, but it provides precision and protection against forgetting. The *outlined talk* is natural-sounding and quick to prepare, but can be imprecise. The *impromptu talk* is best reserved for presentations on subjects you have given talks about several times before.

3. *Integrate visual aids into your presentation.* By using visual aids, you can help your audience follow the organization of your talk. Visual aids can also help you attract and hold your audience's attention, emphasize your key points, and explain your material.

4. *Select the visual medium best suited to your purpose, audience, and situation.* When thinking about which medium to use, consider such things as the size of your audience and room, the time you have to prepare, the desirability of preparing in advance, the flexibility you want the visuals to have, the availability of necessary equipment, and cost.

5. *In your presentation, talk* with *your listeners.* When preparing your talk, imagine your listeners' moment-by-moment reactions to what you are saying—and respond to those reactions. Also, build rapport by using *you* or *your* in the first sentence, using a conversational style, sticking to points that are clearly relevant to your audience, discussing the implications of your main points, and being very clear.

6. *Strongly emphasize your main points.* Your main points are what your *audience* most wants to hear; by emphasizing those points you ensure that your listeners get what they want. Also, your main points are what *you* most want your audience to take away from your talk; emphasizing those points increases the chances that your listeners will notice and remember them.

7. *Make the structure of your talk evident.* Forecast your organization. At places where you shift topics, announce the shift, pause, change visuals, or move from one spot to another. By taking these measures to help your listeners organize your points in their own minds, you increase the likelihood that they will understand and remember your presentation.

8. *Prepare for interruptions and questions.* At work, question periods and even interruptions are normal parts of oral presentations. Prepare for

them by predicting what your listeners are likely to ask and by planning your answers.

9. *Rehearse.* You can gain the full benefit from your rehearsal by practicing in front of other people, using your visuals, and timing your talk so that you can be sure that you will not talk past your time limit. When rehearsing, pay special attention to working out your delivery of your key points.

## EXERCISES

1. Imagine three situations in which you might have to prepare talks at work: one in which the talk would be scripted, one outlined, and one impromptu. Write a paragraph or two describing each of these situations. Using the "Worksheet for Defining the Objectives for an Oral Presentation" (Figure 34-1) as your guide, tell the following:
   - What your topic will be
   - Who will have asked you to talk
   - Why you have been asked to talk
   - What the purpose of your talk will be
   - Who your listeners will be
   - What your listeners will do with the information you provide
   - Where you will speak
   - What your time limit will be
   - What visual aids you will use

2. Make an outline for a talk to accompany a written communication you have prepared or are preparing in one of your classes. The audience for your talk will be the same as for your written communication. The time limit for the talk will be ten minutes. Be sure that your outline indicates the following:
   - The way you will open your talk
   - The overall structure of your talk
   - The main points from your written communication that you will emphasize
   - The visual aids you will use

   Be ready to explain your outline in class.

3. Imagine that you must prepare a five- to ten-minute talk on some mechanism, process, or procedure. For suggested topics, see Exercises 4, 5, and 6 in Chapter 12. Identify your purpose and readers. Then write a script or outline for your talk (whichever your instructor assigns). Be sure to plan what visual aids you will use and when you will display each of them to your listeners.

# Presenting
# Your Talk

## PREVIEW OF GUIDELINES

1. Arrange your stage so you and your visual aids are the focus of attention.
2. Look at your audience.
3. Speak in a natural manner, using your voice to clarify your message.
4. Exhibit enthusiasm and interest in your subject.
5. Display each of your visual aids only when it supports your words.
6. Make purposeful movements, but otherwise stand still.
7. Respond courteously to interruptions, questions, and comments.
8. Learn to accept and work with your nervousness.

ngela sits in one of the soft, leather chairs placed against the wall. In the center of the wood-paneled room is a long table, at which ten men and women sit, all managers or staff advisers in the organization that hired Angela after she graduated eighteen months ago. As soon as they have completed one last piece of business, Angela will speak to them, at their request, concerning a new marketing plan she has developed.

Angela is nervous. For the past three weeks she's concentrated most of her attention on preparing the ten-minute talk in which she will explain her plan. She's carefully written a script, designed a handout, created a set of twelve posters, sought advice and approval from her boss, and practiced over and over again. In a moment, all that work—and the months of labor that preceded it—will be on the line.

As Angela rises from her chair and begins her presentation, what can she do to make her presentation effective? What kinds of things should you try to do when you find yourself in a similar situation? This chapter presents eight guidelines that will help you deliver your oral presentations effectively.

## CONCENTRATE ON COMMUNICATING WITH EACH INDIVIDUAL IN YOUR AUDIENCE

The eight guidelines in this chapter touch on many activities involved with delivering talks, including such things as deciding where to stand, using your voice, and responding to interruptions, questions, and comments. The guidelines concerning all these activities deserve your careful attention. There is, however, one piece of advice that merits special emphasis: when you are presenting a talk, above all else concentrate on communicating with each individual in your audience. Each has come to hear you. Each is looking for something in particular from you. Each wants to feel that you are striving to address his or her individual interests and concerns.

"But," you may be thinking, "why tell me to concentrate on communicating with my audience? What else would I be trying to do?"

Actually, as they stand to speak, many people seem to forget that their purpose is to communicate. They seem to focus mainly on following their notes flawlessly and handling their visual aids without mishap. They speak in a flat, lifeless manner, wear wooden expressions, and make foolish mistakes, like standing in front of their visual aids. Instead of acting like people with something helpful or significant to tell someone else, they act as if they were indifferent toward their audience and subject matter.

You know from your own experience how such speakers affect people. As a listener, you become irritated or bored. You look out the window or at the floor. You begin thinking about something else.

To avoid such reactions from the listeners to your talks, you must strive to make each person feel that you have come for the express purpose of saying something important to him or her. Concentrate on those listeners, not on anything else. Don't simply read your notes; speak to your listeners. Don't simply display your visual aids; use them to help you say something to your listeners.

The various guidelines in this chapter will assist you in many other ways, but all will also help you achieve this very important goal of communicating directly and effectively with each individual in your audience.

## Arrange Your Stage so You and Your Visual Aids Are the Focus of Attention

Guideline

1

When you deliver a talk at work, you are a little like a stage performer, with some words to deliver and some props (your visual aids) to manage. For your performance to go smoothly, you must be sure that your "stage" is set carefully. Any difficulties with your stage can interfere with your listeners' ability to concentrate on and understand your message.

To arrange your stage in a way that will help your listeners focus their attention on your message, you can do three things. First, arrange your stage so that you are the focus of attention. Do not allow yourself to be dwarfed or obscured by your posters or other visual aids. Second, place your visual aids where they are visible to all members of your audience. Third, place your visual aids where you can manipulate them easily.

For most of your talks, these arrangements will require little thought or effort. Usually you will be speaking in familiar rooms that are often used for such presentations. You will be able to set your stage effectively if you imitate the arrangements you have seen others use or those you yourself have used successfully in the past. However, you will benefit from thinking more deliberately about the arrangement of your stage when you find yourself in any of the following situations:

- *You are going to speak in an unfamiliar location.* Visit the room beforehand to arrange your stage.

- *You are speaking in a room that is larger than you are used to.* Be sure that your visual aids will be visible to all and that everyone in your audience will be able to hear you.

- *You are going to use a 35-mm slide projector.* Be sure that you will have a source of light for reading your notes even when the room is darkened to show the slides.

Of course, an essential part of arranging your stage is being sure that your props will be usable. Make arrangements ahead of time if you are going to need a projector, blackboard, easel (for holding posters), or any other piece of equipment for your talk.

## Look at Your Audience

One of the most important ways to let your listeners know that you want to communicate with each of them is to look at them while you talk. By looking at them you create a personal bond. You show them that you are interested in them as individuals, both personally and professionally.

Also, by looking at your listeners you can get them to pay attention to you. Social convention requires people to pay attention to people who are talking to them. If you look at your listeners, they are much more likely to mind their manners by paying close attention to you.

In addition, by looking at your listeners you can see how your talk is going. You can see the eyes fastened on you with interest, the nods of approval, the smiles of appreciation. And you can see the puzzled looks, the wandering attention. As you observe these reactions, you can adjust your talk appropriately. If you are presenting an outlined talk, you can add needed explanations or drop unnecessary elaborations. Even with a scripted talk, you can make some adjustments, changing your rate of speech, speaking more loudly, giving your audience more time to study a visual aid.

Of course, you don't need to look at your audience constantly. If you are using notes (either an outline or a script), you will want to refer occasionally to them. You should not, however, rivet your gaze on your notes so that you never look at your audience.

If you have difficulty looking at your listeners when you speak, here are some strategies you can use:

- *Look at your audience before you start to speak.* Many people have difficulty looking at others when they are nervous or insecure. Perhaps you are one of them. Often these people are afraid that if they look at their listeners they will see signs of disagreement or disinterest. If that's the case, this first look before you begin to speak can help you greatly because it will come at a time when you cannot possibly fear seeing an adverse reaction to your talk. Having built up your confidence at the beginning of your talk, you may be more willing to look at your audience in the midst of it.

- *Follow a plan for looking.* For instance, at the beginning of each paragraph of your talk look at a particular part of your audience—to the right for the first paragraph, to the left for the second, and so on.

- *Target a particular part of your listeners' faces for your glances.* If their eyes are too threatening for you, plan to look at their noses or foreheads.

- *When rehearsing, practice looking at your audience.* For instance, develop a rhythm of looking down at your notes then up at your audience, down then up. Establishing this rhythm in rehearsal will help you avoid holding your head down throughout your talk.

When looking at your audience, try to avoid skimming over their faces without focusing your eyes on any individual. You must focus on an individual to make that person feel that you are paying attention to him or her. Try setting the goal of looking at a person for four or five seconds— long enough for you to state one sentence or idea. Some speakers find that such a pattern of looking helps them establish an even pace of speaking and also helps them avoid using distracting "filler" words ("umm," "you know," and so on).

In sum, by looking at your listeners you can command their attention and interest in your message.

## Speak in a Natural Manner, Using Your Voice To Clarify Your Message

Guideline
3

Another way to persuade your individual listeners that you are speaking directly to each of them is to speak like a person having a conversation. Listen to yourself and your friends converse. Your voices are lively and animated. To emphasize points, you change the pace, rhythm, and volume of your speech. You draw out words. You pause at key points. Your voice rises and falls in the varied cadence of natural speech.

When they make oral presentations, however, many people drain all of the natural life from their voices. They speak in a monotone, at a deadeningly even pace, without pause or variation of any sort. You know from your own experience how difficult it is to listen to such speakers. You rely on variations in the speaker's voice to maintain your interest, to help you distinguish major points from minor ones, to clarify the connections between ideas, and to identify the transitions and shifts in topic that indicate the structure of the speaker's talk.

Here are a few strategies that will help you use your voice to keep your listeners' interest and clarify your message:

- *In rehearsal and when delivering your talk, focus your attention on your listeners.* Keep in mind that you are talking to people, not reciting a well-practiced string of words. When you focus your attention on your listeners you are much more likely to fall into natural speech patterns.

- *In rehearsal, concentrate on your manner of speaking.* This will be easiest to do if you practice in front of people, but you can also do it if you *imagine* that your audience is before you. If you find a passage that is difficult for you to say in a natural voice, rephrase it.

- *In rehearsal, decide what points you want to emphasize, then practice presenting them emphatically.* In your notes (whether an outline or a script), mark these points so that you remember to emphasize them when you stand in front of your audience.

Guideline

4

## Exhibit Enthusiasm and Interest in Your Subject

When rehearsing and presenting your talk, keep in mind that even at work feelings are contagious. People who are enthused about projects often engender enthusiasm in others. At the same time, people who are indifferent often infect others with their indifference. For that reason, be sure to show your interest and enthusiasm when you speak.

What can you do to show your commitment to your subject? Certainly, speaking in a natural, forceful fashion will help. So will your facial expressions. An occasional smile can be your best visual aid. Remember that you must communicate more than information and ideas: you must communicate your *attitude* toward your subject.

Guideline

5

## Display Each of Your Visual Aids Only when It Supports Your Words

When you give a talk, you offer your audience information through two senses: hearing and seeing. This is a real advantage if you provide essentially the same message to both senses. However, if you talk about one thing but display a visual aid about something else, each person in your audience must choose to either listen to your words or read your visual aid. Either way, your audience will probably miss part of your message.

Therefore, you should be very careful to show each of your visual aids only when it supports and reinforces your words. Don't show your visual aids too soon. Don't leave them up too long.

One way you can coordinate your visual aids with your words is to prepare visual aids that you can reveal *gradually*. For instance, you may have a poster or overhead transparency that lists each of the three anticipated outcomes of a certain course of action. You can cover the last two outcomes with a sheet of paper while you state and discuss the first, then uncover the second as you state and discuss it, and so on.

Guideline

6

## Make Purposeful Movements, but Otherwise Stand Still

Another point to remember as you give your talk is that your movements can support or detract from your presentation. In conversation, you naturally make many movements—nodding or shaking your head, holding out your arms to show the size of something, and so on. When making oral presentations, you can also use such movements to make your meaning and feelings clear.

Also helpful are the movements you make to control your visual aids. In fact, one of the advantages of using visual aids is that they give you a good reason for moving, so that you don't stand stiffly and unnaturally throughout your talk.

Unhelpful are movements that express nervousness: pacing, rocking, fidgeting of any sort. These movements distract your audience's attention from your message. You should strive to eliminate them.

## Respond Courteously to Interruptions, Questions, and Comments

Audiences at work often ask questions and make comments to a speaker. In fact, most of the talks you give at work will be followed by discussion periods during which members of your audience will ask you for more information, discuss the implications of your talk, and even argue with you about points you have made. This give and take helps explain the widespread popularity of oral presentations at work: they permit speaker and audience to engage in a mutual discussion of what the speaker has to say.

When you respond to your audience, try to sustain your good relationship with all of your listeners. Be courteous, even if the questions are annoying or hostile. Remember that the questions and comments are important to the people who ask them, even if you don't see why. Give the requested information if you can. If you don't know how much detail the questioner wants, offer some and ask the questioner if he or she wants more. If you don't know the answer to a question, say so.

Although questions and comments are often held until after the talk, in some situations people will interrupt you to say or ask something. Interruptions require special care. First, speak to the person immediately. If you are already planning to address later in your talk the very matter that your questioner has raised, you may want to ask the person to wait for your response. If not, you may want to respond right away. Once you are done with your response and again pick up the thread of your talk, be sure to remind your listeners where you are in your talk: "Well, now I'll return to my discussion of the second of my three recommendations."

## Learn To Accept and Work with Your Nervousness

This final guideline is the most difficult for many novice speakers to follow. But it is very important. Not only is nervousness unpleasant for you to experience, but it can lead to distracting and unproductive behaviors that greatly diminish the effectiveness of your talk. Furthermore, some of these behaviors can cause your listeners to mistakenly believe that you are not interested in them or in your subject. The following list identifies some of the most common nervous behaviors, behaviors you should strive to avoid:

- Looking away from your audience instead of looking into their eyes.
- Speaking in an unnatural or forced manner.

- Exhibiting a tense or uninterested facial expression.
- Fidgeting and pacing.

What can you do to avoid these problems? It would be nice if you could simply banish the nervousness that causes them, making it go away forever. Unfortunately, nervousness cannot be eliminated through a simple act of will-power. Even some highly polished speakers with decades of experience still feel nervous as they rise to make each of their presentations.

If you can't stop your nervousness, what can you do about it? First, accept it. It's natural. It's going to be there whether you want it or not. If you start to worry about being nervous, you merely compound the emotional tension you must deal with when you speak. Keep in mind that your nervousness is not nearly as obvious to your listeners as it is to you. Even if your listeners do notice that you are nervous, they are more likely to be sympathetic than displeased. Furthermore, a certain amount of nervousness can help you. The adrenaline it pumps into your system will make you more alert and more energetic as you speak.

Besides learning to accept your nervousness, you can learn to control the undesirable mannerisms it can foster—the pacing, the fidgeting, and so on. They can be eliminated through conscious effort. Perhaps by asking for help from your friends, classmates, or co-workers, you can identify any distracting mannerisms you have and concentrate on controlling them.

In sum, you should remember that nervousness is natural. Instead of being embarrassed by it you should accept it and work to overcome the unproductive behaviors it fosters.

## SUMMARY

By a conscious effort you can improve your delivery of your oral presentations at work. The primary advice of this chapter is to strive to make each person feel that you have come for the express purpose of saying something important to him or her. The eight guidelines in this chapter tell you how to accomplish that and make your presentation effective in other ways as well.

1. *Arrange your stage so you and your visual aids are the focus of attention.* Make sure that your visual aids are clearly visible to all members of your audience, but take care that the visuals do not dwarf or hide you. Become familiar with the location in which you will speak, and arrange for necessary equipment ahead of time.

2. *Look at your audience.* By doing so, you let your listeners know that you are interested in them, and you subtly encourage them to pay attention to you. Also you are more likely to become aware of your audience's reactions to your talk and you can adjust your remarks accordingly, thereby increasing the effectiveness of your presentation.

3. *Speak in a natural manner, using your voice to clarify your message.* Use your voice in the ways you do when conversing with a friend. Alter the pace, rhythm, and volume of your talk to make your points clearly and emphatically.

4. *Exhibit enthusiasm and interest in your subject.* Feelings are contagious. Use your facial expressions and a natural speaking style to convey your enthusiasm for your subject.

5. *Display each of your visual aids only when it supports your words.* You distract your audience when you talk about one part of your subject but show a visual aid concerning another part. Don't show your visual aids too soon. Don't leave them up too long.

6. *Make purposeful movements, but otherwise stand still.* To make your message and feelings clear, use the sort of body motions that you use in ordinary conversation. Avoid nervous movements such as pacing, rocking, and fidgeting.

7. *Respond courteously to interruptions, questions, and comments.* Strive to maintain your good relationship with your listeners. Remember that the points your listeners raise are important to them, even if you don't see why. After an interruption, remind your listeners where you are in the overall structure of your talk.

8. *Learn to accept and work with your nervousness.* Don't compound your nervousness by worrying about it. Instead, learn to recognize and control the undesirable mannerisms that nervousness sometimes causes.

# EXERCISES

1. At work you will sometimes be asked to contribute to discussions about ways to make improvements. For this exercise, you are to deliver a five-minute impromptu talk describing some improvement that might be made to some organization you are familiar with. Topics you might choose include ways of improving efficiency at a company that employed you for a summer job, ways of improving the operation of some club you belong to, and ways that an office on campus can provide better service to students.

   In your talk, clearly explain the problem and your solution to it. Your instructor will tell you which of the following audiences you should address in your talk:

   - Your classmates in their role as students. You must try to persuade them of the need for and reasonableness of your suggested action.

   - Your classmates, playing the role of the people who actually have the authority to take the action you are suggesting. You should take one additional minute at the beginning of your talk to describe these people to your classmates.

2. At work you will sometimes have to tell your boss and co-workers about your plans for a communication you are writing. For this exercise, you are to describe in a five-minute outlined talk one of the written projects you are preparing for class. Let your talk fall into two parts. In the first, describe the background and objectives of your project by talking about such things as the following:

- What your topic will be.
- Who asked you to write the communication.
- Who your readers will be.
- What task your readers will perform while reading your communication.
- How you want your communication to affect your readers' attitudes.
- What limitations and expectations apply to your communication (including such things as limits on length and expectations about appearance).

In the second part of your talk, describe your plans for your communication in a way that shows how your plans are designed to achieve your objectives.

Be sure to use at least two visual aids. These might include, for instance, a poster outlining the main points of your talk, a handout showing your proposed table of contents, or a sample page showing some of the important features of your design.

Be prepared to answer questions from your audience.

3. Standing before your class, deliver the talk that you prepared in Exercise 2 or 3 of Chapter 34. Let your classmates and instructor play the role of your intended audience. When preparing your talk, be sure to rehearse it more than once, timing it to see that you can stay within your time limit. After your talk, take questions from your audience, who will be asking the types of questions your intended listeners would ask.

# Appendixes

The two appendixes that follow concern writing assignments that your instructor might give you. The first presents seven assignments that he or she might adapt to your course. The second contains a sample student response to one of those assignments—an assignment in which you would write a recommendation to someone who has not asked for your advice. Even if you don't prepare such a recommendation in your course, you will still find that the student sample makes an interesting piece for analysis and discussion from the point of view of the reader-centered approach to writing that you are learning in this book.

## CONTENTS

**Appendix I**
**Writing Assignments**

**Appendix II**
**Sample Unsolicited Recommendation**

# Writing Assignments

## PREVIEW OF ASSIGNMENTS

**Resume and Letter of Application**
**Instructions**
**Unsolicited Recommendation**
**Brochure**
**Project Proposal**
**Progress Report**
**Formal Report or Proposal**

I n this appendix you will find seven writing assignments that your instructor may ask you to complete. These assignments share this important feature: they all ask you to write to particular people for specific purposes that closely resemble the purposes you will have for writing in your career. To succeed with these assignments, you will need to keep both your readers and your purpose continuously in mind. For that reason, the best way to start your work on any of the assignments is to complete the Worksheet for Defining Objectives (page 112). For some or all of the assignments, you may also benefit from filling out the Worksheet for Deciding What To Say (page 142).

You will notice that these assignments contain specifications about such details as length and format. Your instructor may change these specifications to tailor the assignments to your writing course.

*Note to the instructor:* You will find helpful notes about these and other possible assignments in the *Instructor's Manual.* The manual includes suggestions about how you might adapt these assignments to your course.

## RESUME AND LETTER OF APPLICATION

Write a resume and letter of application addressed to some *real* person in an organization with which you might *actually* seek employment. If you will graduate this year, you will probably want to write for a full-time, permanent position. If you aren't about to graduate, you may want to apply for a summer position or an internship. If you are presently working, imagine that you have decided to change jobs, perhaps to obtain a promotion, secure higher pay, or find more challenging and interesting work.

To complete this project, you may need to do some research. Among other things, you will have to find an organization that really employs people in the kind of job you want, and you will need to learn something about the organization so that you can persuade your readers in the organization that you are knowledgeable about it. Many employers publish brochures about themselves; your campus placement center may have copies. Of course, your campus library is also a good source of information. If the publications you find don't give the name of some specific person to whom you can address your letter, call the organization's switchboard to ask for the name of the employment director or the manager of the particular department in which you would like to work. While you are working on this assignment, keep this real person in mind—even if you are not actually going to send your letter to him or her.

Your letter should be an original typed page, but your resume may be a high-quality photocopy of a typed original. Remember that the appearance of your resume, letter, and envelope will affect your readers, as will your attention to such details as grammar and spelling. Enclose your letter and resume in an envelope complete with your return address and your reader's name and address. You will find information about the formats for letters and envelopes in Chapter 26. Throughout your work on this project, you should carefully and creatively follow the advice given in Chapter 3 on resumes and letters of application.

## INSTRUCTIONS

Write a set of instructions that will enable your readers to operate some device or perform some process used in your major. The procedure must involve at least 24 steps. (For additional possible topics, see Exercise 5 in Chapter 12.)

Your instructions should guide your readers through some specific process that your classmates or instructor could actually perform. Do not write generic instructions for performing a general procedure. For instance, do not write instructions for "Operating a Microscope" but rather "Instructions for Operating the Thompson Model 200 Microscope."

Be sure to divide (or segment) the overall procedure into groups of steps, rather than presenting all the steps in a single list.

When preparing your instructions, pay careful attention to the visual design of your finished communication. You must include at least one illustration, and you must use press-on letters, Croy lettering, or some other form of large type for your title, headings, and other appropriate portions of your instructions. You may rely heavily on figures if they are the most effective way for you to achieve your objectives; in fact, your instructions need not contain a single sentence. Finally, you must use a multicolumn page design.

You may use the format for your instructions that you believe will work best—whether it is a single sheet of 8½ × 11″ paper, a small booklet printed on smaller paper, or some other design. Turn in a photocopy made from a typed original. In this way, you can create a professional-looking mixture of text and graphics in the way that many professional writers do: by pasting up a master copy that is then used to print the finished work. For more information about making your pages, see Chapter 21 ("Designing Pages").

Don't forget that your instructions must be accurate.

*Note to the instructor:* If you wish to make this a larger project, increase the minimum length of the procedure about which your students write. In that case, you may also want to specify that they create an instruction manual rather than, for example, an instruction sheet (and have them read the discussion of the book format in Chapter 26).

## UNSOLICITED RECOMMENDATION

This assignment is your chance to improve the world—or at least one small corner of it. You are to write a letter of 400 to 800 words in which you make an unrequested recommendation for improving the operation of some organization with which you have personal contact—perhaps the company that employed you last summer, a club you belong to, or your sorority or fraternity.

There are four important restrictions on the recommendation you make.

1. Your recommendation must concern a *real* situation in which your letter can *really* bring about change. As you consider possible topics, focus on situations that can be improved by the modest measures you can argue for effectively in a relatively brief letter. It is not necessary, however, that your letter aim to bring about the complete solution. In your letter, you might aim to persuade *one* of the key people in the organization that your recommendation will serve the organization's best interests.

2. Your recommendation must be unrequested; that is, it must be addressed to someone who has not asked for your advice.

3. Your recommendation must concern the way an organization operates, not just the way one or more individuals think or behave.

4. Your recommendation may *not* involve a problem that would be decided in an essentially political manner. Thus, you are not to write on a problem that would be decided by elected officials (such as members of Congress or the city council), and you may not address a problem that would be raised in a political campaign.

Of course, you will have to write to an *actual* person, someone who, in fact, has the power to help make the change you recommend. You may have to investigate to learn who that person is. Try to learn also how that person feels about the situation you hope to improve. Keep in mind that most people are inclined to reject advice they haven't asked for; that's part of the challenge of this assignment. From time to time throughout your career, you will find that you want to make recommendations your reader hasn't requested.

In the past, students have completed this assignment by writing on such matters as the following:

- A no-cost way that the student's summer employer could more efficiently handle merchandise on the loading dock.
- A detailed strategy for increasing attendance at the meetings of a club the student belonged to.
- A proposal that the Office of the Dean of Students establish a self-supporting legal-aid service for students.

Bear in mind that one essential feature of a recommendation is that it compares two alternatives: keeping things the way they are now and changing them to the way you think they should be. You will have to make the change seem to be the better alternative *from your reader's point of view.* To do that, you will find it helpful to understand why the organization does things in the present way. By understanding the goals of the present method, you will probably gain insight into the criteria your reader will apply when comparing the present method with the method you recommend.

In Appendix II, you will find an example of an unsolicited recommendation written by a student in response to this assignment.

## BROCHURE

Write a brochure about some academic major or student service on your campus. Alternatively, write a brochure for a service organization in your community.

Begin by interviewing people at the organization to learn about their aims for such a brochure. Then follow the advice given in Chapter 5 ("Understanding Your Readers") to learn about the target audience for the brochure. Remember that to be effective, the brochure must meet the needs of both the organization and the readers.

Use an 8½ × 14″ sheet of paper that is folded so that there are three columns (or panels) on each side of the sheet. When the brochure is folded, the front panel should serve as a cover. Use press-on letters, Croy lettering, or some other form of large type for your cover and headings. In addition to any artwork you may decide to include on the cover, use at least one visual aid (such as a table, flow chart, drawing, or photograph). Note that your cover need not have any artwork; it may consist solely of attractively lettered and arranged words that identify the topic of your brochure.

Turn in a photocopy made from a typed original. In this way, you can create a professional-looking mixture of text and graphics by pasting up a master copy that is then used to print the finished work. For more information about making your pages, see Chapter 21 ("Designing Pages").

Your success in this project will depend largely upon your ability to predict the questions your readers will have about your subject—and upon your ability to answer those questions clearly, concisely, and usefully. Also, think very carefully about how you want your brochure to alter your readers' attitudes about your subject—and remember that, along with the prose, the neatness and visual design of your brochure will have a large effect on your readers' attitudes.

## PROJECT PROPOSAL

Write a proposal seeking your instructor's approval for another project you will prepare later this term.

Your work on this proposal serves three important purposes. First, it provides an occasion for you and your instructor to agree about what you will do for the later project. Second, it gives you experience at writing a proposal, a task that will be very important to you in your career. Third, it gives you a chance to show how well you have mastered the material in Chapters 4 ("Defining Your Purpose") and 5 ("Understanding Your Readers").

Notice that while working on this assignment, you will have to define the objectives of two different communications: (1) the *proposal* you are writing now, which is addressed to your instructor, and (2) the *project* you are seeking approval to write, whose purpose and audience you will have to describe to your instructor in the proposal.

When writing your proposal, you may think of your instructor as a person who looks forward with pleasure to working with you on your final project and wants to be sure that you choose a project from which you can learn a great deal and on which you can do a good job. However, until your instructor learns from your proposal some details about your proposed project, his or her attitude toward it will be neutral. Your instructor's task while reading your proposal will be to seek the answer to many questions, including the following:

- What kind of communication do you wish to prepare?
- Who will its readers be?
- What is its purpose?
    What is its ultimate objective?
    What task will it enable its readers to perform?
    How will it alter its readers' attitudes?
- Is this a kind of communication you will have to prepare at work?
- Can you write the communication effectively in the time left in the term using resources that are readily available to you?

For additional insights into the questions your instructor (like the reader of any proposal) will ask, see Chapter 32 ("Proposals").

Your proposal should be between 400 and 800 words long. Write it in the memo format, use headings, and include a schedule chart (see Chapter 20, "Twelve Types of Visual Aids").

## PROGRESS REPORT

Write a report of between 400 and 800 words in which you tell your instructor how you are progressing on the writing project you are currently preparing. Be sure to give your instructor a good sense not only of what you have accomplished but also of what problems you have encountered or anticipate. Use the memo format. (You may wish to refer to Figure 31-1, which shows a progress report written by a student to the instructor of her technical writing class.)

## FORMAL REPORT OR PROPOSAL

Write an empirical research report, feasibility report, or proposal. Whichever form of communication you write, it must be designed to help some organization—real or imaginary—solve some problem or achieve some goal, and you must write it in response to a request (again, real or imaginary) from the organization you are addressing.

A real situation is one you have actually encountered. It might involve your employer, your major department, or a service group to which you belong—to name just a few of the possibilities. Students writing on real situations have prepared final projects with such titles as:

- *Feasibility of Using a Computer Data Base To Catalog the Art Department's Slide Library.* (The student wrote this feasibility report at the request of the chair of the Art Department.)
- *Attitudes of Participants in Merit Hotel's R.S.V.P. Club.* (The student wrote this empirical research report at the request of the hotel, which wanted to find ways of improving a marketing program that provides parties and prizes for secretaries who book their companies' visitors at that hotel rather than at one of the hotel's competitors.)
- *Expanding the Dietetic Services at the Campus Health Center: A Proposal.* (The student wrote this proposal to the college administration at the request of the part-time dietitian employed by the Health Center.)

An imaginary situation is one you create to simulate the kinds of situations that you will find yourself in once you begin your career. You pretend that you have begun your career working for an employer who has asked you to use your specialized training to solve some problems or answer some questions that face his or her organization. Students writing about imaginary situations have prepared final projects with such titles as:

- *Performance of Three Lubricants at Very Low Temperatures.* (The student wrote this empirical research report about an experiment he had conducted in a laboratory class. He imagined that he worked for a company that wanted to test the lubricants for use in manufacturing equipment used at temperatures below $-100°F$.)
- *Upgrading the Monitoring and Communication System in the Psychology Clinic.* (The student who wrote this report imagined that she had been asked by the Psychology Clinic to investigate the possibility of purchasing equipment that would improve its monitoring and communication system. All of her information about the clinic and the equipment were real.)
- *Improving the Operations of the Gift Shop of Sea World of Ohio, Aurora, Ohio.* (The student who wrote this proposal had worked at this shop for a

summer job; she imagined that she had been hired by the manager to study its operation and recommend improvements.)

Prepare this project as a formal report or proposal, using the book format (see Chapter 26). Remember that your purpose is to help your readers make a practical decision or take a practical action in a real or imaginary organization. The body of your report should between twelve and twenty pages long (not counting cover, executive summary, title page, table of contents, appendixes, and similar parts). Be sure to use headings within your sections (or chapters) where appropriate, and to use press-on letters, Croy lettering, or some similar technique to create larger type for your cover, title page, and chapter titles. You may also want to use larger lettering for headings.

# Sample Unsolicited Recommendation

## PREVIEW OF APPENDIX

**Scott Houck's Letter**
**Exercises**

This appendix presents a recommendation the writer prepared without being asked to do so by his reader.[1] The recommendation was written by a student, Scott Houck, in response to the Unsolicited Recommendation assignment given in Appendix I.

## SCOTT HOUCK'S LETTER

In his letter, Scott addresses Mrs. Stroh, Executive Vice-President of Thompson Textiles, a company that employed him before he came to college. In college, Scott learned many things that made him think that Thompson would benefit if its managers were better educated in modern management techniques. Thompson Textiles could enjoy this benefit, Scott believes, if it offered courses in management to its employees and if it filled job openings at the managerial level with college graduates. However, if Thompson were to follow Scott's recommendations, it would be changing its current practices considerably. Thompson has never offered courses for its employees, and has long sought to keep payroll expenses low by employing people without a college education, even in management positions. (In a rare exception to this practice, the company has guaranteed Scott a position after he graduates.)

To attempt to change the company's policies, Scott decided to write a letter to one of the most influential people on its staff, Mrs. Stroh. Unfortunately for Scott, throughout the three decades that Mrs. Stroh has served as an executive officer at Thompson, she has consistently opposed company-sponsored education and the hiring of college graduates.

Thus, in writing his letter, Scott faced one of the most challenging of all writing situations, one in which the writer—without being asked for advice—tries to persuade the reader to change his or her beliefs, policies, or behavior. Most readers, when presented with advice they haven't requested, instinctively reject it. Mrs. Stroh would have an especially strong motive for rejecting Scott's advice: she was likely to feel that, if she were to agree that Thompson's educational and hiring policies should be changed, she would be admitting publicly that she had been wrong all those years to support the current policies.

In response to this challenge, Scott wrote a skillfully crafted letter, shown in Figure A-1 (page 805).

## NOTE

1. The recommendation letter and background explanation are reprinted by permission of D.C. Heath. Anderson, Paul V. "Unrequested Recommendation," in *What Makes Writing Good: A Multiperspective*, ed. W.E. Coles, Jr., and J. Vopat (Lexington, Massachusetts: D.C. Heath, 1985), 221–3.

## EXERCISES

1. Imagine that a friend of yours has been assigned to write an unsolicited recommendation. Your friend doesn't know how to approach this assignment and so has asked for your advice about the important things to do in such a communication. Respond with a memo in which you tell your friend the important lessons that can be learned about writing unsolicited recommendations by reading Scott Houck's letter: the things Scott does well are things your friend should try to do also. Consider such things as Scott's organization, his phrasing, and his selection of examples. Imagine that you will be able to enclose a copy of Scott's letter for your friend to examine.

2. Think of an unsolicited recommendation that you might write. What are some of the things you would do in your letter to increase its chances of succeeding? Explain your reason for each of the strategies you list.

616 S. College #84
Oxford, Ohio 45056

April 28, 19--

Georgiana Stroh
Executive Vice-President
Thompson Textiles Incorporated
1010 Note Ave.
Cincinnati, Ohio 45014

Dear Mrs. Stroh:

As my junior year draws to a close, I am more and more eager to return to our company where I can apply my new knowledge and skills. Since our recent talk about the increasingly stiff competition in the textile industry, I have thought quite a bit about what I can do to help Thompson continue to prosper. I have been going over some notes I have made on the subject and I am struck by how many of the ideas stemmed directly from the courses I have taken here at Miami University.

Almost all of the notes featured suggestions or thoughts I simply didn't have the knowledge to consider before I went to college! Before I enrolled, I, like many people, presumed that operating a business required only a certain measure of commonsense ability—that almost anyone could learn to guide a business down the right path with a little experience. However, I have come to realize that this belief is far from the truth. It is true that many decisions are common sense, but decisions often only appear to be simple because the entire scope of the problem or the full ramifications of a particular alternative are not well understood. A path is always chosen, but how often is it the BEST path for the company as a whole?

In retrospect, I appreciate the year I spent supervising the Eaton Avenue Plant because the experience has been an impetus to actually learn from my classes instead of just

**Figure A-1.**   Sample Unsolicited Recommendation.

2

receiving grades. But I look back in embarrassment upon some of the decisions I made and the methods I used then. I now see that my previous work in our factories and my military experience did not prepare me as well for that position as I thought they did. My mistakes were not so often a poor selection among known alternatives, but were usually sins of omission. For example, you may remember that we were constantly running low on packing cartons, and we sometimes ran completely out, causing the entire line to shut down. Now I know that instead of haphazardly placing orders for a different amount every time, we should have used a forecasting model to determine demand and establish a reorder point and a reorder quantity. But I was simply unaware of many of the sophisticated techniques available to me as a manager.

I respectfully submit that many of our supervisory personnel are in a similar situation. This is not to downplay the many contributions they have made to the company. Thompson can directly attribute its prominent position in the industry to the devotion and hard work of these people. But very few of them have more than a high school education or have read even a single text on management skills. We have always counted on our supervisors to pick up their management skills on the job without any additional training. While I recognize that I owe my own opportunities to this approach, this comes too close to the commonsense theory I mentioned earlier.

The success of Thompson depends on the abilities of our managers relative to the abilities of our competitors. In the past, EVERY company used this commonsense approach and Thompson prospered because of the natural talent of people like you. But in the last decade many new managerial techniques have been developed that are too complex for the average employee to just "figure out" on his own. For example, mankind had been doing business for several thousand years before developing the Linear Programming Model for transportation and resource allocation problem-solving. It is not reasonable to expect a high school graduate to recognize that his or her particular

**Figure A-1.** *Continued*

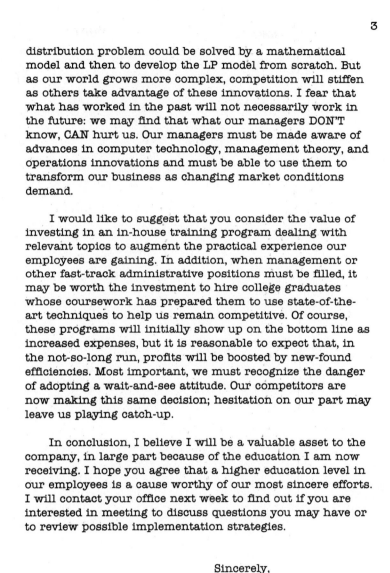

3

distribution problem could be solved by a mathematical model and then to develop the LP model from scratch. But as our world grows more complex, competition will stiffen as others take advantage of these innovations. I fear that what has worked in the past will not necessarily work in the future: we may find that what our managers DON'T know, CAN hurt us. Our managers must be made aware of advances in computer technology, management theory, and operations innovations and must be able to use them to transform our business as changing market conditions demand.

I would like to suggest that you consider the value of investing in an in-house training program dealing with relevant topics to augment the practical experience our employees are gaining. In addition, when management or other fast-track administrative positions must be filled, it may be worth the investment to hire college graduates whose coursework has prepared them to use state-of-the-art techniques to help us remain competitive. Of course, these programs will initially show up on the bottom line as increased expenses, but it is reasonable to expect that, in the not-so-long run, profits will be boosted by new-found efficiencies. Most important, we must recognize the danger of adopting a wait-and-see attitude. Our competitors are now making this same decision; hesitation on our part may leave us playing catch-up.

In conclusion, I believe I will be a valuable asset to the company, in large part because of the education I am now receiving. I hope you agree that a higher education level in our employees is a cause worthy of our most sincere efforts. I will contact your office next week to find out if you are interested in meeting to discuss questions you may have or to review possible implementation strategies.

Sincerely,

Scott Houck

**Figure A-1.**   *Continued*

# Index

## A

Aamdal, Steiner, 420
Abel, Ernest L., 246
Abstract words, 356–358
Abstracts, *see* Summaries
Accuracy
  pairs of words often confused, 359–361
  in wording, 359–361
Active voice
  defined, 32
  converting passive voice to active voice, 332
  identifying, 330
  in instructions, 740
  reasons for using, 331
Activities, describing in resumes, 62–63
Advisers, *see* Readers' roles
Akron Standard, 736, 745
Allegheny Ludlum Steel Corporation, 419
Alternating pattern for comparison, 165–166, 243–244,
  680–681
American Tool, Incorporated, 429
Analogy
  use to help readers understand unfamiliar terms,
    137, 366–368
  *See also subtopic* comparison *under* Segments,
    types
Anderson, John R., 488
Anderson, Paul V., 3, 4, 5, 271
Anderson, Roy E., 643
Anderson, Wayne B., 378
Appearance
  importance of neatness, 25
  professional appearance of resume, 64–65
  readers' expectations about, 120
  *See also* Page design
Appendixes
  cautions and advice, 586
  when to use, 585–586
Apple Computer, 737, 746
Argumentation, *see* Persuasion
Armco, Inc., Midwestern Steel Division, 426
Assumptions
  about readers, 111
  about what employers and readers expect, 116–117
Attitudes of readers, *see* Readers' attitudes
Audience, *see* Readers

## B

Background information
  helps readers understand communications, 275
  in reports, 628, 675–676, 693–694, 706–707,
    711

Bai, Wen-ji, 297
Balde, John W., 452
Bar graphs, *see* Visual aids, types of
Battelle Memorial Institute, 434
Beginnings, *see subtopic* beginning segments *under*
  Segments, types; Introductions
*Better Homes and Gardens,* 451
Bibliographies, *see* Documentation of sources
*Bicycling Magazine's Complete Guide to Bicycle
  Maintenance and Repair,* 432
*BioScience,* 303
Bishop, Georgia A., 288
Blackboards, 774
Book format, *see* Format
Bottom-up processing, 204, 207–208
Brainstorming
  defined, 145, 148–149
  example, 150–151
  use in generating ideas about what to say, 148–149
  use in learning who your readers will be, 100
Bransford, John D., 204
Budget statements, *see* Visual aids, types of
Bush, John, 413

## C

Cause and Effect, *see* Segments, types
Chai, Yao-chu, 297
Chapters, *see* Segments
Checking
  defined, 481, 486
  guidelines
    consider the consequences of your
      communication for your employer,
      486–487
    double-check everything you're unsure about,
      492–493
    let time pass before you check, 488
    play the role of your readers, 486
    read draft aloud, 490
    read draft more than once, concentrating on
      different aspects each time, 488–489
    use checklists, 489–490
    use computers to find but not to cure possible
      problems, 493–494
    when checking for spelling, focus on individual
      words, 490–492
  sample checking worksheet, 491
  why checking is difficult, 488
  *See also* Evaluating
*Chicago Manual of Style,* 119, 597, 605
Citations, *see* Documentation of sources
Claims, *see subtopic under* Persuasion

Clarity
checking for, 488–489
importance of, 7
Classification, *see* Segment, types
Common sense
about people at work, 8
when choosing your voice, 20–21
when writing beginnings of communications, 275
Communication area, 454; *see also* Page design
Comparison, *see* Segments, types
Complex audiences, *see subtopic under* Readers
Computers
as aids to checking drafts, 493–494
use in checking spelling, 492
use for wordprocessing, 553–554
*See also subtopic* use of computers to create visual
aids *under* Visual aids, uses; *subtopic* use of
computers to design pages *under* Page design
Conclusions
in book formats, 582
distinct from interpretation, 630
drawn from visual aids, 396
in empirical research reports, 649–651
in feasibility reports, 682
in general superstructures for reports, 630
in progress reports, 696–697
putting them first, 210–211
*See also subtopic* ending segments *under* Segments,
types
Conley, Larry, 406
Consistency
checking for, 489
in page design, 462–464
Contents, *see* Deciding what to say
Conventions
about appearance of resumes, 64–65
conventional patterns for writing, 126–129; *see also*
Format; Segments; Superstructures
social conventions in ending communications,
289–290
Correctness
checking for, 489
importance of, 7
Costan, C. D., 289
Cox, Joseph L. III, 468

**D**

Davis, Richard M., 3
Davis, William S., 437
Deadlines, 119–120
Deciding what to say
guidelines
generate ideas freely before evaluating them, 140
list additional things your readers will need to
know, 137–138
list persuasive arguments you will have to make,
138–139
list the questions your readers will ask, and
answer them, 132–137

review and revise the contents of your
communication throughout your work on it,
140–141
in job application letters, 67–71
readers' and employers' expectations about what
you'll say, 121
in resumes, 54
say everything needed but nothing more, 27–30
two activities
generating a list of possible contents, 131–132
selecting the items you should include, 132
worksheet, 142
*See also* Generating ideas about what to say;
Planning; Superstructures
Decision-makers, *see* Readers' roles
Definitions
classical, 366–367
help readers understand unfamiliar terms, 137,
366–367
Description of a process, *see subtopic* segmentation
*under* Segments, types
Diagrams, *see* Visual aids, types of
Dijk, Teun A. van, 170, 208, 622
Divided pattern for comparisons, 165–166, 243–244,
680–681
Documentation of sources
author–year citations combined with reference lists
deciding where to place citations, 599
making the reference lists, 599
references to books, 600–601
references to journal articles, 601–602
references to other kinds of sources, 603–604
sample list of references, 604, 606–607
writing author–year citations, 597–599
bibliographies
entries for books, 615–616
entries for journal articles, 616
entries for other sources, 616–617
form of bibliographic entries, 613–615
sample bibliography, 617–619
two purposes, 613
use in book format, 586–587
choosing a format for documentation, 596–597
deciding what to acknowledge, 595–596
footnotes
sample footnotes, 613, 614
second notes, 612–613
use in book format, 586
writing first notes for books, 608–610
writing first notes for journal articles, 610
writing first notes for other sources, 610–612
writing footnote citations, 605–607
writing individual notes, 608
numbered citations combined with reference lists,
605
purposes of documentation, 595
reference lists
in documents in book formats, 586–587
making reference lists, 599
references to books, 600–601

references to journal articles, 601–602
references to other kinds of sources, 603–604
sample list of references, 604, 606–607
Dodge, Richard W., 135
Drafting
  aided by conventional patterns, 223–224
  drafting oral presentations, 776–780
  drafting prose, *see* Guiding readers; Headings; Job
    application letters; Mapping; Resumes;
    Segments; Segments, types; Sentences; Words
  efficiency when drafting, 195–196
  place in the writing process, 9–10
  process of drafting, 195
  use in generating ideas about what to say, 149–163;
    *see also* Throw-away drafts
  visual aids, *see* Page design; Visual aids, types of;
    Visual aids, use of
  *See also* Evaluating
Drawings, *see* Visual aids, types of

**E**

Eastman Kodak Company, 379
Echo words, *see* Words
Education, describing in resume, 61
Elliot, N., 242
Emphasis
  at beginning of a communication, 274
  at end of a communication, 283
  focus on key information, 26–27, 186, 208–211
  in oral presentations, 778–779
  using outline to sharpen focus, 186–190
Empirical research reports
  example, 652–670
  a note about headings, 651–652
  readers' questions, 642
  superstructure
    conclusions, 649–651
    discussion, 648–649
    introduction, 644–646
    method, 647–648
    objectives of research, 646–647
    recommendations, 651
    results, 648
  typical writing situations, 641–642
  *See also subtopic* general superstructure *under*
    Reports
Endings, *see subtopic* ending segments *under*
  Segments, types
Endnotes
  use in book formats, 586–587
  *See also subtopic* footnotes *under* Documentation of
    sources
Engelke, H., 377
Evaluating
  definition, 481
  goals, 481–482
  job application letters, 73
  learning from evaluating, 484
  need for evaluation, 483–484

place in the writing process, 9–10
  questions of evaluation, 482–483
  resumes, 66
  *See also* Checking; Drafting; Reviewing; Testing
Evidence, *see* Persuasion
Ezell, Edward Clinton, 391
Ezell, Linda Neuman, 391
Executive summaries, *see* Summaries
Expectations of listeners in oral presentations, 766
Expectations about your writing
  checking assumptions about expectations, 116–117
  example, 116
  how to find out what's expected
    ask your boss, readers, or other knowledgeable
      people, 118
    look at similar communications written by other
      people, 119
    obtain and study written regulations concerning
      your writing, 118–119
  in job application letters, 66–67
  in resumes, 53, 64–65
  what expectations can be about
    appearance, 25, 120
    deadlines, 119–120
    how you'll treat matters covered by special
      policies, 121–122
    length, 120
    review procedures, 122
    style, 120–121
    what you'll say, 121
  *See also* Conventions
Eye contact, 786–787

**F**

Facts
  discussion, 629–630
  method of obtaining, 628–629
  presentation, 629
  reliability, 626–627
  significance, 626–627, 629
  *See also subtopic* evidence *under* Persuasion
Faigley, Lester, 3
Farrar, William L., 378
Feasibility reports
  example, 683–689
  questions readers ask, 672–673
  superstructure
    conclusions, 682
    criteria, 676–678
    evaluation, 679–682
    introductions, 674–676
    method of obtaining facts, 678–679
    overview of alternatives, 679
    recommendations, 682–683
  typical writing situation, 672
  *See also subtopic* general superstructure *under*
    Reports
Feelings, *see* Readers' feelings
Figures, *see* Visual aids, types of; Visual aids, use of

Filby, Nikola, 331
Filho, Spartaco Astolfi, 288
Firor, Kay, 239–241
Fit among parts of a communication, 187–190
Flesch, Rudolf, 494
Flow charts
  defined, 158
  example, 158–160
  use of to generate ideas about what to say, 158
  *See also* Visual aids, types of
Flower, Linda S., 305, 529–530
Fodstad, Oystein, 420
Fog index, *see* Readability formulas
*Food Engineering,* 435
Footnotes, *see* Documentation of sources
Forecasting statements
  in oral presentations, 280–281
  in written communications, 40, 207–208
Foreman, K. M., 291
Format
  book format
    appendixes, 583–586; *see also* Appendixes
    body, 580–582
    conventions about style, 591–592; *see also*
      *subtopic* conventions in book format *under*
      Style
    cover, 569–570
    definition, 568
    features, 568
    glossary, 587–588
    importance of studying, 568–569
    index, 588–589; *see also* Index
    letter of transmittal, 590–593; *see also*
      Transmittal letters
    list of symbols, 587–588
    list of references; endnotes; bibliography,
      586–587; *see also* Documentation of sources
    page numbering, 590
    summary, 572–578; *see also* Summaries
    table of contents, 578–580; *see also* Tables of
      contents
    title page, 570–572
    typing, 590
  defined, 127, 551–552
  distinguished from superstructures, 127, 621
  how formats help, 127, 551–552
  letter format
    body, 560
    complimentary close, 560–561
    conventions about style, 563–564
    examples, 555–557, 559
    format for additional pages, 563
    format for envelope, 563–565
    heading, 555–558
    inside address, 558
    placement of text on first page, 563
    salutation, 558–560
    signature block, 561
    special notations, 561–562
    subject line, 560

  memo format
    body, 567
    conventions about style, 567
    format for additional pages, 567
    heading, 564–566
    placement of text on first page, 567
    signature, 567
    special notations, 567
    subject line, 566
  readers' expectations about format, 120
  selecting the best format to use, 552–553
  style guides, 552
  use when planning, 127
Frey, Richard L., 643

**G**

General Electric Company, 286, 316
General Motors Corporation, 278, 574
Generalizations
  in conclusion sections of reports, 630, 649–651,
    682, 696–697
  about contents, 208–211
  creation of meaning by readers, 26
  in discussion sections of reports, 621–630, 648–649
  headings that state the main idea of a segment, 305
Generating ideas about what to say
  generate freely before evaluating and selecting, 140
  generating structured and unstructured lists, 145
  as a guide to information gathering, 145–146
  importance of using more than one technique, 145
  as a technique for assembling knowledge already
    acquired, 146
  techniques
    ask readers, 146
    brainstorm, 148–149
    draw a flow chart, 158
    look at conventional superstructures, 148
    look at successful communications used in
      similar situations, 147–148
    make an idea tree, 154–155
    make a matrix, 158–159
    make an outline, 155–158
    talk with someone, 146–147
    when to use these techniques, 145–146
    write a throw-away draft, 149–153
  *See also* Deciding what to say; Planning
Glossaries, 587–588
Grammar, correct, 22, 25
Grids, 454–461, 462, 464–465; *see also* Page design
Grouping related information
  effects on readers' attitudes, 169
  choosing a method of grouping, 169–170
  use in helping readers find information, 167–169
  *See also* Organizing
Guidelines for writing
  distinct from rules, 8
  good judgment required when using, 8–9
  more difficult to apply than to memorize, 8
  use in this book, 8–9, 13

Guiding readers from sentence to sentence
guidelines
avoid shifts in topic from one sentence to the
next, 340–343
use echo words at the beginning of sentences, 36,
344–346
use paragraph breaks to signal major shifts in
topic, 347–348
use transitional words and phrases at the
beginning of sentences, 33, 36, 343–344
readers' versus writers' views, 339
relating new information to old, 339
Gunning, Robert, 494

## H

Handouts, 773
Hargrave, Janice, 60
Hayes, John R., 305, 529–530
Hayes, Robert B., 643–644
Headings
help readers focus on key information, 30
how many levels to use, 310–311
how many to use, 299
indicate organizational hierarchy, 307–311, 457
in instructions, 740
location, 310
parallelism, 305–307
phrasing, 304–305
readers' and employers' expectations about, 120
relationship to topic sentences, 298–299
size, 307–310
in tables of contents, 318
uses, 298
when and how often to use, 298
*See also* Mapping your communication; Page design
Hierarchies
in cause and effect, 254
choosing the best hierarchy, 171–172
in classification, 225–231
as created by readers, 294–295
creating hierarchies in your writing, 172–173
defined, 37, 170
helping readers create
adjust location of prose on page, 311–314
state general points first, 41
use forecasting statements, 40
use headings, 298, 307–311
use page design, 448
use topic sentences, 37
how they help readers, 170
and outlines, 173
in partition, 234
in problem and solution, 255
of related words, 356–358
in segmentation, 236–237
as shown by outlines, 182–183
Hogan, Steve, 239
Huckin, Thomas, 170

## I

IBM, 375, 741, 743
Idea tree
defined, 154
example, 155–156
use in generating ideas about what to say, 154–155
*IEEE Spectrum,* 410
*IEEE Transactions on Systems, Man, and Cybernetics,*
315
Implementers, *see* Readers' roles
Impromptu talks, 770
Indexes
functions, 30, 318–319
how to write, 319
use in book format, 588–589
Indirect readers, *see* Readers
Information, old and new in sentences, 339
Instructions
defined, 727
effects upon readers' attitudes, 84–86, 728
examples, 744, 748–752
readers' tasks while reading, 80–81
superstructure
description of equipment, 735
directions, 738–744
introduction, 731–735
list of materials and equipment, 738
theory of operation, 735–738
troubleshooting, 744
variety, 727
visual design, 728–729
*See also* Testing
Interests, describing in resume, 63
Interruptions, 780, 789
Introductions
in communications written in the book format,
580–582
in empirical research reports, 644–646
in feasibility reports, 674–676
in general superstructures for reports, 627–628
in instructions, 731–735
in progress reports, 693–694
in proposals, 710–711
*See subtopic* beginning segments *under* Segments,
types
Invention, rhetorical, *see* Deciding what to say;
Generating ideas about what to say

## J

Job application letters
defining objectives, 66–67
drafting, 71–72
evaluating, 73
examples, 68–70
importance of taking a reader-centered approach
with, 49
planning, 67–71
revising, 73
writing, 66–73

Job objective, *see* Professional objective
Johnson, Marcia K., 204

# K

Kaalhus, Olav, 420
Kelton, Robert W., 98, 195
Kennedy, Leslie W., 578
Kieras, David E., 209
King, James S., 288
Kintsch, Walter, 622
Klare, George R., 362, 493
Kresovich, S., 230

# L

Lane, Perry, 317
Layout, *see* Page design
Liebman, Bonnie, 421
Length, readers' expectations about, 120
Letters, *see* Format; Job application letters
Lewis, James R., 643
Line graphs, *see* Visual aids, types of
Line of reasoning, *see* Persuasion
Lists
    structured versus unstructured lists in generating
        ideas about what to say, 145
    of symbols, 587–588
    use in instructions, 738
    use in writing, 30
L-Tec Welding and Cutting Systems, 404

# M

Mancusi-Ungaro, Harold R., Jr., 285
Mapping your communication
    guidelines
        adjust location of prose to indicate organizational
            hierarchy, 311–318
        adjust size and location of headings to indicate
            organizational hierarchy, 307–311
        phrase each heading to tell specifically what
            follows, 304–305
        provide an index if it's useful, 318
        provide a table of contents if it's useful, 318
        use headings often, 298–304
        write each heading independently, using parallel
            structure where helpful, 305–307
    importance, 294–295
    in oral presentations, 779–780
    *See also* Page design
Margins
    adjust to show organizational hierarchy, 311–314
    readers' and employers' expectations about, 120
    as white space, 454
    *See also* Page design
Matrix
    defined, 158
    example, 158–161
    use in generating ideas about what to say,
        158–159

Mechanism description, *see subtopic* partition
    segments *under* Segments, types
Memory, short-term, 31–32, 329–330
Memos, *see* Format
Miami University, Manufacturing Engineering
    Department, 426
Mier, Edwin E., 413
Miller, Thomas P., 3
Mills, Alan, 390
*MLA Handbook,* 605
Monarch Marking, 470–471
Monotony
    avoiding in sentences, 342
Monsanto Company, 381
Montalbano, Frank, III, 387
MTD Incorporated, 405
Murray, Raymond L., 285, 434

# N

NASA, 376, 386, 431
National Audubon Society, 308–309
National Cancer Institute, 286
Neatness, importance, 25, 553, 554
The Nelson A. Rockefeller Institute of Government,
    424
Nervousness, 789–790
Nissan Motor Company, Ltd., 430
Numbering
    figures and tables, 393–394
    pages, 590

# O

Oak Ridge Associated Universities, 392, 415
Objectives
    importance, 10
        in empirical research reports, 646–647, 649–650
        in proposals, 713–715
    as standards for evaluating drafts, 10, 482–483
    use in revising writing, 10, 539–541
Objectives, defining
    example, 75, 77
    for job application letters, 66–67
    for oral presentations, 765–768
    place in the writing process, 9–10
    for resumes
        defining your communication's purpose, 51–53
        finding out what's expected, 53
        understanding your readers, 50
    three activities, 75
    *See also* Purpose, defining; Readers; Readers'
        expectations
Ohr, Stephen, 453
Olson, David R., 331
Oral presentations
    defined, 761, 765, 784
    guidelines for preparing
        define your objectives, 765–768
        integrate visuals into your presentation, 770–772

make the structure of your talk evident, 779–780
prepare for interruptions and questions, 780
rehearse, 780
select the form of delivery best suited to your
    purpose and audience, 768–770
select the visual medium best suited to your
    purpose, audience, and situation, 772–776
strongly emphasize main points, 778–779
talk with your listeners, 776–778
guidelines for presenting
    arrange stage so you and your visuals are the
        focus of attention, 785
    display visuals only when they support words, 788
    exhibit enthusiasm and interest, 788
    learn to accept and work with your nervousness,
        789–790
    look at your audience, 786
    make purposeful movements, but otherwise stand
        still, 788–789
    respond courteously to interruptions, questions,
        and comments, 789
    speak in a natural manner, 787
importance of concentrating on each audience
    member, 784–785
listener-centered approach, 761
speaking versus writing, 761–763
types
    impromptu talk, 770
    outlined talk, 769–770
    scripted talk, 768–769
Organization
effects on readers' attitudes, 169
explaining to listeners of oral presentations,
    779–780
explaining to readers, 272–273
forecasting organization of segments, 207–208
helping readers find information, 167–169
notational systems, 182–183
readers' expectations about, 120
showing to readers, 294–296
*See also* Mapping your communication;
    Superstructures
Organization, of this book, 9
Organizational charts
defined, 438
example, 438
use in generating ideas about what to say, 438–439
use in proposals, 719–720
using to learn who your readers will be, 100
*See also* Visual aids, types of
Organizational patterns, *see* Format; Segments, types
Organizing
aided by outlining, 182–184
criteria for organizing
    readers' attitudes, 165–167
    readers' tasks, 165
finding patterns, 176
guidelines
    adapt organizational patterns used successfully by
        others, 175–176

arrange parts of hierarchies into reader-centered
    sequences, 173
build hierarchies, 170–173
group related material, 167–170
organize all parts of your communication,
    174–175
plan visuals, 173–174
effective versus logical organization, 165
evaluating and adapting conventional patterns, 176
importance of thinking about audience and purpose,
    164–167
job application letters, 67–71
organizing versus outlining, 181
readers' expectations about how you'll organize, 120
resumes, 55
signs of poor organization, 184
using patterns used successfully by other writers,
    175–176
various approaches to organizing
    reader-centered approach, 164
    subject-centered approach, 164
    writer-centered approach, 164
what to organize, 174–175
*See also* Organization; Planning; Segments;
    Segments, types; Superstructures
Outlined talks, 769–770
Outlining
guidelines
    change your outline when circumstances warrant,
        190
    don't outline when outlining won't help you,
        190–191
    outline to check the content and organization of
        your draft, 184
    outline to find the best content and organization
        for your communication, 183–184
    outline to obtain approval for your writing plan,
        185
    outline when on a writing team, 185–186
    use a sentence outline to sharpen the focus of
        your writing, 186–190
importance of flexibility, 190
notational systems, 182–183
outline as a sketch, 182
outlining versus organizing, 181
types of outlines
    sentence, 186–190
    topic, 186–187
use to generate ideas about what to say, 155–158
value of outlining, 182–183
what to outline, 184
when outlining won't help, 190–191
*See also* Organizing; Planning
Overhead transparancies, 773

**P**

Page design
arranging prose to reveal hierarchy, 311
defined, 371

Page design *(continued)*
  guidelines
    create a grid to serve as the visual framework for your pages, 454–455
    use the grid to coordinate related visual elements, 456–462
    use the same design for all pages that contain the same types of information, 462–465
  importance, 448
  for instructions, 728–729
  practical procedures for designing pages, 466–469
  readers' expectations about, 120
  for report covers, 569–570
  single-column and multicolumn designs, 449
  use of computers to design pages, 466–467
  *See also* Headings; Mapping your communication; Margins
Paragraphs, *see* Segments
Paragraph breaks to guide readers, 347–348
Partition, *see* Segments, types
Passive voice
  identifying, 330
  when to use, 331–332, 342–343
  why avoid, 331
  *See also* Active voice
Paul, Ronald S., 419
Personal data, describing in resume, 63
Persuasion
  claims
    choosing appealing ones, 138–139
    defined, 138
    implicit claims, 139
    supporting claims, 139
    use in persuasion, 247–251
  defined, 245
  evidence, 247–251
  guidelines
    begin your persuasive segment by stating your claim, 248
    explicitly justify your reasoning if readers might question its validity, 250–251
    present evidence that your readers will agree is sufficient and reliable, 248–250
  how persuasion works, 247
  line of reasoning, 247–251
  persuading about cause and effect, 254–255
  persuading about problems and solutions, 255–258
  skepticism of readers, 245
  structure of a persuasive segment, 251
  when you'll use persuasion, 245
  writing persuasive segments, 245–251
Phantom readers, *see* Readers
Photographs, *see* Visual aids, types of
Pictographs, *see* Visual aids, types of
Pie charts, *see* Visual aids, types of
Pihl, Alexander, 420
Pitter, R. K., 419
Plain words, 361–363
Planning
  as aided by conventional patterns, 223–224

aims, 125
boundaries of planning, 126
considerations, 125–126
forms plans can take, 125–126
importance of making flexible plans, 126
job application letters, 67–71
place in the writing process, 9–10
resumes, 54–55
use of outlines in, 185–186
using conventional patterns to plan, 126–129
visual aids, *see subtopic* guidelines for planning *under* Visual aids, use of
*See also* Deciding what to say; Generating ideas about what to say; Organizing; Outlining
Point of view of reader, 21
Policies about your writing, 121–122; *see also* Style guides
Poltenson, Frank, 417
Posters, 776
Problem and solution, *see* Segments, types
Problems, identifying and explaining as a way of establishing relevance, 267–272
Process description, *see subtopic* segmentation *under* Segments, types
Professional objective, 60
Professionalism in writing
  choosing words commonly used in your professional setting, 22
  discussing your subject from the point of view of your professional role, 21–22
  example, 22–23
  preparing neat communications, 25
  writing grammatically correct sentences and spelling words correctly, 22, 25
  *See also* Grammar, importance; Neatness, importance; Spelling, importance; *subtopic* writing in your professional role *under* Role; *subtopic* use words common in your professional setting *under* Words
Progress reports
  defined, 691
  example, 698, 700–702
  questions readers ask, 692–693
  readers' concerns with the future, 692
  superstructure
    conclusions, 696–697
    facts and discussion, 695–696
    introduction, 693–694
    recommendations, 697
  tone, 697
  typical writing situations, 691
Pronouns
  use as echo words, 36
  use of for sentence variety, 342, 364
Proofreading, *see* Checking; Reviewing
Proposals
  defined, 704
  example, 720–725
  questions readers ask, 706–707
  readers as investors, 706

relationship among problem, objectives, and
    product, 717
strategy of the conventional superstructure
    length, 709–710
    persuasiveness, 707–709
superstructure
    costs, 720
    introduction, 710–711
    management, 719–720
    method, 717–718
    objectives, 713–715
    problem, 711–713
    product, 716
    qualifications, 719
    resources, 718
    schedule, 718
variety of proposal-writing situations, 704–706
Purpose
    guidelines for defining purpose
      describe the final result you want your
        communication to bring about, 77–78
      describe the tasks your readers will try to perform
        while reading, 78–84
      tell how you want your communication to alter
        your readers' attitudes, 84–88
    for job application letters, 67–71
    for oral presentations, 766
    organizing in light of purpose, 164–165
    readers' purposes, 26
    for resumes, 51–53
    uses of definitions of purpose as guide for
      checking and reviewing drafts, 486, 505, 510
      classification, 228–229
      comparison, 244
      designing pages, 464–465
      ending communications, 284–285
      partition, 234
      revising, 539
      segmentation, 237–238
      selecting visual aids, 379–382
      testing, 517–518, 520–521
      writing, 77
    when to use your definition of purpose, 89
    worksheet for defining purpose, 89
    writer's purpose versus the purpose you explain to
      your readers, 77, 87, 89

**Q**

Questions, *see* Readers' questions

**R**

Rappaport, Norman H., 291
Rapport, in oral presentations, 777–778
RCA Corporation, 304
Readability formulas, 493–494
Reader-centered approach to writing
    defined, 7, 13, 339
    example, 7

    general guidelines, 13–43
Readers
    checking assumptions about them, 111
    complex audiences, 19, 98
    create meaning when they read, 25–26
    as a guide for designing pages, 464–465
    as a guide for selecting and designing visual aids,
      384–388
    guidelines for understanding readers
      aggressively seek information about your readers,
        110–111
      create imaginary portraits of your readers in the
        act of reading, 96–97
      find out about the setting in which your readers
        will read your communication, 110
      find out how familiar your readers are with your
        speciality, 108
      find out if your readers have any special
        characteristics, 109–110
      find out your readers' reasons for holding their
        present attitudes, 108–109
      find out whether your readers will be decision-
        makers, advisers, or implementers, 103–108
      learn who your readers will be, 98–103
      portray readers as people who ask questions as
        they read, 97–98
    importance of understanding readers' points of view,
      21
    indirect readers, 100–102
    interact with text, 14–17, 78–79, 212–213, 339,
      528–531
    as investors in reading and evaluating proposals,
      706
    number of, 7, 98; *see also subtopics* complex
      audiences, indirect readers, *and* phantom
      readers *under* Readers
    phantom readers, 99
    procedures for identifying
      brainstorming, 100
      examples of procedures, 102–103
      looking for indirect readers, 100–102
      using organizational charts, 100
    purposes and feelings while reading, 26, 88–89
    readers at work compared with readers at school,
      95–96, 117, 355, 365–366, 501
    as a source of ideas about what you should say, 146
    talking with your readers, 13–19
    understanding readers
      of job application letters, 66–67
      of resumes, 50–51
    use of in testing, 517–518, 518–519, 520–526,
      531–532
    ways that readers read, 26–27, 448
    writer's relationship with readers, 13–25
Readers' attitudes
    altering in resume, 51–52
    as guide for selecting visual aids, 382
    as guide for reviewing drafts, 510–511
    as guide for testing, 520, 526–528
    how readers' attitudes are changed, 84–85, 88

Reader's attitudes *(continued)*
  reasons for holding them, 108–109
  as shaped by instructions, 728
  use in organizing communications, 73
  what readers' attitudes are about, 85–87
Readers' characteristics
  attitudes, *see* Readers' attitudes
  familiarity with your specialty, 108, 137
  as guide for classification, 228–229
  as guide for comparison, 244–245
  as guide for partition, 234–235
  as guide for persuasion, 248
  as guide for segmentation, 237–238
  roles, *see* Readers' roles
  special characteristics, 109–110
Readers' expectations
  as guide for ending communications, 284
  as guide for selecting a form of oral delivery, 768
  for job application letters, 66–67
  for resumes, 53, 64–65
*See also* Expectations about your writing
Readers' feelings
  as affected by page design, 448
  as focus of an ending segment, 286–287
  when reading, 26
  *See also* Readers' attitudes
Readers' preferences, 109–110
Readers' questions
  answering in instructions, 743–744
  answering them where they arise, 213–215
  portraying your readers as people who ask
    questions, 97–98
  related to the various elements of empirical research
    reports, 643–651
  related to the various elements of feasibility reports,
    673–683
  related to the various elements of progress reports,
    693–697
  related to the various elements of proposals,
    707–720
  related to the various elements of reports, 627–631
  typical questions asked by decision-makers, advisers,
    and implementers, 133–137
  use when deciding how much to say in generalizing
    statements, 210
  use when deciding what to say, 132–137
  use when reviewing and revising a draft, 141
  when reading empirical research reports, 642
  when reading feasibility reports, 672–673
  when reading job application letters, 66–71
  when reading progress reports, 692–693
  when reading proposals, 706–707
  when reading reports, 626
  when reading resumes, 52–54
Readers' roles
  advisers
    defined, 107
    questions they typically ask, 134–135
  affects the generalizations they make, 209–210

affects the things they want from a communication,
    268
  decision-makers
    defined, 107
    questions they typically ask, 133–134
  as guide to the contents of summaries, 575
  as guide to material to include in appendixes,
    585–586
  as guide for checking and reviewing drafts, 486,
    505, 510–511
  implementers
    defined, 107
    questions they typically ask, 136–137
  importance of defining, 103–107
  three major roles, 103–107
Readers' tasks
  describing to define purpose, 78–84
  as guide for creating visual aids, 381–382, 422–423
  as guide for testing, 520–526
  helping readers find key information
    begin sections and paragraphs with topic
      sentences, 30, 205
    group related material, 167
    provide indexes and tables of contents in longer
      communications, 30–31, 318–319
    put important information first, 30
    say everything needed but nothing more, 27,
      212–213
    use headings, 30, 298
    use lists, 30
  overlap of, 80
  typical reader tasks while reading, 79–84
  use in organizing, 173
  when reading empirical research reports, 641–642
  when reading reports, 625
  when reading resumes, 52–53
Reading ease scale, *see* Readability formulas
Reasoning, *see* Persuasion
Recommendations
  in beginning segments, 274
  place at beginnings of segments, 212–213
  presentation in reports, 631, 651, 682–683, 697
  in summaries, 575
References, *see* Documentation of sources
References, presented in resumes, 63–64
Rehearsal of oral presentations, 780
Relevance, *see* Significance of your communication to
    your readers
Rensberger, Boyce, 256–257
Reports
  defined, 724
  examples, 632–639
  general superstructure
    conclusions, 630
    discussion, 629–630
    facts, 629
    introduction, 627–628
    method of obtaining facts, 628–629
    recommendations, 631

importance of flexibility, 624
note about summaries, 631–632
questions readers ask, 626
readers want to make practical use of them, 625
varieties of report-writing situations, 624–625
*See also* Empirical research reports; Feasibility
  reports; Progress reports
Resumes
  appearance, 64–65
  examples, 56–59
  importance of taking a reader-centered approach, 49
  parts
    activities, 62–63
    education, 61
    interests, 63
    personal data, 63
    professional objective, 60
    references, 63–64
    work experience, 61–62
  writing a resume
    defining objectives, 50–53
    drafting, 55–65
    evaluating, 65
    planning, 54–55
    revising, 65–66
Reviewing
  defined, 481, 500
  finding out when and whether you'll need your
    communications reviewed, 122
  guidelines for having your writing reviewed
    ask reviewers to explain the reasons for their
      suggestions, 506–507
    stifle your tendency to be defensive, 506
    take notes on your reviewers' suggestions, 507
    tell reviewers about the purpose and readers of
      your communication, 505
    tell reviewers what you want them to do, 505–506
    think of reviewers as partners, 505
  guidelines for reviewing the writing of others
    ask writers about the purpose and readers of their
      communications, 510
    ask writers what they want you to do, 510
    begin suggestions by praising strong points in the
      communication, 511
    determine which revisions will most improve the
      communication, 512–513
    distinguish matters of substance from matters of
      personal taste, 511–512
    explain suggestions fully, 512
    review from the point of view of the readers,
      510–511
    think of writers as people you're helping, not
      judging, 508–510
  importance of good relations between writers and
    reviewers, 504
  reviewing in the classroom, 501–503
  reviewing in the workplace
    formal procedures for reviewing, 501
    why employers require reviews, 500–501

when to review, 140–141
*See also* Evaluating
Revising
  defined, 537
  guidelines
    determine how good your communication is
      before you revise it, 541
    determine how good your communication needs
      to be, 538–541
    determine which revisions will improve your
      communication most, 541–542
    plan to make most important revisions first,
      543–545
    plan when to stop revising, 545
    revise to become a better writer, 545–546
  following guidelines for writing well, 537–538
  investment of time, 537
  job application letters, 73
  place in the writing process, 9–10, 481
  resumes, 65–66
  when to revise, 140–141
Role
  writing in your professional role, 21–24
  your two roles, 4–6
  *See also* Readers' roles; Writing at work as social
    action
Ryther, John H., 252–253

**S**

St. Joe Mineral Company, 386
St. Joe Resources Co., 436
Scene of oral presentations, 766
Schedule charts, *see* Visual aids, types of
Schedule for writing, 119–120, 122
Scope
  communicating scope to readers, 273
  indicating scope of instructions, 731, 732–733
  indicating scope of reports, 628, 676, 693–694
Scripted talks, 768–769
Sears Roebuck and Company, 393
Sections, *see* Segments
Segmentation, *see* Segments, types
Segments
  advice for using conventional segments, 224
  benefits of learning and using conventional
    segments, 223–224
  defined, 127, 199–201
  distinguished from superstructures, 129
  guidelines for writing
    answer readers' questions where they arise,
      213–215
    forecast organization, 207–208
    generalize about contents, 208–211
    provide transitional statements between segments,
      216–217
    put most important information first, 212–213
    use topic sentences, 30, 202–207
  recognizing segments, 199–201, 203

Segments *(continued)*
  use when planning, 127–128, 176
Segments, types
  beginning segments
    determining length of, 275–276
    examples of brief beginnings, 276
    examples of long beginnings, 276–278
    functions, 264
    imagining readers' thoughts, feelings, and
      expectations as they begin reading,
      264–266
    providing background, 275
    telling main points, 274
    telling organization, 272–273
    telling relevance, 267–272
    telling scope, 273
  cause and effect segments
    aims, 251
    defined, 251–254
    example, 255–257
    guidelines for writing, 254
  classification segments
    creating hierarchies in, 225–227
    defined, 225
    examples, 229–233
    formal and informal methods, 227
    guidelines for formal classification, 227–228
    guidelines for informal classification, 228–229
  comparison segments
    alternating and divided patterns, 165–166,
      243–244, 680–681
    defined, 243
    example, 245–246
    guidelines for writing, 244–245
    how comparison works, 243–244
    when you'll use it, 242–243
  ending segments
    aims, 283
    focusing on key feelings, 286–287
    following social conventions, 289–290
    identifying further study, 288
    referring to goal stated earlier, 288–289
    repeating main point, 285
    stopping after last point, 284–285
    summarizing key points, 285–286
    telling readers how to get assistance or more
      information, 287
    telling readers what to do next, 287
  mixing of types, 224
  partition segments
    aims of partitioning, 234
    defined, 231
    example, 235–236
    guidelines for writing, 234–235
    how partitioning works, 231–234
  persuasion segments
    example, 251–253
    guidelines for writing, 248–251
    how persuasion works, 247–248
    structure of a persuasion segment, 251

    when you'll use persuasion segments, 245
    *See also* Persuasion
  problem and solution segments
    aims, 255
    defined, 255
    examples, 258–259
    guidelines for writing, 255
  segmentation segments
    defined, 235–237
    examples, 238–242
    guidelines for writing, 237–238
    how segmenting works, 237
    two purposes for, 235–236
  variety of types, 224
Selzer, Jack, 493
Sentence outlines
  use in establishing tight fit among parts, 187–190
  use in selecting and emphasizing main points,
    186
Sentences
  flexibility, 326
  guidelines
    eliminate unnecessary words, 32, 326–328
    keep related words together, 31, 328–330
    put action in main verbs, 332–334
    use active voice, 32, 330–332
    use words familiar to your reader, 33
  *See also* Guiding readers from sentence to sentence;
    Verbs
Setting, 110
Short-term memory, *see* Memory, short-term
Significance of your communication to your readers,
  explaining
  in beginnings of communications, 267–273
  in conclusions of reports, 626, 632, 642, 649–651,
    673, 682–683, 693, 696–697
  in instructions, 731–732, 734
  in introductions of reports, 625, 627–628,
    641–642, 644–646, 672, 673, 674–676, 691,
    692, 692–693, 693–694, 706–707,711
  in oral presentations, 778–779
  guidelines
    begin by telling your readers why they should
      read, 41
    begin with summaries, 42
    explicitly state significance, 42
Silverman, Robert A., 578
Skepticism of readers at work, 245–247
Slides, 35-mm, 776
Smith, Frank, 25, 490
Sojka, J. J., 387
Society of Automotive Engineers, 427
Souther, James W., 135
Spelling
  importance of correct spelling, 22
  *See also* Checking
Spinks, Nelda, 60
Standard Oil Company, 415
State of Ohio, Department of Highway Safety,
  302

Statewide Tissue and Organ Donor Program,
    Department of Health and Rehabilitative Services,
    Tallahassee, Florida, 306
Storyboards, 771–772
Style
    conventions
        book format, 591–592
        letter writing, 563–564
        memos, 567–568
    in oral presentations, 778
    readers' expectations about, 120–121
    *See also* Guiding readers; Sentences; Words
Style guides
    as aids in documenting sources, 596–597
    defined, 118–119
    types, 596–597
    *See also subtopic* style guides *under* Format
Subjects of sentences, use to indicate topics,
    340–343
Summaries
    abstracts
        defined, 575
        example, 576
    examples, 575–578
    executive summaries
        defined, 631–632
        example, 577
    features, 575
    how they help readers, 572–575
    length, 578
    use in reports for complex audiences, 135
    use to end communications, 285–286
    use to help readers see significance of your
        communication, 42
    when to write, 576–578
Superstructures
    defined, 127, 621
    how knowledge of superstructures helps you,
        127–128, 621–622
    relationship to formats, 127, 129, 621
    use when generating ideas about what to say, 148
    use when organizing, 175–176
    use when planning, 127–128
    *See also* Empirical research reports; Feasibility
        reports; Instructions; Progress reports;
        Proposals; Reports
Swarts, Heidi, 305, 529–530


**T**

Tables, *see* Visual aids, types of
Tables of contents
    examples, 579–580
    functions, 30, 318
    how to write, 318
    lists of figures and tables, 580–581
    purposes, 578–579
Talks, *see* Oral presentations
Tasks, of readers, *see* Readers' tasks
Team writing, use of outlining in, 185–186

Testing
    basic strategy, 517–518
    defined, 481, 517
    elements, 520
    guidelines
        ask test readers to use your draft in the same way
            target readers will, 520–526
        choose test readers from your target audience,
            531
        learn how your draft affects your test readers'
            attitudes, 526–528
        learn how your test readers interact with your
            draft while reading, 528–531
        use drafts that are as close as possible to final
            drafts, 532
        use enough test readers to determine typical
            responses, 531–532
    of instructions, 729–730
    major uses of test results, 518–519
    strategies for learning about readers' interactions
        with your draft
        ask test readers to provide reading aloud
            protocols, 529–530
        interview test readers after they've finished
            reading, 530–531
        sit beside test readers, 529
        videotape test readers, 529
    types
        location tests, 525–526
        performance tests, 521–523
        understandability tests, 523–525
    when to test, 519–520
    *See also* Evaluating
Test laboratories, 522
Thomas, Thomas P., 378
Throw-away draft
    defined, 150
    example, 152–153
    procedure for writing, 151–152
Thumbnail sketch, 464–467
Tone
    of job application letters, 71–72
    in progress reports, 697
Top-down processing, 204, 208–209
Topic outlines, 186–187
Topic sentences, *see* Topic statements
Topic statements
    defined, 30, 37–40
    how they help readers, 202–205
    relationship to headings, 298–299
    use at beginning of segments, 202–207
Toshiba Corporation, 742
Transitions
    between segments, 216–217
    between sentences, 33–36
    in oral presentations, 779–780
    why place at the beginning of sentences, 36,
        343–344
Transmittal letters, 590–593
Tvergyak, Paul J., 416

## U

United States Bureau of the Census, 389, 402, 411, 412, 418, 443
United States Department of Agriculture Extension Service, 739
United States Department of the Interior, 232–233, 444
Unsolicited communications, 271–272, 802-807
USAir, 374
USV Pharmaceuticals, 472–477

## V

Verbs
  putting action in verbs, 332–334
  *See also* Active voice; Passive voice
Visual aids, types of
  bar graphs
    cautions about misleading readers, 413–414
    creating, 411–413
    defined, 410–411
    examples, 4, 5, 135, 381, 383, 410, 411, 412, 413, 414
  budget statements
    creating, 442
    defined, 440–442
    examples, 441
    in proposals, 720
  diagrams
    creating, 434–435
    defined, 433–434
    examples, 303, 377, 386, 387, 434, 435
  drawings
    choosing between photographs and drawings, 430–431
    creating, 428–430
    defined, 428
    examples, 14, 374, 375, 376, 391, 429, 430, 431, 432, 433, 662, 739, 741, 742, 749, 750, 751, 752, 754, 755, 756, 757, 758, 759
  flow charts
    creating, 437–438
    defined, 158, 436–437
    examples, 158, 160, 436, 437
    use in generating ideas about what to say, 158
  line graphs
    creating, 418–421
    defined, 417–418
    examples, 378, 384, 385, 389, 392, 415, 417, 418, 419, 420, 544, 545, 687, 688, 689
  organizational charts
    creating, 439
    defined, 438
    examples, 101, 438
    use in generating ideas about what to say, 438–439
    use in learning who your readers will be, 100
    use in proposals, 719–720
  photographs
    choosing between photographs and drawings, 433
    creating, 426–428
    defined, 423
    disadvantages, 423
    examples, 304, 390, 393, 425, 426, 427, 736
  pictographs
    cautions about misleading readers, 416–417
    creating, 415
    defined, 414–415
    examples, 415, 416
  pie charts
    creating, 422
    defined, 421
    examples, 421, 424
  schedule charts
    creating, 440
    defined, 439–440
    examples, 439
  tables
    creating formal tables, 406–409
    creating informal tables, 404–406
    defined, 401–404
    examples, 379, 382, 385, 394, 402, 404, 405, 406, 407, 408, 409, 660, 661, 665, 666
    ineffective use of tables, 409–410
    rows and columns, 404–406
Visual aids, use of
  choosing between photographs and drawings, 430–431
  choosing the best display for numerical data, 422–423
  defined, 371, 373
  guidelines for designing
    label important content clearly, 389–392
    make visual aids easy to use, 388
    make visual aids simple, 388–389
    provide informative titles, 392–394
  guidelines for integrating with prose
    introduce visual aids, 394–395
    put visual aids where readers can easily find them, 396–397
    state conclusions you want readers to draw from visual aids, 396
  guidelines for planning
    choose visual aids appropriate to your purpose, 379–382
    choose visual aids readers know how to read, 384–388
    while planning communications, look for places to use visual aids, 376–379
  importance, 173–174, 373–374
  for instructions, 729, 740–741
  list of figures and tables, 580–581
  numbering, 393–394
  in oral presentations
    benefits, 771
    coordination with verbal messages, 788
    location, 785
    types, 773–776
  titles and their placement, 392, 394, 401

use of computers to create, 400–401
uses
  describing something physical, 43, 376
  describing something that's not physical, 43, 376–377
  explaining relationships among data, 43, 377–378
  making detailed information easy to find, 43, 378–379
  supporting arguments, 379
  when to plan them, 173–174
  *See also* Page design
Vocabulary, specialized
  explaining technical terms not familiar to readers, 366–368
  using technical terms your readers understand and expect, 365–366
  *See also* Words
Voice
  adjusting, 20
  choosing, 19–20
  using a respectful voice, 19–20
  *See also* Active voice; Passive voice; *subtopic* voice *under* Sentences

**W**

Wang, Bix-xi, 297
Wang, Heng-Sheng, 297
Wang Laboratories, 236, 320, 466–467
Wells, Baron, 60
White Consolidated, 429
Whitehead, Hal, 450
White space, *see* Page design
Wordiness, 327–328
Words
  choosing words suited to your readers and purpose, 22, 355–356
  echo words
    defined, 36, 344
    kinds, 36, 344–345
    importance of placing at the beginning of sentences, 36, 345–346
  guidelines
    avoid unnecessary variation of terms, 363–364

  choose plain words, 361–633
  explain technical terms your readers don't know, 366–368
  use concrete, specific words, 356–358
  use technical terms your readers will understand and expect, 365–366
  use words accurately, 359–361
  transitional, 33
  use words common in your professional setting, 22
  use words familiar to your readers, 33
Work experience, describing in resume, 61–62
Worksheets
  checking, 491
  deciding what to say, 142
  defining objectives for oral presentations, 766–767
  defining purpose, 90–93, 112
Writer-centered writing, 339
Writing as a conversation with readers, 17
Writing process
  examples
    writing job application letters, 66–73
    writing resumes, 50–66
  flexibility, 9–10
  five steps, 9; *see also* Defining; Drafting; Evaluating; Objectives; Planning; Revising
  improving your writing process, 9–10
  order of steps, 10, 126, 140–141
  role of superstructures in, 621–622
  use in drafting visual aids, 371
Writing at work
  as action, 6
  importance in your career, 3–6
  means of sharing ideas and information, 3
  percentage of time spent writing, 3, 5
  as social action, 6–7, 13

**X**

Xtek, Inc., 425

**Y**

Yamaguchi, Jack, 430